MCAT® Biology

2025–2026 Edition: An Illustrated Guide

Copyright © 2024
On behalf of UWorld, LLC
Dallas, TX
USA

All rights reserved.
Printed in English, in the United States of America.

Reproduction or translation of any part of this work beyond that permitted by Sections 107 and 108 of the United States Copyright Act without the permission of the copyright owner is unlawful.

The Medical College Admission Test (MCAT®) and the United States Medical Licensing Examination (USMLE®) are registered trademarks of the Association of American Medical Colleges (AAMC®). The AAMC® neither sponsors nor endorses this UWorld product.

Facebook® and Instagram® are registered trademarks of Facebook, Inc. which neither sponsors nor endorses this UWorld product.

X is an unregistered mark used by X Corp, which neither sponsors nor endorses this UWorld product.

Acknowledgments for the 2025–2026 Edition

Ensuring that the course materials in this book are accurate and up to date would not have been possible without the multifaceted contributions from our team of content experts, editors, illustrators, software developers, and other amazing support staff. UWorld's passion for education continues to be the driving force behind all our products, along with our focus on quality and dedication to student success.

About the MCAT Exam

Taking the MCAT is a significant milestone on your path to a rewarding career in medicine. Scan the QR codes below to learn crucial information about this exam as you take your next step before medical school.

Basic MCAT Exam Information

Scores and Percentiles

MCAT Sections

Registration Guide

Preparing for the MCAT with UWorld

The MCAT is a grueling exam spanning seven subjects that is designed to test your aptitude in areas essential for success in medicine. Preparing for the exam can be intimidating—so much so that in post-MCAT questionnaires conducted by the AAMC®, a majority of students report not feeling confident about their MCAT performance.

In response, UWorld set out to create premier learning tools to teach students the entire MCAT syllabus, both efficiently and effectively. Taking what we learned from helping over 90% of medical students prepare for their medical board exams (USMLE®), we launched the UWorld MCAT Qbank in 2017 and the UWorld MCAT UBooks in 2024. The MCAT UBooks are meticulously written and designed to provide you with the knowledge and strategies you need to meet your MCAT goals with confidence and to secure your future in medical school.

Below, we explain how to use the MCAT UBooks and MCAT Qbank together for a streamlined learning experience. By strategically integrating both resources into your study plan, you will improve your understanding of key MCAT content as well as build critical reasoning skills, giving you the best chance at achieving your target score.

MCAT UBooks: Illustrated and Annotated Guides

The MCAT UBooks include not only the printed editions for each MCAT subject but also provide digital access to interactive versions of the same books. There are eight printed MCAT UBooks in all, six comprehensive review books covering the science subjects and two specialized books for the Critical Analysis and Reasoning Skills (CARS) section of the exam:

- Biology
- Biochemistry
- General Chemistry
- Organic Chemistry
- Physics
- Behavioral Sciences
- CARS (Annotated Practice Book)
- CARS Passage Booklet (Annotated)

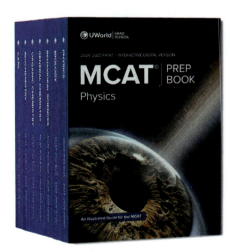

Each UBook is organized into Units, which are divided into Chapters. The Chapters are then split into Lessons, which are further subdivided into Concepts.

MCAT Sciences: Printed UBook Features

The MCAT UBooks bring difficult science concepts to life with thousands of engaging, high-impact visual aids that make topics easier to understand and retain. In addition, the printed UBooks present key terms in blue, indicating clickable illustration hyperlinks in the digital version that will help you learn more about a scientific concept.

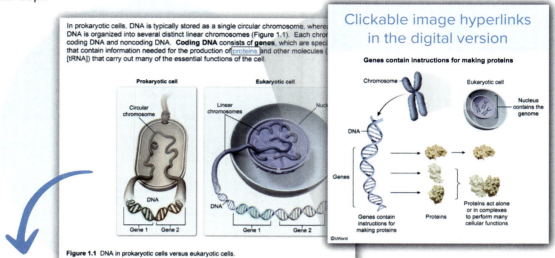

Thousands of educational illustrations in the print book

Clickable image hyperlinks in the digital version

Figure 1.1 DNA in prokaryotic cells versus eukaryotic cells.

Test Your Basic Science Knowledge with Concept Check Questions

The printed UBooks also include 450 new questions—never before available in the UWorld Qbank—for Biology, General Chemistry, Organic Chemistry, Biochemistry, and Physics. These new questions, called Concept Checks, are interspersed throughout the entire book to enhance your learning experience. Concept Checks allow you to instantly test yourself on MCAT concepts you just learned from the UBook.

Short answers to the Concept Checks are found in the appendix at the end of each printed UBook. In addition, the digital version of the UBook provides an interactive learning experience by giving more detailed, illustrated, step-by-step explanations of each Concept Check. These enhanced explanations will help reinforce your learning and clarify any areas of uncertainty you may have.

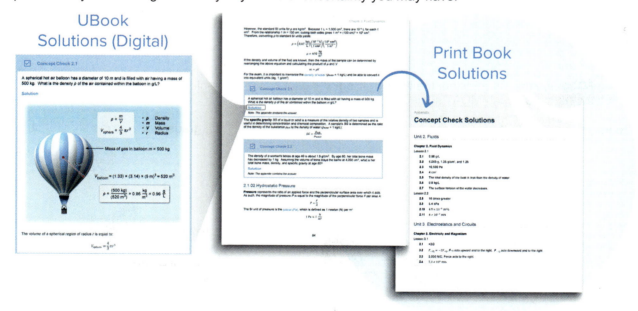

UBook Solutions (Digital)

Print Book Solutions

MCAT CARS Printed UBook Features

For CARS, the main book, or Annotated Practice Book, teaches you the specialized CARS skills and strategies you need to master and then follows up with multiple sets of MCAT-level practice questions.

Additionally, the CARS Passage Booklet includes annotated versions of the passages in the CARS Main Book. From these annotations, you will learn how to break down a CARS passage in a step-by-step manner to find the right answer to each CARS question.

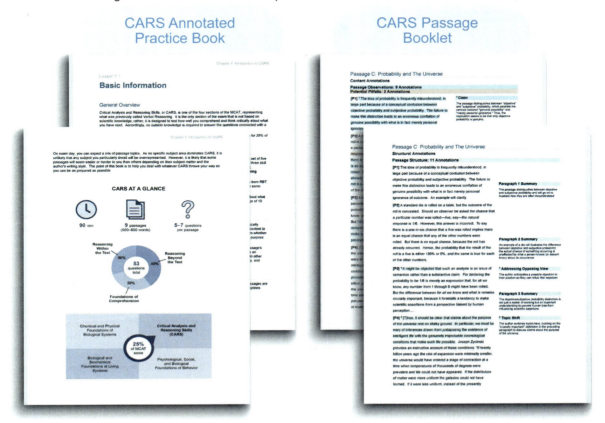

MCAT-Level Exam Practice with the UWorld Qbank

UWorld's MCAT UBooks and Qbank were designed to be used together for a comprehensive review experience. The UWorld Qbank provides an active learning approach to MCAT prep, with thousands of MCAT-level questions that align with each UBook.

The printed UBooks include a prompt at the end of each unit that explains how to access unit practice tests in the MCAT Qbank. In addition, the MCAT UBooks' digital platform enables you to easily create your own unit tests based on each MCAT subject.

To purchase MCAT Qbank access or to begin a free seven-day trial, visit gradschool.uworld.com/mcat.

Boost Your Score with the #1 MCAT Qbank

Why use the UWorld Qbank?

- Thousands of high-yield MCAT-level questions
- In-depth, visually engaging answer explanations
- Confidence-building user interface identical to the exam
- Data-driven performance and improvement tracking
- Fully featured mobile app for on-the-go review

Special Features Integrating Digital UBooks and the UWorld Qbank

The digital MCAT UBooks and the MCAT Qbank come with several integrated features that transform ordinary reading into an interactive study session. These time-saving tools enable you to personalize your MCAT test prep, get the most out of our detailed explanations, save valuable time, and know when you are ready for exam day.

My Notebook

My Notebook, a personalized note-taking tool, allows you to easily copy and organize content from the UBooks and the Qbank. Simplify your study routine by efficiently recording the MCAT content you will encounter in the exam, and streamline your review process by seamlessly retrieving high-yield concepts to boost your study performance—in less time.

Digital Flashcards

Our unique flashcard feature makes it easy for students to copy definitions and images from the MCAT UBooks and Qbank into digital flashcards. Each card makes use of spaced repetition, a research-supported learning methodology that improves information retention and recall. Based on how you rate your understanding of flashcard content, our algorithm will display the card more or less frequently.

Fully Featured Mobile App

Study for your MCAT exams anytime, anywhere, with our industry-leading mobile app that provides complete access to your MCAT prep materials and that syncs seamlessly across all devices. With the UWorld MCAT app, you can catch up on reading, flip through flashcards between classes, or take a practice quiz during lunch to make the most of your downtime and keep MCAT material top of mind.

Book and Qbank Progress Tracking

Track your progress while using the MCAT UBooks and Qbank, and review MCAT content at your own pace. Our learning tools are enhanced by advanced performance analytics that allow users to assess their preparedness over time. Hone in on specific subjects, foundations, and skills to iron out any weaknesses, and even compare your results with those of your peers.

Explore the Periodic Table

You will need to use the periodic table to answer questions on the MCAT for specific sections. Introductory general chemistry concepts constitute 30% of the material tested in the Chemical and Physical Foundations of Biological Systems section of the exam. In addition, General Chemistry constitutes 5% of the Biological and Biochemical Foundations of Living Systems section of the MCAT. Using and understanding the periodic table is a crucial skill needed for success in these sections.

1 H 1.0																	2 He 4.0
3 Li 6.9	4 Be 9.0											5 B 10.8	6 C 12.0	7 N 14.0	8 O 16.0	9 F 19.0	10 Ne 20.2
11 Na 23.0	12 Mg 24.3											13 Al 27.0	14 Si 28.1	15 P 31.0	16 S 32.1	17 Cl 35.5	18 Ar 39.9
19 K 39.1	20 Ca 40.1	21 Sc 45.0	22 Ti 47.9	23 V 50.9	24 Cr 52.0	25 Mn 54.9	26 Fe 55.8	27 Co 58.9	28 Ni 58.7	29 Cu 63.5	30 Zn 65.4	31 Ga 69.7	32 Ge 72.6	33 As 74.9	34 Se 79.0	35 Br 79.9	36 Kr 83.8
37 Rb 85.5	38 Sr 87.6	39 Y 88.9	40 Zr 91.2	41 Nb 92.9	42 Mo 95.9	43 Tc (98)	44 Ru 101.1	45 Rh 102.9	46 Pd 106.4	47 Ag 107.9	48 Cd 112.4	49 In 114.8	50 Sn 118.7	51 Sb 121.8	52 Te 127.6	53 I 126.9	54 Xe 131.3
55 Cs 132.9	56 Ba 137.3	57 La* 138.9	72 Hf 178.5	73 Ta 180.9	74 W 183.9	75 Re 186.2	76 Os 190.2	77 Ir 192.2	78 Pt 195.1	79 Au 197.0	80 Hg 200.6	81 Tl 204.4	82 Pb 207.2	83 Bi 209.0	84 Po (209)	85 At (210)	86 Rn (222)
87 Fr (223)	88 Ra (226)	89 Ac⁺ (227)	104 Rf (261)	105 Db (262)	106 Sg (266)	107 Bh (264)	108 Hs (277)	109 Mt (268)	110 Ds (281)	111 Rg (280)	112 Cn (285)	113 Uut (284)	114 Fl (289)	115 Uup (288)	116 Lv (293)	117 Uus (294)	118 Uuo (294)

	58 Ce 140.1	59 Pr 140.9	60 Nd 144.2	61 Pm (145)	62 Sm 150.4	63 Eu 152.0	64 Gd 157.3	65 Tb 158.9	66 Dy 162.5	67 Ho 164.9	68 Er 167.3	69 Tm 168.9	70 Yb 173.0	71 Lu 175.0
*														
+	90 Th 232.0	91 Pa (231)	92 U 238.0	93 Np (237)	94 Pu (244)	95 Am (243)	96 Cm (247)	97 Bk (247)	98 Cf (251)	99 Es (252)	100 Fm (257)	101 Md (258)	102 No (259)	103 Lr (260)

Table of Contents

UNIT 1 MOLECULAR BIOLOGY

CHAPTER 1 DNA STRUCTURE, SYNTHESIS, AND REPAIR 1
- Lesson 1.1   Nucleic Acids 3
- Lesson 1.2   DNA Replication 9
- Lesson 1.3   DNA Repair 17
- Lesson 1.4   Eukaryotic Chromosome Organization 23

CHAPTER 2 GENE EXPRESSION 27
- Lesson 2.1   The Central Dogma of Molecular Biology 27
- Lesson 2.2   Transcription 29
- Lesson 2.3   Translation 37
- Lesson 2.4   Regulation of Gene Expression in Eukaryotes 57

UNIT 2 BIOLOGICAL RESEARCH TECHNIQUES

CHAPTER 3 DESIGNING AND INTERPRETING EXPERIMENTS 69
- Lesson 3.1   Experimental Design 71
- Lesson 3.2   Statistics 79

CHAPTER 4 BIOTECHNOLOGY 89
- Lesson 4.1   DNA Technology 89
- Lesson 4.2   Analyzing Gene Expression 113
- Lesson 4.3   Determining Gene Function 121
- Lesson 4.4   Practical Applications of Biotechnology 127
- Lesson 4.5   Special Considerations in Biotechnology 137

UNIT 3 CELLULAR BIOLOGY

CHAPTER 5 EUKARYOTIC CELLS 143
- Lesson 5.1   Cells 145
- Lesson 5.2   Plasma Membrane Components and Functions 151
- Lesson 5.3   Eukaryotic Organelles 172
- Lesson 5.4   Cell Growth and Division 193
- Lesson 5.5   Eukaryotic Tissues 208

CHAPTER 6 PROKARYOTES AND VIRUSES 211
- Lesson 6.1   Prokaryotic Cells 211
- Lesson 6.2   Growth and Reproduction of Prokaryotes 224
- Lesson 6.3   Prokaryotic Genetics 237
- Lesson 6.4   Viruses 250
- Lesson 6.5   Viral Life Cycles 257
- Lesson 6.6   Sub-Viral Particles 265

UNIT 4 GENETICS AND EVOLUTION

CHAPTER 7 GENETICS 269
- Lesson 7.1   Meiosis 270
- Lesson 7.2   Mendelian Concepts 278
- Lesson 7.3   Chromosomes and Inheritance 287

CHAPTER 8 EVOLUTION 299
- Lesson 8.1   Factors Affecting Allele Frequency 299
- Lesson 8.2   Evolution of Species 307

UNIT 5 REPRODUCTION

CHAPTER 9 REPRODUCTIVE SYSTEMS ... 315
- Lesson 9.1 Biological Sex and Reproduction ... 316
- Lesson 9.2 Male Reproductive System ... 318
- Lesson 9.3 Female Reproductive System ... 328

CHAPTER 10 PREGNANCY, DEVELOPMENT, AND AGING ... 336
- Lesson 10.1 Pre-Implantation Development ... 336
- Lesson 10.2 Post-Implantation Development ... 342
- Lesson 10.3 Cellular Mechanisms of Development ... 348
- Lesson 10.4 Gestation ... 359
- Lesson 10.5 Cellular Regeneration and Senescence ... 367

UNIT 6 ENDOCRINE AND NERVOUS SYSTEMS

CHAPTER 11 ENDOCRINE SYSTEM ... 371
- Lesson 11.1 Endocrinology ... 372
- Lesson 11.2 Hormones ... 383

CHAPTER 12 NERVOUS SYSTEM ... 418
- Lesson 12.1 Cells of the Nervous System ... 418
- Lesson 12.2 Neural Communication ... 424
- Lesson 12.3 Nervous System Structure and Function ... 440

UNIT 7 CIRCULATION AND RESPIRATION

CHAPTER 13 CIRCULATION ... 449
- Lesson 13.1 Circulatory System ... 450
- Lesson 13.2 Lymphatic System ... 481

CHAPTER 14 RESPIRATION ... 488
- Lesson 14.1 Respiratory System Structure ... 488
- Lesson 14.2 Respiratory System Function ... 494

UNIT 8 DIGESTION AND EXCRETION

CHAPTER 15 DIGESTION ... 511
- Lesson 15.1 Alimentary Canal ... 513
- Lesson 15.2 Accessory Digestive Organs ... 524
- Lesson 15.3 Control of Digestion ... 532

CHAPTER 16 EXCRETION ... 536
- Lesson 16.1 Excretory System Structure ... 536
- Lesson 16.2 Excretory System Function ... 541

UNIT 9 MUSCULOSKELETAL SYSTEM

CHAPTER 17 MUSCULAR SYSTEM ... 555
- Lesson 17.1 General Muscle Characteristics ... 556
- Lesson 17.2 Characteristics of Specific Muscle Types ... 570

CHAPTER 18 SKELETAL SYSTEM ... 582
- Lesson 18.1 Skeletal System Structure ... 582
- Lesson 18.2 Skeletal System Function ... 595

UNIT 10 SKIN AND IMMUNE SYSTEMS

CHAPTER 19 SKIN SYSTEM ... 599
- Lesson 19.1 Skin Structure ... 600
- Lesson 19.2 Skin Function ... 605

CHAPTER 20 IMMUNE SYSTEM 610
    Lesson 20.1   Immune System Components 610
    Lesson 20.2   Immune System Function 617

APPENDIX

CONCEPT CHECK SOLUTIONS 635

INDEX **645**

Unit 1 Molecular Biology

Chapter 1 DNA Structure, Synthesis, and Repair

1.1 Nucleic Acids

   1.1.01   Overview of DNA and RNA Function
   1.1.02   Nucleotides and Nucleic Acids
   1.1.03   Nucleic Acid Structure
   1.1.04   Nucleic Acid Hybridization and Denaturation

1.2 DNA Replication

   1.2.01   Semiconservative Replication
   1.2.02   Origin of Replication
   1.2.03   Mechanism of Replication
   1.2.04   Replication at the Ends of Linear Chromosomes

1.3 DNA Repair

   1.3.01   DNA Proofreading and Mismatch Repair
   1.3.02   DNA Damage and Repair

1.4 Eukaryotic Chromosome Organization

   1.4.01   Histones and Nucleosomes
   1.4.02   Euchromatin and Heterochromatin

Chapter 2 Gene Expression

2.1 The Central Dogma of Molecular Biology

   2.1.01   Flow of Genetic Information
   2.1.02   Gene Expression

2.2 Transcription

   2.2.01   Mechanism of Transcription
   2.2.02   Modifications to mRNA Ends
   2.2.03   RNA Splicing Mechanisms

2.3 Translation

   2.3.01   Interpreting the Genetic Code
   2.3.02   Transfer RNA (tRNA) and Anticodons
   2.3.03   The Degenerate Code
   2.3.04   Ribosomes and Ribosomal RNA (rRNA)
   2.3.05   Mechanism of Translation
   2.3.06   Post-Translational Modification of Proteins
   2.3.07   Mutations and Mutagens
   2.3.08   Types of Mutations

2.4 Regulation of Gene Expression in Eukaryotes

   2.4.01   Differential Gene Expression
   2.4.02   Chromatin Remodeling
   2.4.03   Regulation of Transcription
   2.4.04   Noncoding RNA

Lesson 1.1
Nucleic Acids

Introduction

Deoxyribonucleic acid (DNA) is the heritable material passed from parent to offspring that allows for the transmission of genetic information from one generation to another. DNA stores the information needed for an organism's development and vital processes, and it plays a role in regulating the expression of that information. The expression of the information encoded in DNA is mediated by **ribonucleic acid (RNA)**, a related molecule.

Both DNA and RNA are composed of building blocks called **nucleotides**. The order in which nucleotides are joined together is the mechanism by which genetic information is stored and transmitted. This lesson explores DNA and RNA structure and function.

1.1.01 Overview of DNA and RNA Function

DNA can be transferred from one generation to the next via asexual or sexual reproduction. **Asexual reproduction** involves a single parent organism that produces offspring genetically identical to the parent. In contrast, **sexual reproduction** involves two parent organisms that both contribute genetic material to produce genetically unique offspring.

In prokaryotic cells, DNA is typically stored as a single circular chromosome, whereas in eukaryotic cells, DNA is organized into several distinct linear chromosomes (Figure 1.1). Each chromosome contains coding DNA and noncoding DNA. **Coding DNA** consists of **genes**, which are specific sequences of DNA that contain information needed for the production of proteins and other molecules (eg, transfer RNA [tRNA]) that carry out many of the essential functions of the cell.

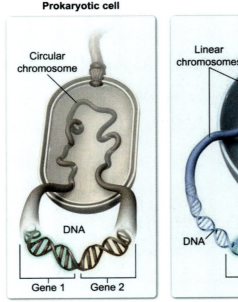
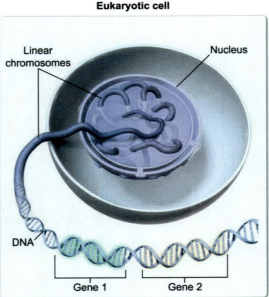

Figure 1.1 DNA in prokaryotic cells versus eukaryotic cells.

The information contained in a gene can be converted to a related molecule known as **ribonucleic acid (RNA)** through a process called transcription. For genes that encode a protein, the RNA is transcribed and processed to become messenger RNA (mRNA), which leaves the nucleus and enters the cytosol after transcription. The mRNA is then translated by a ribosome to produce a protein. Lessons 2.2 and 2.3 explore transcription and translation in more detail.

In addition to genes, each chromosome contains noncoding DNA (Figure 1.2), which accounts for the majority of human genetic material (ie, the genome). **Noncoding DNA** does not code for proteins or other known functional biomolecules. While the function of some noncoding regions remains unknown, many regions are known to be involved in maintenance of chromosomal integrity (eg, telomeres) or regulation of gene expression.

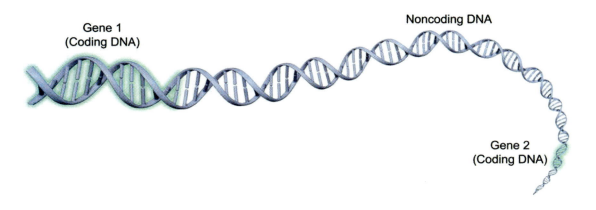

Figure 1.2 Coding versus noncoding DNA.

1.1.02 Nucleotides and Nucleic Acids

DNA and RNA are composed of **nucleotides**, which are naturally occurring molecules that can be classified as either deoxyribonucleotides (found in DNA) or ribonucleotides (found in RNA). Each nucleotide consists of a sugar linked to a nitrogenous base and one, two, or three phosphate (PO_4^-) groups (Figure 1.3).

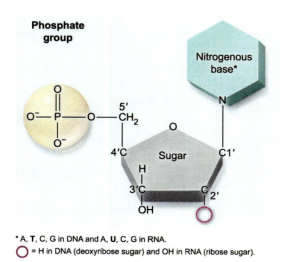

* A, T, C, G in DNA and A, U, C, G in RNA.
○ = H in DNA (deoxyribose sugar) and OH in RNA (ribose sugar).

Figure 1.3 General structure of a nucleotide.

Deoxyribonucleotides differ from ribonucleotides primarily in the type of sugar incorporated into the structure. Ribonucleotides contain ribose, whereas deoxyribonucleotides contain deoxyribose. In both types, each carbon is numbered 1' (1-prime) through 5' (5-prime). In nucleotides, a negatively charged phosphate group is attached to the sugar at the 5' carbon, and a hydroxyl (OH) group is attached to the 3' carbon, as shown in Figure 1.3.

The nitrogenous base is attached to the 1' carbon of the sugar. Nitrogenous bases may be either **purines** (containing two rings) or **pyrimidines** (containing one ring). Both DNA and RNA contain the purines adenine (A) and guanine (G), and the pyrimidine cytosine (C). However, DNA contains thymine (T) and RNA contains uracil (U), which are both pyrimidines.

Nucleotides are the building blocks of **nucleic acids**, which are polymers composed of multiple nucleotide monomers linked together by covalent bonds (Figure 1.4). Linkages between nucleotides form when the 3' end of one nucleotide reacts with the 5' end of another, forming a dinucleotide.

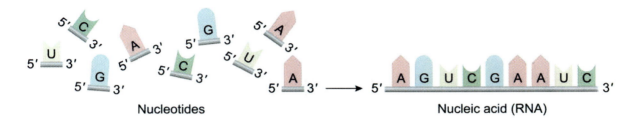

Figure 1.4 Nucleotides join to form nucleic acids.

The 3' end of the resulting dinucleotide can then react with the 5' end of another nucleotide to form a trinucleotide, and so on. This process can eventually lead to the formation of nucleic acids that contain thousands of individual nucleotides linked together by a repeating sugar-phosphate backbone. The linkage of nucleotides is the basis for both DNA synthesis (replication) and RNA synthesis (transcription).

1.1.03 Nucleic Acid Structure

DNA is made up of two distinct nucleic acid strands that wrap around one another to form a **double helix**. The strands are aligned in an antiparallel direction, meaning that the 5' end of one strand aligns with the 3' end of the other strand. As shown in Figure 1.5, the double helix is arranged such that the sugar-phosphate backbone of each strand faces outward and the nitrogenous bases face the inside of the double helix, towards one another.

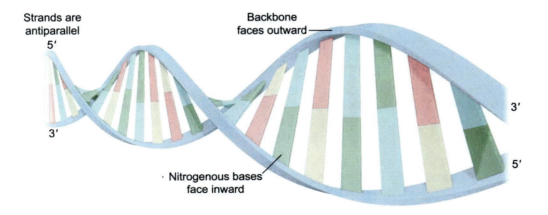

Figure 1.5 General structure of a DNA double helix.

Each nitrogenous base in one strand of the double helix forms hydrogen bonds with a base in the opposite strand to produce a **base pair** (ie, a purine pairs with a pyrimidine). The pairing of nitrogenous bases is highly specific. In DNA, adenine (A) always pairs with thymine (T), and guanine (G) always pairs with cytosine (C) (Figure 1.6). Therefore, A and T bases should always be equal in number, and G and C bases should always be equal in number. This **complementarity** guides DNA replication.

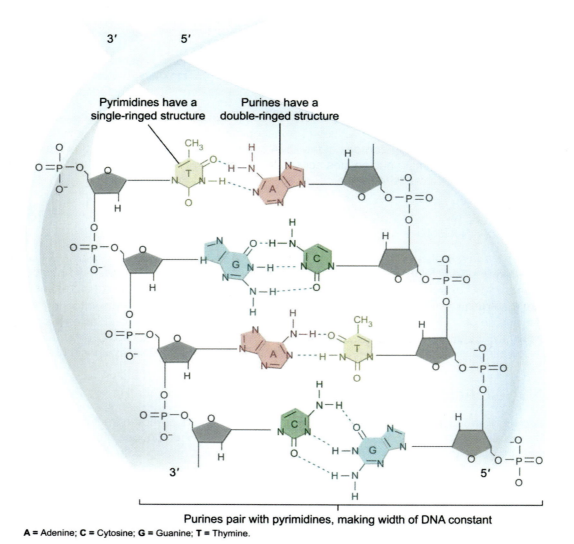

Figure 1.6 The pairing of nitrogenous bases is highly specific.

Unlike DNA, which is made up of two nucleic acid strands, RNA is composed of a single nucleic acid strand. Both double- and single-stranded nucleic acids may exhibit complementary base pairing. For example, during transcription, ribonucleotides in the growing mRNA strand pair with deoxyribonucleotides in the portion of the DNA strand being transcribed (see Lesson 2.2).

The complementarity of this pairing is the same as in DNA, with the exception that A in the DNA strand pairs with uracil (U), instead of T in the mRNA strand. During transcription, A in the RNA strand continues to pair with T in the DNA strand. RNA base pairing with itself occurs primarily when an RNA strand forms a loop called a hairpin by bringing complementary bases within the strand into proximity of one another.

> ## ☑ Concept Check 1.1
>
> A fragment of DNA contains 157 adenine (A) bases and 225 guanine (G) bases. What is the total number of nucleotides present in this DNA fragment?
>
> ### Solution
>
> *Note: The appendix contains the answer.*

1.1.04 Nucleic Acid Hybridization and Denaturation

When two nucleic acid strands are joined in a double helix, they are said to be **hybridized**, or annealed. Within a cell, DNA not being used for replication or transcription is fully hybridized. However, for replication or transcription to occur, portions of the DNA double helix must separate to provide the necessary enzymes and incoming nucleotides access to nitrogenous bases in the DNA. The process of separating hybridized DNA strands is known as denaturation or melting.

DNA can be denatured by enzymes or due to environmental factors such as high temperatures, pH levels well outside of physiological levels, and changes in salt concentrations. These environmental factors cause denaturation by disrupting hydrogen bonds and other interactions between bases.

The temperature required to separate half of the double helices in a sample into single strands is referred to as the **melting temperature T_m** of the DNA being assessed (Figure 1.7). An increased number of hydrogen bonds between bases increases the stability of a double helix and therefore increases the T_m for that double helix.

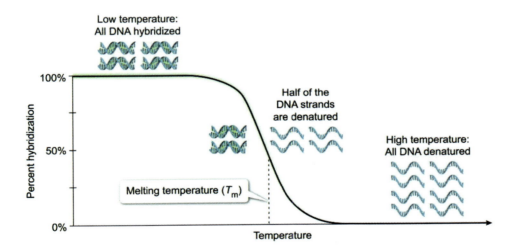

Figure 1.7 Effect of temperature on DNA hybridization.

G-C base pairing involves more hydrogen bonds than A-T or A-U base pairing, so a double helix with a higher percentage of G-C pairs tends to have a higher T_m than a double helix with a lower percentage of G-C pairs. Similarly, longer double helices have more total intermolecular interactions than shorter helices with similar G-C levels, so longer double helices tend to have a higher T_m than shorter helices.

When denatured DNA strands in a solution are returned to temperatures well below their T_m, the strands quickly reanneal. The speed with which reannealing occurs can be influenced by certain factors, including length of the DNA (ie, annealing takes longer when strands are longer), pH of the solution (ie, annealing takes less time when pH is near 7.4 [ie, physiological pH in most cell nuclei]), and salt concentration of the solution (ie, annealing is more stable when cations in a salt interact with the negatively charged phosphate groups in DNA).

These factors must be considered when optimizing conditions for molecular techniques that rely on DNA annealing and denaturation, such as PCR (see Concept 4.1.02).

Lesson 1.2

DNA Replication

Introduction

For a complete and accurate copy of genetic information (DNA) to be passed to daughter cells during cell division, DNA must first be copied (replicated). **DNA replication** takes place during the cell cycle phase known as S phase. This lesson expands on the process of DNA replication, as well as special mechanisms for beginning and ending DNA replication.

1.2.01 Semiconservative Replication

The replication of DNA is **semiconservative**, which means that each new (ie, daughter) strand is synthesized using one of the original (parent or parental) strands as a template. During replication, the parent strands are separated from each other (denatured), and each parent strand is used as a template to synthesize a new, complementary daughter strand (Figure 1.8). When replication is complete, each new double helix is composed of one parent strand (the template) and one newly synthesized daughter strand.

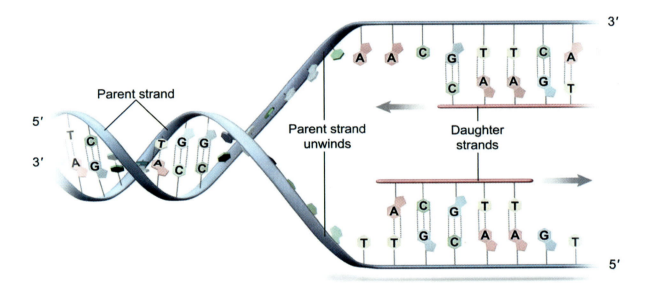

Figure 1.8 Semiconservative DNA replication.

1.2.02 Origin of Replication

DNA replication begins at sites within the DNA called **origins of replication**, which are specific sequences of nucleotides. Enzymes that initiate replication attach to the origin of replication, which triggers separation of the strands and formation of a replication bubble. Replication proceeds in both directions from the origin.

Prokaryotic organisms generally have circular chromosomes with a single origin of replication, and replication occurs in both directions until the entire chromosome is copied. In eukaryotic organisms, chromosomes are linear and may have hundreds of origins of replication, which open simultaneously so replication can occur more quickly. As in prokaryotes, replication in eukaryotes proceeds in both directions from each origin (Figure 1.9).

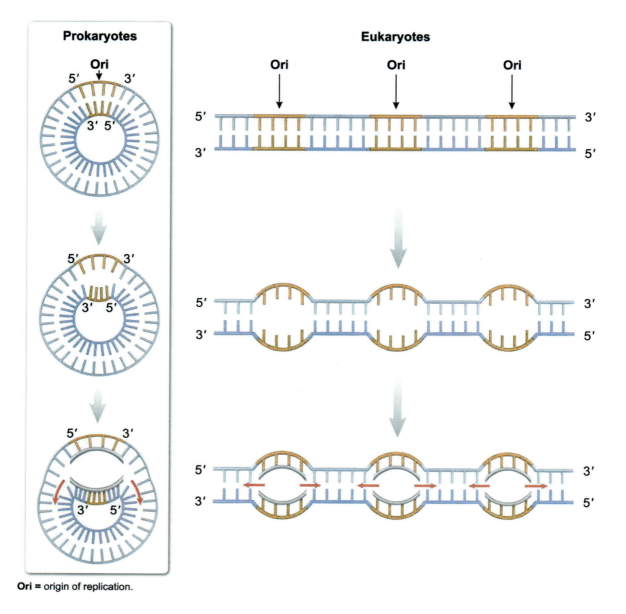

Ori = origin of replication.

Figure 1.9 Origins of replication in prokaryotes versus eukaryotes.

1.2.03 Mechanism of Replication

The mechanism of DNA replication is very similar in prokaryotes and eukaryotes. For simplicity, the prokaryotic model of DNA replication is discussed in this concept.

At the beginning of DNA replication, the enzyme **helicase** unwinds the parent DNA helix at the origin of replication, and **single-stranded DNA-binding proteins** hold the two strands apart (Figure 1.10). This creates a **replication bubble** and two **replication forks**. Synthesis of the daughter strands proceeds from the forks in both directions from each parent strand, lengthening the replication bubble as DNA synthesis continues.

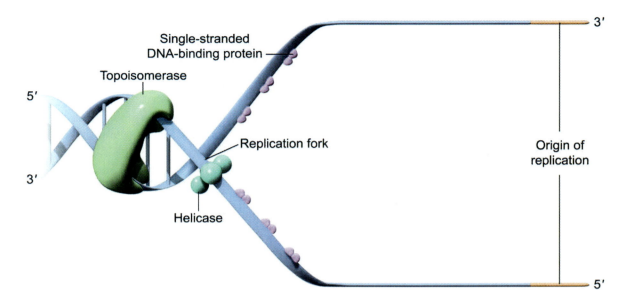

Figure 1.10 Formation of the replication fork.

As helicase unwinds DNA at the replication fork, the DNA ahead of the replication fork becomes overwound (negatively supercoiled). **Topoisomerase** reduces the strain caused by DNA supercoiling in front of the replication fork by transiently cleaving one DNA strand.

Because all DNA polymerases can synthesize DNA only in the 5' → 3' direction, one daughter strand is synthesized continuously toward the replication fork (**leading strand**). The other strand must be synthesized discontinuously in a direction away from the replication fork (**lagging strand**), with new segments being added as more DNA is unwound, as shown in Figure 1.11.

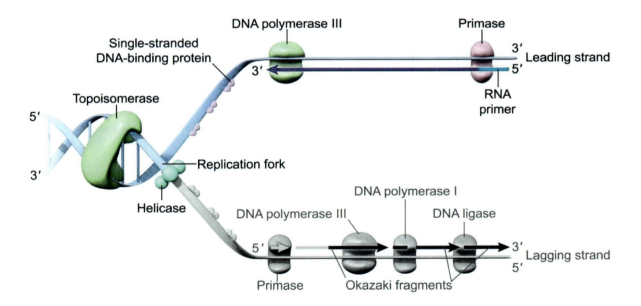

Figure 1.11 Synthesis of DNA from the leading strand.

The initiation of DNA synthesis requires the synthesis of an RNA primer on both the leading and lagging strands because all DNA polymerase enzymes are able to attach new nucleotides only to an existing strand. This short RNA primer is synthesized by the enzyme **primase** in the 5' → 3' direction.

Once the RNA primer is in place, **DNA polymerase III** attaches uncoupled deoxynucleotide triphosphates (dNTPs) to the growing DNA strand. Each dNTP is composed of a nitrogenous base (eg, A, G, C, T) and three phosphate groups. Based on the complementary A-T/G-C base pairing rules, a free dNTP enters the catalytic site of DNA polymerase III and forms hydrogen bonds with a complementary nucleotide on the parent (template) DNA strand. The 3′ OH from the last nucleotide of the growing strand "attacks" the 5′ PO_4^- group of the incoming dNTP.

A pyrophosphate is released from the dNTP, and the nucleotide is attached to the growing strand via a condensation reaction (Figure 1.12). Condensation is an exergonic process, and the energy released from this reaction is used to form a covalent phosphodiester bond between the growing strand and the incoming dNTP.

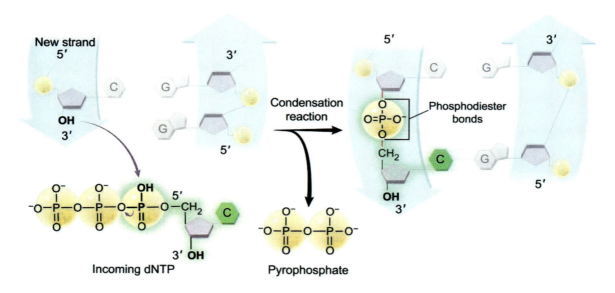

Figure 1.12 Coupling of free nucleotides during DNA replication.

When DNA synthesis is complete, **DNA polymerase I** removes RNA primers and replaces them with DNA nucleotides, and **DNA ligase** catalyzes the formation of phosphodiester bonds to join all DNA fragments.

The replication of the lagging strand is problematic due to the antiparallel orientation of the two DNA strands (Figure 1.13). Because DNA polymerase enzymes are able to synthesize DNA only in the 5′ → 3′ direction, new RNA primers must be added periodically as the replication fork advances. As a result, replication on the lagging strand is discontinuous, and short fragments of newly synthesized DNA called **Okazaki fragments** are formed. RNA primers must be removed from each Okazaki fragment and replaced with DNA before the fragments can be joined by DNA ligase.

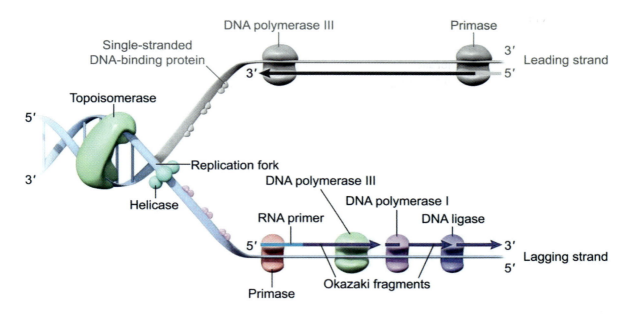

Figure 1.13 Synthesis of DNA from the lagging strand.

The rate of DNA replication in prokaryotes is ~1,000 nucleotides per second, but in eukaryotes it is much slower, at ~50–100 nucleotides per second. While synthesis of the leading and lagging strands occurs simultaneously, synthesis of the lagging strand occurs at a slightly slower rate, as the template strand must be continuously revealed at the replication fork for continued replication of the lagging strand. The various proteins involved in DNA replication are summarized in Table 1.1.

Table 1.1 Proteins involved in DNA replication.

Protein	Function
Helicase	Unwinds DNA double helix to form replication fork
Single-stranded DNA-binding protein	Prevents DNA double helix from reannealing
Topoisomerase	Relieves DNA supercoiling in front of the replication fork
Primase	Synthesizes RNA primer
DNA polymerase III	Performs 5' → 3' DNA synthesis
DNA polymerase I	Replaces RNA primers with DNA
DNA ligase	Joins DNA fragments after replacement of RNA primers

1.2.04 Replication at the Ends of Linear Chromosomes

DNA replication machinery requires a free 3' end to incorporate a new dNTP into the daughter strand, so replicating the ends of a linear chromosome poses a particular problem. When the 3' end of the template is too short to engage with DNA polymerase, DNA synthesis cannot continue, and chromosome ends are incompletely replicated.

Because these uneven, overhanging ends cannot be copied, the chromosome becomes shorter with each successive round of replication (Figure 1.14). Shortening of chromosomes is not an issue in prokaryotes, since prokaryotic chromosomes are generally circular; however, shortening of chromosomes is problematic in eukaryotes, because eukaryotes have linear chromosomes.

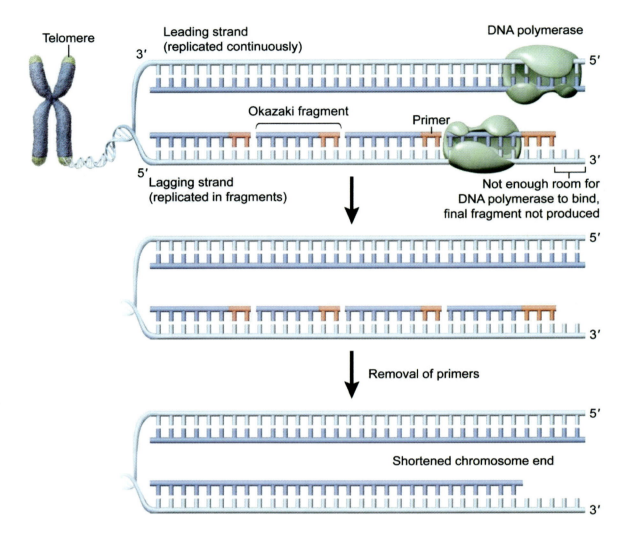

Figure 1.14 Shortening of linear chromosome ends during replication.

The progressive shortening of chromosomes is addressed in eukaryotes by the addition of noncoding DNA sequences called **telomeres** to the ends of chromosomes (see Figure 1.15). Telomeres consist of copies of a six-nucleotide sequence (5'-TTAGGG-3') repeated 100 to 1,000 times at the end of each chromosome. Telomeres do not contain coding information, so the portions of the telomeres lost with each replication cycle do not result in a loss of genetic information.

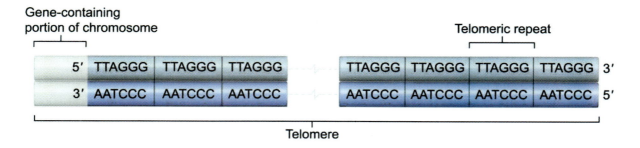

Figure 1.15 Telomere sequences at the end of a linear chromosome.

Telomeres function to protect the chromosome from loss of important DNA sequences as the chromosome becomes increasingly shorter with each round of replication. The protection of chromosome ends via looping and telomere-specific proteins also prevents activation of DNA repair mechanisms (discussed in Lesson 1.3), because overhanging ends may trigger a response that could lead to cell cycle arrest or cell death.

Normal shortening of chromosomes may protect organisms from cancer by limiting the number of somatic cell divisions that can occur. The successive shortening of telomeres has also been linked with cellular and organism aging, since this shortening results in a finite number of possible cell divisions. In cells that must retain the potential to divide indefinitely (ie, stem cells and germ cells), and in some cancer cells, the enzyme telomerase is expressed and functions to replenish shortened telomere sequences on the ends of chromosomes (see Concepts 5.4.04 and 10.5.02).

Lesson 1.3
DNA Repair

Introduction

DNA replication must be highly accurate to avoid introduction of mistakes into newly copied DNA strands. The error rate for DNA polymerases is estimated to be once every ~10^5 nucleotides. One factor that allows for such high accuracy is the nucleotide selectivity of DNA polymerases. DNA polymerases also have a proofreading function, which further improves the accuracy of replication by a factor of 100–1,000.

In addition to their ability to add nucleotides to a growing strand, DNA polymerases possess exonuclease activity, which allows them to remove bases from the end of a DNA strand. When a mismatch is detected directly after replication or when DNA damage occurs outside of the replication process, distinctive DNA repair pathways are activated. This lesson discusses various mechanisms of DNA proofreading and repair.

1.3.01 DNA Proofreading and Mismatch Repair

During DNA replication, DNA polymerases may incorporate an incorrect nucleotide into the newly synthesized strand, creating a base pair mismatch between the daughter and parent strands (eg, G-T and A-C mispairing instead of proper G-C and A-T pairing). DNA polymerases are equipped with 3' → 5' **exonuclease** activity, which acts as a proofreading function, allowing incorrectly paired nucleotides to be replaced during DNA replication (Figure 1.16).

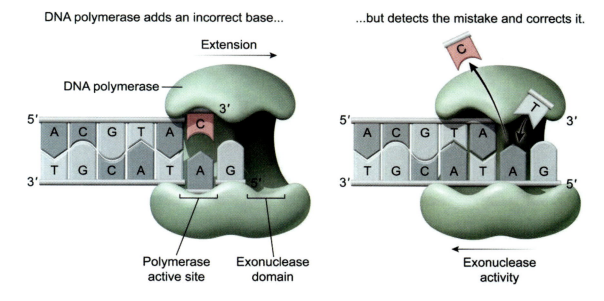

Figure 1.16 DNA polymerases can correct nucleotide mismatches during replication.

When a base pair mismatch is detected immediately after replication, the DNA **mismatch repair (MMR)** system is activated, as shown in Figure 1.17. In MMR, an endonuclease enzyme removes (excises) the mismatched base and several nucleotides on either side of it from the daughter strand.

Prokaryotes can distinguish the template from the daughter strand by the state of strand methylation. The template strand has methyl groups attached to some of its bases, whereas the newly synthesized daughter strand does not. In eukaryotes, the process of determining which strand contains the mismatched base involves recognition of single-stranded breaks found only in newly synthesized DNA.

MMR enzymes must have **endonuclease**, rather than exonuclease, activity because mismatched nucleotides must be excised from within an existing DNA strand, and only endonucleases can cleave phosphodiester bonds in the middle of a nucleic acid strand. After mismatched nucleotides are excised, DNA polymerase incorporates appropriate nucleotides and DNA ligase catalyzes the formation of new phosphodiester bonds to rejoin the strands.

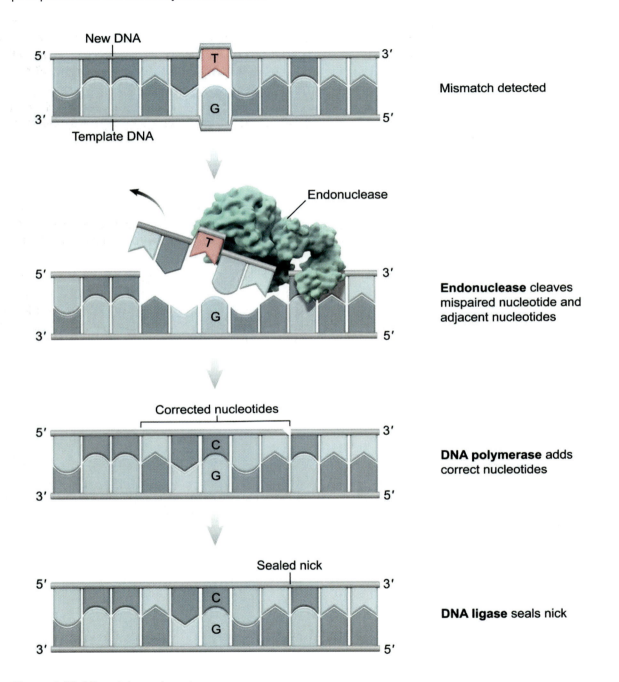

Figure 1.17 Mismatch repair system.

When both the inherent accuracy of DNA polymerase, its proofreading ability, and the MMR system are taken into account, the error rate of DNA replication is approximately one in every 10^{10}–10^{11} nucleotides.

1.3.02 DNA Damage and Repair

Errors in replication are a common cause of DNA mismatches, but at other times during the cell cycle nucleic acids may also undergo spontaneous changes or damage due to exposure to harmful conditions (eg, UV light, high energy radiation, X-rays, chemicals). When DNA damage is detected, specific DNA repair pathways are activated, depending on the type of damage.

The **base excision repair** system (see Figure 1.18) corrects damage that does not cause significant distortion of the double helix, such as oxidation, deamination, and alkylation of bases. A DNA glycosylase enzyme recognizes the site of DNA damage and excises the damaged base. An endonuclease enzyme then cleaves the phosphodiester bond, and DNA polymerase fills the gaps while DNA ligase rejoins the DNA strands.

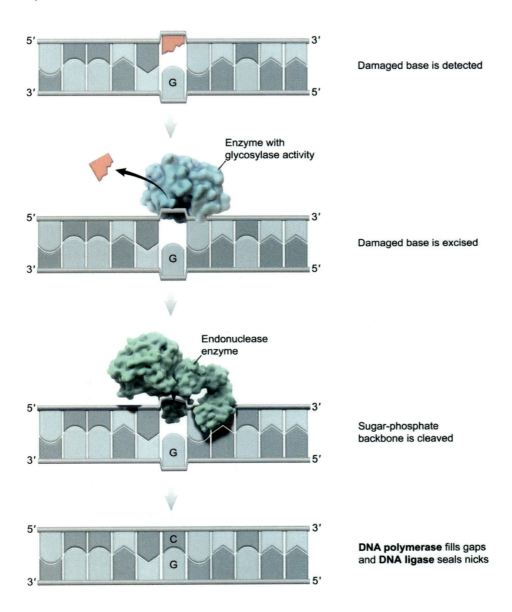

Figure 1.18 Base excision repair system.

If damage to DNA is more extensive or bulky, the **nucleotide excision repair (NER)** pathway is activated, shown in Figure 1.19. For example, UV radiation can cause thymine dimers, or linkages of two thymine bases, which leads to a distortion in the double helix. When this type of damage is detected,

NER endonuclease enzymes are mobilized to remove the damaged region. DNA polymerase and DNA ligase subsequently fill in the gaps and rejoin the repaired DNA strands.

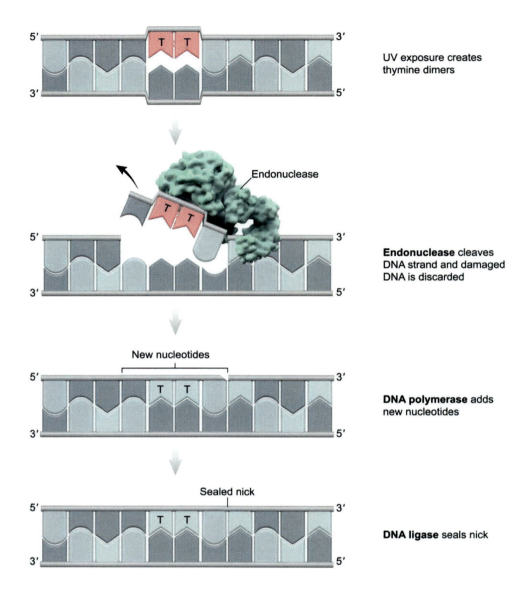

Figure 1.19 Nucleotide excision repair system.

In some situations, double-stranded breaks in DNA can separate parts of the chromosome. If these breaks are not repaired, large portions of the chromosome can be lost. There are two major mechanisms for repairing double-stranded breaks: homologous recombination and nonhomologous end joining. These two mechanisms are summarized in Figure 1.20.

If an organism is diploid or a sister chromatid is nearby (such as during the S and G_2 phases of the cell cycle), the corresponding chromosome is used as a template to repair the broken chromosome, a process known as **homologous recombination**. However, if a suitable homologue is not present (eg, in haploid organisms or when no duplicated chromosome is present), broken strands can be rejoined by enzymes via **nonhomologous end joining**. Because there is no template involved in nonhomologous end joining, mutations are often incorporated at the site where the DNA is rejoined.

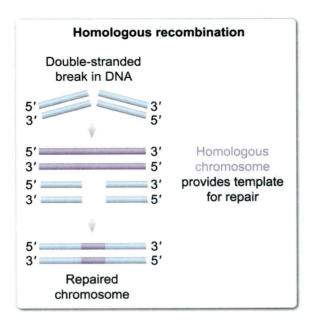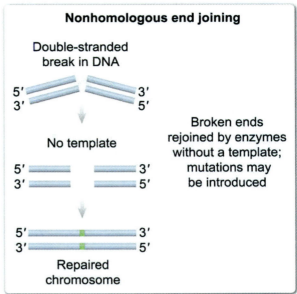

Figure 1.20 Homologous recombination and nonhomologous end joining.

Lesson 1.4
Eukaryotic Chromosome Organization

Introduction

In a typical eukaryotic cell, each linear chromosome consists of a DNA molecule that contains an average of 1.5×10^8 nucleotide pairs. A completely unwound chromosome would be about 4 cm long, which poses a problem because many chromosomes must fit into the nucleus of a cell.

To address this problem, eukaryotic cells organize and condense DNA, packaging it densely with proteins to create **chromatin**. Chromatin packaging changes dramatically during cell division, and modification of chromatin structure is intricately linked with gene regulation. This lesson explores the organization of eukaryotic chromosomes and different states of chromosome organization.

1.4.01 Histones and Nucleosomes

During the initial steps of chromatin formation, the DNA double helix wraps around a complex of eight **histone** proteins (an octamer) two times to form a structural subunit called a **nucleosome** (Figure 1.21). There are five major histone types (ie, H1, H2A, H2B, H3, and H4). A DNA-wrapped octamer core is composed of eight subunits (two each of H2A, H2B, H3, and H4), and a linker histone (H1) is present outside of the core.

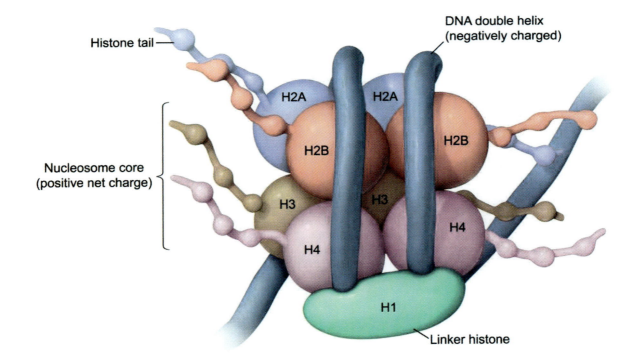

Figure 1.21 Components of a nucleosome.

Because DNA is negatively charged, histone proteins must have a net positive charge to facilitate DNA binding. The rich presence of arginine and lysine (positively charged, basic amino acids) in histones confers this necessary positive charge.

In unwound chromatin, the nucleosomes resemble beads on a string, and the DNA between each "bead" is called linker DNA (see Figure 1.22). During replication and transcription, when access to the DNA helix is required, the histones leave the DNA for a short period of time.

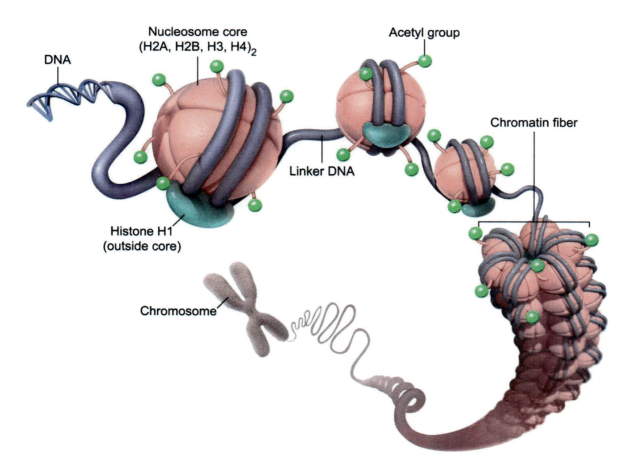

Figure 1.22 Eukaryotic DNA organization.

1.4.02 Euchromatin and Heterochromatin

When DNA is not actively being replicated or transcribed, chromatin may be configured either loosely (**euchromatin**) or densely (**heterochromatin**), as shown in Figure 1.23. Whether chromatin is configured as euchromatin or heterochromatin impacts its accessibility and the expression of genes located in that region of the chromosome. During cell division, chromatin becomes even more condensed in preparation for mitosis (see Concept 5.4.02).

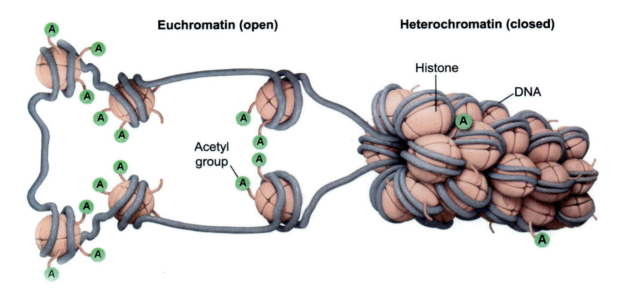

Figure 1.23 Euchromatin and heterochromatin.

Heterochromatin consists of DNA tightly coiled around histone proteins, bound by ionic interactions between negatively charged phosphates on the DNA backbone and positively charged lysine residues on the histones. Because DNA in heterochromatin is densely packed, it is not readily accessible to transcription machinery. Many key noncoding regions of the chromosome, such as telomeres and centromeres, are composed of heterochromatin.

Euchromatin forms when histones are modified, often due to acetylation of lysine residues. Acetylation neutralizes the positive charge on the lysine residue, which reduces interactions between histones and DNA. These reduced interactions yield a more open form of chromatin that allows better accessibility for transcription machinery.

Lesson 2.1
The Central Dogma of Molecular Biology

Introduction

As presented in Chapter 1, the information needed for an organism's development and vital processes is stored in its genetic code (DNA), and the transmission of genetic information from one generation to the next cannot proceed without DNA replication. However, the *expression* of an organism's genetic code is dependent on the flow of information from DNA to RNA to proteins, mediated by the processes of transcription and translation. This lesson provides an overview of the mechanism by which genes are expressed, otherwise known as the **central dogma of molecular biology**.

2.1.01 Flow of Genetic Information

The expression of genes is the highly regulated manifestation of a set of genetic instructions (DNA) into a physical form. It is the link between the genetic code of an organism (its **genotype**) and the organism's physical and biochemical attributes (its **phenotype**), as illustrated in Figure 2.1.

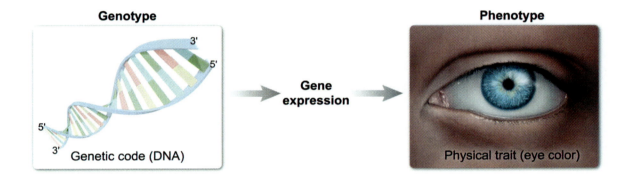

Figure 2.1 An organism's physical and biochemical attributes are a consequence of gene expression.

The **central dogma of molecular biology** explains how genetic information, stored as DNA, is used to regulate the synthesis of proteins (polypeptides). In turn, the proteins made using this genetic information control the activities of the cell and organism.

2.1.02 Gene Expression

According to the central dogma of molecular biology, gene expression can be divided into two stages: transcription and translation (Figure 2.2). During **transcription**, information stored in DNA is copied, or transcribed, into a more mobile form known as RNA. Multiple types of RNA are involved in gene expression, but only **messenger RNA** (mRNA) contains the information needed to synthesize a protein via **translation**. During translation, a ribosome and two other forms of RNA (transfer RNA [tRNA] and ribosomal RNA [rRNA]) are used to decode the information in mRNA, resulting in a protein product.

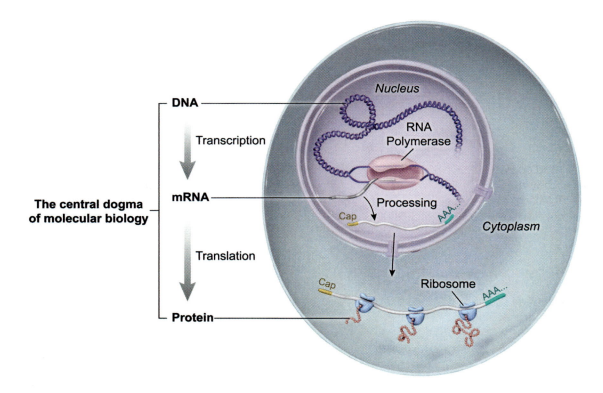

Figure 2.2 The central dogma of molecular biology.

In general, the information encoded in an individual mRNA results in the synthesis of one protein (polypeptide), consisting of a single chain of amino acids. However, some proteins consist of multiple associated polypeptide chains (subunits), which may be derived from more than one gene. Each unique protein subunit is translated from its own mRNA, associating with additional subunits once translation is complete.

The basic mechanisms of transcription and translation are the same in all organisms. In eukaryotic organisms, transcription and translation are separated in both space and time. Transcription occurs first, in the nucleus, and translation occurs later, outside of the nucleus. One important difference between prokaryotic and eukaryotic gene expression is that prokaryotic organisms generally do not have membrane-bound organelles, so DNA is not separated from the rest of the cell in a nucleus.

While prokaryotic and eukaryotic gene expression have many similarities, this chapter focuses on eukaryotic gene expression and regulation. Additional information about prokaryotic gene regulation is presented in Concept 6.3.05.

Lesson 2.2
Transcription

Introduction

Genomic DNA can be classified based on characteristics such as nucleotide composition, sequence structure, and coding potential (ie, whether a region of DNA contains information for generating a product). Only a small fraction of eukaryotic DNA contains coding information, giving rise to different types of RNA (eg, mRNA, rRNA, tRNA). While all types of RNA may be created via transcription, this lesson focuses on the transcription of mRNA, which is the only type of RNA that contains the information needed to synthesize proteins.

2.2.01 Mechanism of Transcription

The process of gene expression (ie, DNA → mRNA → protein) begins when the cell receives signals indicating that transcription should take place. Because the majority of DNA is noncoding, certain boundaries exist that define a **transcription unit** (ie, where a coding region begins and ends). These boundaries are marked by specific sequences of nucleotides called **promoters** and **terminators**, which identify where transcription should begin and end, respectively (Figure 2.3).

The promoter and the DNA sequences before the transcription unit are considered *upstream*, and the region to be transcribed is considered *downstream*. These terms also describe the position of nucleotides within a transcription unit. For example, the terminator is *downstream* of the coding region.

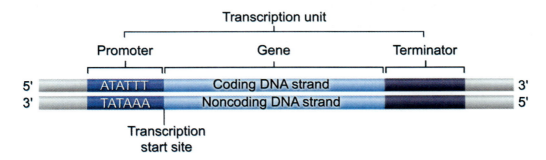

Figure 2.3 A transcription unit.

Transcription is similar to DNA replication (see Lesson 1.2) in that the enzyme that synthesizes mRNA molecules (**RNA polymerase II**) can assemble nucleotides in the 5′ → 3′ direction only. However, unlike DNA polymerase III, RNA polymerase II does not need a primer to begin assembling nucleotides.

Transcription occurs in three broad stages: initiation, elongation, and termination. **Initiation** begins with the recognition of the promoter sequence, which is generally located approximately 25–50 nucleotides upstream of the transcription start site (Figure 2.4). A **transcription initiation complex** is formed when RNA polymerase II and other regulatory proteins known as **general (basal) transcription factors** bind to a specific nucleotide sequence called the **promoter**. There are a variety of promoter sequences, but the most well-known is the TATA box, which consists of a sequence of thymine (T) and adenine (A) nucleotides.

Once the transcription machinery is assembled at the promoter, RNA polymerase II unwinds the DNA helix and transcription begins at the transcription start site. One strand is known as the **coding** (sense) strand (5′ → 3′), and the other is the **noncoding** (antisense) strand (3′ → 5′).

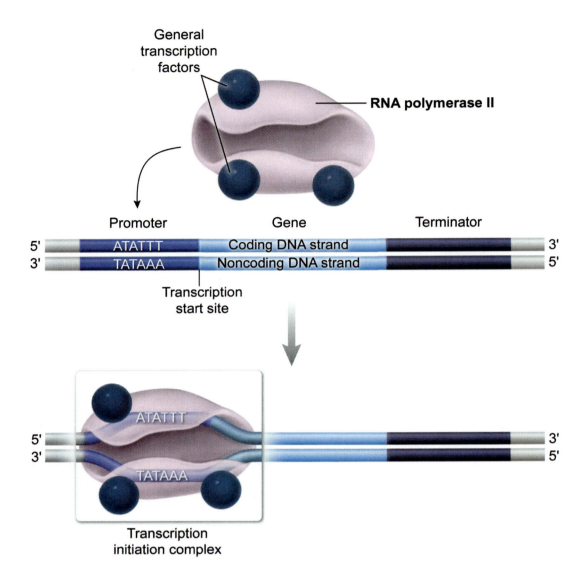

Figure 2.4 Transcription initiation.

During **elongation**, the noncoding DNA strand (ie, the 3' → 5' DNA strand) is used as a template to build the mRNA molecule, as shown in Figure 2.5. The mRNA transcript is synthesized in the 5' → 3' direction by RNA polymerase II through complementary base pairing with the noncoding DNA strand. Accordingly, the DNA coding strand has a similar sequence and directionality (5' → 3') as the newly synthesized mRNA transcript, except that the mRNA transcript has uracil (U) nucleotides where the DNA has thymine (T) nucleotides.

As RNA polymerase II progresses along the template strand, the DNA double helix reforms behind the enzyme, and multiple RNA polymerase II molecules can transcribe a single gene at the same time.

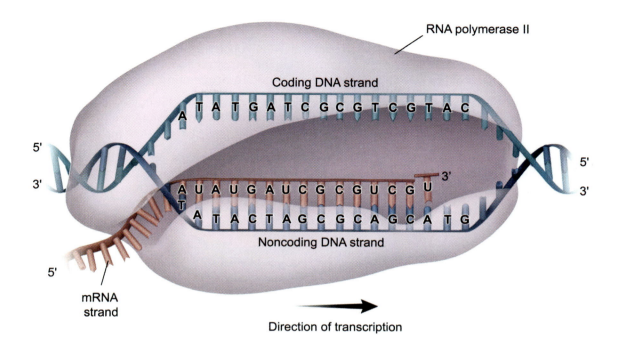

Figure 2.5 Transcription elongation.

Termination of transcription occurs when the transcription complex reaches the terminator sequence in the DNA (Figure 2.6). In prokaryotes, the mRNA transcript then detaches from the DNA and is available for translation with no further modification.

In eukaryotes, RNA polymerase II transcribes a sequence in the DNA known as the **polyadenylation signal sequence** (3'-TTATTT-5'), which is then bound by proteins that separate the mRNA transcript from RNA polymerase II. At this time, the mRNA transcript is considered **pre-mRNA** (ie, primary transcript) and must undergo further processing before it can be exported from the nucleus as a **mature mRNA** transcript.

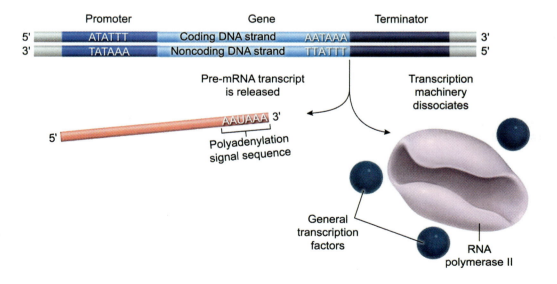

Figure 2.6 Transcription termination.

2.2.02 Modifications to mRNA Ends

While prokaryotic mRNA is immediately available for translation once transcription is complete, eukaryotic pre-mRNA must undergo certain modifications before it can be exported from the nucleus and translated. Modifications to both ends of pre-mRNA are the first steps in **RNA processing**, shown in Figure 2.7.

The 5' end of pre-mRNA is modified almost immediately after it is transcribed with the addition of a **5' cap**, which is a modified guanosine triphosphate (GTP) nucleotide (7-methylguanosine [m⁷G]). This 5' cap is recognized by the ribosome during translation and prevents degradation of mRNA in the cytoplasm.

When transcription is complete, a chain of adenine nucleotides known as the **poly-A tail** is added to the 3' end of mRNA, directly downstream of the transcribed polyadenylation signal sequence. Like the 5' cap, the poly-A tail also functions to prevent mRNA degradation. In addition, the poly-A tail facilitates export of mature mRNA from the nucleus to the cytoplasm.

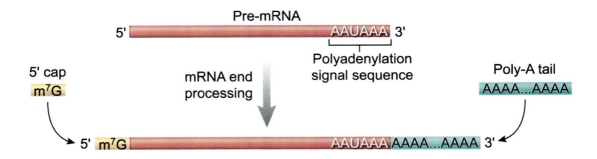

Figure 2.7 5' and 3' mRNA end processing.

2.2.03 RNA Splicing Mechanisms

In eukaryotes, gene structure is considerably more complex than in prokaryotes. The vast majority of DNA in eukaryotes is noncoding, and DNA sequences that code for a particular protein may be interrupted by noncoding regions. During transcription, these noncoding regions are transcribed, resulting in a pre-mRNA transcript that contains coding and noncoding information.

To produce mature mRNA transcripts, the intervening noncoding regions (**introns**) must be removed, and coding regions (**exons**) must then be linked together. Therefore, the pre-mRNA transcript is often significantly longer than the mature mRNA transcript (Figure 2.8).

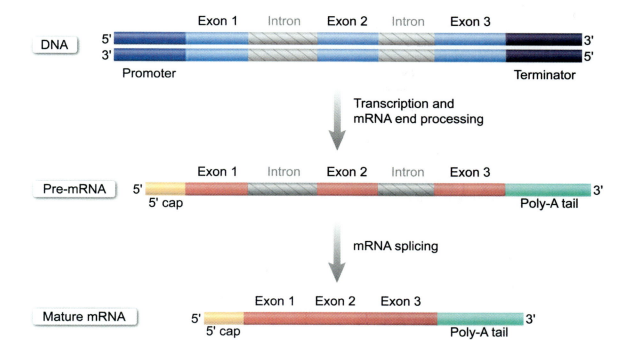

Figure 2.8 mRNA processing is required to form a mature mRNA transcript.

The process of removing noncoding regions, called **RNA splicing**, is carried out by a type of molecular machine called the **spliceosome**, illustrated in Figure 2.9. Spliceosomes are composed of specific proteins and small nuclear riboproteins (snRNPs) containing small nuclear RNA (snRNA). Spliceosomes recognize specific sites in introns called **splice donor sites** and **splice acceptor sites**.

Splice donor sites are located at the 5' end of introns, adjacent to the 3' end of the upstream exon, and splice acceptor sites are found at the 3' end of introns, adjacent to the 5' end of the exon directly downstream. Guided by snRNAs, spliceosomes remove intronic sequences and join exons to produce mature mRNA transcripts with no introns.

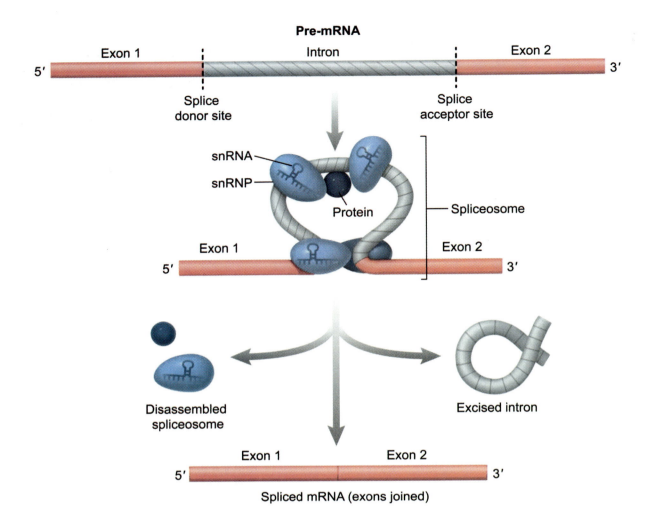

Figure 2.9 Spliceosomes remove introns and join adjacent exons.

During RNA splicing, exons and introns can be removed in a differential manner, which means that multiple unique mRNA transcripts can be made from the same pre-mRNA. This **alternative mRNA splicing** allows for the synthesis of multiple distinct proteins (**isoforms**) from a single mRNA transcript based on differential inclusion or exclusion of exons during splicing (Figure 2.10).

Each differentially spliced mRNA contains a subset of exons in the pre-mRNA and may also contain portions of introns. Differentially spliced mRNAs and their resulting protein isoforms are often used in regulatory roles, due to their different and sometimes unique properties. New combinations of exons due to alternative splicing may provide an evolutionary benefit (see Concept 8.1.06), because new proteins with unique structures and functions can be generated from a single gene.

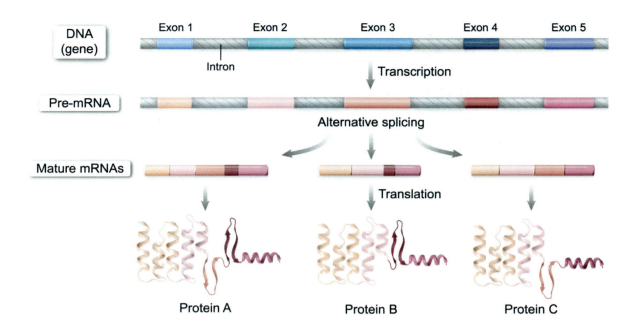

Figure 2.10 Alternative mRNA splicing.

Chapter 2: Gene Expression

Lesson 2.3
Translation

Introduction

The process of transcription (described in Lesson 2.2) involves using DNA as a template to transcribe a copy of the genetic code in the form of messenger RNA (mRNA), a similar molecule. This process is not overly complex because the two molecules (DNA and mRNA) are both nucleic acids, and complementary base pairing is an efficient and natural way to transfer the information from a DNA molecule into mRNA form.

Translation is a more complex enterprise, by which the information encoded in mRNA (a nucleic acid) must be transformed into a completely different type of macromolecule (a protein). This lesson details the mechanism by which translation is carried out by a ribosome and how mutations or changes in the genetic code may result in changes to the protein that is produced.

2.3.01 Interpreting the Genetic Code

As established in Lesson 1.1, nucleotides are the building blocks of nucleic acids (ie, DNA and RNA), and there are four types of nucleotides in RNA (A, G, C, and U). The building blocks of proteins are amino acids, of which there are 20 different types. In **translation**, a cell converts genetic information encoded within the nucleotide sequence into a protein.

Because there are 20 types of amino acids in proteins and only four types of nucleotides in mRNA, the code cannot read as 1:1 (one nucleotide specifying one amino acid). Rather, a combination of three nucleotides (termed a **codon**) specifies a particular amino acid in the protein. A **codon table** can be used to determine which amino acid a particular codon specifies, as shown in Figure 2.11.

Therefore, translation is the process by which the order of nucleotides (in the form of codons) in the mRNA transcript specifies the order in which amino acids are added to the growing protein chain. During translation, each codon in the mRNA transcript is read by the translation machinery in the 5′ → 3′ direction.

Figure 2.11 The codon table for mRNA.

As codons are read by translation machinery in groups of three, it is crucial that the *correct* groups of three are used to decode the information in the mRNA transcript. The translation machinery recognizes and begins translation at a specific set of three nucleotides (AUG) known as the **start codon**. Starting translation with a defined start codon ensures that the translation machinery sets the correct **reading frame**, as illustrated in Figure 2.12.

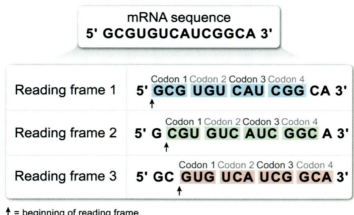

↑ = beginning of reading frame.

Figure 2.12 An mRNA sequence has three possible reading frames.

Incorrect groupings of nucleotides can result in the formation of completely different proteins, similar to how incorrect groupings of letters change a sentence's meaning. For example, if read using the correct starting point, the statement, "the car was red" is a complete sentence with recognizable words. If the statement is read in groups of three letters starting just one letter to the right, the statement becomes, "hec arw asr ed," which is no longer a coherent sentence.

By always starting with the correct group of three nucleotides (ie, the start codon), the translation machinery ensures that the reading frame is accurate, and the correct protein is made. There must also be a way to specify when the translation machinery should stop attaching amino acids to the protein chain. Just as start codons designate where translation begins, **stop codons** (UGA, UAG, or UAA, see Figure 2.11) specify where translation ends.

2.3.02 Transfer RNA (tRNA) and Anticodons

To convert information encoded in the mRNA into the correct sequence of amino acids in the protein, each codon (made up of three mRNA nucleotides) must correspond with a specific amino acid. The key to matching the correct amino acid with its specific codon is found in another type of RNA, called **transfer RNA (tRNA)**.

A tRNA molecule is a small, single-stranded RNA molecule that base pairs with itself to form a complex three-dimensional L-shaped structure, which is often depicted as a cloverleaf shape for simplicity. When folded, the 5' and 3' ends of the tRNA are both located near the same end of the molecule. At the opposite end of the molecule, a loop called the **anticodon** is formed. Figure 2.13 depicts both the cloverleaf and three-dimensional form of tRNA structure.

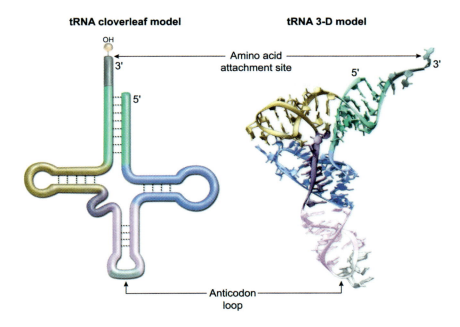

Figure 2.13 tRNA structure.

The anticodon is a three-nucleotide sequence found in each tRNA which is complementary to a particular mRNA codon. Although multiple codons may correspond to the same amino acid (see Concept 2.3.01), each tRNA is paired with a specific amino acid based on the unique sequence of the tRNA's anticodon.

For example, an mRNA codon for the amino acid threonine is 5'-ACA-3'. The corresponding tRNA has the complementary anticodon 3'-UGU-5' and should always carry the amino acid threonine on its 3' end, as shown in Figure 2.14. The tRNA molecule is the necessary link to "decode" or convert the information found in the nucleotide sequence of the mRNA, to specify the correct order of amino acids during translation.

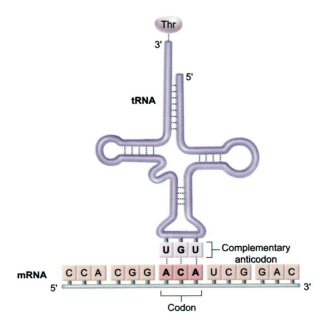

Figure 2.14 tRNA anticodons base pair with mRNA codons.

Nucleotides in the first or second position within a codon require traditional Watson-Crick base pairing with the tRNA anticodon. However, the nucleotide in the third position may undergo less stringent base pairing in a non-Watson-Crick manner known as **wobble pairing**. For example, a tRNA with the anticodon 3'-UCA-5' can base pair with the mRNA codons 5'-AGU-3' and 5'-AGC-3', both of which code for the amino acid serine, as shown in Figure 2.15.

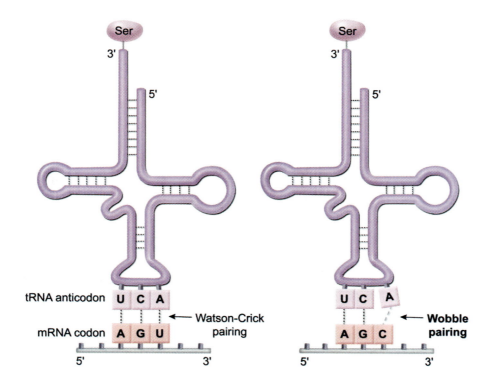

Figure 2.15 Example of wobble pairing.

Like mRNA molecules, tRNA molecules must be transcribed from a DNA template in the nucleus; however, tRNA molecules are transcribed by a different RNA polymerase (RNA polymerase III). Subsequently, tRNA molecules are transported to the cytoplasm, where the correct amino acid is attached and used to build a protein during translation (see Concept 2.3.05). While it is important to maintain accuracy during transcription from DNA to RNA, it is equally important that the correct amino acid is attached to its designated tRNA to prevent the wrong amino acid from being incorporated into a protein.

To pair amino acids with the correct tRNA molecules, a family of enzymes known as **aminoacyl-tRNA synthetases** are employed. For each of the 20 amino acids, there is a designated aminoacyl-tRNA synthetase, which joins a specific amino acid to its appropriate tRNA via an ester linkage. Once the amino acid has been joined to its corresponding tRNA, the tRNA is charged and is ready to deliver its amino acid to the growing protein chain (Figure 2.16).

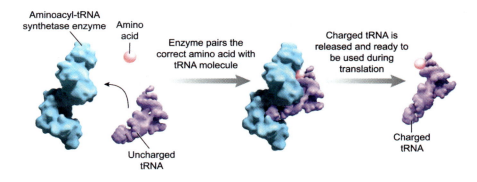

Figure 2.16 Charging of tRNAs by aminoacyl-tRNA synthetases.

2.3.03 The Degenerate Code

There are 64 possible combinations of three nucleotides (ie, codons), but only 20 amino acids. Therefore, the genetic code is considered *degenerate*, or *redundant*, meaning that multiple codons code for the same amino acid (Figure 2.17).

The degenerate nature of the genetic code should not be mistaken for *ambiguity* within the code. During translation, each tRNA is assigned to a single, specific amino acid. In many cases, codons that specify the same amino acid differ only at the nucleotide in the third position of the codon. Although multiple codons can code for the same amino acid, certain codons can be utilized preferentially to increase the speed of translation and the amount of protein produced.

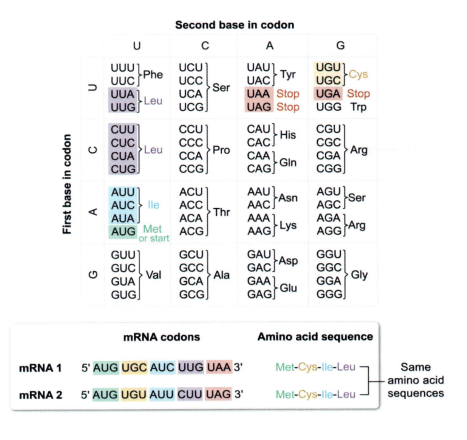

Figure 2.17 Degeneracy in the genetic code.

2.3.04 Ribosomes and Ribosomal RNA (rRNA)

Ribosomes participate in translation by providing a location where mRNA and tRNA molecules can interact and by catalyzing the formation of peptide bonds between amino acids to create a protein chain.

Eukaryotic ribosomes are composed of two parts: a large (60S) subunit and a small (40S) subunit, which are assembled to form a complete (80S) ribosome. Each subunit is made up of a complex of proteins and another specialized type of RNA, called **ribosomal RNA** (rRNA, Figure 2.18). Eukaryotic (80S) and prokaryotic (70S) ribosomes are remarkably similar in their structure and mechanism of action. Prokaryotic large (50S) and small (30S) subunits are distinguishable from eukaryotic subunits both in size and composition, but all ribosomes perform the same essential functions.

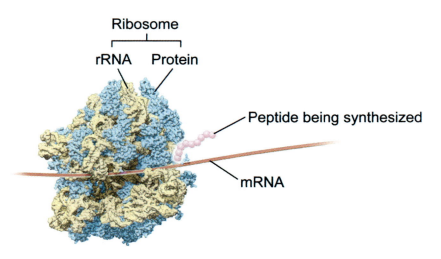

Figure 2.18 Ribosomes are composed of proteins and rRNA.

Like mRNA and tRNA, rRNA must be transcribed from DNA (ie, ribosomal DNA) in the nucleus. Because of its central structural role in the formation of ribosomes, rRNA is the most abundant form of RNA in the cell. In eukaryotes, rRNA genes are clustered in a specialized region of the nucleus called the **nucleolus**. rRNA genes are mostly transcribed by RNA polymerase I and undergo processing in the nucleolus; subsequently, transcribed rRNA is assembled into ribosomal subunits with proteins imported from the cytoplasm. Completed subunits are exported to the cytoplasm to perform their role in translation.

Within a fully assembled ribosome (large + small subunits), there are locations where mRNA and tRNA can interact. The small ribosomal subunit contains an mRNA binding site, and the large ribosomal subunit contains three tRNA binding sites and catalyzes peptide bond formation via its peptidyl transferase activity (Figure 2.19).

These three tRNA binding sites include the peptidyl-tRNA binding site (**P site**), which holds the tRNA carrying the growing protein chain; the aminoacyl-tRNA binding site (**A site**), which holds the tRNA carrying the next amino acid to be added to the chain; and the exit site (**E site**), which is where the discharged (ie, empty) tRNA leaves the ribosome after the amino acid has been transferred to the growing protein.

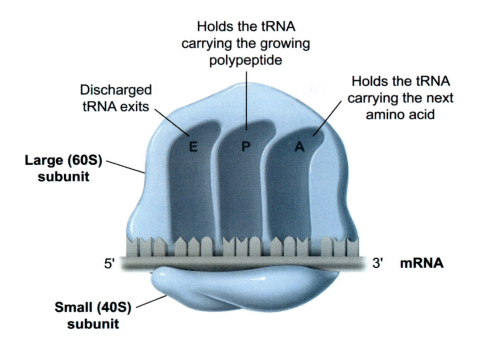

Figure 2.19 Interior sites within an assembled ribosome.

2.3.05 Mechanism of Translation

Like transcription, translation can be divided into three phases: initiation, elongation, and termination. Each phase is regulated by additional factors, which are usually proteins. Some steps of translation require the hydrolysis of guanosine triphosphate for energy.

Eukaryotic translation **initiation** begins when the small (40S) ribosomal subunit binds to a specific **initiator tRNA** carrying the amino acid methionine. Along with initiation factors, the initiator tRNA binds to the small ribosomal subunit. The small ribosomal subunit then binds to the mRNA 5′ cap (see Concept 2.2.02). The small subunit (bound by initiator tRNA), glides along the mRNA in the 5' → 3' direction until it detects the first AUG codon (ie, the start codon). When the initiator tRNA anticodon recognizes and base pairs with the mRNA start codon, the reading frame is established.

The portion of mRNA that is translated by ribosomes is called the **open reading frame (ORF)**. It begins with a start codon (AUG) and ends with a stop codon (UGA, UAG, UAA). The regions upstream and downstream of the ORF are known as the 5' and 3' untranslated regions (UTRs), respectively. Once the mRNA, initiator tRNA, and small ribosomal subunit are engaged, the binding of the large (60S) ribosomal subunit completes the **translation initiation complex** (Figure 2.20).

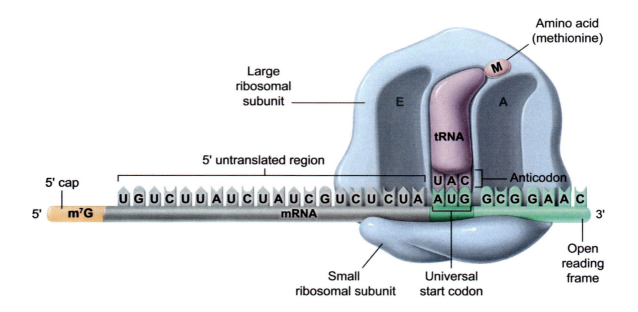

Figure 2.20 Translation initiation complex.

Translation initiation complex formation requires protein initiation factors and energy obtained via GTP hydrolysis. At the end of the initiation phase, the initiator tRNA occupies the P site and the A site is unoccupied, ready to accept the next charged tRNA.

During the **elongation** phase, tRNA molecules deliver amino acids to the ribosome for incorporation into the growing protein chain in a cycle, depicted in Figure 2.21. The steps in the cycle proceed as follows:

1. The mRNA codon following the start codon dictates the next tRNA that will bind. Base pairing between the mRNA codon and the anticodon of the next tRNA occurs within the ribosome's A site.

2. The **peptidyl transferase center** within the ribosome's large subunit catalyzes the formation of a peptide bond between the carboxyl end of the growing protein (attached to the tRNA in the P site) and the amino end of the amino acid attached to the tRNA in the A site. This step results in the attachment of the growing protein to the amino acid of the tRNA in the A site.

3. The empty tRNA in the P site is translocated to the E site, and the tRNA containing the growing protein is translocated from the A to P site. The mRNA moves along with the bound tRNAs, advancing the next codon into the A site.

4. The discharged tRNA exits from the E site and the ribosome is now ready to accept the next charged tRNA at the A site.

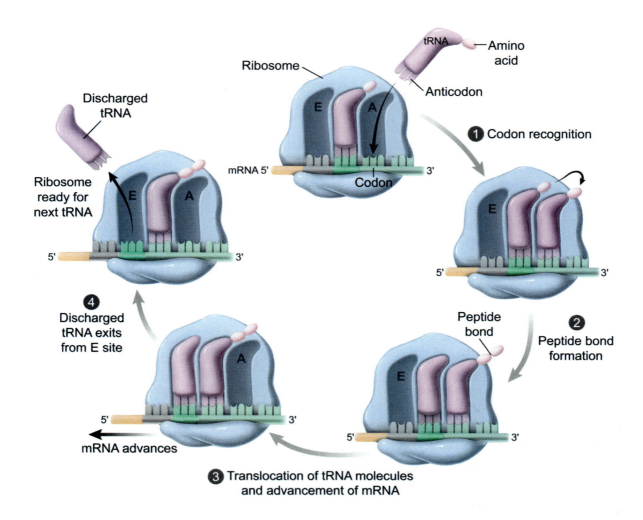

Figure 2.21 Elongation phase of translation.

The elongation process continues until the ribosome reaches a stop codon (UGA, UAG, UAA) within the mRNA at the A site of the ribosome. When a stop codon is encountered, instead of a tRNA, a **release factor** binds to the codon in the A site, leading to the **termination** of translation. The release factor promotes hydrolysis of the ester linkage between the protein and the tRNA in the P site, separating the protein from the ribosome. After the protein is released, dissociation of the ribosome requires additional protein release factors and GTP hydrolysis (Figure 2.22).

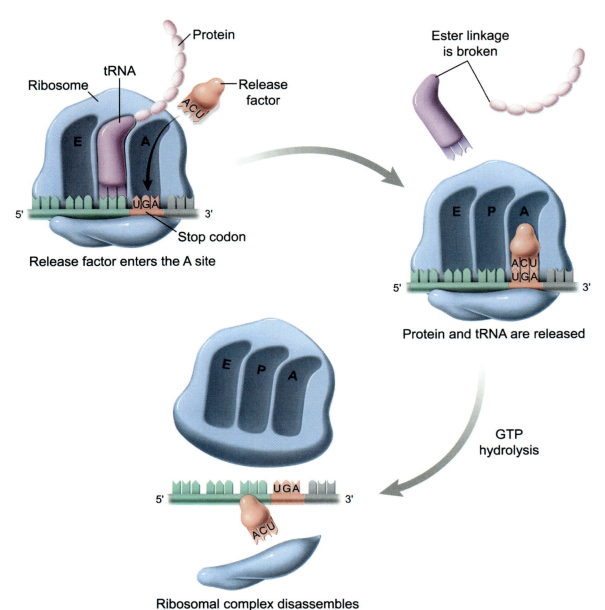

Figure 2.22 Termination phase of translation.

Ribosomes can be found in the cytosol and attached to the cytosolic side of the endoplasmic reticulum (ER). Proteins destined for secretion or for specific locations within the cell contain a **signal peptide**. The signal peptide is a sequence of ~20 amino acids near the protein's amino (N) terminus. The signal peptide directs the ribosome to a translocation complex on the surface of the ER membrane, where protein synthesis continues. As the growing protein passes into the ER lumen, the signal peptide is cleaved.

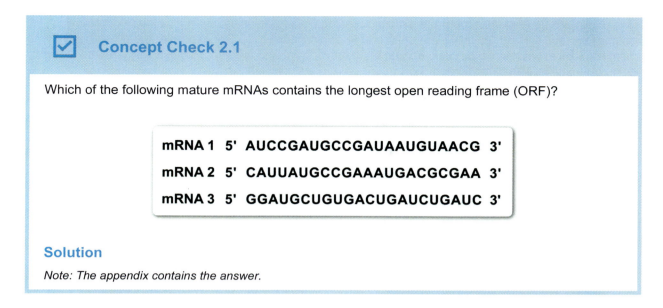

2.3.06 Post-Translational Modification of Proteins

As a protein emerges from a ribosome, it adopts its final three-dimensional structure through protein folding. Protein folding often requires the assistance of other proteins called molecular **chaperones**. Additional chemical modifications may be necessary for the protein to become functional (Figure 2.23). Chemical modifications may include the addition of carbohydrates (**glycosylation**), lipids (**lipidation**), certain proteins (eg, **ubiquitination**), phosphate groups (**phosphorylation**), and other chemical groups.

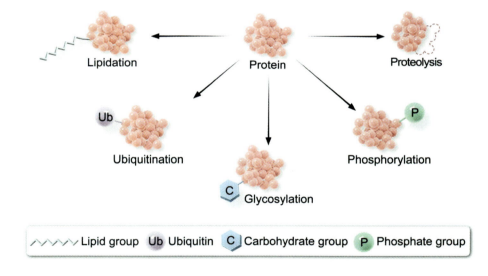

Figure 2.23 Examples of post-translational modifications to proteins.

Chemical modifications may affect the localization or biological activity of a protein. For example, protein glycosylation has numerous biological functions, including roles in stability, cellular localization, and molecular trafficking. Protein phosphorylation also plays multiple cellular roles, most notably in the activation and deactivation of proteins during cell signaling. Ubiquitin is a small regulatory protein tag that plays a role in membrane trafficking of proteins and the timing of protein degradation (discussed in Concept 2.4.03).

Some proteins may also undergo proteolytic processing (ie, proteolysis), in which a longer, inactive precursor protein is enzymatically cleaved to create an active protein. For example, the digestive enzyme pepsin is initially synthesized as pepsinogen, an inactive form, and is later proteolytically cleaved to form pepsin, an enzyme with catalytic activity.

2.3.07 Mutations and Mutagens

Gene expression relies on passing the information encoded in DNA accurately through transcription and translation to create a protein. Maintaining the accuracy of genetic information in DNA is crucial because it is within DNA that genetic information is maintained on a long-term basis. As discussed in Lesson 1.3, there are mechanisms for DNA proofreading and repair to prevent and correct errors in the DNA code. However, sometimes these errors are not detected or are unable to be repaired by the cell. These errors are called **DNA mutations**.

Mutagens are chemical, physical, or biological agents that can lead to mutations in DNA (Table 2.1). Chemical agents include chemicals similar in structure to DNA nucleotides that can cause incorrect pairing during replication, insert themselves into DNA or otherwise distort the double helix, or cause chemical changes to nucleotides, all of which may alter base pairing. Physical agents include UV light, X-rays, and other forms of high-energy radiation. Biological agents can insert mobile DNA elements or viral DNA into the genome. Agents that lead to cancer-causing mutations are also called **carcinogens**.

Table 2.1 Types of mutagens and their effects.

Type of mutagen	Effect
Physical agents (eg, ionizing radiation, UV light)	• Nucleic acid strand breaks • Pairing of noncomplementary bases
Chemical agents (eg, base analogs, intercalating agents)	• Direct interaction with nucleic acids leading to chemical alteration of bases • Insertion of agents between bases • Strand breaks
Biological agents (eg, viruses, transposons)	• Incorrect proofreading by DNA polymerase during genome replication • Nucleic acid strand breaks caused by reactive oxygen species • Insertion of transposons (mobile DNA elements) or viral DNA into the genome

The effects of mutations in coding DNA can range from inconsequential to profound, depending on how the mutation affects the product that is formed. While mutations in regulatory and noncoding DNA regions do not affect the amino acid sequence of the product formed, these mutations may have critical effects on the timing and level of gene expression.

Whether a mutation is passed to subsequent generations depends on both the timing of the mutation and the cell type in which the mutation occurs. Mutations that occur in single-celled organisms are always passed to the next generation. In multicellular organisms, mutations are passed to the next generation only if they occur in gametes or precursors to gametes (**germline mutations**).

Spontaneous mutations that are not present in either parent may also arise during the early stages of embryonic development and affect the offspring only. After the early stages of embryogenesis and throughout the remainder of life, mutations that occur in somatic cells (non-gametes) cannot be passed down in a hereditary fashion but may affect the individual in which they occur. **Somatic mutations** contribute to the normal aging process of an organism and are often involved in the development of cancer and other degenerative conditions.

2.3.08 Types of Mutations

Mutations can range from a change in a single nucleotide to large-scale disruptions of the chromosome. Larger-scale changes in chromosomes can result in duplications, deletions, inversions, and translocations of large chunks of DNA (discussed in Lesson 7.3). This concept focuses on mutations of a smaller scale.

Mutations known as **point mutations** are the result of the replacement of a single nucleotide and its complementary partner with a different pair of nucleotides. **Missense mutations** are substitutions that occur within the open reading frame (ORF) of a gene and cause a different amino acid to be placed into the protein during translation. For example, the mRNA codon 5'-GCG-3' specifies the amino acid alanine. If a mutation in the DNA causes the codon to be transcribed as 5'-UCG-3', the amino acid serine will be inserted instead, as shown in Figure 2.24.

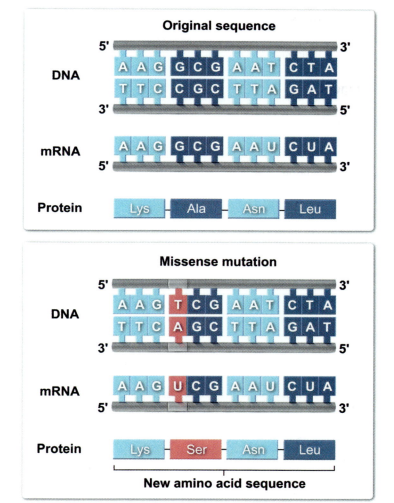

Figure 2.24 Missense mutation.

The effects of a missense mutation on the protein produced may be small if the substituted amino acid has similar properties to the one it replaces or if the amino acid is not essential to the protein's function. However, if the substituted amino acid has significantly different properties or is essential for protein function, the effect can be substantial.

In many cases, a substitution is detrimental to the protein's function in the cell. For example, sickle-cell disease results from a single amino acid substitution in the β-globin subunit of hemoglobin, leading to dramatic effects at the protein, cell, and organism levels (Figure 2.25). In some cases, a missense mutation can lead to a substitution that results in improved protein function (ie, a beneficial mutation).

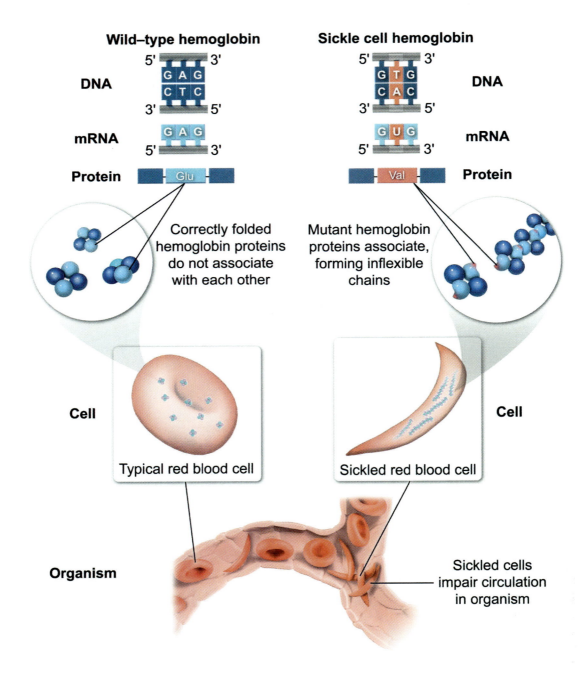

Figure 2.25 A single missense mutation in the β-globin coding region leads to dramatic effects.

Alternatively, a nucleotide substitution may lead to a premature stop codon within the ORF of a gene. This is called a **nonsense mutation** and results in the formation of a shortened protein during translation. For example, the mRNA codon 5'-UCG-3' specifies the amino acid serine. If a mutation in the DNA causes the codon to be transcribed as 5'-UAG-3' (ie, a stop codon), then a release factor binds in place of a tRNA carrying serine, which terminates translation immediately (Figure 2.26). Proteins that result from nonsense mutations are shortened and usually nonfunctional.

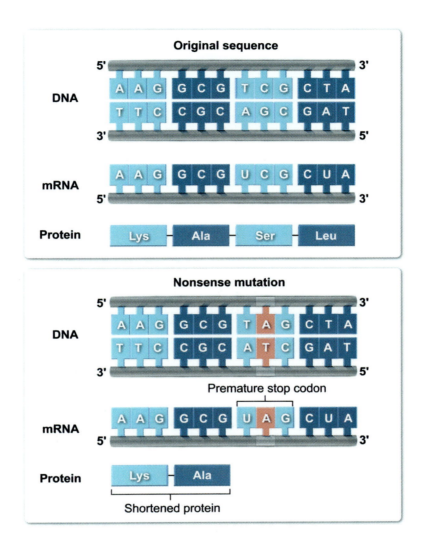

Figure 2.26 Nonsense mutation.

Due to the degeneracy of the genetic code, there are some cases in which a mutation does not result in an amino acid change to the protein produced. These are known as **silent mutations**. In many cases, silent mutations involve a mutation to the nucleotide in the third position of a codon in an ORF. For example, the mRNA codon 5'-GCG-3' specifies the amino acid alanine. If a mutation in the DNA causes the codon to be transcribed as 5'-GCC-3', the amino acid specified is still alanine (Figure 2.27). While the proteins produced in both cases are identical, the genetic code itself is different.

During translation, specific codons are sometimes preferred over others that code for the same amino acid. Although the protein product is the same, when the preferred codon is not present, the speed of translation and level of expression can be affected.

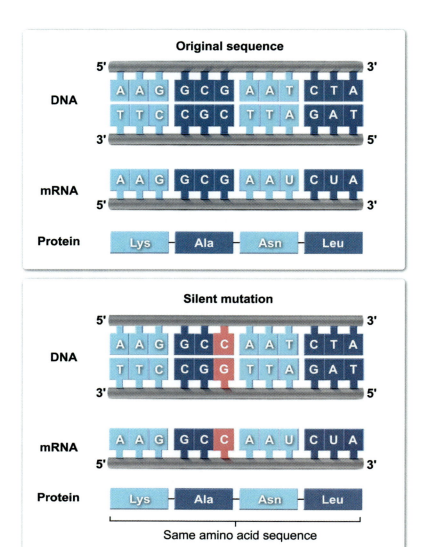

Figure 2.27 Silent mutation.

In contrast to point mutations, in which one nucleotide is substituted for another, insertion or deletion of nucleotides within an ORF is likely to have a more extensive effect on the protein produced. Because nucleotides are read in groups of three (ie, codons), an alteration of the reading frame caused by adding or deleting nucleotides often results in substantial changes to the protein produced.

A **frameshift mutation** occurs when nucleotides are deleted or inserted within an ORF in a way that alters the groups of three nucleotides recognized as codons by tRNA during translation. The reading frame of the mRNA codons directly downstream of the deletion or insertion is shifted, which results in the translation of a completely different amino acid sequence from that point on (Figure 2.28). The protein product formed is often nonfunctional.

Chapter 2: Gene Expression

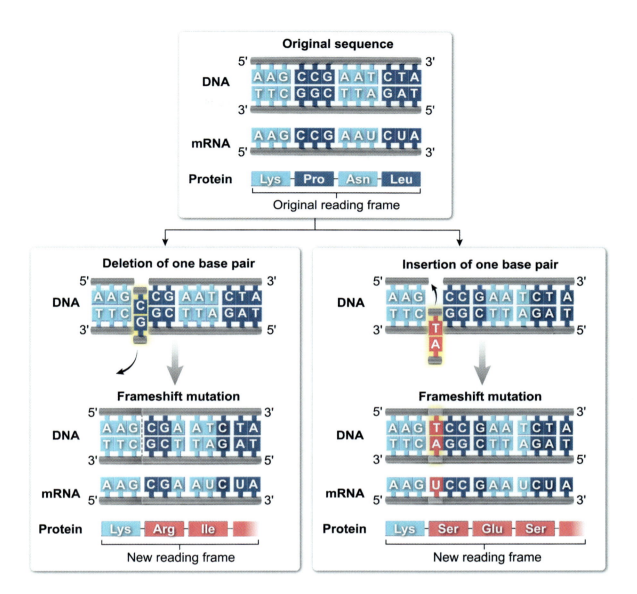

Figure 2.28 Deletion and insertion frameshift mutations.

If a deletion or insertion involves a group of three nucleotides, it could occur *within the correct reading frame* (**in-frame**). An in-frame deletion of three nucleotides (or multiple of three nucleotides) results in the deletion of the corresponding amino acids from the protein but leaves the rest of the protein intact (Figure 2.29). An in-frame insertion of three nucleotides or a multiple of three nucleotides results in one or more extra amino acids being incorporated into an otherwise correct protein sequence. Depending on the position and length of the in-frame deletion or insertion, these types of mutations may range in effect from minimal to significant.

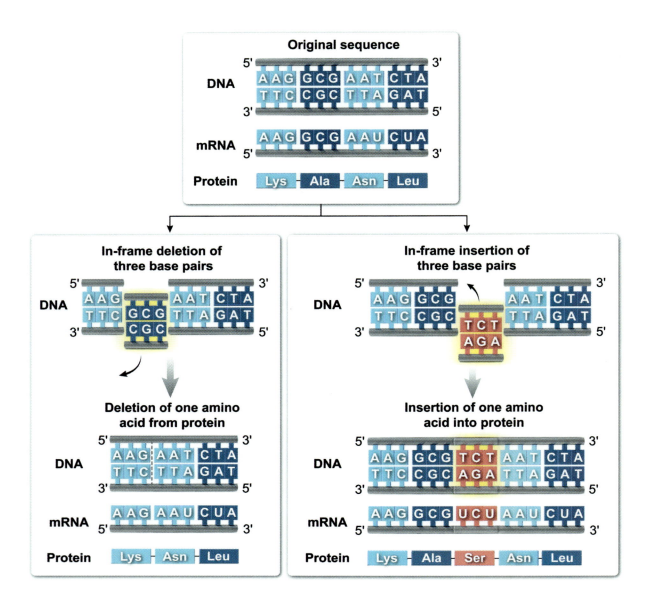

Figure 2.29 In-frame deletion and insertion mutations.

The types of mutations covered in this lesson are summarized in Table 2.2.

Table 2.2 Types of mutations.

Type of DNA mutation	Effect of mutation
Missense	Single base pair change that gives rise to a different amino acid
Nonsense	Single base pair change that gives rise to a stop codon
Silent	Single base pair change that does not change the amino acid sequence
Insertion	Addition of one or more base pairs into the DNA sequence
Deletion	Removal of one or more base pairs from the DNA sequence
Frameshift	Addition or removal of one or more base pairs that results in a new reading frame

Concept Check 2.2

The following shows a portion of a gene containing an open reading frame (ORF):

5' AAGTATGCCAAAAGGATGACGC 3'

3' TTCATACGGTTTTCCTACTGCG 5'

If the nucleotide at the tenth position in the *coding strand* is mutated to a C, what effect would this have on the protein produced? Use a codon table to answer this question.

Solution
Note: The appendix contains the answer.

Lesson 2.4

Regulation of Gene Expression in Eukaryotes

Introduction

Each cell in an organism contains a copy of that organism's entire genome, which includes all coding and noncoding DNA. However, it would not be favorable to express every gene simultaneously. Whether an organism is unicellular or multicellular, regulation of gene expression allows cells to respond appropriately and differentially to external and internal signals.

While there are many similarities between prokaryotic and eukaryotic gene regulation, this lesson focuses on eukaryotic gene regulation, and prokaryotic gene regulation is covered in more detail in Concept 6.3.05.

2.4.01 Differential Gene Expression

While all cells in a multicellular organism contain the same set of genes, only some genes are expressed at any given time. A subset of genes, called housekeeping genes, are **constitutively** (ie, always) expressed, and are required for the maintenance of basic cellular functions. The remaining genes are expressed only when they are required for a specific function.

The difference between two individual cell types is ultimately a result of which specific genes are expressed in each. For example, during embryonic development, genes specific to neuronal differentiation (eg, genes that specify neurotransmitter types) are expressed in cells that will become neurons, whereas genes specific to muscle differentiation (eg, genes that specify muscle fiber types) are expressed in cells that will become muscle cells (Figure 2.30).

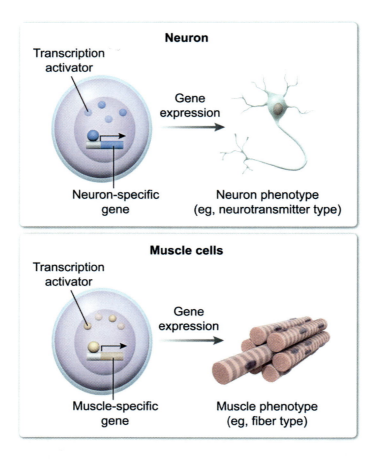

Figure 2.30 Differential gene expression leads to cell specialization.

In multicellular organisms, differential gene regulation is the basis for the initiation and maintenance of cell specialization. During embryogenesis, gene expression is regulated in a differential manner to give rise to all varieties of cell types found in the mature organism (see Lesson 10.3 for more information).

2.4.02 Chromatin Remodeling

As discussed in Lesson 1.4, eukaryotic DNA is packaged with histone proteins into chromatin, which is condensed to fit inside the nucleus of a cell. The packaging of chromatin inside the nucleus affects gene expression on a large scale (ie, large regions of a chromosome encompassing many genes) and at the individual gene level.

Chromatin exists in two different conformations: a closed form known as **heterochromatin** and an open form called **euchromatin** (Figure 2.31). Heterochromatin is more densely arranged, and genes present within regions of heterochromatin are rarely expressed. Euchromatin is more openly arranged and is associated with higher levels of gene expression.

Histone proteins in chromatin have tails that are accessible to modifying enzymes, which can add or remove chemical groups to change chromatin structure, resulting in induced or repressed gene expression. For example, the addition of acetyl groups to histone tails is catalyzed by the enzyme **histone acetylase**.

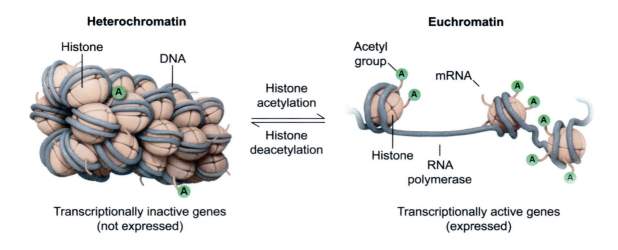

Figure 2.31 Histone acetylation.

Histone acetylation influences gene transcription by promoting the formation of more openly arranged euchromatin, making that region of DNA more readily accessible to transcription machinery. Conversely, **histone deacetylase** enzymes downregulate gene expression by removing acetyl groups from histones, which promotes a denser heterochromatin conformation and restricts access of transcriptional machinery (see Figure 2.31).

Histone tails may also be modified by the addition of methyl (–CH$_3$) or phosphate groups (–PO$_4^{-3}$), which may also affect chromatin conformation and transcriptional activity. In addition to affecting the density of chromatin conformation, chromatin modifications may obscure or create more accessible binding sites for proteins that affect transcription. Although the modification of chromatin does not alter the sequence of nucleotides present in the DNA, it can have a profound effect on the timing and level of gene expression.

In addition to histone tail modifications, specific DNA nucleotides may be modified to affect gene expression. For example, addition of a methyl group to cytosine nucleotides by a family of enzymes known as **DNA methyltransferases** is associated with decreased transcriptional activity (Figure 2.32). In contrast, removal of such methyl groups by DNA demethylation enzymes can restore expression levels.

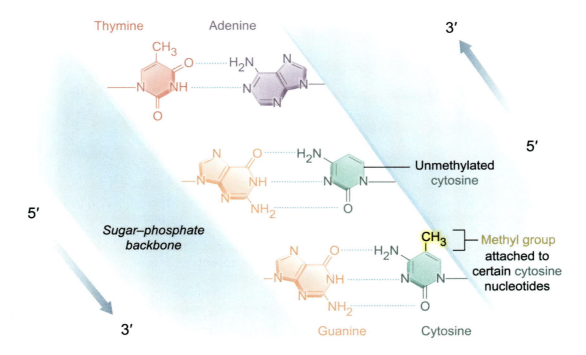

Figure 2.32 DNA methylation.

Gene methylation patterns are usually maintained through DNA replication and are passed to daughter cells. Maintenance of certain methylation patterns in offspring contributes to genomic imprinting in mammals (discussed in Concept 10.3.02).

Mutations in the DNA sequence itself are maintained and can be inherited by offspring. However, modifications to chromatin and DNA methylation patterns may also be passed to future generations, a phenomenon known as **epigenetic inheritance**.

One important difference between inheritance of DNA mutations and epigenetic changes is that epigenetic modifications have the potential to be reversed, depending on environmental conditions. The regulation of gene expression through epigenetic mechanisms is now widely accepted and can be considered a link between *nature* (ie, the genome) and *nurture* (ie, the environment).

2.4.03 Regulation of Transcription

Gene expression can be regulated at multiple levels. For example, controlling whether a gene is *transcribed* affects whether mRNA is synthesized. Likewise, controlling when, where, or even if *translation* occurs affects protein expression and function. While there are many mechanisms of transcriptional and translational control, the most common method of gene regulation occurs by controlling the initiation of transcription.

The assembly of the transcription initiation complex at a gene's promoter is discussed in Concept 2.2.01. RNA polymerase II and its associated protein factors are essential for transcription of all protein-coding genes and are considered part of the general transcription machinery. The assembly of such factors at a gene's promoter is required for transcription to occur but leads to a relatively low (ie, **basal**) level of transcription in the absence of additional factors. To achieve a higher level of expression under appropriate conditions, additional proteins called **transcription factors** are required.

In addition to the promoter region, specific regulatory DNA sequences may be located both proximal (near) to and distal (far) from the promoter and may also be intronic (ie, within an intron). These control elements, called **enhancers**, contain clusters of DNA binding sites for transcription factors (Figure 2.33).

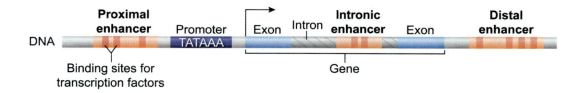

Figure 2.33 Enhancers are noncoding regulatory regions.

Enhancers are found mostly in noncoding regions (eg, between two genes or within introns) and are sometimes found at great distances both upstream and downstream of a target gene. The presence of multiple enhancers allows for differential gene regulation, with each enhancer contributing to gene expression in different cell types and under different conditions.

Transcription factors often have two types of structural domains: a **DNA-binding domain**, which interacts with DNA binding sites in the enhancer, and an **activation domain**, which interacts with other regulatory proteins or transcription machinery to facilitate transcription initiation (Figure 2.34). Transcription factors may either activate (ie, enhance) or inhibit gene expression.

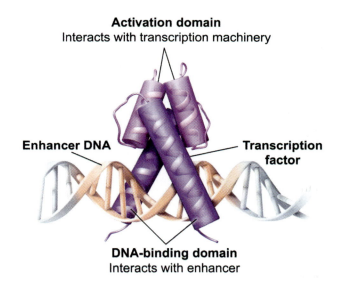

Figure 2.34 Transcription factor domains.

Even though enhancers may be located at a great distance from the promoter, under the correct conditions, the DNA may be bent (ie, looped) to bring **transcription activators** into contact with the general transcription machinery assembled at the promoter (Figure 2.35). When transcription activators bound to regulatory sites in the enhancer interact with the general transcription machinery, levels of transcription are enhanced above a basal level.

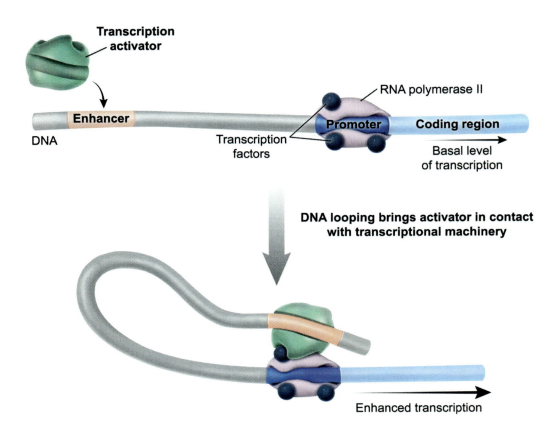

Figure 2.35 Transcription activator binding to enhancer stimulates transcription.

In contrast to transcription activators, **transcription repressors** inhibit transcription initiation. Repressors achieve this by competing with transcription activators for binding sites within an enhancer or otherwise blocking activator binding. Alternatively, repressors may bind directly to activators to prevent the action of transcription activators, as shown in Figure 2.36.

In eukaryotes, transcriptional control of gene expression is complex and depends in large part on the presence of tissue-specific regulatory elements or enhancers. In any given cell type, the unique combination of regulatory elements and transcription factors controls whether transcription occurs. While gene expression is largely controlled through regulating transcription initiation, additional methods of control can occur once transcription is completed. For example, by controlling which splice variants are formed via alternative splicing (see Concept 2.2.03), a cell can alter which protein isoforms are expressed.

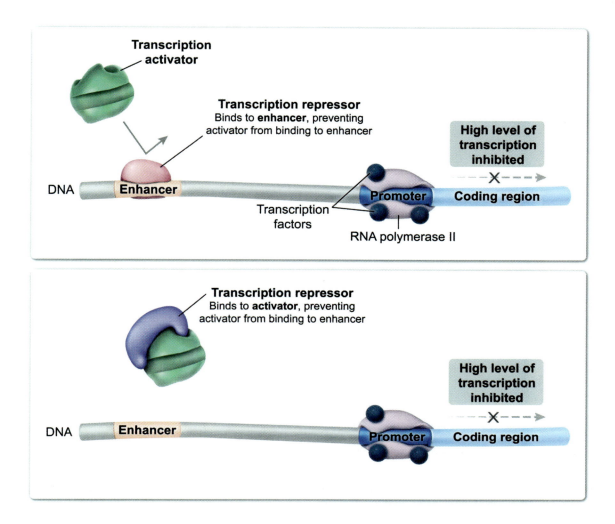

Figure 2.36 Transcription repressors inhibit high levels of transcription.

Another form of post-transcriptional control includes regulating access to translational machinery. If a cell needs to respond quickly to certain environmental conditions, the cell's mRNA may be transcribed and stored, being translated only when needed. Likewise, controlling the timing of mRNA degradation after transcription affects the extent and duration of protein synthesis (discussed further in Concept 2.4.04).

Post-translationally, a cell may also control the length of time a protein is present. While not strictly considered regulation of gene expression, this is a method used to regulate how long a gene product (ie, protein) is available in a cell. Individual proteins are generally considered to have a characteristic lifespan and are flagged for destruction when it is time for that lifespan to end. In addition, proteins may also be flagged for destruction due to an abnormality (eg, misfolding, inappropriate aggregation).

Flagging a protein for destruction involves the addition of a series of **ubiquitin** proteins that are covalently linked to form a **polyubiquitin chain**. Polyubiquitination occurs via the transfer of ubiquitin to the target protein via the enzyme **ubiquitin ligase**, as illustrated in Figure 2.37. Thus, ubiquitin ligases play an important regulatory role in determining how long a protein remains present in the cell.

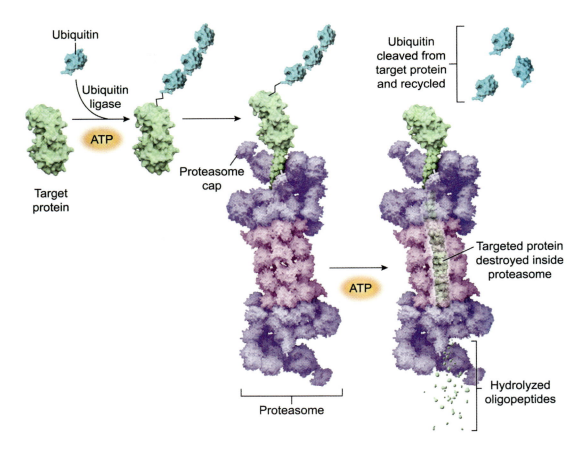

Figure 2.37 Polyubiquitinated proteins are targeted for destruction by proteasomes.

When a protein becomes polyubiquitinated, it can be recognized by a type of molecular machine known as a **proteasome**. Proteasomes are composed of a cylindrical stack of protein rings, which are capped at each end. The polyubiquitinated protein is recognized by a regulatory proteasome cap, which proceeds to unwind the protein and passes the protein into the interior of the cylinder. Once inside, the protein is hydrolyzed into small oligopeptides, which are released into the cytosol, where the oligopeptides can be further degraded into amino acids and recycled for future use.

2.4.04 Noncoding RNA

The term **noncoding RNA** (ncRNA) has traditionally been used to refer to RNA that does not code for a tRNA, rRNA, or protein. However, ncRNAs contain information and serve a purpose. Recent evidence suggests that there is a large and diverse population of ncRNAs which play major roles in gene regulation. There are two major classes of small ncRNAs: **microRNAs (miRNAs)** and **small interfering RNAs (siRNAs)**. miRNAs and siRNAs are RNA molecules capable of self-pairing or binding to other RNA molecules via complementary base pairing to form double-stranded RNA structures.

miRNAs and siRNAs are processed from longer RNA precursors and form a complex with one or more proteins. However, miRNAs typically originate as single-stranded RNAs which exhibit self-complementarity and are processed to form shorter double-stranded mature miRNAs. Generally, siRNAs are first observed as double-stranded RNAs which are further processed into a mature form.

Processed miRNAs and siRNAs are similar in length and interact with the same machinery to form gene silencing complexes. The major functional difference between these two types of ncRNA is that siRNA molecules are highly specific and regulate a single mRNA target, whereas miRNA molecules are less specific and regulate multiple mRNA targets.

Mature ncRNAs of either class may then interact with gene silencing machinery in the cell to regulate gene expression (see Concept 4.4.01). An ncRNA-protein complex interacts with an mRNA molecule to either trigger mRNA degradation or block mRNA translation, as shown in Figure 2.38. Ultimately, the role of ncRNAs in gene regulation is post-transcriptional, preventing or delaying the translation of specific mRNAs.

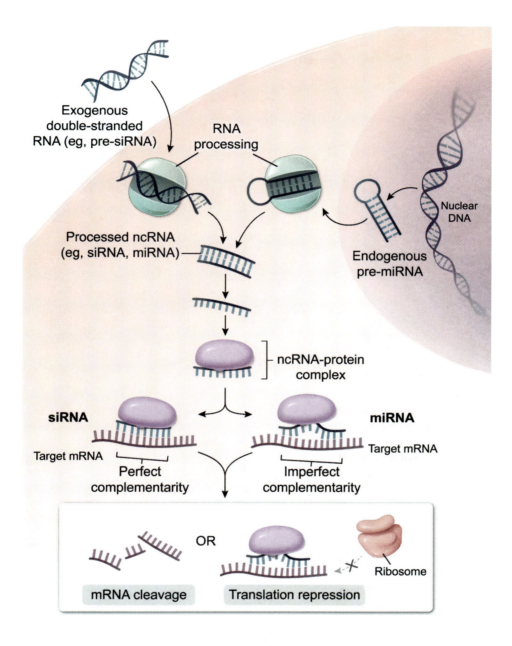

Figure 2.38 Noncoding RNA regulation of gene expression.

END-OF-UNIT MCAT PRACTICE

Congratulations on completing **Unit 1: Molecular Biology**.

Now you are ready to dive into MCAT-level practice tests. At UWorld, we believe students will be fully prepared to ace the MCAT when they practice with high-quality questions in a realistic testing environment.

The UWorld Qbank will test you on questions that are fully representative of the AAMC MCAT syllabus. In addition, our MCAT-like questions are accompanied by in-depth explanations with exceptional visual aids that will help you better retain difficult MCAT concepts.

TO START YOUR MCAT PRACTICE, PROCEED AS FOLLOWS:

1) Sign up to purchase the UWorld MCAT Qbank
   IMPORTANT: You already have access if you purchased a bundled subscription.
2) Log in to your UWorld MCAT account
3) Access the MCAT Qbank section
4) Select this unit in the Qbank
5) Create a custom practice test

Unit 2 Biological Research Techniques

Chapter 3 Designing and Interpreting Experiments

3.1 Experimental Design

    3.1.01   Experimental Approach
    3.1.02   Types of Variables
    3.1.03   Relationships Among Variables
    3.1.04   Experimental Conclusions

3.2 Statistics

    3.2.01   Types of Hypotheses
    3.2.02   Significance
    3.2.03   Errors
    3.2.04   Power
    3.2.05   Confidence
    3.2.06   Validity
    3.2.07   Statistical Conclusions

Chapter 4 Biotechnology

4.1 DNA Technology

    4.1.01   DNA Sequencing
    4.1.02   Polymerase Chain Reaction
    4.1.03   Restriction Enzymes
    4.1.04   Gene Cloning
    4.1.05   Generation of cDNA
    4.1.06   DNA Libraries
    4.1.07   DNA Gel Electrophoresis and Southern Blotting
    4.1.08   Hybridization
    4.1.09   Radiography

4.2 Analyzing Gene Expression

    4.2.01   Detecting and Manipulating RNA
    4.2.02   Detecting and Manipulating Proteins

4.3 Determining Gene Function

    4.3.01   Knockouts
    4.3.02   Complementation

4.4 Practical Applications of Biotechnology

    4.4.01   Antisense Drugs
    4.4.02   Gene Therapy
    4.4.03   Forensics
    4.4.04   Agriculture
    4.4.05   Environmental Cleanup

4.5 Special Considerations in Biotechnology

    4.5.01   Safety Concerns
    4.5.02   Ethical Concerns

Lesson 3.1
Experimental Design

Introduction

Biological research is conducted via a process commonly referred to as the scientific method, which involves systematic observation and experimentation. Appropriate experimental design is an essential aspect of biological research that produces reliable and valid data, which can be analyzed to reach defensible conclusions. This lesson explores fundamental components of experimental design, including types of variables, controls, and conclusions based on experimental outcomes.

Although this lesson focuses primarily on biological research, the fundamentals of experimental design apply to all areas of scientific research. As such, knowledge of the principles of appropriate experimental design may be tested in multiple sections of the MCAT.

3.1.01 Experimental Approach

Biological research typically entails conducting systematic investigations that allow cause-and-effect relationships to be determined among **variables** through the analysis of data generated via experimentation. This experimental approach typically involves the generation of testable **hypotheses**, which then guide the process of experimental design.

The results of an initial experiment typically guide future experiments, either for replicating the initial results (ie, strengthening an underlying scientific theory) or testing a revised hypothesis if the results of the initial experiment did not support the original hypothesis. Figure 3.1 summarizes a common way in which scientific experimentation is used to address a scientific problem.

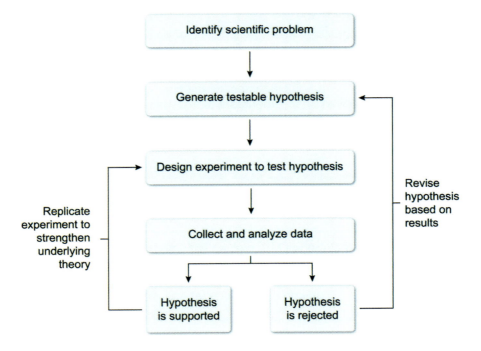

Figure 3.1 Typical experimental approach.

Scientific hypotheses are tested by performing experiments in which researchers deliberately manipulate one or more variables while measuring the effect of this manipulation on a different variable. In properly designed experiments, factors other than the deliberately manipulated variable(s) that could affect the outcome of an experiment (ie, extraneous variables) are **controlled**, or the effects of such variables are minimized by **randomization**. To that end, scientific experiments are characterized by the random assignment of test subjects (eg, research animals, human participants) to **treatment groups** and **control groups**.

A treatment group (also referred to as an **experimental group**) is exposed to the treatment under study in an experiment, whereas a control group is not subjected to the factor being investigated. By comparing the results observed in the treatment group with those observed in the control group, researchers can determine the extent to which the observed experimental outcome is attributable to the manipulated variable.

Certain aspects of experimental design can be optimized to avoid the introduction of systematic error (ie, bias) in experiments involving human subjects. In a **single-blind study**, the research subjects do not know whether they have been assigned to a treatment group or control group. In a **double-blind study**, neither the research subjects nor the researchers conducting the experiment know which subjects have been assigned to treatment and control groups. This concealment (blinding) of the allocation of subjects to treatment and control groups helps prevent bias caused by human expectations, as does the use of placebos in control groups.

3.1.02 Types of Variables

An experimental variable is any factor that can change (ie, take on different values) in an experiment. An **independent variable (IV)** in an experiment is a factor that is purposefully varied (ie, manipulated) by a researcher to determine its effect on another variable. An IV is *independent* in that its variation does not occur in response to another variable (ie, does not depend on another variable). IVs are also referred to as **explanatory variables** or **predictor variables** because IV variation is presumed to explain and/or predict change in another variable that is measured to determine the outcome of the experiment.

A **dependent variable (DV)** in an experiment is a factor measured for the purpose of observing the experiment's outcome. Consequently, DVs are also called **outcome variables** or **response variables**, and changes in a DV typically happen only *after* manipulation of an IV has occurred in the experiment.

In addition to IVs and DVs, experiments typically have multiple **control variables** (which are also known as controlled variables). A control variable is a factor, which could affect the DV, that is held constant in the experiment by the researcher. Figure 3.2 summarizes the characteristics of the main experimental variable types as applied in the context of an experiment to test memory.

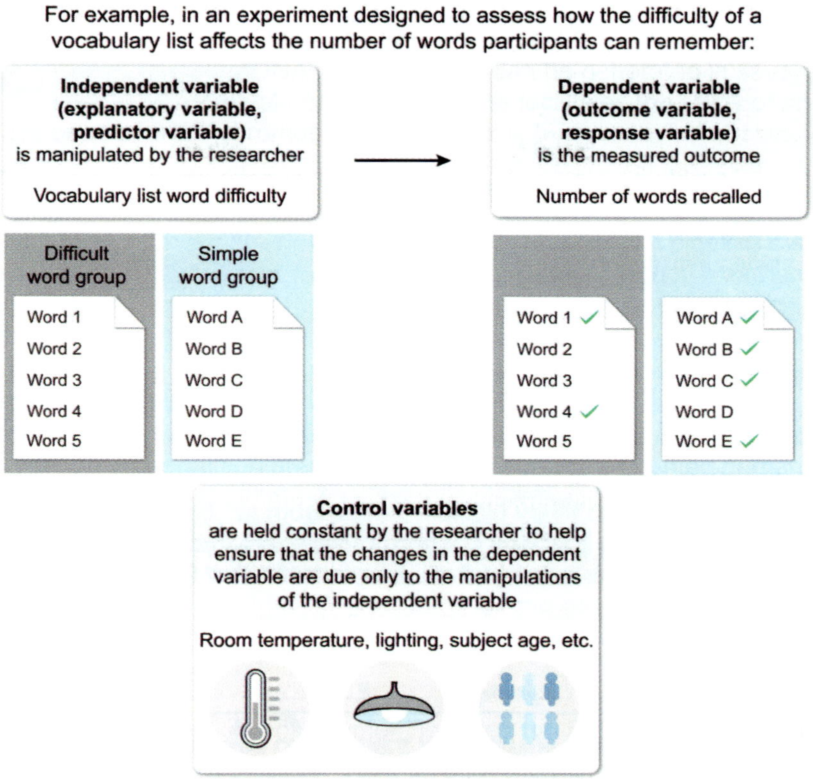

For example, in an experiment designed to assess how the difficulty of a vocabulary list affects the number of words participants can remember:

Independent variable (explanatory variable, predictor variable) is manipulated by the researcher

Vocabulary list word difficulty

Difficult word group: Word 1, Word 2, Word 3, Word 4, Word 5
Simple word group: Word A, Word B, Word C, Word D, Word E

Dependent variable (outcome variable, response variable) is the measured outcome

Number of words recalled

Word 1 ✓, Word 2, Word 3, Word 4 ✓, Word 5
Word A ✓, Word B ✓, Word C ✓, Word D, Word E ✓

Control variables are held constant by the researcher to help ensure that the changes in the dependent variable are due only to the manipulations of the independent variable

Room temperature, lighting, subject age, etc.

Figure 3.2 Independent, dependent, and control variables.

Experimental variables can be classified into two main groups by statistical type. **Quantitative variables** have numerical values that represent quantities (ie, amounts or counts), such as the length of an object or the number of offspring produced. **Categorical variables** have values assigned to a limited number of distinct categories (ie, groups) based on some characteristic, such as blood type or educational level. Both quantitative variables and categorical variables can be further classified into subtypes.

Quantitative variables are classified as being either continuous or discrete. **Continuous quantitative variables** can assume a potentially infinite number of numerical values obtained practically only by performing measurements (eg, height). Alternately, **discrete quantitative variables** take on only certain numerical values (eg, integers) that are determined by counting (eg, number of vertebrae).

Categorical variables may be nominal or ordinal. **Nominal categorical variables** describe characteristics grouped into categories that do not have a natural order but are simply distinguished by arbitrary names (eg, round, wrinkled). Alternately, **ordinal categorical variables** describe characteristics grouped into three or more categories that exhibit an intrinsic order (eg, low, medium, high). Figure 3.3 depicts experimental variable categorization by statistical type.

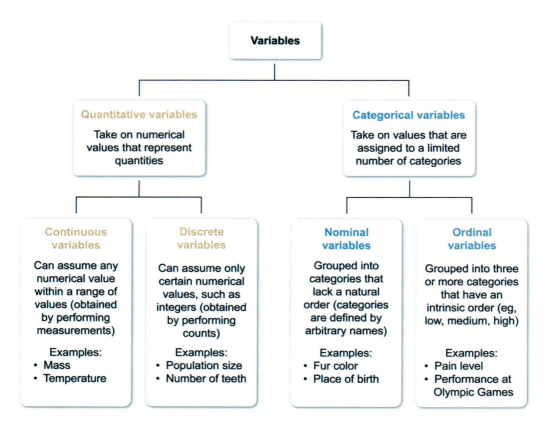

Figure 3.3 Categorization of experimental variables by statistical type.

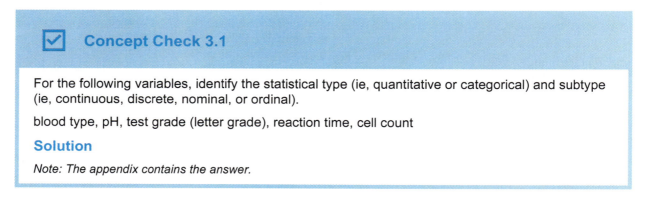

3.1.03 Relationships Among Variables

The purpose of most biological research is to investigate relationships among variables. In particular, biological research that involves experimentation seeks to determine cause-and-effect relationships (ie, **causality**) among variables.

If a relationship exists between two variables, then the two variables correspond to each other in some way. Correspondence that is in the form of a **linear relationship** between variables (ie, the variables change together at a uniform rate) is called **correlation**, which may be positive or negative. In a positive correlation (ie, direct correlation), the change in both variables is in the same direction, whereas, in a negative correlation (ie, inverse correlation), the direction of change of one variable is opposite that of the other variable.

The strength of a correlation between two variables is represented by a **correlation coefficient** (eg, Pearson's correlation coefficient), which ranges from −1 to +1, with −1 indicating a perfect negative correlation between the variables and +1 indicating a perfect positive correlation. A correlation coefficient of 0 indicates that no linear relationship exists between the variables. Visualization of correlations between variables requires that experimental data be commonly displayed in **scatterplots**. Figure 3.4 shows representative scatterplots and associated correlation coefficients (r), which quantify the strengths and directions of the correlations.

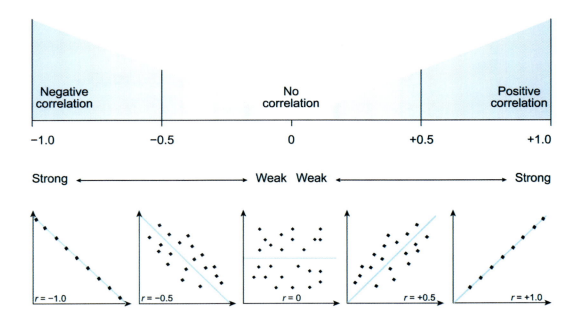

Figure 3.4 Representation of correlation strength and direction.

The existence of a correlation between two variables indicates that the variables are statistically associated; however, the existence of a correlation does *not* indicate whether a change in one variable *causes* the change in the other variable. In other words, the existence of a causal relationship (ie, cause-and-effect relationship) between variables *cannot* be inferred solely based on an observed correlation, even if the correlation is strong. Figure 3.5 shows an example of two variables that are strongly correlated but that do not exist in a causal relationship.

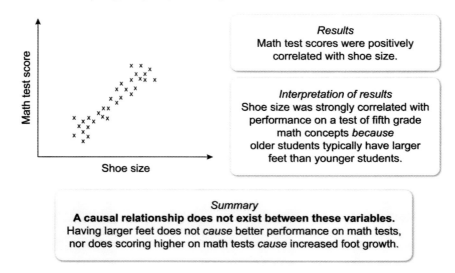

Figure 3.5 Example of correlated variables that do not have a causal relationship.

Because a cause logically must precede an effect, demonstrating that a causal relationship exists between correlated variables requires knowledge of the order in which the variables change. Furthermore, a causal relationship cannot be established unless other plausible explanations, which could account for the observed relationship between the variables, are ruled out. Consequently, scientific experiments, in which manipulation of independent variables precedes change in dependent variables and extraneous variables are controlled, provide the most direct means of investigating causal relationships.

3.1.04 Controls

Controls are an essential component of scientific experiments. **Experimental controls** enable researchers to evaluate dependent variable responses to the manipulation of independent variables in an unbiased, objective manner to reach reliable conclusions. Multiple types of controls, including negative and positive controls (see Figure 3.6), may be included in an experimental design.

A **negative control** provides a basis of comparison that can be used to determine the extent to which manipulation of the independent variable in an experiment is responsible for producing an observed change in the dependent variable. Negative controls are used to reveal the experimental outcome that occurs when the independent variable has *no* effect on the dependent variable. Ideally, the negative control (ie, control group) is exposed to all conditions to which the experimental group is exposed, with the exception of the single factor being investigated in the experiment.

The purpose of a **positive control** is to demonstrate that the experimental system being used is capable of measuring changes in the experiment's dependent variable and that this system is functioning as expected at the time of the experiment. A successful positive control produces the results anticipated to be brought about by the experiment's independent variable. Consequently, if a positive control does *not* produce the expected result in an experiment, such an outcome indicates that the experimental system was likely not functioning properly at the time of the experiment.

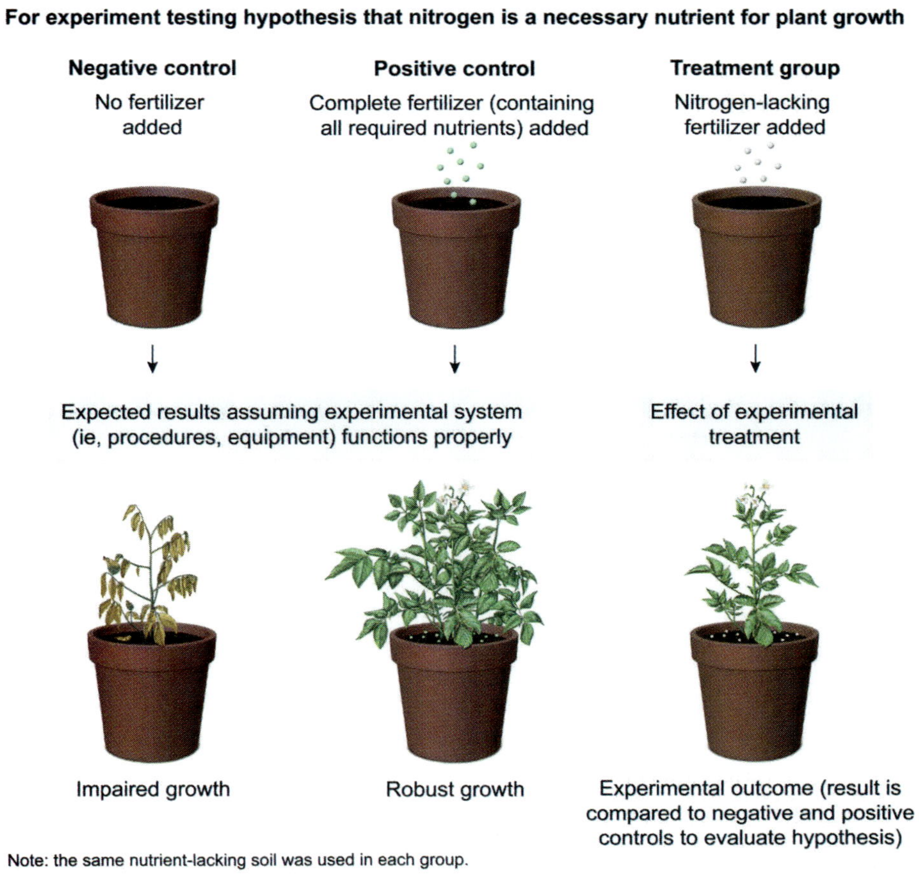

For experiment testing hypothesis that nitrogen is a necessary nutrient for plant growth

Negative control	Positive control	Treatment group
No fertilizer added	Complete fertilizer (containing all required nutrients) added	Nitrogen-lacking fertilizer added

↓ ↓ ↓

Expected results assuming experimental system (ie, procedures, equipment) functions properly | Effect of experimental treatment

Impaired growth | Robust growth | Experimental outcome (result is compared to negative and positive controls to evaluate hypothesis)

Note: the same nutrient-lacking soil was used in each group.

Figure 3.6 Purpose of negative and positive controls in an experiment.

3.1.05 Experimental Conclusions

Valid **experimental conclusions** can be reached through the analysis of data obtained in properly designed and executed scientific experiments. Analysis of experimental data typically involves the use of **statistics** (see Lesson 3.2).

Analysis of experimental data allows researchers to draw conclusions regarding relationships between independent and dependent variables. More specifically, data analysis provides the means by which researchers can directly evaluate hypotheses that guide the design of, and are tested by, scientific experiments. By determining whether experimental results support research hypotheses, researchers contribute to the creation or refinement of scientific theories on which the hypotheses are based (Figure 3.7).

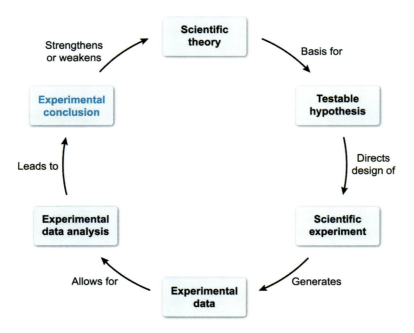

Figure 3.7 Causes and effects of experimental conclusions.

Chapter 3: Designing and Interpreting Experiments

Lesson 3.2
Statistics

Introduction

Statistics is the branch of mathematics used to guide experimental design and data collection. Statistics is also used to organize, present, analyze, and interpret numerical data for describing group characteristics, drawing conclusions, and making predictions.

Descriptive statistics include measures of a data set's central tendency (ie, average), such as **mean**, **median**, and **mode**. In addition, statistical measures such as **range** and **standard deviation (SD)** provide information regarding the variability (ie, dispersion) of a data set. Figure 3.8 summarizes how mean, median, mode, and range are calculated, and Figure 3.9 illustrates the concept of standard deviation.

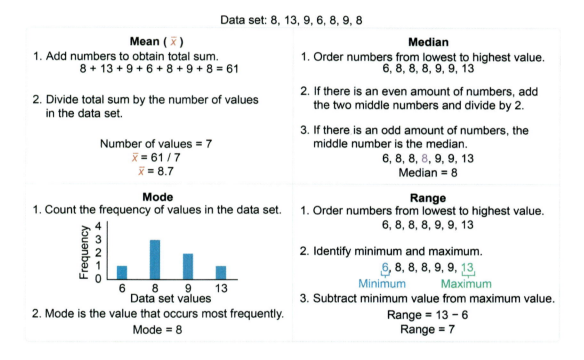

Figure 3.8 Calculation of mean, median, mode, and range.

Standard deviation is a measure of how **variable** or **dispersed** a set of data points are around the mean

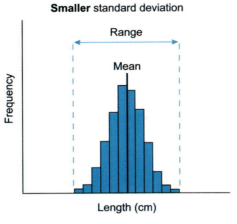
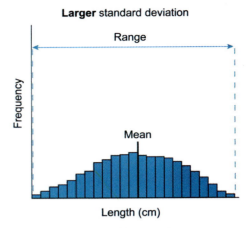

Smaller standard deviation
- **Smaller** range (data points more clustered around the mean)
- **Less** variation among data points

Larger standard deviation
- **Larger** range (data points more spread out around the mean)
- **More** variation among data points

Figure 3.9 Standard deviation.

This lesson examines the components of statistics and the ways that statistics can be used to reach experimental conclusions. Although this lesson pertains to biological research in particular, statistics are used in all areas of scientific research. As such, fundamental knowledge of statistics may be tested in multiple sections of the MCAT.

3.2.01 Types of Hypotheses

Biological research typically involves carrying out experiments designed to test particular **hypotheses**, which, in general terms, are proposed answers to the research questions that underlie experiments. More specifically, a hypothesis is an assumption that is made regarding a relationship between variables being investigated in an experiment. The purpose of an experiment is to generate data by which hypotheses may be tested to determine whether the hypotheses should be rejected or not. A **null hypothesis (H_0)** and an **alternative hypothesis (H_A)** represent opposite assumptions about the relationship being investigated.

The assumption made in a null hypothesis is that no difference exists between two or more groups with respect to the factor being studied in the experiment. In other words, the null hypothesis assumes that the independent variable in an experiment has *no effect* on the dependent variable. In contrast, the alternative hypothesis assumes that manipulation of the independent variable *causes an effect* on the dependent variable, resulting in a difference between groups in the experiment. Figure 3.10 summarizes the characteristics of null and alternative hypotheses.

Experiment: expose plant seeds to darkness or sunlight and compare plant growth

Null hypothesis (H_0)

Sunlight has *no effect* on plant growth

- Typically states the opposite of what researchers predict will occur
- Generally presumed to be true prior to data collection
- Predicts that no statistical difference exists between groups being studied

Alternative hypothesis (H_A)

Sunlight *promotes* plant growth

- States what researchers predict will occur
- Negates H_0
- Predicts that a statistical difference exists between groups being studied

Figure 3.10 Null and alternative hypotheses.

Because the null hypothesis is presumed to be true before conducting an experiment, researchers analyze experimental results to determine whether there is sufficient evidence to reject the null hypothesis. If such evidence is lacking, the null hypothesis *cannot* be rejected (ie, the researchers *fail to reject* the null hypothesis).

3.2.02 Significance

Statistical significance pertains to the probability that an experiment's null hypothesis is true given an acceptable level of uncertainty about the actual situation. Experimental results that are statistically significant are *unlikely* to be due to chance (ie, sampling error). As such, the probability of obtaining a statistically significant result is very low if an experiment's null hypothesis (which asserts that manipulation of the experiment's independent variable has *no* effect) is indeed true.

The ***p*-value** is widely used to determine the significance of experimental results. A *p*-value is formally defined as the probability of observing a given result due to chance alone, assuming that the null hypothesis is true. The *p*-value ranges from 0 to 1 and is generally interpreted as follows, which assumes a **significance level (α)** of 0.05:

- $p > 0.05$ signifies a greater than 5% probability that the observed result is due to chance alone (ie, *not* due to an actual association between the variables under study). Values in this range are generally not considered statistically significant.
- $p \leq 0.05$ signifies a 5% or lower probability that the observed result is due to chance alone. Values in this range are typically considered statistically significant.

By determining the *p*-value for an experimental result, researchers can evaluate the experiment's null hypothesis. That is, if the calculated *p*-value is *greater* than the α chosen before the experiment (eg, 0.05), the null hypothesis *cannot be rejected*, whereas if the *p*-value is *less than or equal to* the predetermined α, the null hypothesis *is rejected*. Furthermore, rejection of the null hypothesis provides convincing evidence in favor of the experiment's alternative hypothesis. Figure 3.11 summarizes the interpretation of *p*-values.

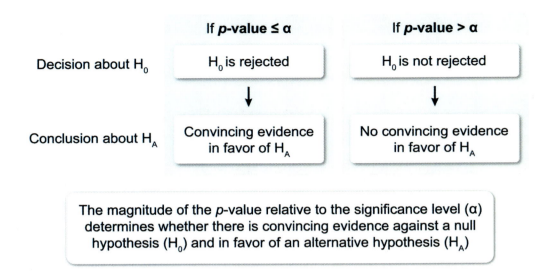

Figure 3.11 Interpretation of *p*-values.

3.2.03 Errors

Analysis of experimental data can sometimes lead researchers to make erroneous judgments regarding null hypotheses. That is, researchers sometimes conclude that the data collected in an experiment demonstrate that a null hypothesis should be rejected when it is true (ie, the experimental treatment indeed has no effect on the dependent variable). Likewise, researchers may fail to reject a null hypothesis when it is true that the experimental treatment *does* produce an effect (ie, the null hypothesis is false).

When researchers mistakenly reject the null hypothesis (ie, reach a **false-positive conclusion**), this type of error is called a **type I error**. Conversely, when researchers mistakenly fail to reject the null hypothesis (ie, reach a **false-negative conclusion**), they have committed a **type II error**. The probability of committing a type I error is represented by α (ie, the significance level chosen by the researchers for the hypothesis test), whereas the probability of committing a type II error is represented by β.

A researcher can reduce the risk of reaching a false-positive conclusion (ie, committing a type I error) by choosing a lower significance level (eg, 0.01 instead of 0.05). However, the risk of committing a type I error is inversely related to the risk of committing a type II error. Consequently, use of a lower significance level results in an increased risk of reaching a false-negative conclusion (ie, committing a type II error). Figure 3.12 summarizes the characteristics of type I and type II statistical errors.

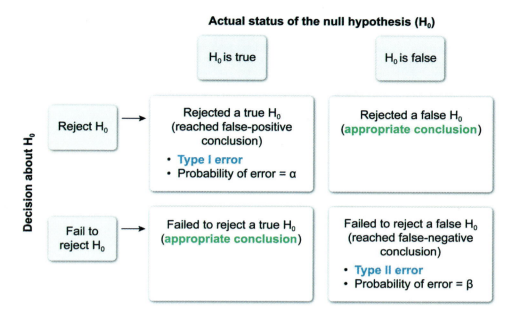

Figure 3.12 Type I and type II errors.

3.2.04 Power

Statistical **power** refers to the probability that a test for statistical significance will detect an effect resulting from manipulation of an independent variable in an experiment, when such an effect exists. In other words, power is the probability that the null hypothesis will be rejected in an experiment when a difference truly exists between groups (ie, probability that the null hypothesis is correctly rejected).

Because a type II error occurs when researchers *fail* to reject the null hypothesis when appropriate (ie, when the experimental treatment *does* produce an effect), power can also be defined as the probability that type II error will be *avoided*. As such, statistical power is mathematically equal to $1 - \beta$ (ie, 1 minus the probability of a type II error occurring), and a hypothesis test with high power (ie, power close to 1) detects false null hypotheses with high sensitivity (Figure 3.13).

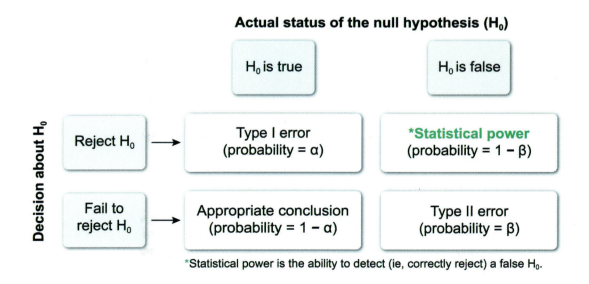

Figure 3.13 Statistical power.

Statistical power increases with increased significance level (α). However, because α represents the probability of type I error, increasing α results in more false-positive conclusions. Researchers can increase statistical power in an experiment *without* increasing the probability of false-positive conclusions by increasing the **sample size** of the treatment and control groups in the experiment.

> ✓ **Concept Check 3.2**
>
> Fill in the blanks in the sentences below with the words "decreases," "increases," "type I," and "type II."
>
> The probability of committing a _____ error _____ as statistical power increases, which occurs when the significance level (α) _____ and/or when the sample size increases. Increased α results in an increased probability of a _____ error.
>
> **Solution**
>
> *Note: The appendix contains the answer.*

3.2.05 Confidence

Biological research entails performing experiments in which samples are investigated to reach conclusions regarding populations from which the samples were taken. Numerical values that describe *samples* are referred to as **statistics** (eg, sample mean), whereas numerical values that describe *populations* are called **parameters** (eg, population mean). Parameters are often unknown in biological research because data from entire populations are typically not available.

A **confidence interval**, which is calculated using a specified **confidence level** (eg, 95%), gives a range of values within which an unknown parameter is likely found and is determined using data obtained from a sample. The confidence interval contains a statistic that describes the sample (ie, **point estimate**) and a **margin of error** on either side of the statistic that reflects the uncertainty associated with using a sample to estimate a population parameter (see Figure 3.14).

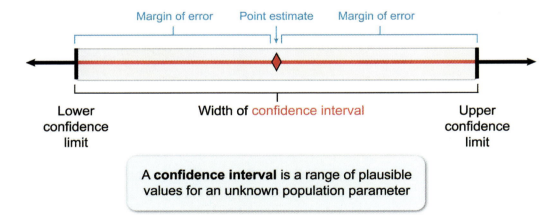

Figure 3.14 Components of a confidence interval.

The confidence level used to calculate a confidence interval represents the percentage of repeated random samples taken from the population that would generate confidence intervals containing the true population parameter. That is, if confidence intervals were calculated from random samples taken numerous times from the same population, the percentage of these intervals that contain the true population parameter would be approximately equal to the confidence level. Figure 3.15 illustrates confidence intervals calculated at the 90% confidence level.

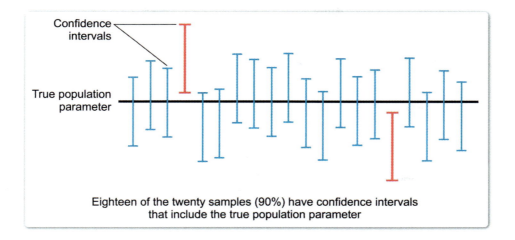

Figure 3.15 90% confidence intervals.

Confidence intervals can be calculated using measurements of **standard error (SE)**, which pertain to the **reliability** (ie, precision) of a statistic, such as a sample mean. For example, standard error of the mean (SEM) is a measure of the variability of means obtained from multiple equally sized random samples drawn from a population. In other words, SEM indicates the *spread* of the sample means (ie, how far from the population mean the sample means tend to fall). As such, SEM provides a measurement of *uncertainty* around an estimate of a population mean.

Confidence intervals can also be used to determine statistical significance. Mathematically, the confidence level is equal to 1 minus the significance level (ie, 1 − α). Therefore, if α = 0.05, the confidence level is 95% (ie, 1 − 0.05 = 0.95). As such, if error bars representing the 95% confidence interval on a graph do *not* overlap, the results are statistically significant (see Figure 3.16). Likewise, error bars representing ±2 SEM on a graph can be interpreted in the same way because, for large samples, the 95% confidence interval is calculated as the sample mean plus or minus 1.96 (ie, ~2) times the standard error of the mean.

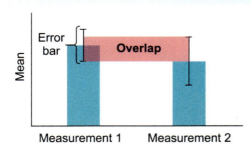
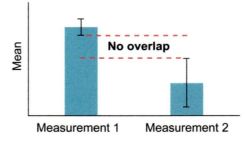

Figure 3.16 Error bars for the 95% confidence interval.

3.2.06 Validity

Experimental validity is a measure of the extent to which an experiment tests what it is intended to test and the extent to which the experimental results represent the true state of the scientific issue being studied. If a scientific experiment is valid, its conclusions are likely to be trusted (ie, accepted by the scientific community) and applied in scientific decision-making and further research.

Various types of validity, including the following general types, affect the acceptance and applicability of biological research findings:

- **Internal validity** is the extent to which the observed effects in an experiment can be attributed to the factor being studied (ie, the independent variable) as opposed to extraneous factors (eg, confounding variables) that can influence the outcome. Experiments with high internal validity provide convincing evidence that causal relationships exist between the experiments' independent and dependent variables.
- **External validity** pertains to the suitability of applying experimental findings to a more generalized context. The results of an experiment with high external validity can justifiably be applied to a broader population than that sampled in the experiment.

The internal and external validity levels of an experiment tend to be negatively correlated; that is, as one increases the other decreases. This negative correlation exists because factors that increase internal validity (eg, highly controlled laboratory conditions) tend to decrease the generalizability of the experimental results to real-world situations. Figure 3.17 summarizes the characteristics of internal validity and external validity.

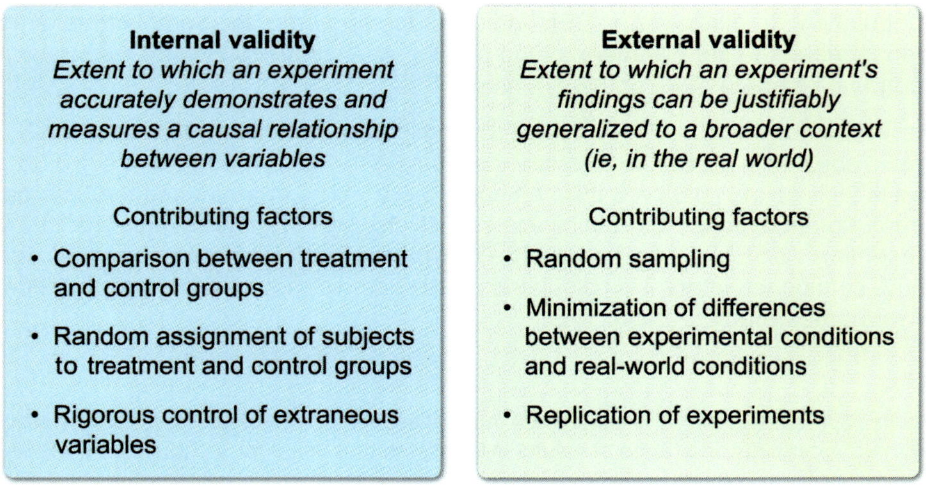

Internal validity	External validity
Extent to which an experiment accurately demonstrates and measures a causal relationship between variables	*Extent to which an experiment's findings can be justifiably generalized to a broader context (ie, in the real world)*
Contributing factors	Contributing factors
• Comparison between treatment and control groups	• Random sampling
• Random assignment of subjects to treatment and control groups	• Minimization of differences between experimental conditions and real-world conditions
• Rigorous control of extraneous variables	• Replication of experiments

Figure 3.17 Internal and external validity of scientific experiments.

3.2.07 Statistical Conclusions

The goal of scientific research is to arrive at valid conclusions regarding scientific problems. Scientific conclusions are based on statistical analyses of data generated in scientific experiments. The degree to which these statistics-based conclusions are reasonable (ie, accurate, appropriate) is referred to as **statistical conclusion validity (SCV)**.

SCV is influenced by various factors, including an experiment's statistical power and the degree to which statistical test assumptions are met. If an experiment has low statistical power, SCV is negatively affected. Likewise, if multiple statistical tests are applied to a data set in an attempt to find a test that provides a favorable result (ie, data fishing) without accordingly adjusting the error rate (ie, significance level), SCV decreases. Figure 3.18 summarizes factors that contribute to SCV.

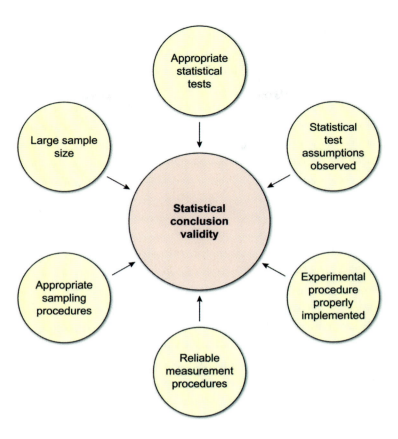

Figure 3.18 Factors that contribute to statistical conclusion validity.

Lesson 4.1

DNA Technology

Introduction

As described in Chapters 1 and 2, certain fundamental characteristics of nucleic acids, such as high-fidelity replication and the ability to base pair, make DNA and RNA well suited to serve as information reservoirs and mediators of gene expression. When combined with knowledge of the cellular machinery involved in nucleic acid handling (eg, enzymes that cut nucleic acids at specific sites) and mathematical consideration of base pairing, these characteristics also enable powerful methods of manipulation and analysis. This lesson focuses on a few concepts related to such methods as applied to DNA.

Please note that much of the information presented in this chapter is presented primarily as an introduction to the concepts involved. In most cases, multiple variations exist for the techniques presented. Therefore, more important than memorizing the specifics presented here is the ability to think critically about the concepts behind these techniques, as they are often the basis of questions and passages.

4.1.01 DNA Sequencing

The order of nucleotides in a DNA strand or fragment is called the DNA sequence, and **DNA sequencing** refers to the process of determining this order. DNA sequencing can be used for several purposes, including the determination of gene or genomic sequences and for the construction of nucleic acid sequences for experimental purposes.

Multiple sequencing techniques exist, including the chain termination method (also called the Sanger dideoxy method after its inventor, Frederick Sanger) as well as newer techniques collectively referred to as **next-generation sequencing**. Many sequencing methods involve tracking the incorporation of labelled nucleotides into new DNA strands; for this reason, such methods are sometimes said to utilize *sequencing by synthesis*.

The **Sanger dideoxy method** was the first widely used sequencing approach and employs dideoxy nucleotides (ddNTPs). Typical deoxyribonucleotide triphosphates (dNTPs) lack a hydroxyl group at the 2' carbon of the ribose ring but possess the 3' hydroxyl group required for phosphodiester bond formation. In contrast, ddNTPs lack a hydroxyl group at both the 2' and the 3' carbons (Figure 4.1).

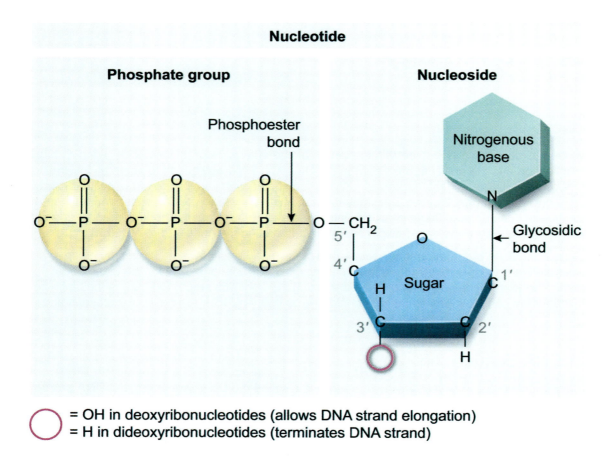

Figure 4.1 Deoxyribonucleotides and dideoxyribonucleotides.

ddNTPs can be incorporated into growing DNA strands via phosphodiester bonds involving the ddNTP 5′ phosphate groups. Incorporation of ddNTPs in this way results in the termination of DNA strand elongation due to the lack of 3′ hydroxyl groups to which new nucleotides can bind. By labelling the specific ddNTPs, the identity of the ddNTP terminating each DNA strand can be determined.

In the Sanger dideoxy method, small amounts of specific ddNTPs are mixed with denatured (ie, single-stranded) sample DNA, an excess of dNTPs, DNA polymerase, and a primer complementary to the denatured sample DNA. The polymerase synthesizes complementary DNA using the dNTPs and ddNTPs, thereby generating strands of various lengths, some terminating with a ddNTP. These new double-stranded DNA molecules are subsequently denatured to single-stranded DNA for further analysis.

The synthesized strands terminating with ddNTPs are loaded onto a gel and separated according to size using electrophoresis. Radiography (see Concept 4.1.09) was originally used to detect each ddNTP separately, but radioactive ddNTPs were eventually replaced by fluorescent ddNTPs (deemed safer and enable the use of multiple labels at once). Since each nucleotide position in the original DNA sample is represented, the nucleotide sequence in the original DNA sample can be determined by combining the ddNTP identities with the relative position of each DNA segment in the gel (Figure 4.2).

Chapter 4: Biotechnology

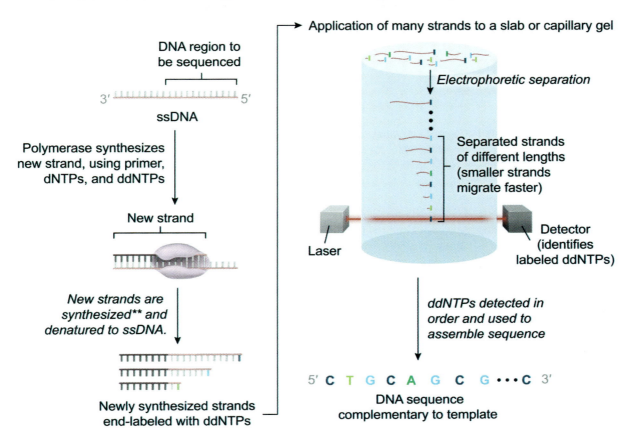

Figure 4.2 Sanger (dideoxy) sequencing.

Several next-generation methods of DNA sequence analysis have been developed. Some of these methods utilize sequencing by synthesis, but on a much larger scale than Sanger sequencing. Next-generation methods typically result in a greatly accelerated rate of sequence determination at a lower cost, making these methods well suited for large-scale genomic analyses.

In one next-generation method, large DNA samples are cut into smaller fragments and attached to the bottom of special wells on a plate. The fragments are amplified to produce clusters of ~1,000 template strand copies of each DNA fragment. Labeled versions of each dNTP are then sequentially added along with DNA polymerase, as shown in Figure 4.3.

Similar to Sanger sequencing with fluorescent ddNTPs, each labeled dNTP in this next-generation method contains both a tag that fluoresces upon incorporation and a 3' blocking group that terminates elongation. However, the fluorescent tag and blocking group are both removable, thereby imparting a temporary indicator of dNTP incorporation into the new DNA strand as well as the ability to add additional dNTPs after removal of the blocking group.

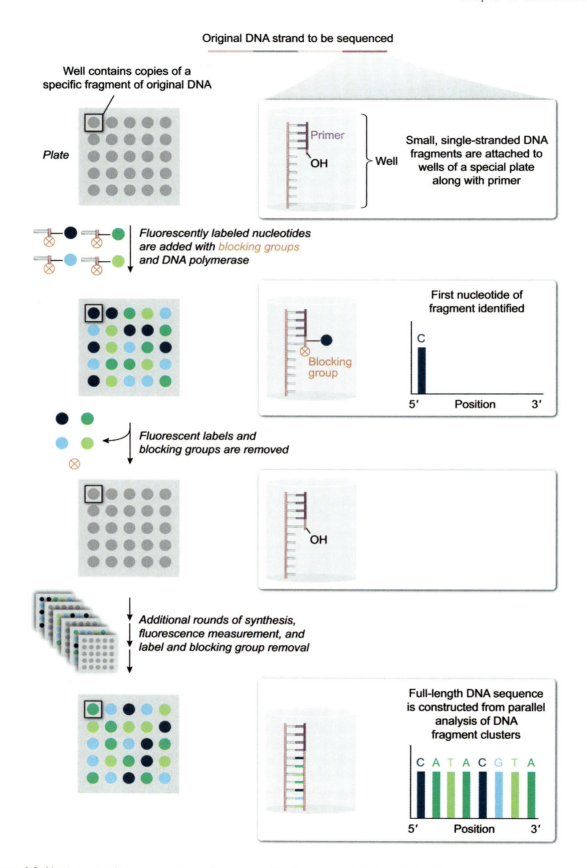

Figure 4.3 Next-generation sequencing using removable fluorescent labels and blocking groups.

Because the fluorescent tag and blocking group are removable, the template strands in each well can be reused over many rounds of elongation. This ability to recycle template strands, as well as the capacity to monitor multiple sequences simultaneously in separate wells, greatly increases the efficiency of this next-generation method compared to Sanger sequencing. Monitoring light emission during successive rounds of dNTP addition and label and blocking group removal enables determination of the DNA sequence of the strand synthesized in each cluster.

> ☑ **Concept Check 4.1**
>
> Based on the pattern of radioactive bands in the given gel, created during Sanger sequencing of a DNA fragment, state the order of nucleotides in the DNA fragment that was run on the gel.
>
>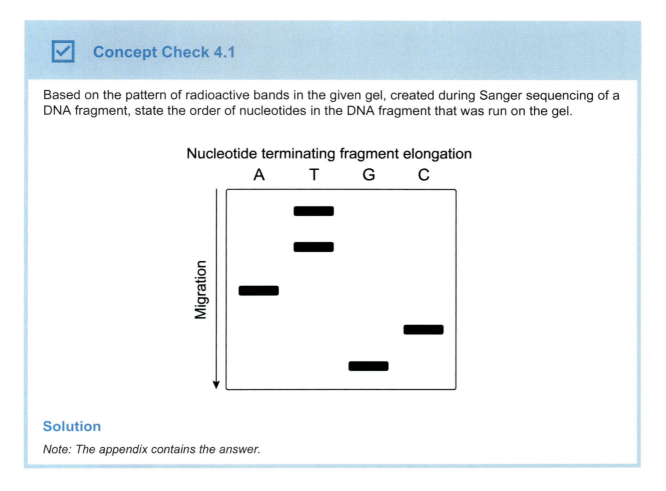
>
> **Solution**
>
> Note: The appendix contains the answer.

4.1.02 Polymerase Chain Reaction

The **polymerase chain reaction (PCR)** is a technique for generating multiple copies of a DNA sequence. The amplification of DNA in PCR allows even very small amounts of DNA to be detected and studied. The reagents required for PCR include the following:

- A **source DNA template** that encompasses both the target region to be amplified and its adjacent flanking sequences, to which primers bind
- Short, single-stranded **primer pairs** consisting of *forward* (ie, complementary to the 3' end of the antisense strand of the DNA sample) and *reverse* (ie, complementary to the 3' end of the sense strand) primers designed to bind the regions flanking the target sequence
- A **thermostable DNA polymerase** (ie, not denatured at the high temperatures used in PCR) to replicate the DNA template
- A pool of supplied **deoxyribonucleoside triphosphates (dNTPs)**

PCR occurs in a repeating cycle of three steps, as depicted in Figure 4.4:

1. Denaturation: heat-induced separation of double-stranded DNA into single-stranded DNA

2. **Annealing:** cooling-induced hybridization of single-stranded DNA into double-stranded DNA by binding of the primers to the single-stranded flanking ends of the target DNA region
3. **Elongation:** nucleotide addition to primer strands in a 5′ to 3′ direction by the thermostable DNA polymerase

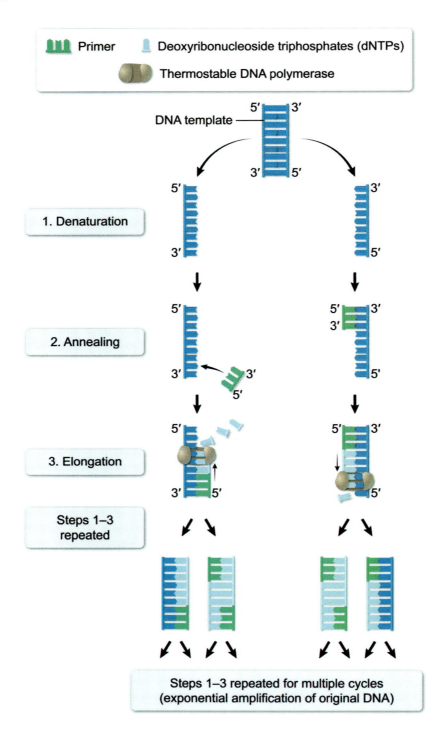

Figure 4.4 The polymerase chain reaction (PCR).

Each PCR cycle amplifies the double-stranded DNA through semiconservative DNA replication. After denaturation, each strand in a DNA duplex serves as the template to form a new duplex; therefore, each PCR cycle theoretically doubles the amount of DNA in a sample, as shown in Figure 4.5. When repeated over many cycles, millions of copies of the target DNA segment can be produced in a short time. Standard PCR quantifies DNA amplification after all the cycles are complete, and the DNA produced is often utilized for another procedure or analysis (eg, DNA sequencing, synthesis of a specific protein).

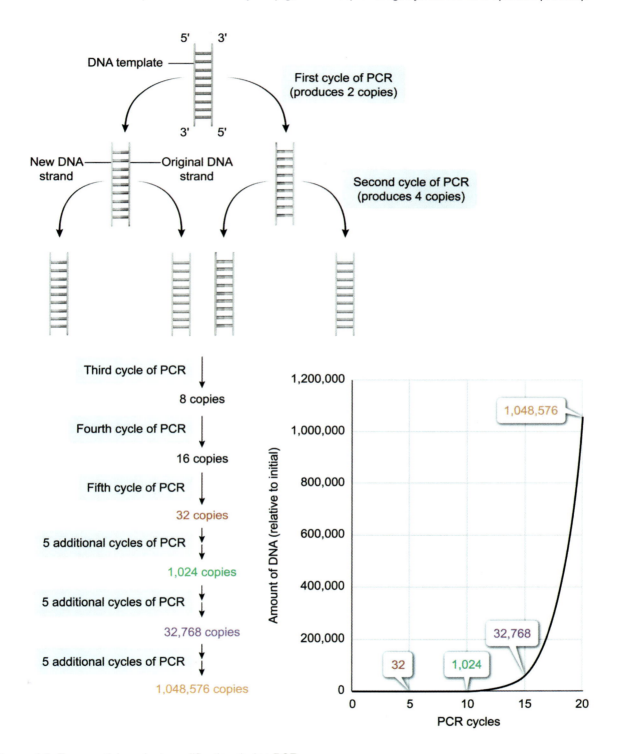

Figure 4.5 Exponential product amplification during PCR.

Due to the exponential nature of the increase in DNA levels during PCR, when DNA production by PCR is plotted using a linearly scaled *y*-axis, production during early cycles becomes more difficult to distinguish as the number of cycles plotted increases. In situations such as this, a logarithmic scale can make actual changes more apparent, as shown in Figure 4.6.

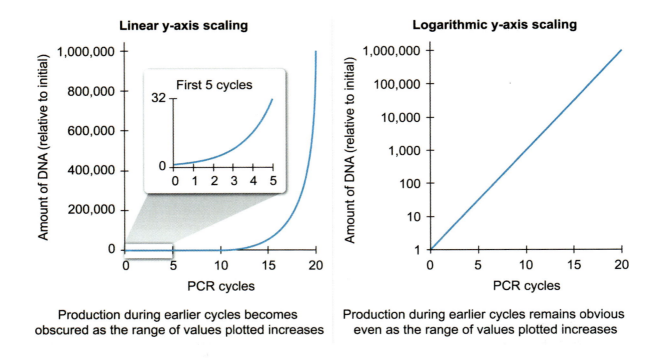

Production during earlier cycles becomes obscured as the range of values plotted increases

Production during earlier cycles remains obvious even as the range of values plotted increases

Figure 4.6 Linear versus logarithmic scaling of DNA production during PCR.

Real-time or quantitative PCR (qPCR) is PCR in which product amplification is quantified in real time as the reaction progresses, typically by using fluorescent DNA markers. Of note, qPCR is sometimes called RT-PCR; however, the exam uses *RT-PCR* to designate PCR analysis of RNA, which is discussed in detail in Concept 4.2.01.

In qPCR, the theoretical relationship between cycle number and product amount can be capitalized upon to estimate the amount of DNA in a sample. For example, the number of cycles necessary to reach a certain threshold amount of DNA varies according to the number of DNA molecules in the sample, as shown in Figure 4.7.

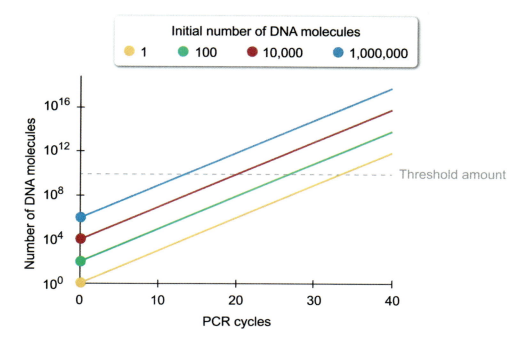

Figure 4.7 The number of PCR cycles necessary to reach a particular threshold varies with the initial amount of DNA.

In practice, the amount of DNA produced per PCR cycle eventually levels off as the number of cycles increases, and the fluorescence or other signal used for quantifying the amount of DNA by qPCR is not initially detectable. The first cycle number at which fluorescence exceeds background levels is designated as the cycle threshold (C_t) and is related to the initial amount of DNA in a sample, as depicted in Figure 4.8.

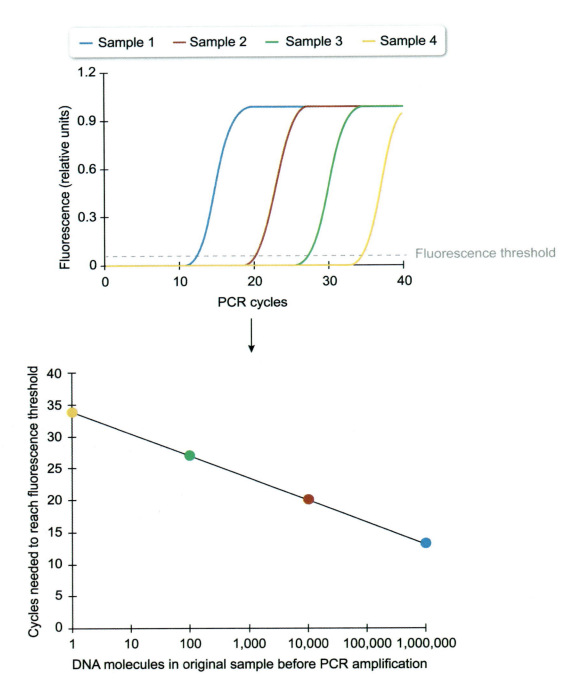

Figure 4.8 The PCR cycle number at which the fluorescence threshold (ie, C_t) is reached varies with the starting amount of DNA.

4.1.03 Restriction Enzymes

Restriction enzymes are endonucleases that typically originate in bacteria and cut double-stranded DNA (dsDNA) at nonterminal sites (ie, at places other than dsDNA ends). In nature, these endonucleases protect bacteria from bacterial viruses (ie, bacteriophages) by cutting viral DNA into small pieces, thereby restricting transfer of foreign DNA into bacterial cells.

In laboratory settings, restriction enzymes are used to create cuts at specific sites in dsDNA. Each restriction enzyme recognizes a characteristic short sequence of nucleotides (typically four to eight pairs);

this sequence is called the target sequence, recognition sequence, or restriction site for the enzyme. Most target sequences for restriction enzymes used in laboratory settings are palindromic sequences (ie, paired nucleotide sequences on the two strands that are identical when read in the same 5' to 3' orientation), as illustrated in Figure 4.9.

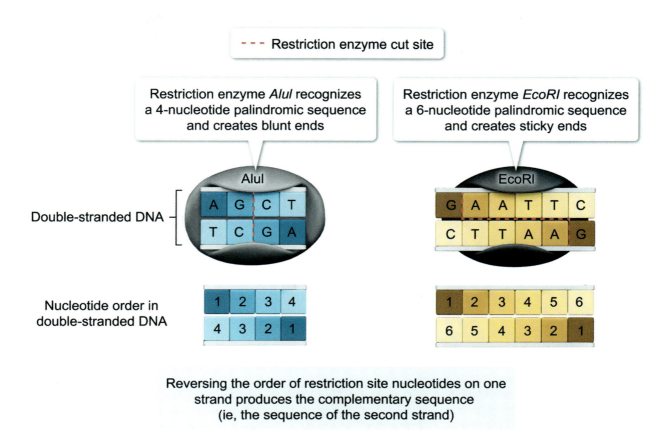

Figure 4.9 Examples of palindromic restriction site sequences.

Restriction enzymes differ in the type and number of cuts they make in dsDNA. Some enzymes (eg, AluI) cut evenly across dsDNA to create *blunt ends*, or cuts in dsDNA that do not create single-stranded regions. Others (eg, EcoRI) cut unevenly across dsDNA, thereby creating short single-stranded segments known as *sticky ends* on each of the cut strands.

4.1.04 Gene Cloning

Gene cloning is the insertion of a previously isolated gene into a DNA sequence, typically to make multiple copies of the gene. Gene cloning is often carried out for the ultimate purpose of creating a large amount of a specific gene's transcribed or translated products (RNA or protein, respectively). The gene or gene products formed during cloning are called **recombinant**, meaning that they are from more than one source (eg, different species or bacterial strains).

One approach to gene cloning consists of cutting plasmid DNA with a restriction enzyme (Concept 4.1.03) and adding the DNA fragment to be cloned along with DNA ligase. Plasmid DNA is small, circular, nonchromosomal DNA, typically found in bacteria (which have a single origin of replication [see Concept 6.3.02]). Cutting the plasmid DNA linearizes it and introduces a site for the DNA fragment to be inserted. DNA ligase is added to join the ends of the DNA fragment and plasmid together, thereby converting the DNA back into circular form. The general process of gene cloning is shown in Figure 4.10.

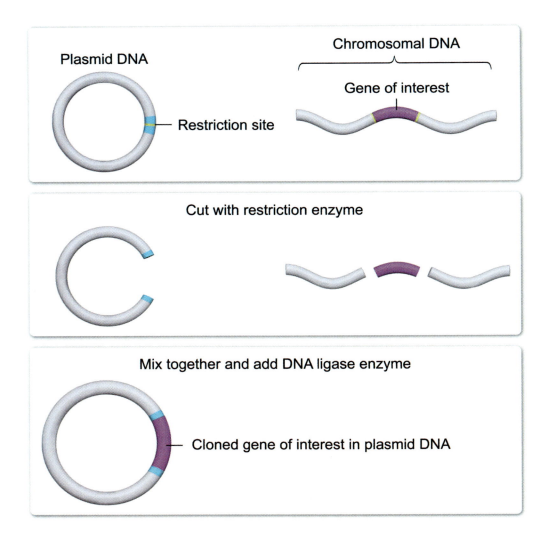

Figure 4.10 The general cloning process.

The ends of the DNA fragment and the plasmid must match (ie, be complementary) for successful incorporation of the fragment into the plasmid. One way this is accomplished is by using the same restriction enzyme in both the isolation of the DNA fragment and the cutting of the plasmid, such that the ends of the plasmid and fragment are complementary. Another approach is to use PCR to add restriction sites at each end of the sequence to be cloned.

When the use of different restriction enzymes produces a combination of noncomplementary blunt and sticky ends on the DNA fragment and the plasmid, dNTPs and DNA polymerase can be added to fill in the staggered ends. The resulting fragment and plasmid, both blunt-ended, are then joined by DNA ligase (Figure 4.11).

Chapter 4: Biotechnology

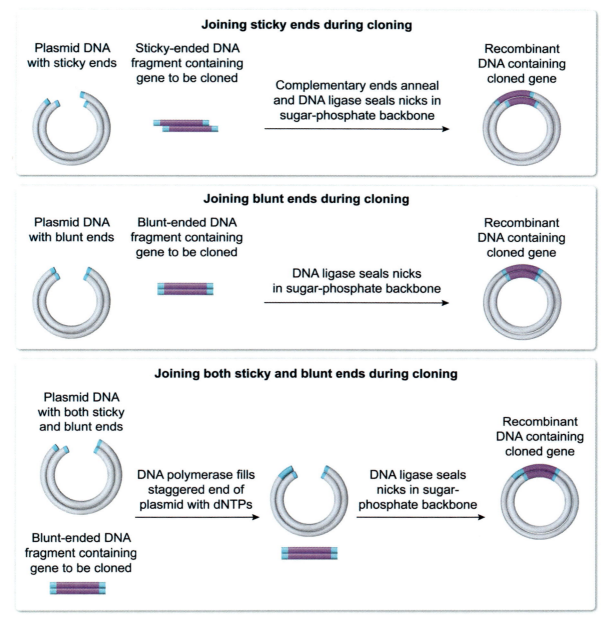

Figure 4.11 Joining of plasmid and cloning fragment DNA to form recombinant DNA.

After addition of the DNA fragment, the plasmids are added to bacterial cells, and the cells are stimulated to take up the plasmids. This stimulation often requires heating the cells, and the process of taking up the plasmids is called transformation (as discussed in Concept 6.3.03).

Notably, not all stimulated cells take up the experimental plasmid, and not all transformed bacteria contain the correct plasmid. Some cells do not undergo transformation, and the successful insertion of the DNA fragment containing the gene to be cloned is only one of several possibilities, as shown in Figure 4.12. For example, in some cases, complementary ends of the linearized plasmid DNA are ligated back together without the target gene having been inserted. In addition, the gene may be inserted, but in the wrong orientation (ie, reversed) or position (ie, somewhere in the plasmid other than the intended location).

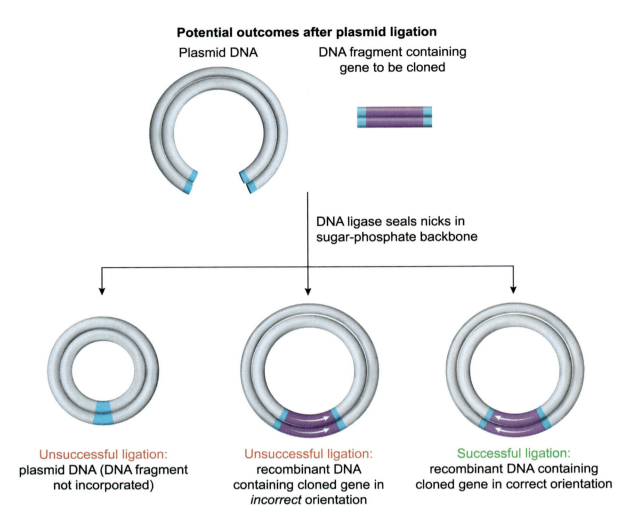

Figure 4.12 Successful insertion of the gene to be cloned does not always occur.

Because gene insertion is not always successful, additional measures are often taken to aid in selecting bacterial cells containing the gene of interest. An antibiotic resistance gene is typically inserted into the plasmid with the DNA sequence containing the gene to be cloned, and the bacterial colonies are then grown in media containing the antibiotic. This process facilitates the selection of bacterial colonies containing the experimental plasmid, as bacteria lacking the antibiotic resistance gene die.

Further screening via PCR or DNA sequencing is used to verify whether the gene sequence is found in the surviving colonies, which can then be isolated and expanded (Figure 4.13).

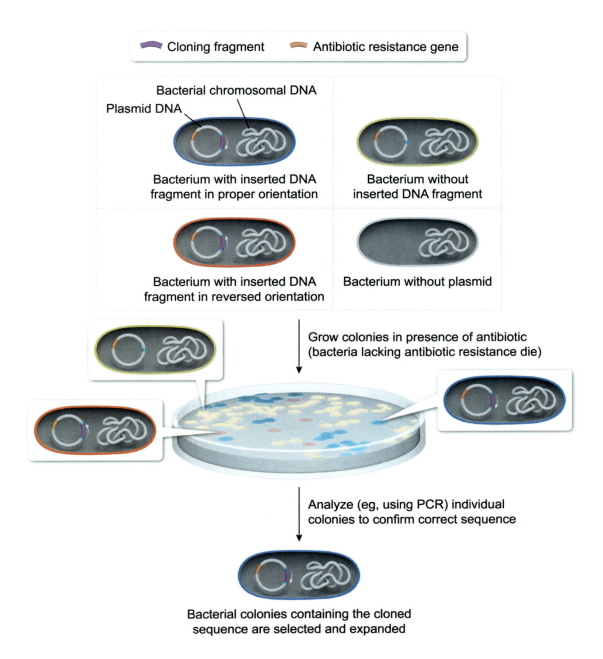

Figure 4.13 Selection of bacterial colonies containing a cloned gene of interest.

4.1.05 Generation of cDNA

In the context of gene cloning and experimental nucleic acid generation, **complementary DNA (cDNA)** refers to DNA generated from an RNA template, typically mRNA. Because mature mRNA lacks introns, cDNA represents the minimum amount of genetic information necessary for gene expression: a contiguous sequence of nucleotides (ie, exons) coding for a protein.

To generate cDNA, mRNA is first used as a template for the synthesis of a single cDNA strand. This single cDNA strand is then used as a template to synthesize a complementary second cDNA strand, thereby completing the synthesis of double-stranded cDNA, as shown in Figure 4.14. Once formed, cDNA can be amplified by insertion into plasmids and subsequent transformation into cells or by PCR (see Concept 4.1.02).

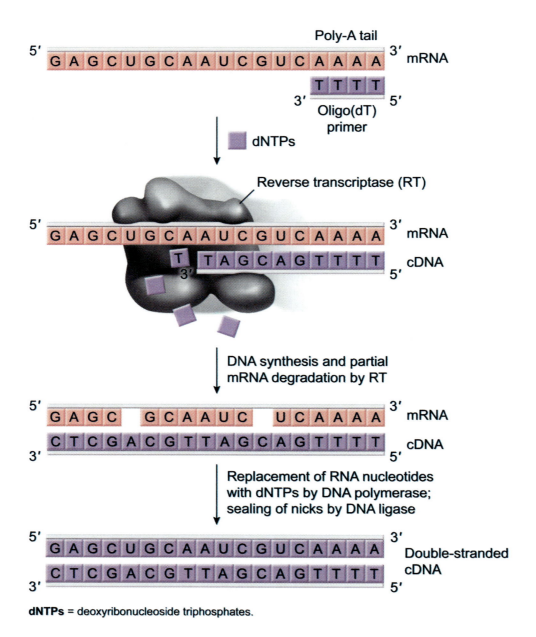

Figure 4.14 Generation of cDNA.

To initiate synthesis of the first cDNA strand, isolated mRNA is typically mixed with primers composed of thymine nucleotides. These primers, known as poly-T or oligo(dT) primers, are complementary to the 3' poly-A tails of eukaryotic mRNA molecules (other primers can be used for other RNA types). Next, the enzyme **reverse transcriptase** uses the primer-bound mRNA as a template for the synthesis of DNA, adding dNTPs to generate single-stranded cDNA. The mRNA strand is then partially degraded by the reverse transcriptase, leaving the first cDNA strand bound to the original RNA in some regions.

The second (ie, complementary) cDNA strand is then synthesized by DNA polymerase, using the first cDNA strand as a template. The RNA fragments that remain bound to the first cDNA strand serve as primers to which DNA polymerase successively adds dNTPs. Remaining RNA fragments are removed during synthesis of the second cDNA, as the 5'→3' exonuclease activity of DNA polymerase displaces NTPs (ie, RNA nucleotides) as it adds dNTPs (ie, DNA nucleotides). DNA ligase is then added to seal any nicks in the DNA, completing the synthesis of the double-stranded cDNA.

4.1.06 DNA Libraries

DNA libraries are collections of cloned DNA fragments and can be generated in several ways. Separation of DNA fragments or individual genes into libraries facilitates the study of specific, smaller regions of DNA as opposed to entire chromosomes.

One approach to DNA library construction is to enzymatically (eg, using restriction enzymes) or mechanically (eg, using sudden, forceful movements) break chromosomal DNA into smaller pieces of double-stranded DNA. Each of these smaller pieces is then inserted into a plasmid that is taken up by a bacterium or fungal cell. Amplification of the cloned fragment occurs as the recipient cell undergoes replication. Because the entire genome is represented in the collection of DNA fragments, a library produced in this way is called a **genomic DNA library** (Figure 4.15).

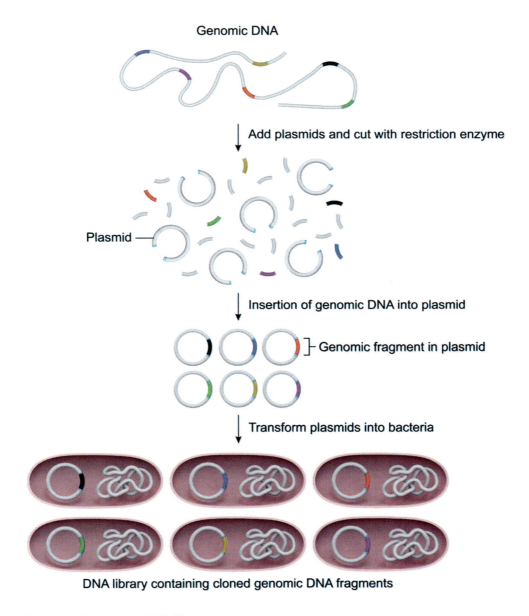

Figure 4.15 Creation of a genomic DNA library.

A second approach for constructing a DNA library begins with mRNA, which is reverse transcribed into double-stranded cDNA (see Concept 4.1.05) before being inserted in plasmids and transferred to recipient cells for amplification. Because the molecules from which the library is derived are mRNA, this type of library represents only the portion of the genome that codes for proteins. Therefore, these **cDNA libraries** do not contain intronic segments that would be present in a genomic library.

A third approach uses PCR to amplify cDNA of individual genes or genomic fragments. These fragments can be transferred into plasmids and expressed in bacteria.

4.1.07 DNA Gel Electrophoresis and Southern Blotting

DNA electrophoresis is a means of separating DNA molecules of different sizes by using an electric current to induce the molecules to migrate through a special gel. The gel is formed by particles that link to each other, and one end of the gel has rectangular wells into which DNA samples can be added. The linking of the particles in the gel results in the formation of tiny spaces through which DNA molecules can pass.

An electrophoresis system includes the gel, a tank containing an electrophoresis buffer solution (which conducts charge and into which the gel is submerged), and electrodes attached to a power supply (Figure 4.16). The gel electrophoretic system operates as an electrolytic cell in which the current supplied by the power source causes one electrode to become negative (the cathode) and the other electrode to become positive (the anode).

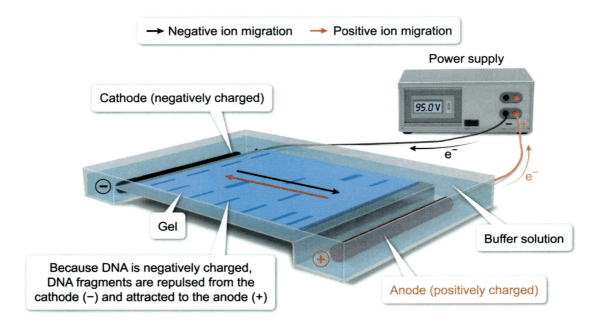

Figure 4.16 A DNA electrophoresis system.

During DNA electrophoresis, samples in a special loading buffer are loaded into wells on the side of the gel next to the cathode while the power supply is off. When the power supply is turned on, the negatively charged DNA molecules in the samples migrate away from the negatively charged cathode toward the positively charged anode at the opposite end of the gel. The generally straight migration path from each sample well to the other end of the gel is called a *lane*.

Smaller DNA molecules migrate longer distances during electrophoresis, and larger molecules migrate shorter distances (Figure 4.17). DNA samples are sometimes treated with restriction endonucleases before loading to break the DNA strands into smaller DNA fragments that migrate farther along the lane than the intact DNA molecule would.

Typically, one lane of the gel is devoted to a standard sample containing DNA fragments of known lengths. This lane is sometimes called a *ladder* because the regular pattern of bands from the standard resembles rungs on a ladder. The ladder allows the sizes of the sample DNA molecules to be estimated by comparing the location of sample bands with those in the ladder.

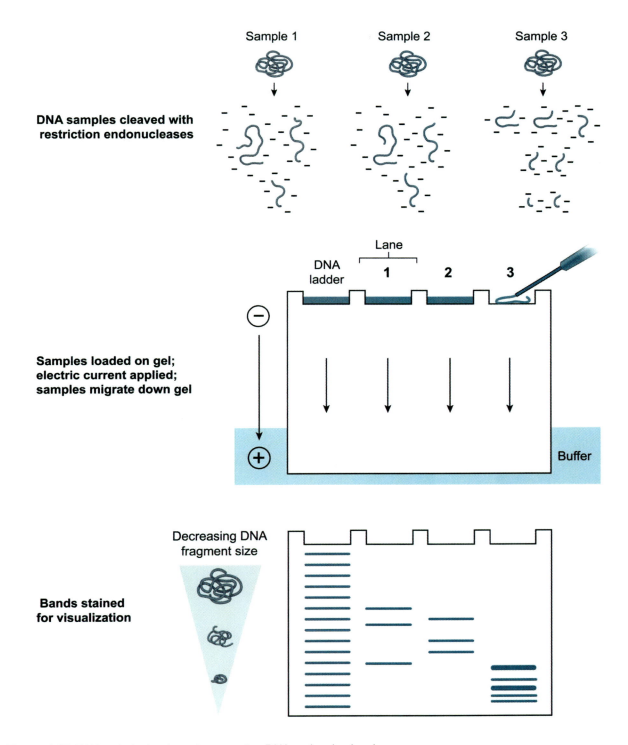

Figure 4.17 DNA gel electrophoresis separates DNA molecules by size.

The bands produced during electrophoresis can be visualized using stains such as ethidium bromide, which inserts between the nucleic acid bases and fluoresces under ultraviolet light. When ethidium bromide is applied to a gel, the bands on the gel are illuminated upon application of ultraviolet light. A general overview of ethidium bromide staining is presented in Figure 4.18. Because nucleic acids absorb ultraviolet light, DNA bands can also be detected without ethidium bromide by measuring ultraviolet light absorption along each lane in the gel.

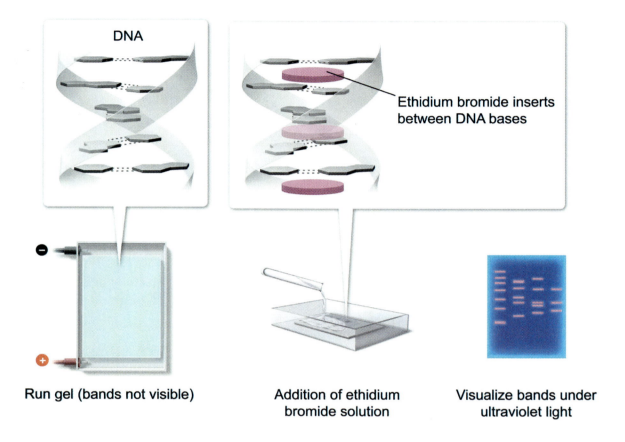

Figure 4.18 Ethidium bromide can be used to detect bands on DNA electrophoresis gels.

Following electrophoresis, DNA bands can also be detected using a technique called **Southern blotting** (named after its inventor, Edwin Southern). Rather than detecting all the DNA bands on a gel, Southern blotting uses a nucleic acid probe to detect only bands that contain a specific DNA sequence. In Southern blotting, the bands from the gel must be transferred to special blotting membranes prior to detection, and the nucleic acid probe is labeled in some way that facilitates detection (eg, radioactivity, ability to fluoresce). The general steps of Southern blotting are shown in Figure 4.19.

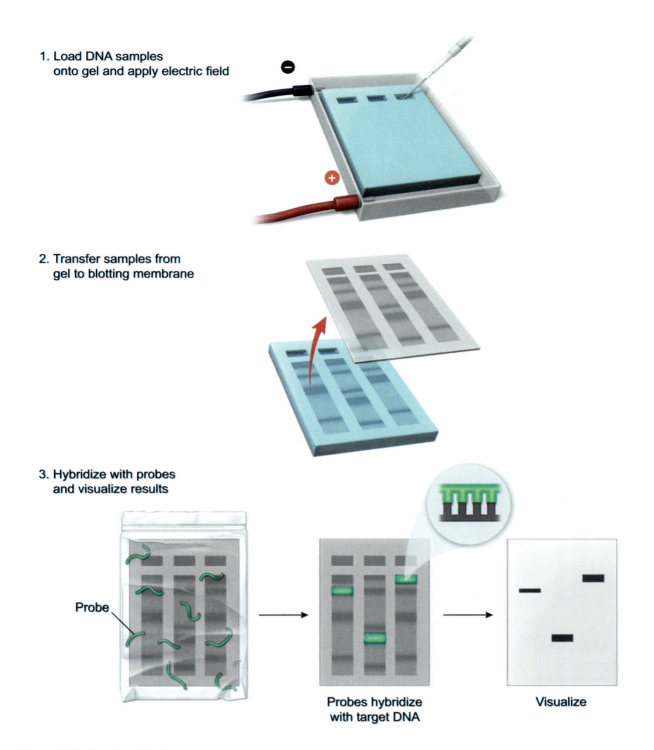

Figure 4.19 Southern blotting.

4.1.08 Hybridization

Hybridization is the base pairing of complementary strands of nucleic acids via hydrogen bonding. Because hydrogen bonds are relatively weak, two hybridized polynucleotide strands can be easily separated, or denatured, through heating or by chemical means. In addition to its role in nature (detailed in Concept 1.1.04), hybridization has been harnessed for numerous biotechnological applications.

There is a notable combinatorial character to hybridization: Perfect pairing between two DNA strands becomes less likely to occur simply by chance as the number of nucleotides involved increases, as shown in Figure 4.20.

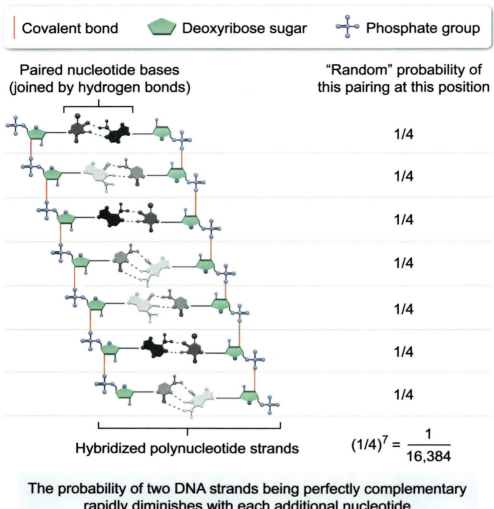

The probability of two DNA strands being perfectly complementary rapidly diminishes with each additional nucleotide

Figure 4.20 Hybridization is very specific because of its combinatorial nature.

By generating a specific sequence of nucleotides that have been modified in such a way as to facilitate detection (eg, by using radioactive nucleotides or attaching a chemical group to which an antibody can bind), a hybridization probe can be created. The probability of the complementary sequence occurring simply by chance in a genome drops by one-fourth for each additional nucleotide in the probe sequence. Therefore, binding of a typical 30-nucleotide hybridization probe to a nucleotide sequence in a sample is highly specific.

Hybridization probes are used to determine which part or parts of a sample contain the complementary nucleic acid sequence. For example, hybridization probes are used in conjunction with electrophoretic separation of DNA samples to determine whether any band on the gel corresponds to a particular sequence of interest (as in Southern blotting, see Concept 4.1.07).

In another type of hybridization, called **in situ hybridization**, hybridization probes are used to localize nucleic acid sequences within a chromosome, cell, tissue, or organism. A specific example of this technique known as *chromosome painting* is illustrated in Figure 4.21.

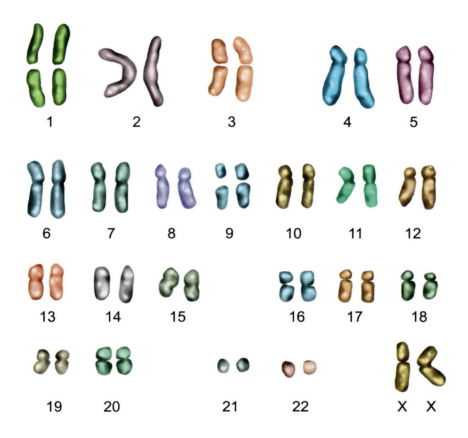

Figure 4.21 Chromosome painting, in which fluorescently labelled DNA probes that hybridize to sequences associated with each chromosome, is used to locate or track chromosomes.

4.1.09 Radiography

Radiography is the use of ionizing electromagnetic radiation for imaging or imaging-related measurements. Ionizing radiation is of high enough energy to cause electron loss from an atom, thereby forming an ion, and can be produced during radioactive decay or by other means. In some contexts, the source of radiation is external (ie, produced outside of the cell, molecule, or organism under study). In other cases, procedures are employed that result in high-energy radiation being produced within the structure being studied.

For example, brief bombardment of a sample, body region, or entire person or model organism with externally produced X-rays (a type of ionizing radiation) allows certain structures to be visualized because different materials (eg, fat, bone) absorb such radiation to varying degrees. By using photographic film or other detection methods sensitive to electromagnetic radiation, the differential absorption of X-rays by different regions creates an image of the object being irradiated (Figure 4.22).

Such X-ray absorption can also be used to estimate the composition or density of anatomical structures (eg, bone density, lipid content of muscle).

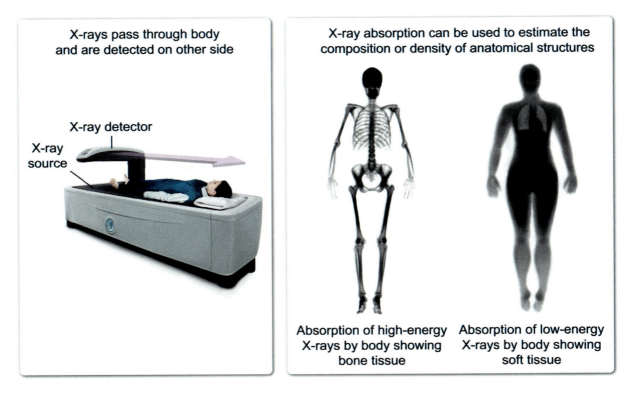

Figure 4.22 The absorption of X-rays by body tissues can be used for assessing body composition.

Autoradiography is a type of radiography in which internally released ionizing radiation is used for imaging or measurement purposes. Such radiation can be incorporated into molecules metabolically *in vivo* (eg, by supplying radioactive amino acids to label proteins) or during *in vitro* synthesis (eg, of a radioactive DNA probe) prior to injection into an organism or application to a sample. In both cases, the radioactivity is detected by placing a film or similar radiation-sensitive detector surface close to the sample to create an image, as depicted in Figure 4.23.

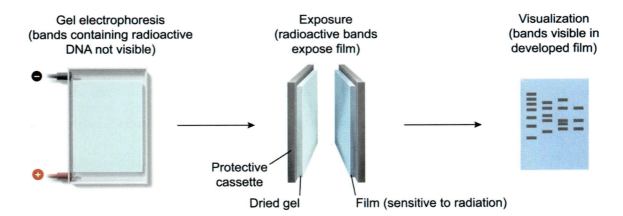

Figure 4.23 Autoradiography of DNA bands on an electrophoresis gel.

Lesson 4.2
Analyzing Gene Expression

Introduction

Given the few differences between DNA and RNA, it is perhaps not surprising that many methods for RNA detection and manipulation resemble those employed when working with DNA (see Lesson 4.1). As with DNA, the ability of nucleic acids to base pair (ie, hybridize) forms the basis for most measurement and manipulation techniques related to RNA.

Although proteins do not hybridize, they do undergo highly specific interactions with other molecules. These interactions are harnessed in numerous methods for studying protein expression and function. In addition, techniques employed in the study of DNA and RNA, such as electrophoresis and radiography, are also suitable for studying proteins.

This lesson explores a few ways scientists detect and manipulate RNA and proteins to analyze gene expression.

4.2.01 Detecting and Manipulating RNA

RNA detection methods include hybridization to an RNA or DNA probe (Figure 4.24), autoradiography, and UV absorption. In addition, the polymerase chain reaction (PCR) can be used for detection and manipulation of RNA. Additional hybridization-related methods are extensively employed for manipulating RNA.

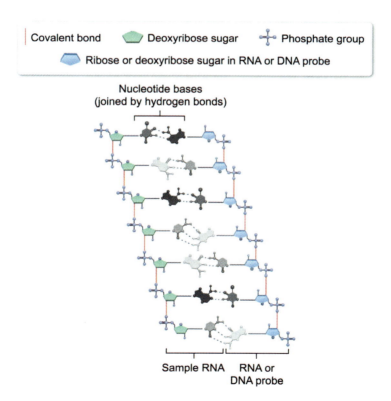

Figure 4.24 Hybridization of sample RNA to an RNA or DNA probe.

RNA can be assessed using nonspecific techniques that measure overall RNA without differentiating between RNA molecules with different sequences. For example, RNA (like DNA) absorbs light in the ultraviolet part of the spectrum, at 260 nm. Making a standard curve relating RNA concentration to absorbance allows the amount of RNA in a sample to be determined (Figure 4.25).

Another nonspecific technique used to analyze RNA is radioactive labeling. As described for DNA in Concept 4.1.09, **radioactive labeling** of RNA during synthesis similarly allows all the RNA bands on a gel to be detected using autoradiography.

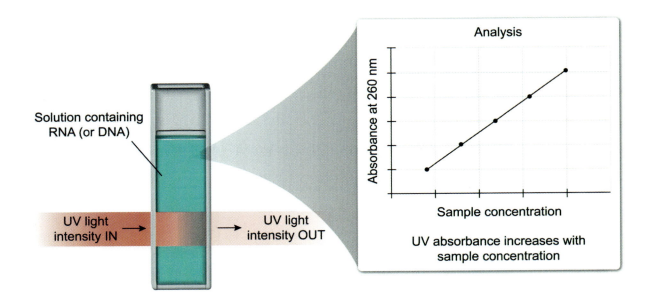

Figure 4.25 Determination of RNA (or DNA) concentration from UV absorbance.

Northern blotting is the electrophoretic separation of RNA molecules followed by detection using a single-stranded nucleic acid (typically DNA) probe that hybridizes with a single specific RNA sequence. The probe can be radioactively labeled, thereby facilitating detection using autoradiography, or labeled in some other way for ease of detection. As with Southern blotting (see Concept 4.1.07), from which the name *northern blotting* is playfully derived, the RNA bands are transferred (blotted) to a special membrane prior to detection.

Another method of detecting RNA is known as **microarray** analysis. Although also based on hybridization between an RNA molecule and a complementary DNA probe, microarrays differ from northern blots in that microarrays consist of probes for many distinct sequences rather than a single sequence. In addition, microarrays do not involve electrophoresis.

The DNA probes for microarray analysis are spatially arranged on a plate (sometimes called a chip) such that measuring the hybridization signal over the many distinct regions on the plate allows thousands of RNA sequences to be assessed simultaneously. If labeled with distinct tags (eg, fluorescence of a particular color), multiple samples can be assessed in the same plate (Figure 4.26).

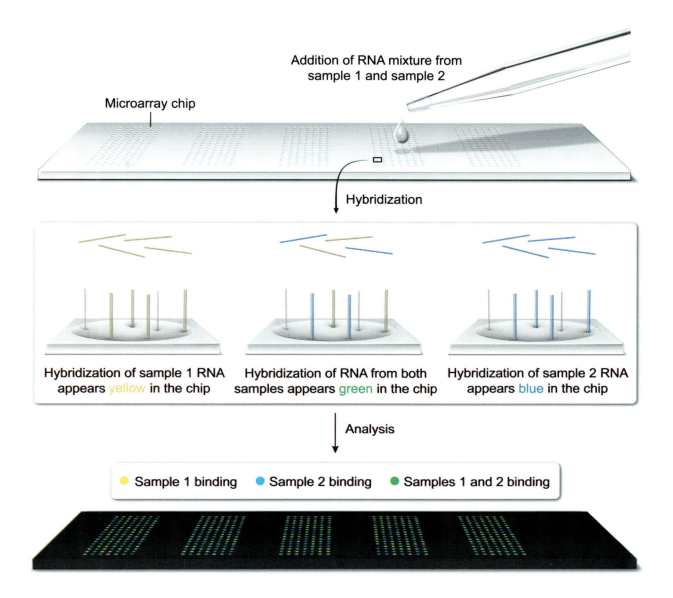

Figure 4.26 Using a microarray to measure amounts of specific RNA sequences.

In situ RNA hybridization, in which a complementary hybridization probe hybridizes with RNA molecules in a tissue sample, allows gene expression (ie, RNA molecules transcribed from DNA) to be localized within a tissue or cell. As with in situ DNA hybridization, the probe is labelled in some way that facilitates detection. For example, if the hybridization probe is labeled with a fluorescent tag, then fluorescence within the sample reveals the location(s) of complementary RNA molecules and the method is called **fluorescence in situ hybridization (FISH)**.

With slight modification to the procedure for amplifying DNA (see Concept 4.1.02), PCR can also be used to study RNA. In this modified approach, called **RT-PCR**, reverse transcriptase (RT) is used to produce complementary DNA (cDNA) molecules from the RNA in a sample, and then PCR is used to amplify the cDNA (Figure 4.27). Due to exponential amplification in PCR, RT-PCR is extremely sensitive for detecting specific RNA sequences. When the mathematical relationship between amplification cycle number and the amount of cDNA synthesized is used to quantify the amount of RNA in a sample, the process is referred to as **RT-qPCR**.

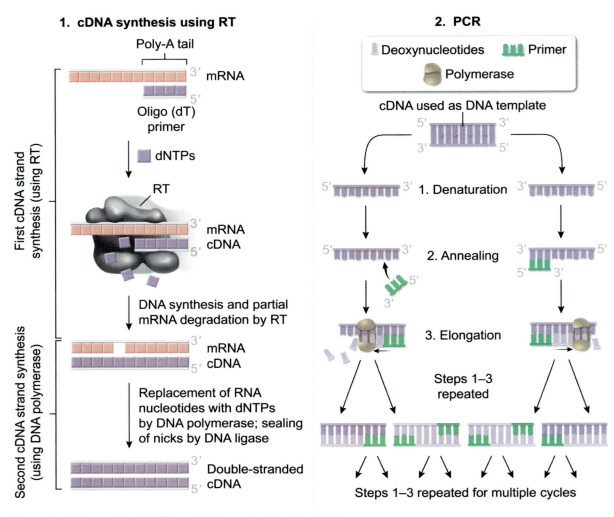

Figure 4.27 RT-PCR uses reverse transcriptase to generate cDNA, which is then used as the template for PCR.

Antisense strategies are another means of RNA manipulation; these strategies are covered in detail in Concept 4.4.01.

4.2.02 Detecting and Manipulating Proteins

The ability of proteins to undergo highly specific interactions with other proteins forms the basis of numerous techniques for studying protein abundance and function. In addition, some of the techniques used for nucleic acid analysis (eg, electrophoresis) can be applied in the study of proteins.

The **enzyme-linked immunosorbent assay (ELISA)** is a technique for measuring the amount of a specific protein in a sample using an immune-based detection mechanism (ie, antibodies). Typically, the sample is applied to a surface upon which the sample either sticks (eg, due to adsorption to a microplate well) or migrates (eg, due to wicking on paper). The specific protein of interest is detected by the binding of an antibody that is either directly or indirectly (through a secondary antibody) attached to a reporter enzyme. After antibody binding has been allowed to occur, unbound antibodies either are washed away or migrate away.

Next, the substrate of the reporter enzyme is added to the antibody-bound protein complex, allowing the substrate to react with any protein-antibody complexes. If a color change is detected, the intensity of the color change is proportional to the amount of bound protein. Protein expression levels are ultimately quantified by comparing the color change in the well or on the paper to the color change observed from a series of known concentration standards (Figure 4.28).

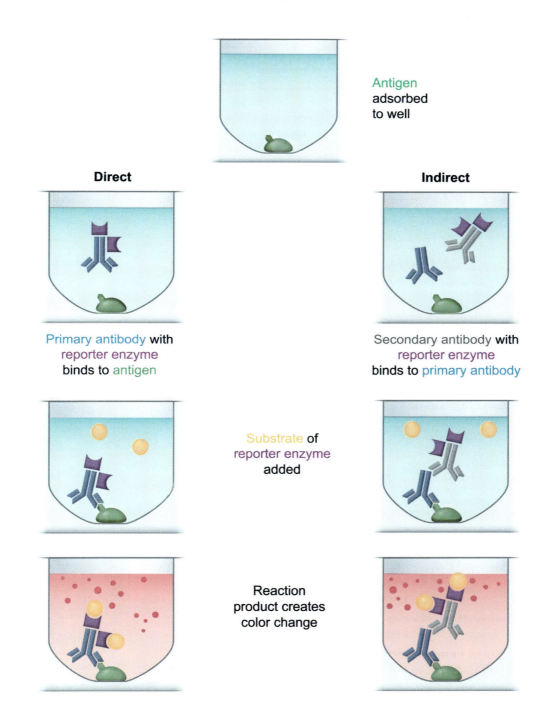

Figure 4.28 Enzyme-linked immunosorbent assay (ELISA).

Proteins can also be studied using **polyacrylamide gel electrophoresis (PAGE)**. Proteins separated on **nondenaturing (native) gels** retain their native structure and charge, both of which affect electrophoretic migration. In contrast, **denaturing gels** employ denaturants such as sodium dodecyl sulfate (SDS) to impart a uniform negative charge to proteins.

This negative charge is proportional to size, minimizing charge differences between proteins, so separation of proteins by denaturing gel electrophoresis (eg, SDS-PAGE) occurs primarily on the basis of size. The distance proteins migrate on the gel is inversely proportional to the logarithm of molecular weight (unless the proteins are being separated on special gradient gels in which the pore size changes from top to bottom).

Like Southern blotting for DNA or northern blotting for RNA, proteins in a sample can be electrophoretically separated, transferred to a membrane, and visualized in a process known as western blotting. The protein bands are detected via binding to a primary antibody, to which a secondary antibody binds for visualization (eg, via chemiluminescence, fluorescence, or autoradiography). Prior to adding antibodies, membranes are typically exposed to milk or another substance (eg, bovine serum albumin) to block nonspecific binding sites, a step that reduces background binding. The western blotting process is depicted in Figure 4.29.

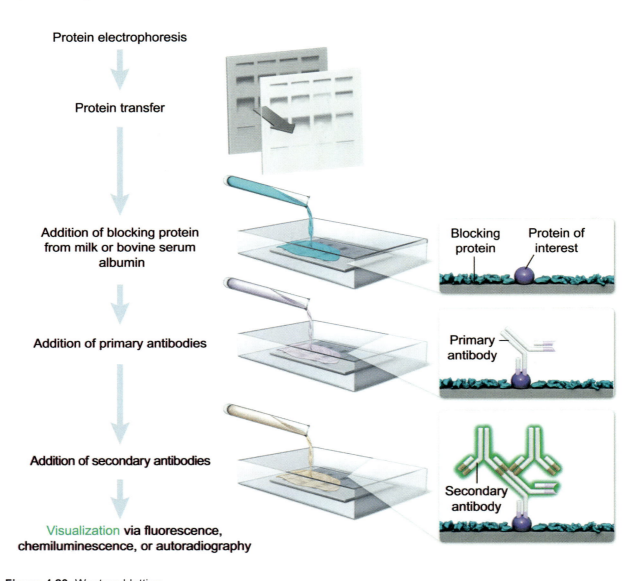

Figure 4.29 Western blotting.

Antibody-based detection of specific proteins can also be used for visualization of proteins in histological sections, a technique broadly known as **immunohistochemistry**. As in western blotting, the antigen (protein) is detected with a primary antibody, and a secondary antibody with a specific characteristic (eg, participation in an enzyme-linked color change reaction) facilitates visualization. For imaging using electron microscopy, secondary antibodies are linked to gold particles that facilitate detection.

Pulldown assays (Figure 4.30) are a means of studying protein-protein interactions using the affinity of a protein for a binding partner. A protein of interest is first bound to a stationary support structure (eg, a bead). This bound protein serves as "bait" for binding by other protein binding partners from a subsequently added sample, and nonbinding proteins are washed away. The bait–binding partner complexes are then eluted and collected via centrifugation (the bait "pulls down" its binding partners from the sample). Analysis of the collected complexes via western blotting or other methods allows identification of the bait––protein binding partners.

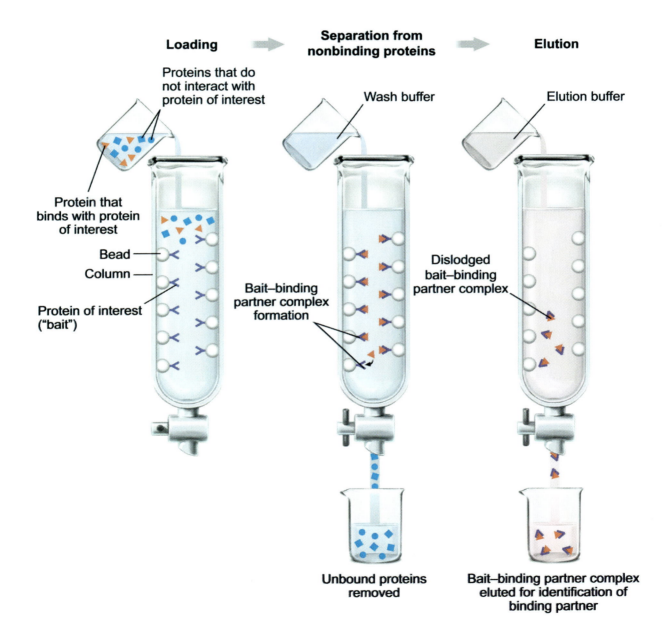

Figure 4.30 A pulldown assay.

Protein levels can be manipulated via various knockout or silencing strategies, discussed in Concepts 4.3.01 and 4.4.01.

> ### ☑ Concept Check 4.2
>
> Fill in the blanks in the paragraph below using the words "denaturing," "molecular weights," "nondenaturing," and "protein interactions."
>
> _____ polyacrylamide gels preserve the native state of proteins and are therefore well suited for determining _____ of proteins with higher-order structure. In contrast, the presence of SDS makes _____ gels better suited for determining the _____ of proteins in a sample.
>
> ### Solution
>
> *Note: The appendix contains the answer.*

Lesson 4.3
Determining Gene Function

Introduction

Protein synthesis depends on the information contained in DNA. Accordingly, one means of studying the function of proteins is to inactivate the gene responsible for a protein and observe the resulting phenotype. Likewise, a phenotype can provide information regarding the genetic mechanisms responsible for producing that phenotype. This lesson focuses on the study of protein function through genetic alteration and inferring genetic alterations through observation of phenotypes.

4.3.01 Knockouts

Knockout models (sometimes referred to simply as **knockouts**) are organisms or cells in which one or more genes have been removed or made nonfunctional (ie, "knocked out"). Knockouts are created by researchers to understand the function of the gene that has been removed or silenced.

One approach for developing knockouts is derived from a bacterial defense system that collects short DNA segments from infecting viruses, enabling each bacterium to accumulate previously encountered viral DNA sequences. The viral sequences are inserted at regular intervals between clusters of palindromic sequences of bacterial DNA, forming clustered, regularly interspaced, short palindromic repeats (CRISPR).

RNA transcribed from the viral DNA, known as short guide RNA (sgRNA), associates with CRISPR-associated (Cas) proteins and through hybridization targets foreign DNA sequences for destruction by a specific Cas endonuclease called Cas9. A portion of sgRNA is complementary to the target DNA sequence, and an additional portion forms a scaffold allowing sgRNA-Cas9 complex formation. After formation, the Cas9-sgRNA complex pairs with target DNA sequences, causing DNA cleavage.

In **CRISPR-Cas9** gene editing, a synthetic sgRNA is employed to target a sequence of interest. Imperfect repair of the cleavage or the use of multiple sgRNA sequences results in the silencing or deletion of the gene (Figure 4.31).

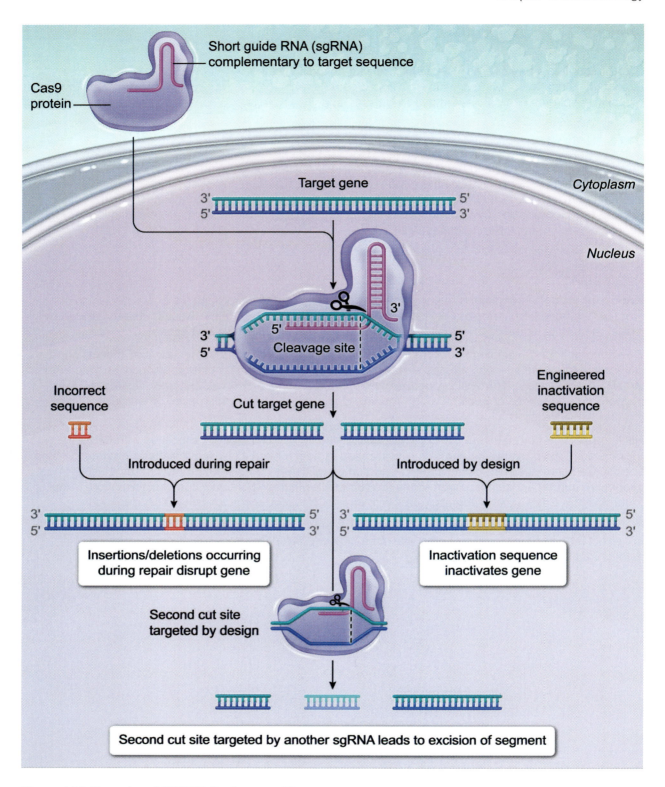

Figure 4.31 Examples of CRISPR-Cas9 gene editing.

Another approach for generating knockouts, called the **Cre-loxP** strategy, employs a recombinase enzyme called Cre, which recognizes special DNA sequences called loxP sites. Cre cuts out the portion of the DNA between two loxP sites as well as one of the two loxP sites, then it joins the two cut ends to leave a single loxP site. In this way, Cre can remove a gene flanked by loxP sites, a so-called *floxed* gene.

Crossing animals engineered to express Cre with animals in which a gene has been experimentally floxed gives rise to offspring in which the floxed gene is removed (ie, knockouts for the gene). Tissue-specific knockouts can be created by using tissue-specific promoters to control Cre expression, as shown in Figure 4.32.

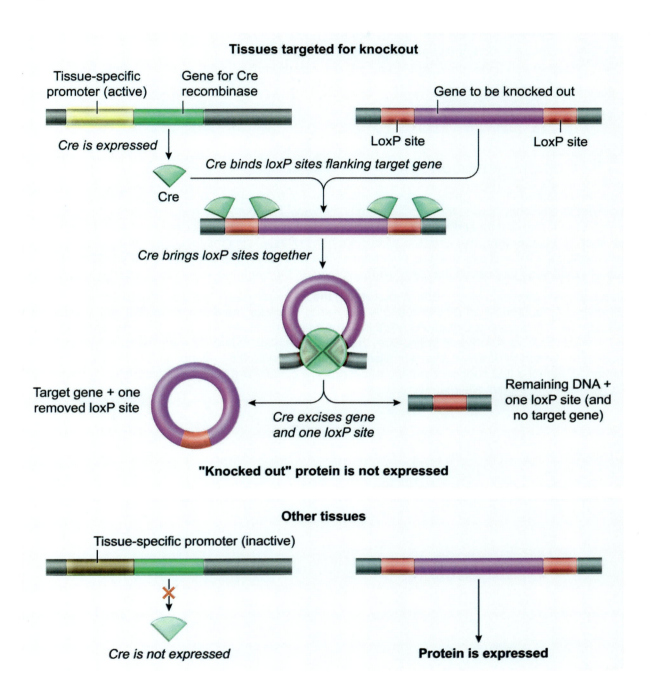

Figure 4.32 Tissue-specific knockout using the Cre-loxP strategy.

Concept Check 4.3

Which of the following patterns of Cre expression would cause germline deletion of a floxed gene (more than one may apply)?

- Cre expression under the control of a general promoter active in all tissues
- Cre expression under the control of a muscle-specific promoter
- Cre expression under the control of a liver-specific promoter
- Cre expression under the control of an egg-specific promoter

Solution

Note: The appendix contains the answer.

4.3.02 Complementation

When studying the effect of an experimental intervention in a population of organisms, tests are often employed to efficiently screen the population in some way. For example, in some cases it may be preferable in terms of time and resources to use phenotypic responses as a proxy for genetic changes. One such test is the **complementation** assay, a means of assessing the location of the gene(s) responsible for a recessively inherited phenotype. The test addresses the question of whether two separate mutations that cause the same phenotype are in the same gene or different genes (Figure 4.33).

In the complementation assay, mutants expressing the same recessive trait are mated, and the phenotype of first generation (F1) offspring is assessed. Because the trait is recessive and manifested in both parents, it is assumed that each parent is homozygous recessive at a genetic locus for the trait. If the F1 offspring have the mutant phenotype, then the two mutations are likely on the same gene, meaning the offspring received a mutant allele from each parent. If F1 offspring do not have the mutant phenotype, then the test suggests that the mutations are on different genes and the offspring have one wild-type and one mutant allele at each locus.

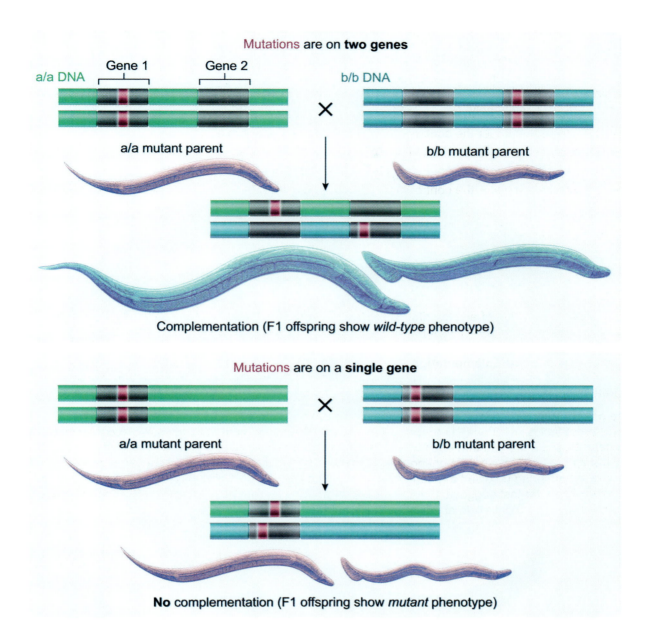

Figure 4.33 The complementation assay.

Lesson 4.4
Practical Applications of Biotechnology

Introduction

Biotechnological advances are often sought for the promise they are envisioned to hold for application in specific settings. For example, biotechnological advances could theoretically lead to reversal or prevention of diseases or augmentation of the food supply. This lesson focuses on potential practical applications of biotechnology in the fields of antisense drugs, gene therapy, forensics, agriculture, and environmental cleanup.

4.4.01 Antisense Drugs

Antisense strategies are designed to post-transcriptionally modulate gene expression by manipulating mRNA function using short (~20–25 nucleotides), synthetic oligonucleotides that undergo Watson-Crick base pairing with mRNA. This pairing (ie, hybridization) results in inhibited translation or altered handling of the complementary mRNA molecule(s). Some of these oligonucleotides have counterparts in nature and are collectively referred to as **RNA interference (RNAi)**, whereas others represent fundamentally altered synthetic versions of natural oligonucleotides.

Short, interfering RNA (siRNA) and microRNA (miRNA) are examples of RNAi. Both are originally duplex (ie, at least partially double-stranded), noncoding RNA molecules (see Concept 2.4.04). Although processed somewhat differently by cells, both siRNA and miRNA are recognized by a cytosolic protein complex that ultimately removes one RNA strand, known as the passenger strand, allowing the other guide strand to hybridize with complementary target mRNA. One difference in processing is that siRNA passenger strands are cleaved, whereas miRNA passenger strands are typically released intact.

Once the guide strand has hybridized to the target mRNA, siRNA and miRNA act by largely different mechanisms. siRNA pairing is usually complete and specific (ie, targeting only one mRNA sequence). This specificity makes the guide strand–target strand pair a target for destruction (ie, cleavage) by specific endonucleases, ultimately resulting in the enzymatic destruction of the target mRNA, as shown in Figure 4.34.

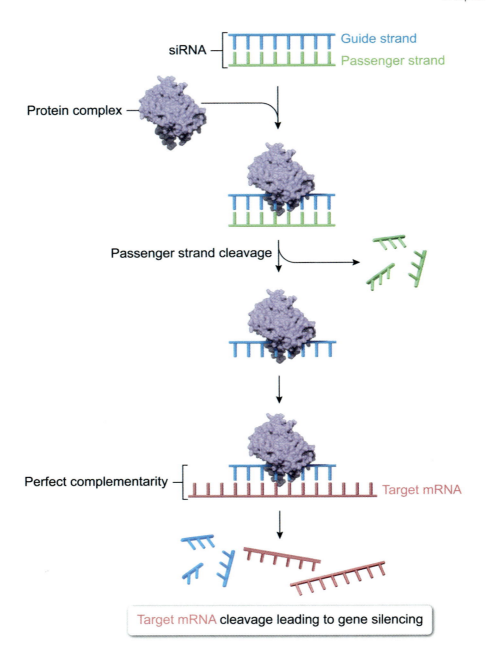

Figure 4.34 siRNA interference of mRNA function.

In contrast to siRNA, miRNA undergoes partial (ie, imperfect) mRNA pairing, allowing miRNA to interact with multiple mRNA sequences rather than a single specific mRNA sequence. miRNA typically either represses translation of the bound mRNA (Figure 4.35) or alters splicing of a pre-mRNA, although with extensive pairing mRNA destruction can be triggered.

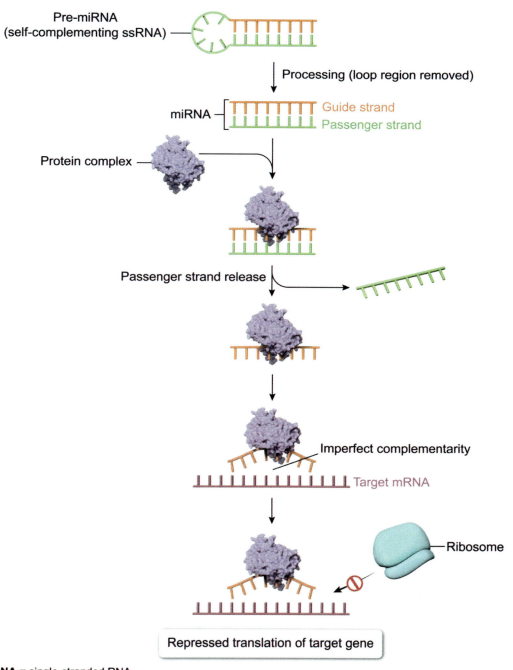

ssRNA = single-stranded RNA.

Figure 4.35 miRNA interference of mRNA function.

Modified versions of nucleic acids are also used for antisense targeting. **Morpholinos** are single-stranded synthetic antisense RNA molecules that hybridize with target mRNA and sterically block translation or alter splicing, similar to miRNA's role in RNAi. Unlike native RNA or DNA (or synthetic siRNA or miRNA), morpholinos are uncharged and, rather than a five-membered ribose or deoxyribose ring, possess a six-membered morpholine ring (from which the name *morpholino* derives). These characteristics make morpholinos less susceptible to degradation and are thought to reduce the likelihood of off-target effects.

> ### ✓ Concept Check 4.4
>
> Consider the use of synthetic miRNA, siRNA, and morpholinos for post-transcriptional inhibition of gene expression.
>
> 1) Which of these antisense molecules hybridizes with target mRNA as part of its mechanism of action (more than one may apply)?
> 2) Would an siRNA molecule be more or less likely to cause target mRNA degradation than a morpholino?
>
> #### Solution
>
> *Note: The appendix contains the answer.*

4.4.02 Gene Therapy

Gene therapy is an attempt to treat a condition or augment health by altering the complement of genes possessed or expressed. The alteration induced by a gene therapy approach may consist of the replacement or correction of a missing or defective gene. Alternatively, a functional gene not responsible for a health condition but thought to be beneficial (eg, an enzyme capable of converting a prodrug into an active drug targeting cancer cells) may be introduced.

While holding great potential, the development of a successful gene therapy approach faces numerous obstacles. A target gene must be identified and cloned, and a viable means of uptake into the tissue(s) in need of therapy is necessary. Various viruses have been employed as vectors to mediate the systemic delivery of gene therapies, as shown in Figure 4.36, as have therapeutic liposomes. To date, obstacles to developing gene therapies have proven formidable, and gene therapies have largely been unsuccessful or of limited success.

1. Clone WT *Hb* gene

2. Insert WT *Hb* gene into viral genome

3. Infect patient stem cells with viral vectors

4. Viral genome integrates into stem cell genome

5. Infuse stem cells back into patient (cells migrate to bone marrow)

6. Stem cells differentiate into RBCs expressing WT Hb and eject nucleus once mature

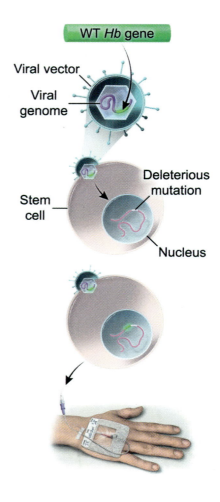
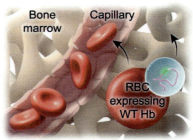

Hb = hemoglobin; **RBCs** = red blood cells; **WT** = wild-type.

Figure 4.36 Gene therapy using a viral vector.

4.4.03 Forensics

In the context of biology, **forensics** is the application of biotechnological methods or knowledge to questions about evidence in legal proceedings. For example, if the blood cell antigen profile (ie, blood type) in a suspect's blood is different from that of a blood sample taken from a crime scene, then the crime scene blood sample likely cannot be used as evidence to link the suspect to the crime. Although blood typing is one means of forensic analysis, much more sophisticated methods, many based on analysis of DNA characteristics, have been in use for several decades.

Chapter 4: Biotechnology

DNA fingerprinting is a type of DNA analysis that focuses on differences in DNA fragment lengths generated by the polymerase chain reaction or by treatment of DNA samples with an endonuclease. Such differences can arise due to variation in the number and locations of restriction sites or due to differences in the number of repeated units in a fragment containing repetitive DNA. Like the variable pattern of ridges comprising an individual's fingerprint, the variable pattern of bands on a Southern blot (see Concept 4.1.07) of these fragments can be used to identify unique biological samples (Figure 4.37).

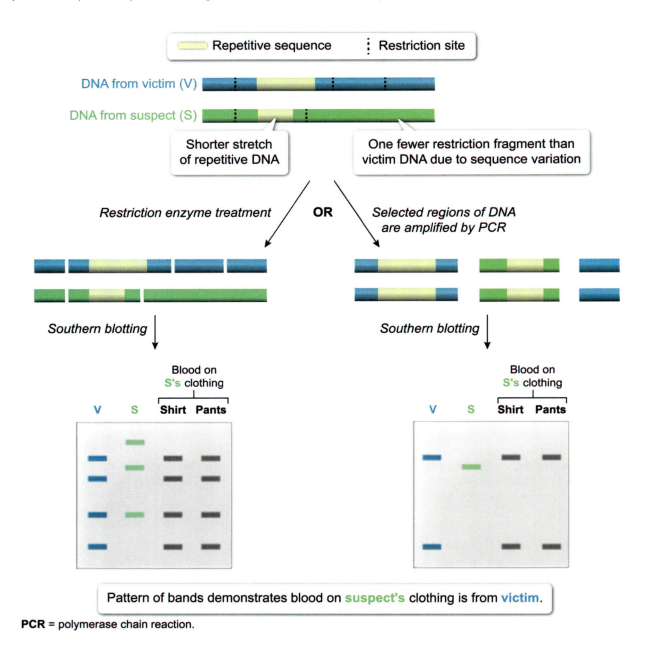

PCR = polymerase chain reaction.

Figure 4.37 DNA fingerprinting.

4.4.04 Agriculture

Biotechnology has been widely applied in **agricultural science**, including in identity verification (eg, of species, as shown in Figure 4.38) as well as in attempts to improve crop yields.

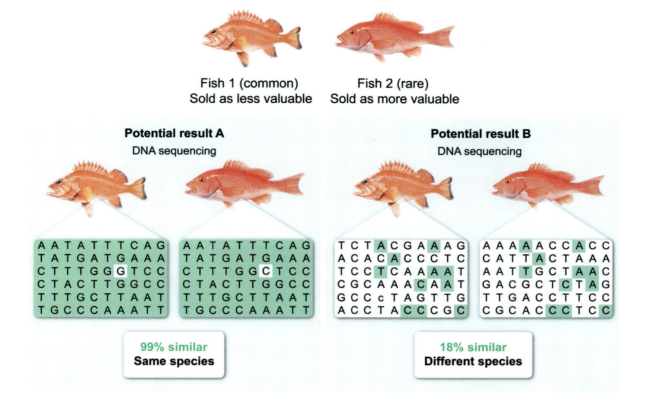

Figure 4.38 DNA sequencing can help identify fish species sold to consumers.

An example of the use of biotechnology in agricultural science is the use of DNA fingerprinting (see Concept 4.4.03) to verify individual livestock animals for breeding purposes. DNA fingerprinting can also be used to track meat from individual animals for safety reasons, for example, in response to an outbreak of prion-mediated disease or contamination during meat processing.

Biotechnology, through the genetic modification of some plants to engineer a favorable trait into a crop, has proved useful in increasing favorability or production of several plant species important to humans. For example, transgenic crops have been created to increase the vitamin content or alter the ripening characteristics of edible plants. In addition, genes for toxins that kill pests have been inserted into plant genomes to improve yields. For example, some cotton plans have been genetically modified to express a pesticidal gene that results in the deal of bollworms, a pest species that consumes cotton plants (Figure 4.39).

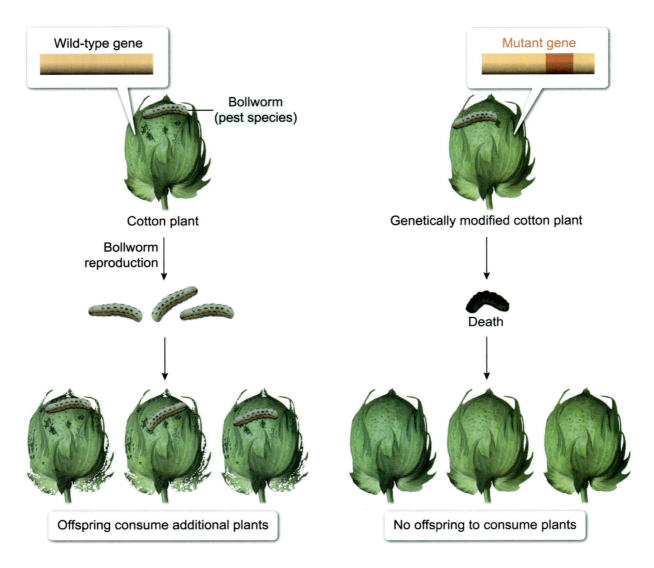

Figure 4.39 Introduction of a pesticidal mutant gene into cotton plants.

4.4.05 Environmental Cleanup

As the size of the human population has grown, so has the impact of humans on the environment (Figure 4.40). In many cases, human environmental impact has been modified by discoveries or technologies that have caused the use of one environmental resource to be replaced by another. In some cases, new technologies have created a demand for resources previously ignored or avoided.

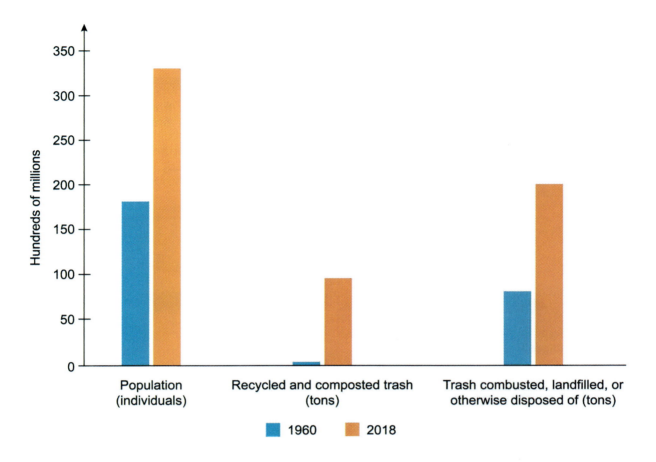

Figure 4.40 Population and trash production in the U.S. in 1960 and 2018.

For example, whales were hunted extensively through the early to mid-twentieth century to supply raw materials (eg, blubber, bones) used in the manufacture of numerous products such as oil for lamps, lubricants, ink, soaps, and vitamin D supplements. However, as new resources and technologies became available, whale hunting sharply declined and has now largely ceased. In recent decades, demand for rare earth elements (eg, palladium and neodymium) used in many modern technologies (eg, cell phones, electric cars, wind turbines) and xenobiotics (eg, some chemotherapies), has resulted in new environmental impacts.

Ongoing research efforts are aimed at developing methods of **bioremediation**, the use of biotechnology to mitigate human environmental impacts. Bioremediation sometimes involves the genetic modification of organisms either disproportionately contributing to an environmental impact or possessing biological characteristics known to reduce a certain environmental impact. In other instances, bioremediation can take the form of biostimulation, the addition of nutrients or other molecules known to stimulate growth or activity of bioremediating organisms.

One example of bioremediation involves mitigating the effect of pig manure on natural bodies of water (Figure 4.41). Manure from wild-type pigs is rich in phosphorus because much of the dietary phosphorus in pigs' diets is in the form of phytate, which is indigestible by pigs. When phosphorus from pig manure enters natural bodies of water, it can lead to algal blooms that deprive other aquatic life of oxygen. One approach to prevent these destructive algal blooms has been to engineer pigs whose salivary glands produce phytase, the enzyme catalyzing breakdown of phytate, leading to less phosphorus excretion in the transgenic pigs' manure.

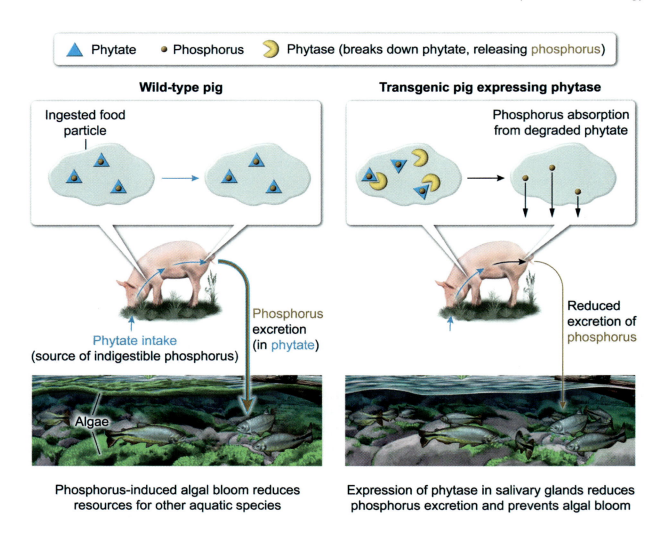

Figure 4.41 A bioremediation technique to reduce the impact of phosphorus from pigs on bodies of water.

Another approach to bioremediation employs organisms capable of sequestering greater-than-typical amounts of harmful molecules. Such organisms, called hyperaccumulators, possess enhanced capacities for transport and/or storage of harmful molecules (eg, heavy metals). Some hyperaccumulators are extremophiles well adapted for survival in a specific type of harsh environment (eg, extremely acidic or salty).

Many plants, including hundreds classified as hyperaccumulators, remain healthy despite taking up and sequestering heavy metals in vacuoles. Plants, however, do not thrive in all environments, and many types of bacteria present in various environments do not possess the requisite cellular machinery (eg, membrane-bound organelles) for bioremediation.

To provide an alternative for hyperaccumulation, baker's yeast (*Saccharomyces cerevisiae*) has been genetically engineered to possess several characteristics of plant hyperaccumulators. For example, *S. cerevisiae* has been engineered to increase metal transporters present in the membranes of cells and vacuoles as well as to increase production of antioxidant molecules to prevent death from metal-catalyzed oxidant production.

Lesson 4.5

Special Considerations in Biotechnology

Introduction

As detailed in Lesson 4.4, the application of biotechnology holds promise for alleviating or eradicating many problems facing humanity. However, biotechnological interventions introduced into complex biological systems do not occur in isolation, and potential or real tradeoffs can give rise to concerns about the overall safety of an intervention. This lesson presents several safety and ethical concerns associated with the development and use of biotechnology.

4.5.01 Safety Concerns

Because **safety** is typically considered as the degree to which risk is minimized, safety is inextricably tied to risk. Safety is often framed using numbers (eg, risk of death or harm), but the *perceived* safety of a condition or intervention can be context-dependent and definitional (eg, what level of risk can be reasonably defined as *safe*?). Faced with identical risk estimates, two individuals may reach different conclusions regarding the safety of a disease or intervention. Similarly, one individual may view identical risks differently in different contexts.

For example, a parent might perceive a certain estimated risk as personally acceptable but consider the same risk as unacceptable for a child. Likewise, an intervention deemed safe enough for one group may be considered too risky for another group with different characteristics. For instance, exposure to an intervention with a risk of slowly developing harm (eg, cancer) may be considered acceptable for older individuals, who have a shorter time for adverse conditions to develop, but unacceptable for younger individuals.

Another type of risk-benefit consideration concerns the constantly evolving nature of scientific data or opinion. The perceived safety of a biotechnological intervention at a particular point in time is arrived at by weighing currently known or hypothesized risks and benefits. However, over time this perception often changes because additional data reflecting more individuals and groups are considered (Table 4.1).

Table 4.1 Evolving perceptions of the risks and benefits of several historical biotechnological advances.

Intervention	Initial perceptions and data	Later perceptions and data
Injectable insulin	• Profound immediate positive impact treating patients with diabetes, a condition with 100% mortality at the time	• Initial insulin preparations were of highly variable potency • Early forms of insulin therapy increased the risk for complications (eg, blindness, amputation) • Still considered by some as the greatest scientific achievement of the twentieth century
Thalidomide	• Released in late 1950s as an effective anti-inflammatory treatment • Highly marketed for reducing nausea and promoting sleep • Available without prescription outside the U.S., including as a treatment for morning sickness in pregnant individuals	• By the early 1960s, known to promote severe birth defects in infants, with reports of increased miscarriage rates • Considered by some as worst human-caused medical disaster • Later successfully used for treating other conditions (eg, leprosy, HIV)
Selective cyclooxygenase-2 (COX-2) inhibitors	• FDA-approved in 1999 • Highly marketed and prescribed as painkillers with lower risk of gastrointestinal side effects than comparable therapies	• In 2004, known to increase risk for cardiovascular complications • Still in use for treating inflammation and pain, with some viewing the risks as acceptable
Antibiotics	• Viewed as an immediate lifesaver and miraculous cure for bacterial infections	• Alters/eradicates beneficial gut flora • Overuse has led to development of multi-drug resistant bacterial strains • Now used more cautiously, but still considered by some as the greatest scientific achievement of the twentieth century

An area of concern regarding the safety of biotechnology is its use for creating genetically modified organisms (GMOs), such as crops consumed by humans. Some people view the development of GMOs as a safe means to increase resource availability through the introduction of beneficial traits (eg, enhanced pest or pesticide resistance, increased nutritional content, reduced water requirements). However, some people express concern over the development of pesticide resistance in pests and the effects of potentially beneficial or harmful alleles being transferred from GMOs to native plants.

Gene therapy is another application of biotechnology for which perceptions of safety vary considerably. Although the viral vectors used to introduce potentially therapeutic genes into cells are typically considered safe, these viral vectors can cause inflammation. In some cases, strong immune reactions to these vectors are thought to have caused the death of clinical trial participants receiving gene therapies.

The biotechnology-driven potential for germ line editing in humans, which would produce heritable changes in the collection of alleles available, is a cause of concern for some individuals. Apart from ethical considerations (see Concept 4.5.02), the inadvertent removal of potentially helpful alleles is possible. In other instances, known benefits and risks associated with certain alleles must be weighed before moving forward with germ line editing. As noted earlier in this concept, some might view the risks of germ line editing as acceptable given the potential benefit, whereas others might reach the opposite conclusion.

4.5.02 Ethical Concerns

Ethical considerations in the context of medicine are traditionally traced back to the writings of Hippocrates, from which the philosophy of the Hippocratic Oath (ie, first, do no harm) is derived, or even the concept of a golden rule to do to others as you would have done to you. However, formal consideration of the ethics of medical treatment and research, and a shift from physician-centered to subject- or patient-centered ethics, began with the Nuremberg trial following World War II. During the so-called Doctors' Trial, 16 Nazi physicians were found guilty of crimes related to unethical experimentation on prisoners of war.

During the war, certain prisoners (primarily Jews) were subjected to torturous scientific experiments, the outcomes of which were often painful harm or death. After the conclusion of the war, these atrocities led to international consideration of research ethics. A set of principles, called the Nuremberg Code for its connection with the Nuremberg trials of war criminals, was borne from these deliberations and is considered a landmark in the development of research guidance and the protection of human research subjects.

The Nuremberg Code lays out foundational guidelines for research involving human subjects (Table 4.2), including studies of new biotechnologies. This code, and the human research declarations that have followed it, are based on respect for individuals and individual autonomy. Over time, the importance of oversight committees has been increasingly emphasized. In addition to guidelines for human research subjects, guidelines have similarly been developed for the humane treatment of animals used for research purposes.

Table 4.2 A brief summary of the principles of the Nuremberg Code.

• Consent of the individual patient or subject is absolutely required and can be withdrawn at any time
• Experiments should be well designed and performed to minimize risk to individuals under study
• Experiments should lead to necessary and useful knowledge
• The lead experimenter should end an experiment before completion if undue risk is believed to be present

In a notable oversight success, a physician for the Food and Drug Administration, Frances Kelsey (in her first assignment), slowed approval of thalidomide in the U.S. Being unconvinced of its purported safety for pregnant individuals and other patients, she prevented the disaster of thalidomide-induced birth defects and miscarriages in the U.S. that occurred in many other countries.

In an infamous example of violating the principles of the Nuremberg Code, the Tuskegee Syphilis Study, started in 1932, studied the natural history of syphilis in African American men. The men were enrolled without formal consent, and after penicillin became available for treating syphilis, those participants with syphilis were denied treatment. After the details of the study came to light in 1972, the study was ended. This study, among others, led to the Belmont Report, which stated that potential societal benefit cannot be used as justification for doing harm to research subjects and called for oversight of research studies by independent committees.

Scientific advances can lead to consideration of the ethics of a biotechnology's application, including who has the right or responsibility to determine whether the use of the technology is warranted. For example, human germ line editing could potentially eradicate certain genetic conditions, alter the human genome for cosmetic reasons, or allow a program of eugenics (increasing the prevalence of genes deemed favorable). Whether parents, governments, or oversight committees should decide whether gene editing (currently condemned) is ethical and, if so, under what circumstances, will likely continue to be debated.

END-OF-UNIT MCAT PRACTICE

Congratulations on completing **Unit 2: Biological Research Techniques**.

Now you are ready to dive into MCAT-level practice tests. At UWorld, we believe students will be fully prepared to ace the MCAT when they practice with high-quality questions in a realistic testing environment.

The UWorld Qbank will test you on questions that are fully representative of the AAMC MCAT syllabus. In addition, our MCAT-like questions are accompanied by in-depth explanations with exceptional visual aids that will help you better retain difficult MCAT concepts.

TO START YOUR MCAT PRACTICE, PROCEED AS FOLLOWS:

1) Sign up to purchase the UWorld MCAT Qbank
   IMPORTANT: You already have access if you purchased a bundled subscription.
2) Log in to your UWorld MCAT account
3) Access the MCAT Qbank section
4) Select this unit in the Qbank
5) Create a custom practice test

Unit 3 Cellular Biology

Chapter 5 Eukaryotic Cells

5.1 Cells

- 5.1.01 Cell Theory
- 5.1.02 Domain Classification
- 5.1.03 Comparing Prokaryotic and Eukaryotic Cells

5.2 Plasma Membrane Components and Functions

- 5.2.01 Components of the Plasma Membrane
- 5.2.02 Membrane Protein Structure and Function
- 5.2.03 Transport Across Membranes

5.3 Eukaryotic Organelles

- 5.3.01 The Nucleus
- 5.3.02 Nucleolus and Ribosomes
- 5.3.03 The Endomembrane System
- 5.3.04 Mitochondria
- 5.3.05 Peroxisomes
- 5.3.06 Cytoskeleton
- 5.3.07 Extracellular Matrix and Cell Junctions

5.4 Cell Growth and Division

- 5.4.01 The Cell Cycle
- 5.4.02 The Mitotic Process
- 5.4.03 Cell Cycle Control
- 5.4.04 Cancer

5.5 Eukaryotic Tissues

- 5.5.01 Types of Animal Tissues

Chapter 6 Prokaryotes and Viruses

6.1 Prokaryotic Cells

- 6.1.01 Classification of Prokaryotes
- 6.1.02 Cell Wall
- 6.1.03 Bacterial Structures

6.2 Growth and Reproduction of Prokaryotes

- 6.2.01 Binary Fission
- 6.2.02 Bacterial Growth
- 6.2.03 Aerobic and Anaerobic Metabolism
- 6.2.04 Symbiotic Relationships
- 6.2.05 Antibiotic Resistance

6.3 Prokaryotic Genetics

- 6.3.01 Prokaryotic Genome
- 6.3.02 Plasmids
- 6.3.03 Horizontal Gene Transfer
- 6.3.04 Transposons
- 6.3.05 Gene Regulation in Prokaryotes

6.4 Viruses
- 6.4.01 Viral Structures
- 6.4.02 Viral Genomes

6.5 Viral Life Cycles
- 6.5.01 Bacteriophages
- 6.5.02 Animal Viruses
- 6.5.03 Retroviruses

6.6 Sub-Viral Particles
- 6.6.01 Viroids
- 6.6.02 Prions

Lesson 5.1
Cells

Introduction

With the development of the microscope, scientists were able to observe a previously invisible world, and the groundwork for the cell theory was laid. The cell theory states that all living organisms are composed of cells and that all cells come from pre-existing cells. In addition to understanding that cells are building blocks of larger organisms, scientists were also able to observe free-living unicellular microorganisms.

The lessons in this chapter highlight the development of the cell theory and provide a review of the vital components of eukaryotic cells and their functions, with an emphasis on animal cells. The unique components and cellular functions of prokaryotic cells are covered in Chapter 6.

5.1.01 Cell Theory

The modern version of the **cell theory** includes the principles outlined in Figure 5.1.

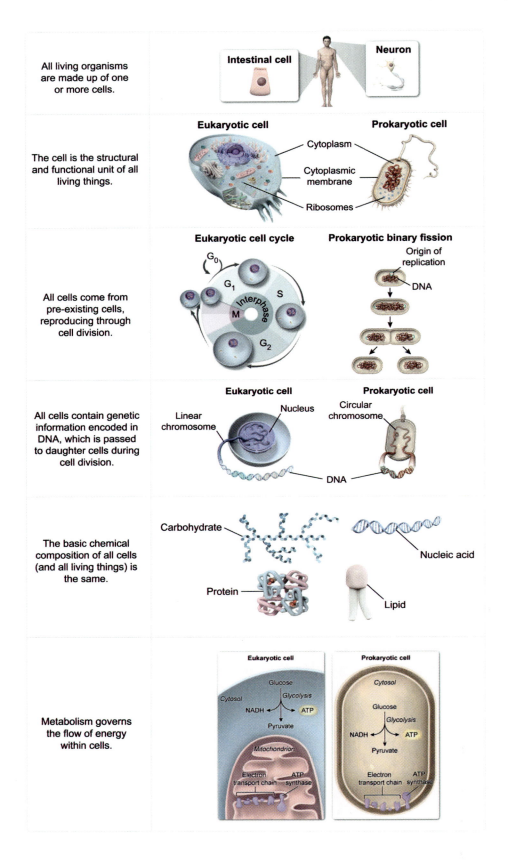

Figure 5.1 Principles of the cell theory.

While not considered cells, infectious agents such as **viruses** share some features with living organisms. Like cells, viruses contain genetic material and are able to evolve and adapt to different environments through natural selection. However, a virus differs from a cell because a virus depends on a host cell to conduct viral metabolic activities, including reproduction (Figure 5.2). In other words, viruses are considered **obligate intracellular parasites** and must utilize the metabolic machinery of a host cell to reproduce.

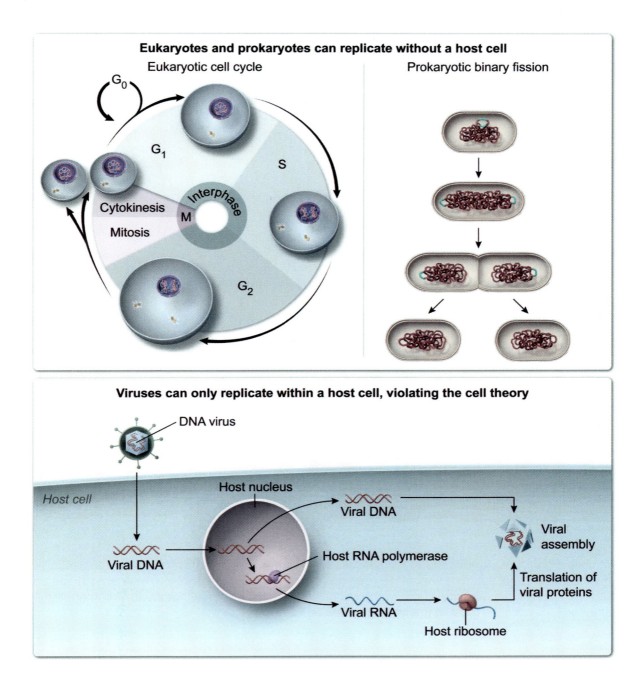

Figure 5.2 Viruses are unable to reproduce or carry out any metabolic activities without a host cell.

While viruses *do* possess genetic information (ie, DNA or RNA), they do not conform completely to the principles of the cell theory and are therefore not cells. Because of this, viruses are generally considered non-living. Viruses are covered in greater detail in Chapter 6.

5.1.02 Domain Classification

Cells exist in many forms and sizes; some live as independent organisms, and others live in groups (eg, a colony) or make up part of a multicellular organism. Based on shared characteristics, cells can be divided into one of two basic types: **prokaryotic** or **eukaryotic**.

Regardless of cell type, all living organisms can be categorized into one of the three domains of life: **Archaea**, **Bacteria**, and **Eukarya**, depicted in Figure 5.3. The three-domain system places domain as the broadest taxonomic rank within the biological classification system. Organisms classified within domains Archaea and Bacteria have a prokaryotic cell structure, whereas organisms classified within domain Eukarya have a eukaryotic cell structure.

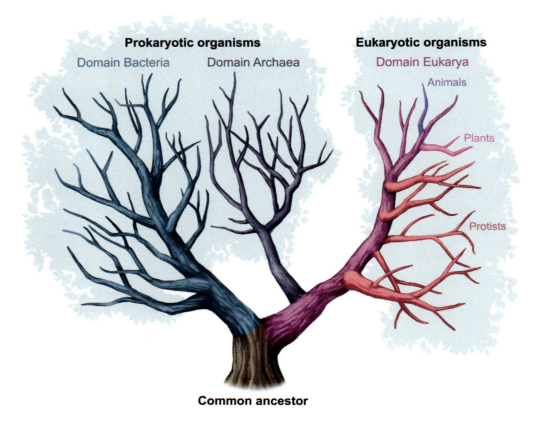

Figure 5.3 The three domains of life.

Prokaryotic organisms are **unicellular** (single-celled) and are usually much smaller than eukaryotic organisms and internal membrane-enclosed compartments are largely absent in prokaryotes (Figure 5.4). Prokaryotic species can adapt to a broad range of habitats, including those that are environmentally extreme, and they are the most abundant organisms on Earth.

Initially, bacteria and archaea were placed within one large group, domain Bacteria, which included all prokaryotic cells. With advancing technology, a subpopulation of prokaryotic organisms intrinsically different from bacteria was discovered, and domain Archaea was established.

While similar to bacteria in appearance, archaea are separated into a distinct subgroup of prokaryotes due to differences in ribosomal RNA (rRNA) genes. Additionally, archaea share other common distinguishing features from bacteria, including differences in cell walls, plasma membranes, and other biochemical signatures. Some archaean metabolic pathways are more closely related to eukaryotic cells than to other prokaryotic cells.

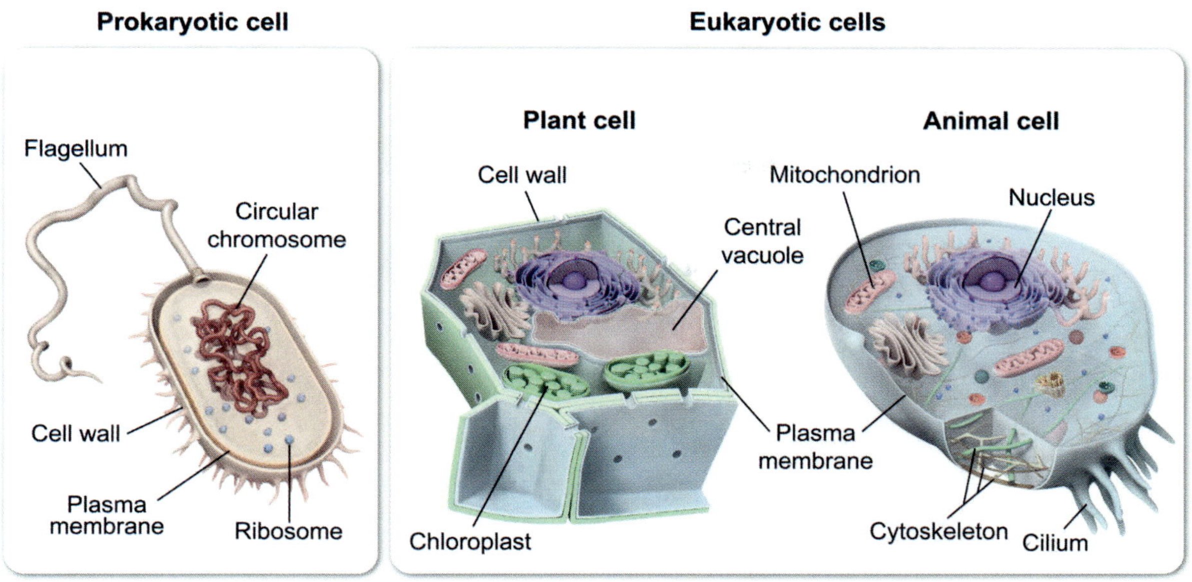

Figure 5.4 Prokaryotic versus eukaryotic cells.

Eukaryotic organisms are classified as part of domain Eukarya, and they encompass all other forms of life, including plants and animals. Eukaryotic cells contain membrane-enclosed compartments such as the nucleus, which contains the cell's DNA. Eukaryotic organisms may be unicellular (eg, some fungi and protists) or **multicellular** (eg, plants, animals).

5.1.03 Comparing Prokaryotic and Eukaryotic Cells

The cell theory establishes that the cell is the basic structural and functional unit of all organisms; therefore, all cells must have certain features in common. Whether classified as prokaryotic or eukaryotic, all cells:

- are contained by a selective barrier known as the **plasma membrane**.
- contain a semifluid substance known as the **cytosol**, in which all internal components of the cell are suspended.
- contain genetic information in the form of **DNA**, which is organized into structures called chromosomes.
- contain **ribosomes**, which translate the information encoded in DNA into functional proteins.

A key difference between eukaryotic and prokaryotic cells is that the majority of the DNA in a eukaryotic cell is enclosed in an **organelle** (ie, functional unit) known as the **nucleus**, which is surrounded by a double membrane. In prokaryotic cells, DNA is concentrated in a cellular region known as the **nucleoid**, which is not enclosed by a membrane. The interior of a eukaryotic cell is called the **cytoplasm**, which is the area contained by the plasma membrane, excluding the nucleus.

Eukaryotic cells contain membrane-bound organelles in addition to the nucleus, each with a specialized function, while membrane-bound organelles are absent in nearly all prokaryotic cells. Both prokaryotic and eukaryotic cells may contain some shared features, such as **flagella** for locomotion and **cell walls** to protect and maintain cell shape, but each cell type has unique components as well (Figure 5.5). For example, some prokaryotic cells have **pili**, structures that can be used to exchange genetic information with other prokaryotic cells, and some eukaryotic cells have **cilia**, structures that help move substances along the outer cell surface.

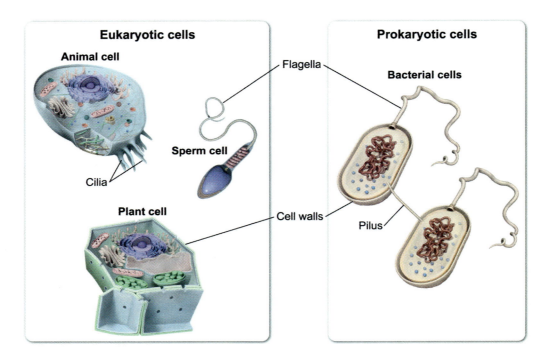

Figure 5.5 Eukaryotic and prokaryotic cells have both shared and distinctive features.

Prokaryotic cells are generally smaller than eukaryotic cells. On average, the diameter of a prokaryotic cell is 0.1–5.0 μm, whereas a eukaryotic cell has a diameter 10–100 times larger, averaging 10–100 μm. Regardless of whether a cell is prokaryotic or eukaryotic, there is an upper limit on cell size due to the limitations that a larger size puts on cellular transport efficiency.

Lesson 5.2

Plasma Membrane Components and Functions

Introduction

All cells are surrounded by a structure known as the **plasma membrane**, which serves as a barrier between the cell's **extracellular** (external) and **intracellular** (internal) compartments. The plasma membrane is the interface between the cell and its environment, and by regulating external signals and molecular traffic, it moderates the cell's response to the outside world. This lesson focuses on the components of the plasma membrane and their functions, with an emphasis on cell-cell signaling and transport across membranes.

5.2.01 Components of the Plasma Membrane

The plasma membrane is composed of a **phospholipid bilayer**. The two phospholipid layers are oriented much like a sandwich, with the hydrophobic (ie, water-repelling) fatty acid tails of the phospholipids in contact with each other and the hydrophilic (ie, water-attracting) head groups in contact with either the intracellular or extracellular aqueous environment, as shown in Figure 5.6.

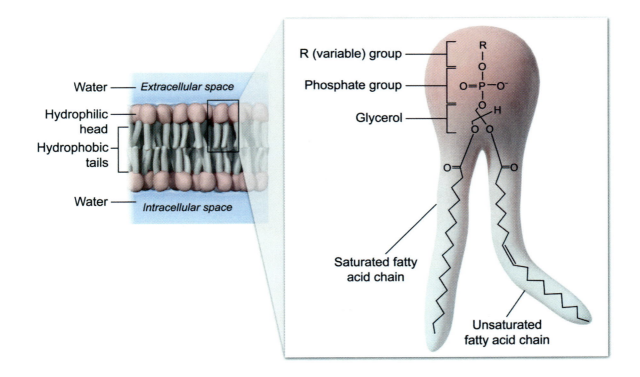

Figure 5.6 Composition of a phospholipid bilayer.

Phospholipids with **saturated fatty acid** tails tend to associate closely with each other and make the membrane less fluid, while phospholipids with **unsaturated fatty acid** tails tend to associate with each other more loosely, which results in a more fluid membrane state, as shown in Figure 5.7.

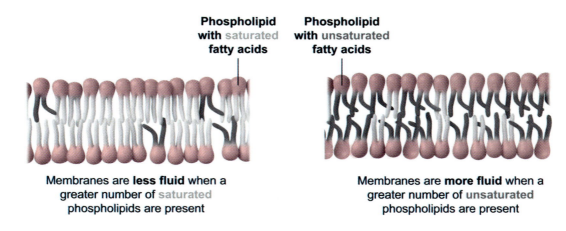

Figure 5.7 Phospholipid composition affects membrane fluidity.

Membrane fluidity is also affected by the length of phospholipid fatty acid tails and environmental factors such as temperature. In addition to the specific ratio of saturated to unsaturated phospholipids, the length of fatty acid tails affects membrane fluidity. Shorter fatty acid tails are associated with greater membrane fluidity and longer tails are associated with less membrane fluidity.

Temperature determines the point at which a membrane solidifies. Higher temperatures result in greater membrane fluidity due to increased kinetic energy of the membrane components. Conversely, lower temperatures result in less membrane fluidity. When the temperature of the environment changes, cells may respond by changing their membrane composition (eg, length of phospholipid fatty acid tails, relative abundance of saturated versus unsaturated phospholipids), as depicted in Figure 5.8.

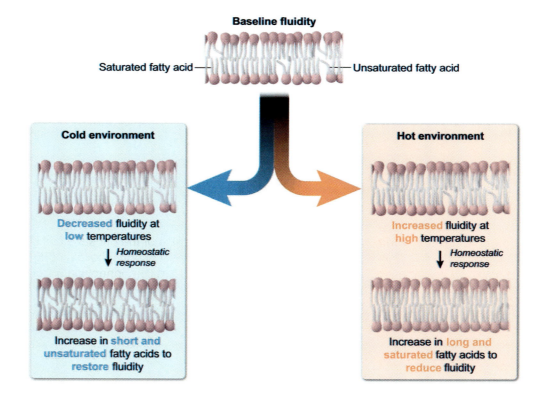

Figure 5.8 Effect of phospholipid composition and temperature on membrane fluidity.

While phospholipids are the main component of the plasma membrane, the steroid cholesterol is another associated lipid. Cholesterol molecules can be inserted between phospholipids and interact with the nonpolar fatty acid tails. Sterols such as cholesterol are not typically found in prokaryotic membranes.

Cholesterol also participates in maintaining membrane fluidity when temperatures fluctuate (Figure 5.9). At human body temperature, cholesterol restricts phospholipid movement, making membranes less fluid. At lower temperatures, cholesterol makes membranes more fluid by interfering with phospholipid packing. The presence of cholesterol allows cells to maintain optimal membrane fluidity when exposed to changing temperatures.

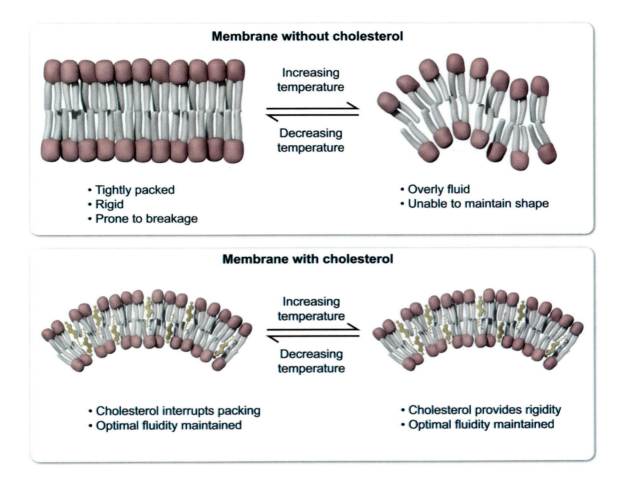

Figure 5.9 Effect of cholesterol on membrane fluidity.

According to the **fluid mosaic model,** which describes certain properties of the phospholipid bilayer, proteins and other components reside in a fluid bilayer composed largely of phospholipids (Figure 5.10). The plasma membrane is said to be fluid, as its various non-phospholipid components can migrate laterally (ie, from side to side) within the membrane.

The "mosaic" aspect of the fluid mosaic model refers to the nonlipid components embedded within the phospholipid bilayer. A variety of proteins are associated with the plasma membrane and are responsible for most of the membrane's functions. Different cell types have different combinations of associated membrane proteins, allowing for a variety of cellular functions.

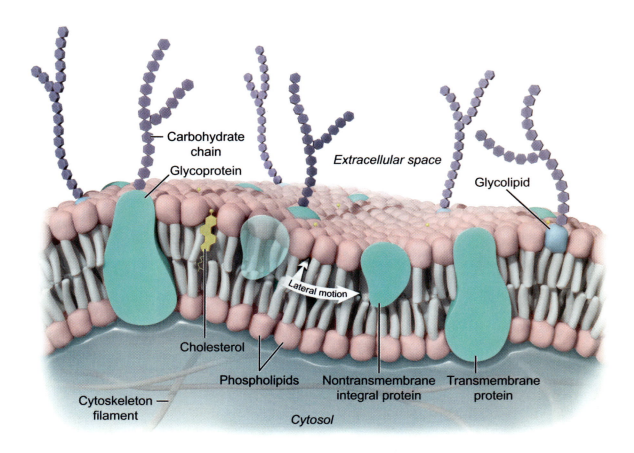

Figure 5.10 The fluid mosaic model of membrane structure.

Membrane-associated proteins can be categorized as integral, lipid-anchored, or peripheral. **Integral proteins** penetrate the phospholipid bilayer either partially (nontransmembrane) or completely (transmembrane).

To be embedded within the membrane, integral membrane proteins must contain regions made up of nonpolar amino acids, which interact with the hydrophobic interior of the bilayer. In addition, some transmembrane proteins contain one or more interior hydrophilic channels, which allow charged or polar molecules to travel through the phospholipid bilayer.

Lipid-anchored proteins undergo a modification known as lipidation, in which certain lipids are covalently attached to the protein post-translationally. These lipid attachments allow the protein to be anchored within the phospholipid bilayer on either the intracellular or extracellular side of the membrane. For example, G proteins, which participate in cell signaling, are anchored to the membrane by a glycosylphosphatidylinositol (GPI) anchor.

Peripheral proteins are not embedded in the bilayer but are loosely (ie, noncovalently) associated with the membrane, attached to integral membrane proteins or to other membrane-associated structures. For example, peripheral proteins of the extracellular matrix (ie, a network of extracellular molecules) are attached to transmembrane proteins (eg, integrins). Figure 5.11 illustrates the types of membrane proteins.

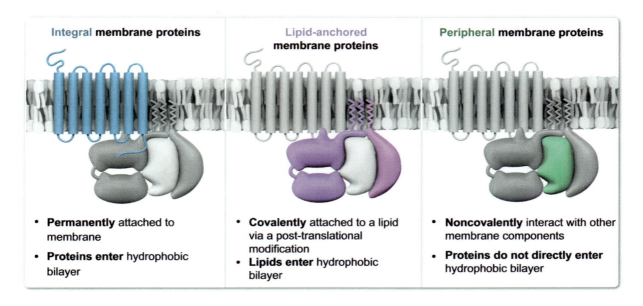

Figure 5.11 Membrane-associated proteins.

Some proteins and lipids within the membrane are modified by the addition of oligosaccharides, which are short, branched chains of fewer than 15 sugars. These glycoproteins and glycolipids are most often found on the extracellular side of the plasma membrane and vary among species, individuals, and among cell types within a single organism. Often, these surface carbohydrates function as markers that distinguish one cell from another. For example, human blood types (A, B, AB, and O) are a result of variations in different glycoproteins on the surface of red blood cells (see Figure 5.12).

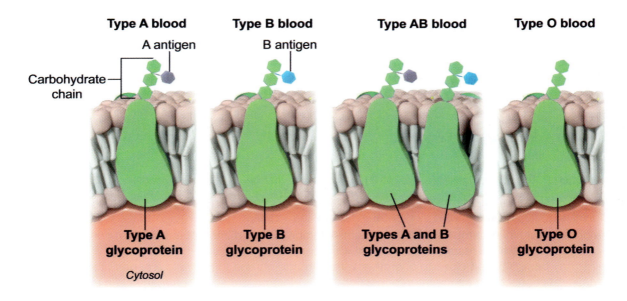

Figure 5.12 Glycoproteins present on the surface of red blood cells determine human blood type.

Within a membrane, certain lipids and cholesterol can become concentrated in small, specialized areas of the membrane known as **lipid rafts** (Figure 5.13). In a lipid raft, lipids are more tightly packed than in adjacent regions of the bilayer, and they form a platform, which retains the ability to move laterally within the membrane. Membrane proteins may aggregate within the raft to maximize protein interactions and function. For example, clustering of membrane protein receptors within a lipid raft may result in enhanced cell signaling capabilities.

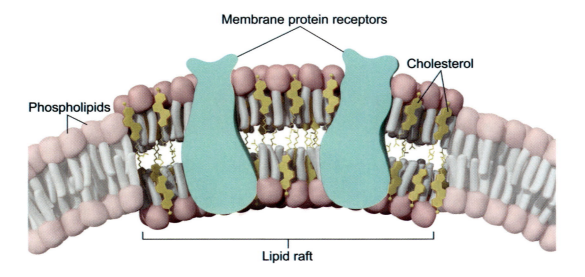

Figure 5.13 A lipid raft.

5.2.02 Membrane Protein Structure and Function

In addition to functioning as barriers, both the plasma membrane and internal membranes perform a variety of functions for the cell. Many of these functions are carried out by various proteins associated with the membrane. Individual cells express a unique subset of membrane proteins, which allows for individual cellular responses within the same environment.

Membrane proteins participate in a variety of cellular activities including:

- transport
- connecting cells via junctions
- attachment to the cytoskeleton and extracellular matrix
- cell-cell recognition
- coordination of enzymatic activities
- cell signaling

In this concept, the role of membrane proteins in the coordination of enzymatic activities and cell signaling is highlighted. Other roles of membrane proteins are discussed throughout this lesson and in Concept 5.3.07.

Some proteins within a membrane (ie, the plasma membrane or an internal membrane) may be enzymes that catalyze biochemical reactions associated with membrane activities. In some cases, multiple proteins may be positioned close to one other within the membrane to facilitate a series of sequential enzymatic reactions, such as the interconnection of the electron transport chain and citric acid cycle at the inner membrane of the mitochondrion.

Transmembrane proteins (ie, proteins that completely cross the membrane) may also serve as **receptors** that receive and integrate signals from the extracellular environment. Cells receive information from the external environment in a variety of ways. Signals may be chemical (eg, hormones, growth factors, neurotransmitters) or mechanical (eg, changes in pressure, sound waves). While this concept focuses on plasma membrane receptors, there are also intracellular receptors found in the cytosol and nucleus (discussed further in Concept 11.1.04).

Receptors are specific for particular signaling molecules, or **ligands**, that initiate a specified set of internal cellular responses. For example, insulin receptors bind and respond specifically to insulin and insulin-like growth factors but generally do not respond to unrelated ligands. Every cell expresses a unique combination of receptors that allows that cell to respond to specific ligands and ignore others, ultimately resulting in specialized activities for each cell (Figure 5.14).

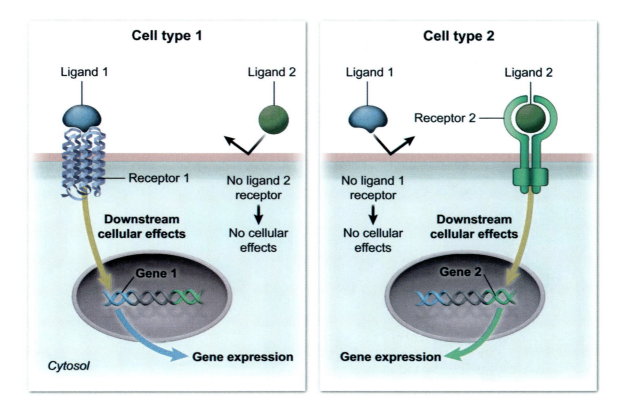

Figure 5.14 Differential expression of receptors leads to different cellular effects.

In general, when an external ligand and transmembrane receptor make contact, a ligand-receptor complex is formed. This connection triggers a conformational change in the intracellular portion of the receptor, activating a series of intracellular events (eg, interactions with other proteins or molecules associated with the cytosolic face of the plasma membrane, other enzymatic activities). The receptor type determines which specific intracellular pathway is engaged (Figure 5.14).

There are various classes of transmembrane protein receptors, including G protein–coupled receptors, enzyme-linked receptors, and ligand-gated ion channel receptors. Because membrane receptors interact with both the extracellular ligand and intracellular molecules, these receptors allow the cell to respond to external cues without the ligand entering the cell. After ligand binding, a signal may be amplified through intracellular effectors, collectively known as a signal transduction cascade. This cascade of events leads to a cellular response, which often includes altered gene expression.

G protein–coupled receptors (GPCRs) are the most abundant type of cell surface receptor and are associated with a variety of cellular functions in eukaryotic organisms. Receptors in the GPCR family are similar in structure and are composed of a single polypeptide with seven α-helices that cross the plasma membrane. The intra- and extracellular GPCR domains consist of loops between each of the α-helices, where ligands and other effectors may interact with the receptor.

In the classical G protein signaling cascade, when a specific ligand binds to the extracellular domain of a GPCR, the ligand acts as an agonist, triggering a conformational change on the intracellular side of the receptor. This conformational change activates a heterotrimeric lipid-anchored protein known as a **G**

protein. The classical G protein signaling cascade is shown in Figure 5.15 and consists of the following steps:

1. A ligand binds to the GPCR at the cell surface.
2. Ligand binding activates the G protein, causing a GDP molecule to be replaced with GTP within the α-subunit of the G protein.
3. The activated α-subunit dissociates from the GPCR and interacts with a second membrane protein known as **adenylate cyclase**. Meanwhile, the βγ subunit complex of the G protein proceeds to activate other signaling cascades.
4. Adenylate cyclase (ie, a peripheral membrane protein) catalyzes the conversion of ATP to cyclic AMP, which is a second messenger.
5. cAMP activates the cytosolic enzyme **protein kinase A**, which subsequently phosphorylates other downstream proteins in the cascade, leading to ligand-specific cellular effects.

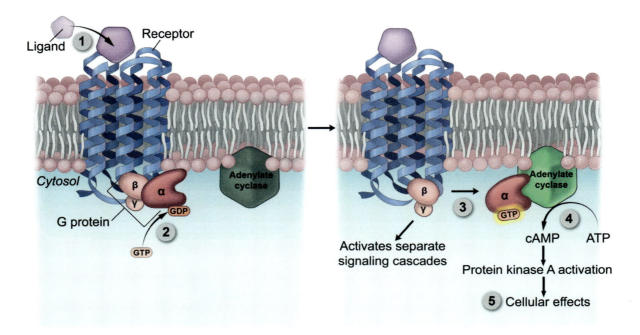

Figure 5.15 Classical G protein signaling cascade.

Enzyme-linked receptors are transmembrane proteins that catalyze chemical reactions in response to extracellular ligands. Upon ligand binding to the extracellular domain of a receptor protein, a conformational change occurs that allows the intracellular domain to become catalytically active.

Receptor tyrosine kinases (RTKs) are one of the most common types of enzyme-linked receptors (Figure 5.16). Upon ligand binding, RTKs dimerize. Each RTK in the dimer then transfers a phosphate group from ATP to multiple tyrosine residues on the intracellular domain of the other RTK via the receptor's intrinsic kinase activity.

This receptor autophosphorylation initiates a signaling cascade in which cytosolic effector proteins are activated, leading to the activation of a small lipid-anchored G protein known as **Ras**. The activation of Ras sets off a phosphorylation cascade in the cytosol, which ultimately leads to specific cellular effects.

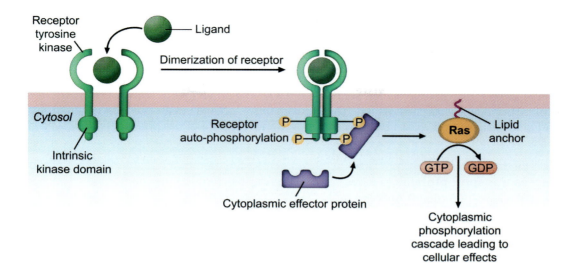

Figure 5.16 Receptor tyrosine kinase signaling cascade.

A **ligand-gated ion channel receptor** is a type of membrane channel receptor that requires the binding of a ligand before the channel opens or closes, as shown in Figure 5.17. Unlike the previously discussed receptors, ligand-gated ion channels may be activated from either side of the membrane. Ligand-gated ion channels are widespread within the nervous system and mediate the effects of various neurotransmitters (eg, acetylcholine, GABA, glutamate).

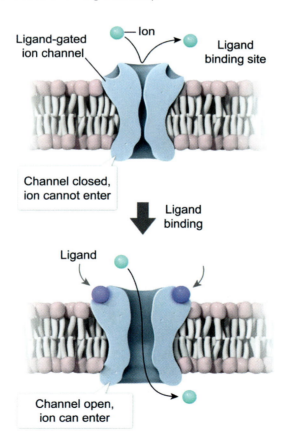

Figure 5.17 Ligand-gated ion channel receptor.

5.2.03 Transport Across Membranes

The cell membrane regulates incoming and outgoing molecular traffic on a differential basis, a concept known as **selective permeability**. In other words, the cell membrane controls precisely which substances are permitted to enter and exit the cell. To some extent, the phospholipid bilayer is permeable to small, nonpolar molecules such as O_2, CO_2, and small hydrocarbons. However, the direct passage of hydrophilic cargo, such as ions (eg, sodium [Na^+], potassium [K^+]) and small polar molecules (eg, water, glucose), is largely prohibited by the hydrophobic interior of the phospholipid bilayer (see Concept 5.2.01).

To facilitate the movement of hydrophilic substances across the membrane, some transmembrane proteins act as **transport proteins** to move specific substances through the membrane. There are two major types of transport proteins: channel proteins and carrier proteins, as shown in Figure 5.18.

Channel proteins are embedded in the membrane, and specific molecules or ions may pass through a hydrophilic "tunnel" within the protein lined with charged and/or polar amino acids. **Carrier proteins** are also embedded in the membrane but undergo reversible conformational changes that move specific solutes from one side of the membrane to the other.

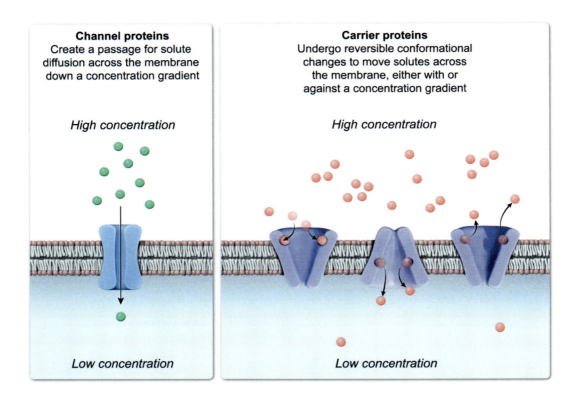

Figure 5.18 Channel proteins versus carrier proteins.

Transport proteins are usually specific for the substance they are meant to transport. Highly specific transport proteins exist for a variety of different solutes, including ions and small molecules such as glucose. The selective permeability of a membrane is further refined based on whether specific membrane transporters are present or absent from the membrane.

Transport across membranes may be passive or active (Figure 5.19). **Passive transport** is energetically favored and occurs without the expenditure of energy. For example, diffusion of molecules down their concentration gradient (from high to low concentration) is a form of passive transport. Conversely, the **active transport** of molecules against their concentration gradient (from low to high concentration) requires energy.

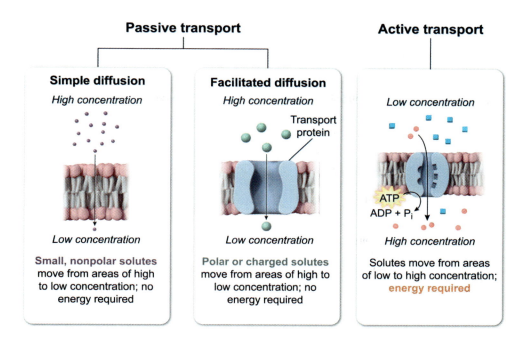

Figure 5.19 Passive transport versus active transport.

Passive Transport

Passive transport across the membrane may take place via two mechanisms: **simple diffusion** or **facilitated diffusion**, as shown in Figure 5.19. Energy is not necessary in either case, but a concentration difference on either side of the membrane is required for net movement of a solute. A solute moves down its concentration gradient until a dynamic equilibrium, in which the concentration on both sides of the membrane is equal and solute movement occurs in both directions at an equal rate, is reached.

In passive transport, the mechanism by which molecules cross the membrane depends on their size and chemical properties. Only small, nonpolar molecules (eg, O_2, CO_2) can cross the membrane via simple diffusion, whereas in facilitated diffusion, ions and small polar molecules most often utilize specific transport proteins to cross the membrane. While transport proteins can be specific for particular molecules, the driving force for movement in passive transport is the presence of a concentration gradient. Therefore, in facilitated diffusion, energy is not required for transport to occur.

While the transport of water molecules down their concentration gradient through aquaporin channels (as shown in Figure 5.20) can be considered a type of facilitated diffusion, there are significant consequences to the rapid movement of water in and out of cells. As such, the net movement of water across a selectively permeable membrane is often considered separately from the movement of other molecules.

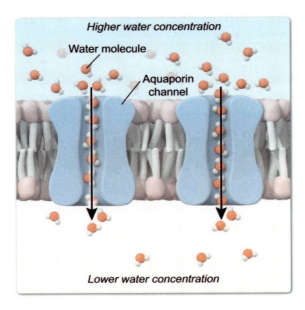

Figure 5.20 Water movement across the cell membrane (osmosis) via aquaporin channels.

Osmosis is the movement of water molecules across a membrane down a concentration gradient from a high to low concentration of water molecules. However, to determine in which direction water will move in the context of a cellular environment, it is generally more useful to consider the concentration of solute molecules (rather than the concentration of water molecules) on either side of the membrane.

Differences in solute concentrations on either side of a membrane lead to different concentrations of free water molecules. When there are *fewer* solutes, *more* free water molecules are available, and when there are *more* solutes, *fewer* free water molecules are available. Therefore, free water molecules diffuse down their concentration gradient (from high to low *water molecule* concentration) across the membrane from a low to high *solute* concentration. A dynamic equilibrium occurs when the concentration of free water molecules is equalized on both sides of the membrane, as shown in Figure 5.21.

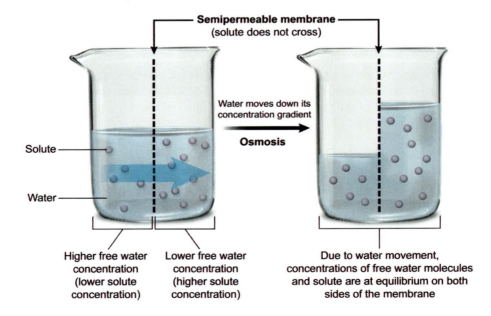

Figure 5.21 Osmosis across a semipermeable membrane.

The **tonicity** of a solution is a measure of its ability to cause water to move into or out of a cell, and it is used to describe the solution surrounding a cell compared to the cell's contents. In an **isotonic** solution, solute concentration is equal inside and outside of the cell; therefore, no net movement of water occurs.

A solution is considered **hypotonic** if its solute concentration is *lower* than the solute concentration inside the cell. In this case, there is a net movement of water into the cell (toward the higher solute concentration), causing the cell to swell and potentially **lyse** (ie, burst). Conversely, a solution is considered **hypertonic** if its solute concentration is *higher* than the solute concentration inside the cell. In this situation, there is a net movement of water out of the cell (toward the higher solute concentration), causing the cell to shrivel (ie, **crenate**). The effect of each type of solution on red blood cells is depicted in Figure 5.22.

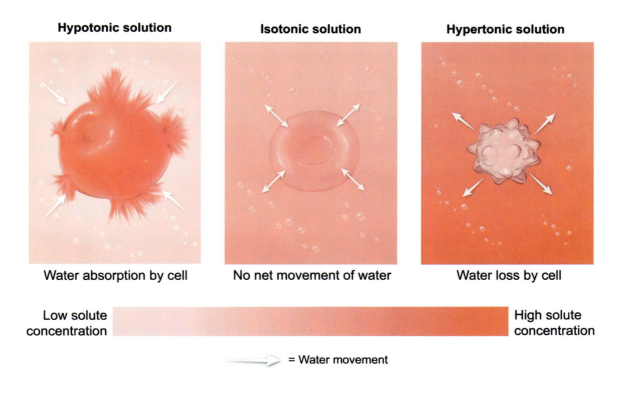

Figure 5.22 A solution's tonicity determines osmotic effects.

The tendency of a solution to draw water through a semipermeable membrane is known as **osmotic pressure**. Osmotic pressure must be well controlled to prevent dramatic consequences to the cell or organism. For organisms with cell walls (eg, bacteria, plants, fungi, some protists), a hypotonic environment typically does not result in lysis of the cell because the cell wall is rigid. This limits the amount of water that can be taken in, resulting in turgor pressure.

The regulation of water balance is crucial to prevent an excessive gain or loss of water, particularly for animal cells, which lack cell walls. Different organisms possess a variety of osmoregulatory adaptations to maintain the correct balance of water and solutes. For example, the kidneys regulate the amount of water reabsorbed from the blood to maintain the correct water balance in the body (discussed in Concept 16.2.01).

Active Transport

Active transport occurs when a cell must use energy to move cargo. Cells may move small solutes or ions against their concentration gradients using carriers similar to those used in facilitated diffusion; however, in active transport, these membrane transporters require the expenditure of energy. In **primary active transport**, energy is most commonly obtained from the hydrolysis of ATP. The direct transfer of a phosphate group from an ATP molecule to a transport protein can induce a conformational change in the transporter, which allows the solute to cross the membrane.

In eukaryotic organisms, an important primary active transport system is the **sodium-potassium pump** (Na^+/K^+-ATPase), which exchanges sodium for potassium against each ion's concentration gradient. For each ATP used, the sodium-potassium pump moves three Na^+ *out of* the cell and two K^+ *into* the cell. This exchange maintains a high concentration of Na^+ outside the cell and a high concentration of K^+ inside the cell.

The sodium-potassium pump exists in two conformational states, as shown in Figure 5.23. The cycle, which can be repeated provided that ATP is readily available, includes the following steps:

1. When the Na^+ binding sites are facing the cytosolic side of the membrane, these sites have a higher affinity for Na^+ than K^+.
2. The binding of three Na^+ ions triggers the transfer of a phosphate group from ATP to the pump.
3. Phosphate transfer provokes a conformational change in the pump that reduces the affinity for Na^+ binding, and Na^+ ions are translocated to the extracellular space.
4. Now facing the extracellular space, the pump's new conformation has binding sites with a high affinity for two K^+ ions to bind. K^+ binding triggers the release of the phosphate group.
5. Phosphate group release results in a conformational change to the pump and K^+ ions are translocated to the cytosol.

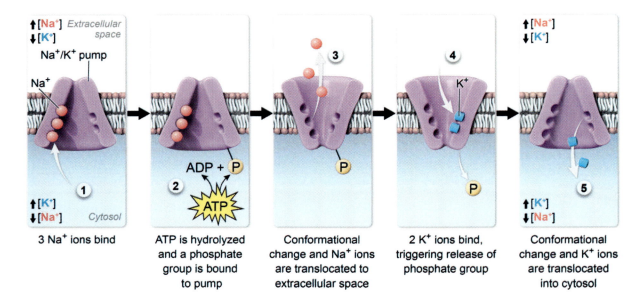

Figure 5.23 The sodium-potassium pump is a primary active transport system.

If a membrane transporter moves only one type of molecule, it is called a **uniport**. However, some transporters can move more than one type of molecule at a time. Transporters that move two or more molecules across the membrane in the same direction are called **symports**. If the molecules are moved in opposite directions, the transporter is called an **antiport**. The three types of port systems are depicted in Figure 5.24.

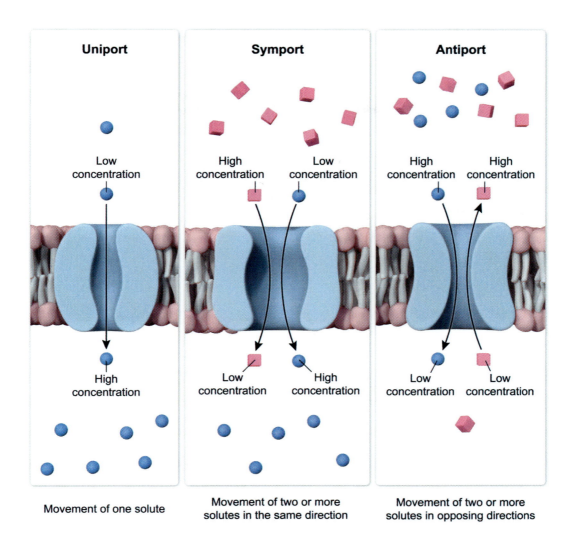

Figure 5.24 Types of port systems.

Secondary active transport is a coupled transport process that uses energy released by the movement of one substance down its concentration gradient (ie, passive transport) to move another substance against its concentration gradient (ie, active transport). In this type of transport, potential energy stored in the concentration gradient of the passively transported molecule is used for the active transport of another.

One example of a secondary active transport system is the **sodium-glucose linked transporter (SGLT)**. As previously described, the sodium-potassium pump maintains a high extracellular Na^+ concentration. As Na^+ ions passively travel back into the cell *down* their concentration gradient, the potential energy stored in the Na^+ concentration gradient is used by the SGLT transporter, which acts as a symport to transport a glucose molecule into the cell *against* the glucose concentration gradient. Figure 5.25 illustrates the difference between primary and secondary active transport, as well as the link between the two types of transport.

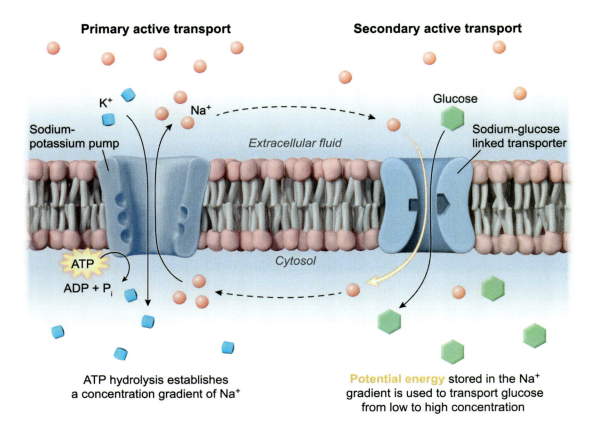

Figure 5.25 The sodium-glucose linked transporter is a secondary active transport system.

When ions are transported across the membrane, an unequal distribution of the overall charge across the membrane may develop, creating an **electrical gradient**. For example, when the overall charge inside of the cell is more negative than the overall charge outside the cell, the passive transport of positively charged ions into the cell and the passive transport of negatively charged ions out of the cell is favored. This transport occurs through passive membrane channels known as leak channels.

There are two factors that drive the diffusion of ions across the membrane: the ion's concentration (chemical) gradient and the effect of the electrical gradient. The combined effect of these two factors is known as the electrochemical gradient. An electrical gradient may work to oppose diffusion in the direction favored by the concentration gradient, thereby slowing or even reversing the direction of diffusion.

For example, the K^+ concentration gradient tends to drive K^+ out of the cell, but the electrical gradient tends to drive K^+ into the cell. The combined effect of both gradients results in a *net* electrochemical gradient favoring the movement of K^+ out of the cell (Figure 5.26).

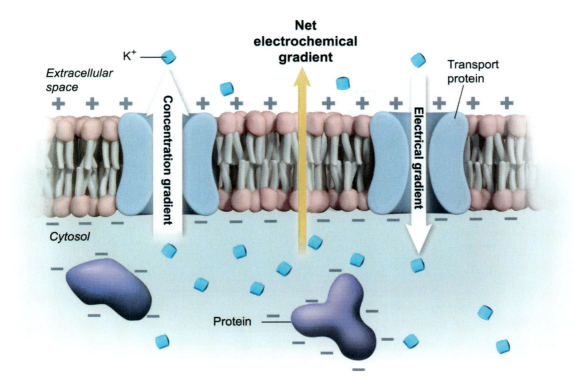

Figure 5.26 The electrochemical gradient.

Largely due to continued active transport of Na⁺ and K⁺ ions via the sodium-potassium pump, the concentration of positively charged ions is consistently greater outside of the cell. Comparatively, there is a net negative charge inside the cell. The voltage difference caused by the unequal distribution of charges on either side of the membrane is called the **membrane potential**.

The sodium-potassium pump plays an important role in the maintenance of membrane potential by keeping the Na⁺ and K⁺ gradients steady. Modulation of the electrochemical gradient facilitates processes such as transport of glucose into the epithelial cells of the small intestine (eg, SGLT) and the transmission of nerve impulses (see Lesson 12.2).

Bulk Transport

In general, membrane transporters are not used to move larger molecules such as polysaccharides and proteins. These types of molecules are typically transported in bulk into the cell via small membrane-bound vesicles in a process known as **endocytosis**.

During endocytosis, the substance being transported enters a small pocket formed by the surrounding membrane. In an energy-dependent process (eg, utilizing ATP and/or GTP), the sides of the pocket are pinched together and sealed off, forming a small membrane-bound compartment (ie, vesicle) called an **endosome** inside of the cell. The contents of the endosome are then trafficked to their final destination via the endocytic pathway (see Concept 5.3.03).

Endocytosis of extracellular fluid containing small solutes is known as **pinocytosis** (cellular "drinking"). Pinocytosis is relatively nonspecific as to the types of particles taken in. Endocytosis of relatively large structures (eg, cells, cell fragments) is known as **phagocytosis** (cellular "eating"). Phagocytosis generally involves receptor binding as well as membrane extensions (**pseudopodia**) that surround a particle to form a vesical called a **phagosome**, which draws the particle into the cell (Figure 5.27). Specialized phagocytic cells play an important role in innate immunity (see Lesson 20.2).

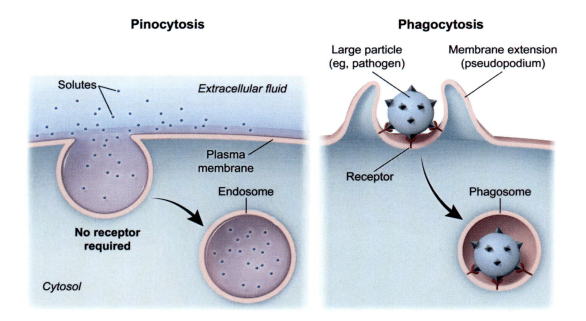

Figure 5.27 Pinocytosis and phagocytosis, two types of endocytosis.

Clathrin-mediated endocytosis utilizes cell surface receptors to target specific molecules for incorporation into the cell. In this type of endocytosis, cell surface receptors may be clustered within specialized regions of the plasma membrane called **clathrin-coated pits**, which are lined with the protein **clathrin** on the cytosolic face of the membrane. When these vesicles pinch off and enter the cytosol, the vesicle formed is completely coated with clathrin, which helps mediate delivery of cargo to its final intracellular destination.

Cells use clathrin-mediated endocytosis to take up cholesterol from the bloodstream, as shown in Figure 5.28. Because cholesterol molecules are insoluble in water, they must be associated with proteins while traveling in the bloodstream. Cholesterol is packaged into particles called low-density lipoproteins (LDLs), which are composed of a phospholipid monolayer embedded with apolipoprotein B (ApoB). ApoB binds to a cell surface LDL receptor in the clathrin-coated pits, which triggers the endocytosis of LDL within a clathrin-coated vesicle. Subsequently, cholesterol can be trafficked to the correct location in the cell.

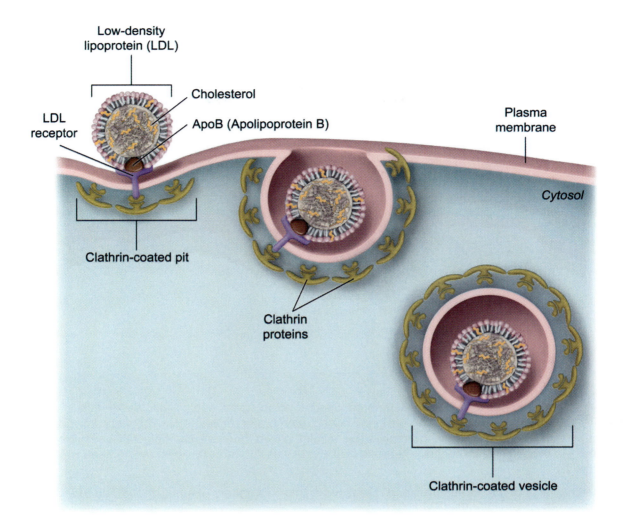

Figure 5.28 Clathrin-mediated endocytosis of a low-density lipoprotein.

Cells may also export large molecules via an energy-dependent bulk transport process called **exocytosis**. While the process is slightly different, exocytosis shares many features with endocytosis. During exocytosis, a secretory vesicle inside the cell approaches the plasma membrane and fuses with the membrane, releasing the vesicle contents into the extracellular space. After fusion, the vesicle membrane and any embedded membrane proteins and lipids become part of the plasma membrane (Figure 5.29).

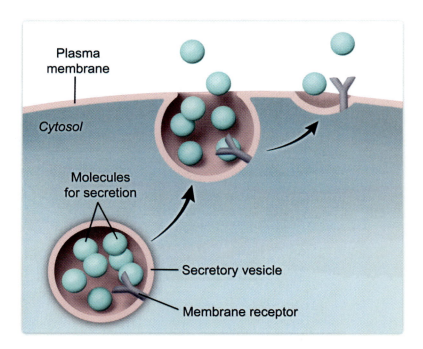

Figure 5.29 Exocytosis.

Exocytosis can be used to secrete cellular products or to expel toxins and cellular waste. It is also the mechanism by which a cell can integrate proteins (eg, membrane receptors), lipids, and carbohydrates into its plasma membrane. Many hormones (eg, insulin, growth hormone) and digestive enzymes are secreted via exocytosis. In addition, neurons utilize exocytosis to release neurotransmitters, which signal other cells.

The mechanisms of membrane transport discussed in this concept are summarized in Table 5.1

.Table 5.1 Mechanisms of membrane transport.

Type of transport	Direction of transport	Requires cellular energy	Examples of substances transported
Passive transport (simple diffusion)	High to low concentration (down the gradient)	No	Small nonpolar molecules
Passive transport (facilitated diffusion)	High to low concentration (down the gradient)	No	Water, ions, polar molecules
Active transport	Low to high concentration (against the gradient)	Yes	Ions, polar molecules
Bulk transport (endocytosis)	Into the cell	Yes	Large particles (eg, macromolecules, viruses)
Bulk transport (exocytosis)	Out of the cell	Yes	Large particles (eg, macromolecules, viruses)

☑ Concept Check 5.1

A scientist uses fluorescently labeled antibodies as a tool to detect the transport of low-density lipoproteins (LDLs) into a cell via clathrin-mediated endocytosis. The scientist chooses specific antibodies that bind to the molecules involved in clathrin-mediated endocytosis of LDL and attaches a different fluorescent marker (fluorophore) to each antibody. When excited by the scientist during the experiment, each fluorophore emits light of a specific color.

The following types of labeled antibodies are available to the scientist for the experiment.

Labeled antibody detects:	Fluorophore emits light:
LDL receptor	~690 nm (red)
Clathrin protein	~500 nm (green)
ApoB protein (detects LDL)	~455 nm (blue)

At which locations in the cell should the scientist expect to detect the red, green, and blue fluorescence prior to and after transport of LDL into the cell?

Solution
Note: The appendix contains the answer.

Lesson 5.3

Eukaryotic Organelles

Introduction

Eukaryotic cells are considerably larger and more complex in structure and function than prokaryotic cells. While both types of cells possess internal structures known as **organelles**, typically only eukaryotic cells contain membrane-bound organelles. This lesson focuses on the structure and function of eukaryotic organelles and other cellular features, with a concentration on animal cells.

5.3.01 The Nucleus

The **nucleus** is the largest eukaryotic organelle and serves as the storage site for the majority of the cell's DNA and is the location of DNA replication and gene transcription (see Lesson 1.2 and Lesson 2.2). It is surrounded by a **nuclear envelope**, a double membrane continuous with another membranous organelle, the endoplasmic reticulum (Figure 5.30). Within the nucleus, DNA and its associated proteins are organized into discrete chromosomes, which are further organized and compacted as chromatin.

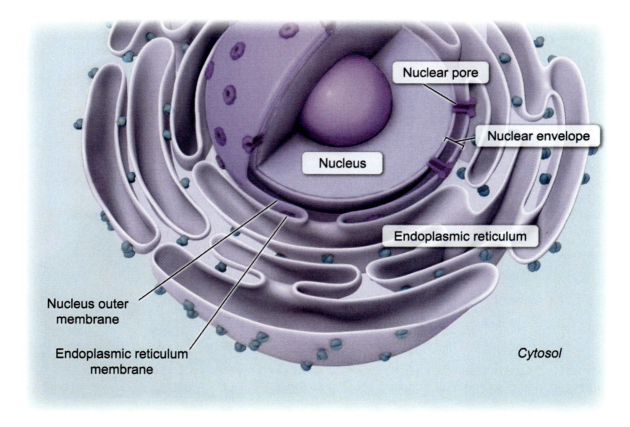

Figure 5.30 The nucleus is surrounded by a nuclear envelope and contains nuclear pores.

Nuclear pores are channels within the nuclear envelope that regulate the passage of materials and place limits on the size and type of molecules that can enter or exit the nucleus. Nuclear pores must be large enough to accommodate the translocation of relatively large molecules such as ribosomal subunits and other proteins. Nuclear pores contain a structure known as the **nuclear pore complex (NPC)**, which is made up of proteins called **nucleoporins**.

Proteins destined for the nucleus are targeted to nuclear pores and are translocated into the nucleus, due to the recognition of a sequence of amino acids called the **nuclear localization sequence**. To exit the nucleus, a protein must move through the NPC back into the cytosol.

5.3.02 Nucleolus and Ribosomes

As discussed in Lesson 2.3, a fully assembled ribosome is composed of two ribosomal subunits and serves as the molecular machinery that translates messenger RNA (mRNA) sequences into proteins. Present in both eukaryotic and prokaryotic organisms, ribosomes are composed of **ribosomal RNA (rRNA)** and proteins and are not membrane-bound organelles. **Nucleoli** (singular: **nucleolus**) are dense, round bodies within the nucleus of eukaryotic cells that serve as the sites of ribosomal synthesis and assembly.

Within the nucleolus, **RNA polymerase I** transcribes rRNA genes from ribosomal DNA (rDNA) into a pre-rRNA template. Ribosomal proteins (synthesized in the cytoplasm from mRNA) are transported into the nucleolus, where these proteins combine with newly transcribed pre-rRNA. Subsequent processing of pre-rRNA forms mature rRNA, which, along with the associated ribosomal proteins, form precursors to the 40S (small) and 60S (large) mature ribosomal subunits (Figure 5.31).

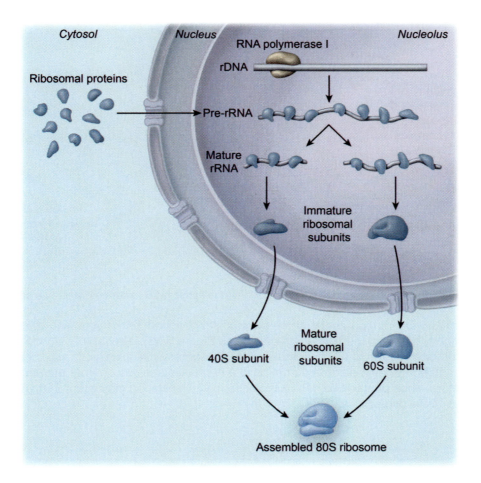

Figure 5.31 Ribosome assembly.

Mature ribosomal subunits are shuttled out of the nucleus via nuclear pores, and the different subunit types combine to form a fully assembled 80S ribosome in the cytosol. Ribosomes synthesize proteins either in a free state (within the cytosol) or in a bound state when attached to the rough endoplasmic reticulum (RER).

The coding region of some mRNA molecules begins with a signal sequence, which is translated to form a **signal peptide**. The signal peptide interacts with a signal recognition particle at the ribosome and induces transport of the ribosome (still joined with the mRNA) to a translocation complex on the RER membrane where translation continues. During or after translation, the signal peptide is cleaved and the polypeptide is either deposited into the RER lumen or embedded in the RER membrane, where further modifications may occur (Figure 5.32).

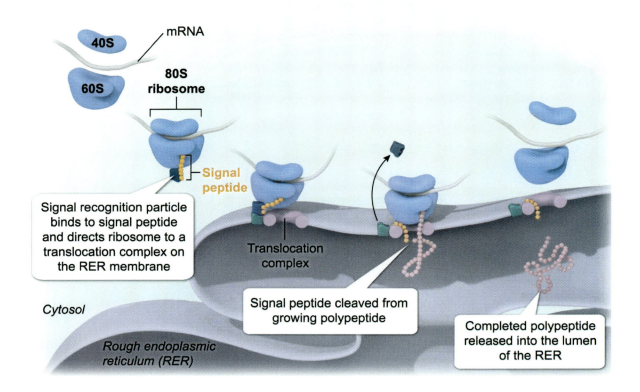

Figure 5.32 Localization of protein synthesis to the rough endoplasmic reticulum membrane by a signal peptide.

While all cells contain ribosomes with the same mechanism of action, prokaryotic and eukaryotic ribosomes are slightly different in size and structure (see Figure 5.33). Prokaryotic ribosomal subunits (30S and 50S) combine to form a fully assembled 70S ribosome, while eukaryotic subunits (40S and 60S) form an 80S ribosome. Prokaryotic cells do not have a nucleus/nucleolus or other membrane-bound organelles; therefore, once prokaryotic mRNA is transcribed, translation may occur immediately via nearby ribosomes in the cytosol.

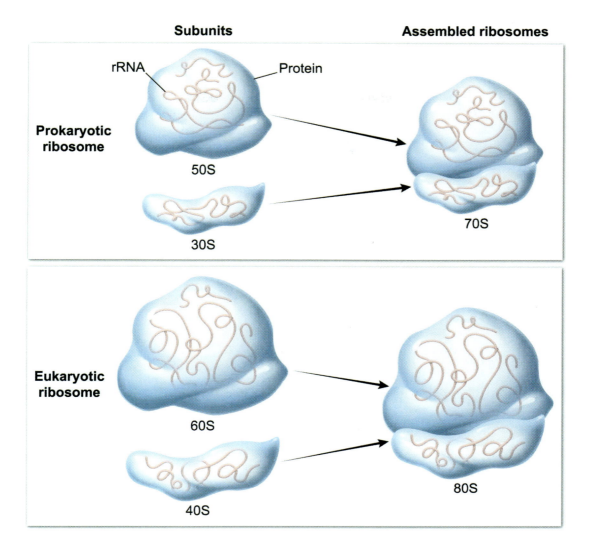

Figure 5.33 Prokaryotic and eukaryotic ribosomes.

5.3.03 The Endomembrane System

The **endomembrane system** is a collection of membranous organelles that carry out various tasks within the cell, including modification and transport of proteins to various locations within or outside the cell. In addition, the endomembrane system plays a role in the metabolism of carbohydrates and lipids and in the neutralization of cellular toxins.

The membranes associated with this system are related because they either come into direct physical contact with one another or they are connected via the transfer of membranous sacs known as **vesicles**. The components of the endomembrane system include the nuclear envelope, endoplasmic reticulum, Golgi apparatus, lysosomes and other vesicles, and the plasma membrane, as depicted in Figure 5.34.

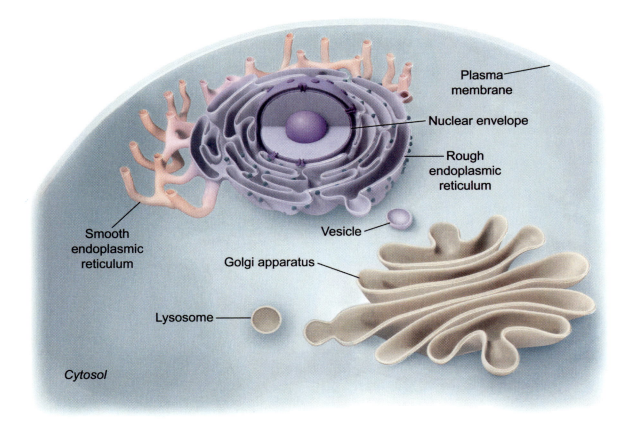

Figure 5.34 Components of the endomembrane system.

Endoplasmic Reticulum

The **endoplasmic reticulum (ER)** is an organelle continuous with the nuclear envelope and composed of a network of connected membranous sacs and tubules. The ER is divided into two parts with different functions: the **rough ER (RER)**, which is studded with ribosomes, and the **smooth ER (SER)**, which has an outer surface free of ribosomes.

Because the SER does not have ribosomes on its surface, it does not participate in protein synthesis. Instead, the SER participates in a variety of cellular functions depending on cell type. These functions include synthesis of lipids such as cholesterol and cholesterol-derived molecules (eg, steroid hormones), triglycerides, and phospholipids destined for new membranes. The SER is also involved in carbohydrate metabolism, detoxification of drugs and poisons, and storage of calcium ions used for contraction in muscle cells (see Concept 17.1.04).

Ribosomes located along the RER cytosolic surface translate proteins destined for other components in the endomembrane system or for secretion from the cell (Figure 5.35). These translated proteins may undergo post-translational modifications (eg, glycosylation) catalyzed by enzymes in the RER (see Concept 2.3.06). Proteins in the RER are transported via vesicles to other locations in the cell, including the Golgi apparatus, where additional modifications may occur.

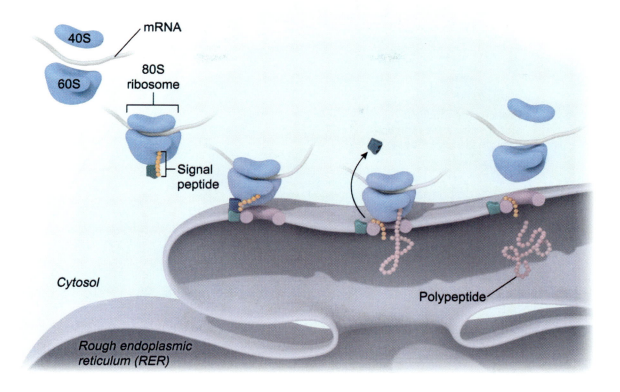

Figure 5.35 Translation of proteins via ribosomes at the rough endoplasmic reticulum.

Golgi Apparatus

After protein-containing transport vesicles exit the RER, many travel to the **Golgi apparatus**, where proteins are further processed, sorted, and packaged for transport to the next destination. The Golgi apparatus is composed of a stack of flat, membranous sacs known as **cisternae**. Each cisterna is a distinct compartment, and materials are transferred between cisternae. There is a directionality to the movement of the proteins between cisternae: proteins move from the RER towards the *cis* (ie, receiving) face of the Golgi apparatus and depart the Golgi apparatus from the *trans* (ie, shipping) face.

As proteins travel though the Golgi apparatus, different chemical groups (eg, carbohydrate, phosphate) may be added or modified, and proteins are sorted before reaching the *trans* face. From the *trans* Golgi, protein-filled transport vesicles are directed to their final destination (Figure 5.36).

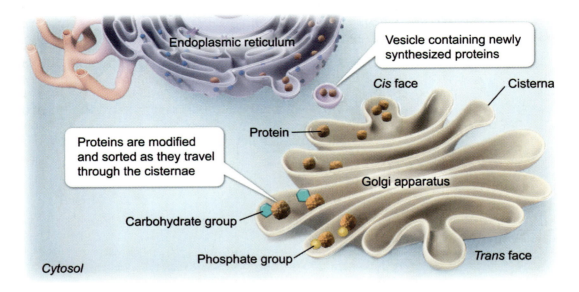

Figure 5.36 The Golgi apparatus functions in the modification and sorting of proteins.

Although the **plasma membrane** is not technically a compartment within the cell, it is considered a part of the endomembrane system because it interacts with other membrane-bound organelles in the endomembrane system. The endomembrane system is the mechanism by which the **secretory pathway** directs the embedding of proteins in the plasma membrane and the secretion of proteins from a cell.

The secretory pathway begins with protein deposition in the RER. Proteins destined to be incorporated into the plasma membrane are embedded in the RER membrane during synthesis, and proteins destined to be secreted from the cell are deposited into the RER lumen. After further modifications in the Golgi apparatus, the resulting proteins are packaged in vesicles that undergo exocytosis. Upon vesicle fusion with the plasma membrane, proteins embedded in the vesicle membrane become part of the plasma membrane, and secreted proteins are released to the extracellular space (Figure 5.37).

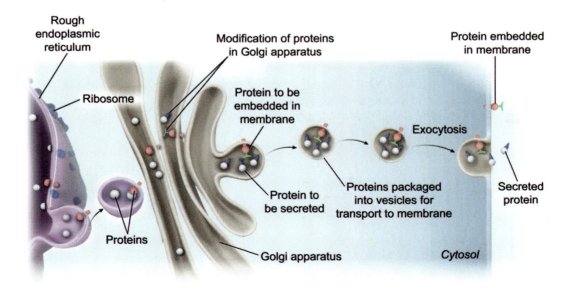

Figure 5.37 Protein trafficking through the secretory pathway.

Chapter 5: Eukaryotic Cells

Lysosome

Lysosomes are specialized vesicles that serve as a type of cellular digestive system. The lysosome interior is maintained as an acidic environment (pH ~4.5), and hydrolytic enzymes packaged within the lysosome facilitate the degradation of various biomolecules. These enzymes are synthesized in the RER and then transferred to the Golgi apparatus via vesicles, where lysosomes are formed by budding from the *trans* face of the Golgi apparatus.

When molecules enter the cell by endocytosis, they are often transported to a lysosome via the **endocytic pathway**. Lysosomes participate in the digestion of food particles, other organic matter, and small organisms engulfed during endocytosis (see Concept 5.2.03). Following the internalization of extracellular materials, **early endosomes** (vesicles) are formed. As an early endosome matures, its contents are sorted and it becomes a **late endosome**.

If targeted for destruction, the late endosome fuses with a lysosome, forming a structure known as an **endolysosome**. Hydrolytic enzymes in the endolysosome digest the trapped materials into organic products (eg, sugars, amino acids, other monomers) which are recycled in the cell. Vesicles containing larger materials (eg, microbes) are called **phagosomes**, and these fuse with lysosomes to form **phagolysosomes**. The contents are then degraded within the phagolysosomes, and waste products are eliminated via exocytosis (Figure 5.38).

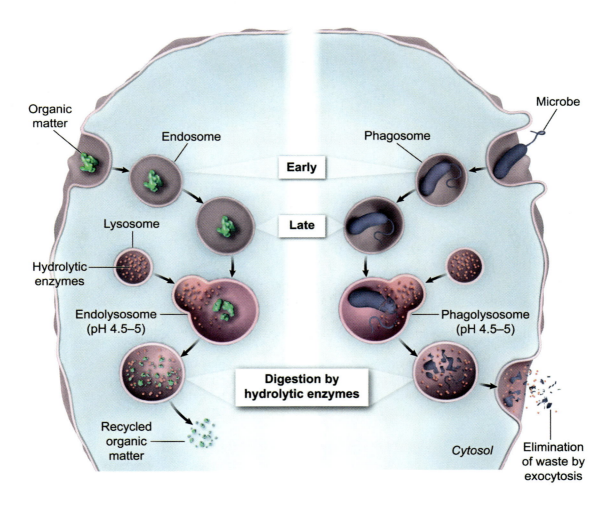

Figure 5.38 Recycling and elimination of waste by lysosomes.

Endosomes may also be trafficked through an alternate pathway to the Golgi apparatus or RER. This pathway is most often used to retrieve specific membrane receptors and lipids from the cell surface. Endocytosed cell surface molecules can be targeted to either the *trans* Golgi or RER for recycling purposes, bypassing degradation by the lysosome, as shown in Figure 5.39. Some microbes may exploit this pathway to avoid destruction by the lysosome.

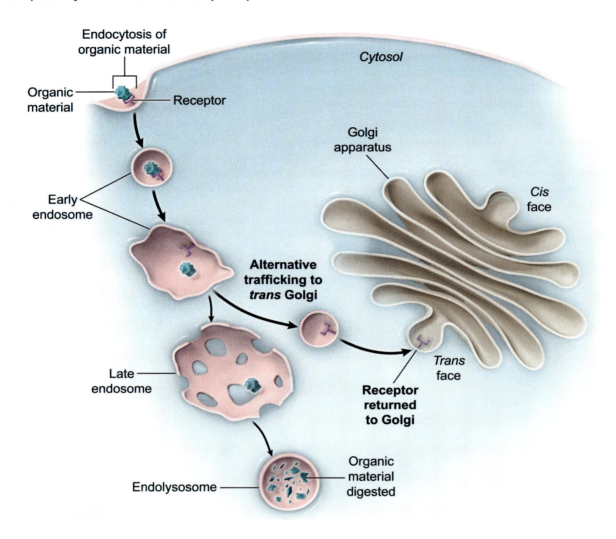

Figure 5.39 Alternative trafficking through the endocytic pathway.

In addition, cells can utilize lysosomes to recycle intracellular organic material in a process known as **autophagy**. Old or damaged organelles can be recycled by becoming enclosed in a vesicle that fuses with a lysosome. Lysosomal enzymes digest the trapped organelle, and organic materials are released for reuse within the cell, as depicted in Figure 5.40.

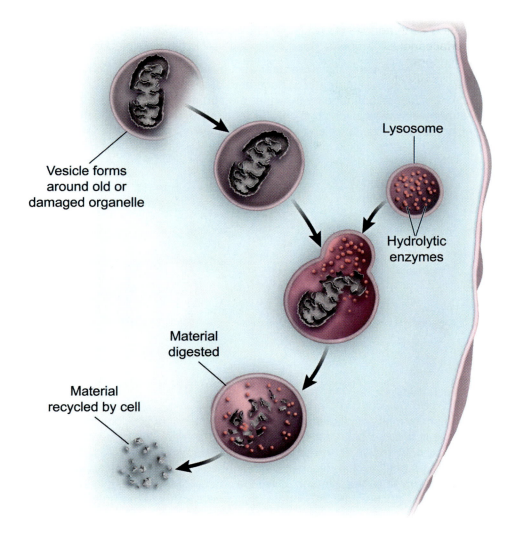

Figure 5.40 Autophagy is a lysosome-mediated intracellular degradation pathway.

> **Concept Check 5.2**
>
> A scientist wants to determine if a protein is localized to the cytosol or is trafficked through the secretory pathway. How could the scientist determine how this protein is localized?
>
> **Solution**
>
> *Note: The appendix contains the answer.*

5.3.04 Mitochondria

To carry out cellular functions, a cell must acquire energy from its environment. In eukaryotic cells, the conversion of raw materials into usable cellular energy occurs in an organelle known as the **mitochondrion**. Mitochondria are the sites of cellular respiration, in which energy extracted from sugars, fats, and other molecules is converted into ATP, a usable energy form. Each cell may contain anywhere from one to thousands of mitochondria depending on that cell's metabolic needs. For example, muscle cells contain more mitochondria than many other cell types due to higher metabolic activity.

Mitochondria are enclosed by an **outer membrane** and an **inner membrane**. Compared to the outer membrane, the inner membrane has a greater surface area due to convolutions (ie, infoldings) known as **cristae** (Figure 5.41). The space bounded by the inner membrane is known as the mitochondrial **matrix**. The matrix contains many enzymes involved in cellular respiration, as well as mitochondrial DNA and ribosomes. The region between the outer and inner membranes is known as the **intermembrane space**.

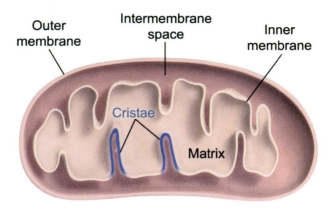

Figure 5.41 Mitochondrial structure.

Because mitochondria share many similarities with bacteria, mitochondrial origin has long been considered to be the result of a symbiotic relationship between an engulfed bacterium and a primitive eukaryotic host cell, a concept known as the **endosymbiotic theory** (Figure 5.42). This theory proposes that an oxygen-using (ie, aerobic) prokaryotic cell was engulfed by a primitive eukaryotic ancestor and that the prokaryotic cell survived its engulfment. Over the course of evolution, it is thought that the engulfed bacterium and host cell became dependent upon one another, evolving into a single organism.

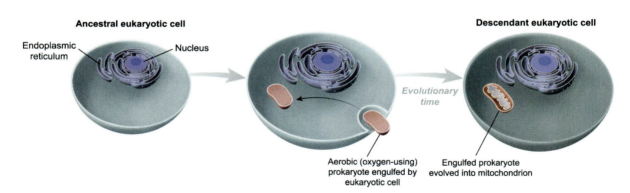

Figure 5.42 Endosymbiotic theory.

The structural and biochemical features of mitochondria provide support for the endosymbiotic theory. For example, mitochondrial ribosomes are more similar to prokaryotic ribosomes than eukaryotic ribosomes. In addition, mitochondria contain their own circular DNA molecules, which share similarities in sequence and organization with prokaryotic DNA. Finally, mitochondria grow and reproduce independently through mitochondrial fission, a process related to prokaryotic reproduction (ie, binary fission, described in Concept 6.2.01).

5.3.05 Peroxisomes

Peroxisomes are small, membrane-bound organelles that carry out a variety of metabolic reactions. Functions of peroxisomes include the facilitation and containment of oxidative reactions, which produce **reactive oxygen species (ROS)**, as well as involvement in fatty acid metabolism and synthesis of certain lipids and bile acid intermediates (Figure 5.43). Peroxisomes are spherical in shape and often contain a crystalline core composed of a dense collection of oxidative enzymes.

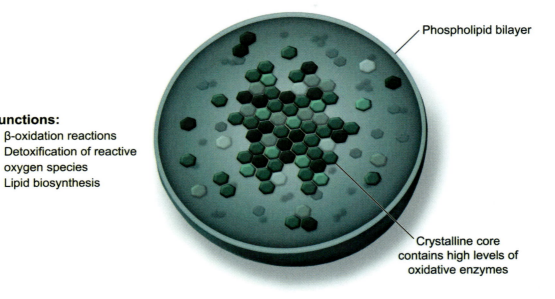

Figure 5.43 Peroxisome structure and functions.

Because ROS (eg, hydrogen peroxide) are harmful to cells, peroxisomes also contain high levels of the enzyme **catalase**, which neutralize the harmful effects of ROS. Fatty acid oxidation can occur in both peroxisomes and mitochondria, and products of fatty acid oxidation in peroxisomes can later be used as fuel for cellular respiration in mitochondria. In addition, reactions in peroxisomes can facilitate detoxification of other harmful substances (eg, alcohol detoxification within liver cells). Peroxisomes are not considered part of the endomembrane system.

Unlike mitochondria, peroxisomes do not contain their own DNA or ribosomes. While the evolutionary origin of peroxisomes is still undetermined, there is some evidence that peroxisomes are able to grow and replicate in a similar fashion to mitochondria (ie, a process similar to binary fission).

5.3.06 Cytoskeleton

The cytoskeleton is a network of intracellular scaffolding fibers deposited throughout the cytosol, as shown in Figure 5.44. Together, these fibers function to influence cell shape, support cellular motility, and help organize intracellular compartments.

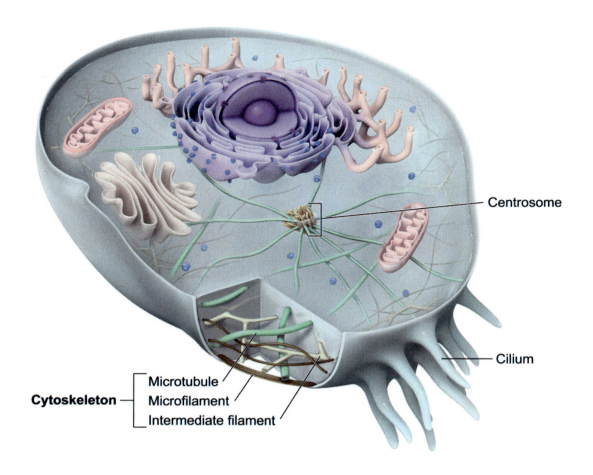

Figure 5.44 Cytoskeletal structures within a cell.

The three major cytoskeletal components are microfilaments (ie, actin), intermediate filaments, and microtubules, as depicted in Figure 5.45.

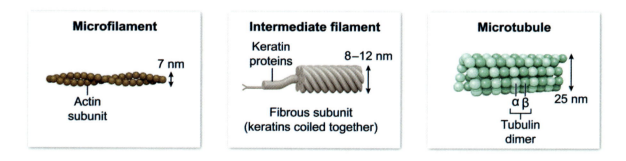

Figure 5.45 Major components of the cytoskeleton.

Microfilaments are composed of a twisted chain of actin protein subunits. The smallest of the cytoskeletal fibers (7 nm diameter), microfilaments are organized into a three-dimensional network that maintains tension and supports cellular shape. Microfilaments also function in cell motility; for example, actin filaments, with the motor protein **myosin**, play a role in muscle contraction, and they participate in cell migration via cellular extensions known as pseudopodia.

Actin filaments (along with myosin) also participate in cell division during cytokinesis, by forming a contractile ring which cleaves the cell in two. Microfilaments make up the core of microvilli, cellular projections that increase surface area in some cells (eg, intestinal lining, kidney tubules).

Intermediate filaments are a diverse group of cytoskeletal fibers made up of a variety of proteins (eg, keratins, lamins, vimentin, desmin) expressed differentially by various cell types. Intermediate filaments are so named because their diameter (8–12 nm) is larger than microfilaments but smaller than microtubules. Functions of intermediate filaments include supporting cell shape, reinforcing the nuclear lamina (inner lining of the nuclear envelope), anchoring organelles to specific cellular locations, and helping the cell resist mechanical forces (eg, compression).

Microtubules, the largest of the cytoskeletal fibers (25 nm diameter), are composed of alternating α- and β-tubulin protein subunits that assemble into hollow tubes. Most microtubules originate near the nucleus from small organelles called **centrosomes**. A centrosome consists of two smaller **centrioles**, from which short **aster** microtubules radiate out towards the plasma membrane.

Microtubules are involved in maintaining cell shape, as well as with various forms of movement within the cell and are essential for cell motility. For example, the mitotic spindle, which is formed from microtubules, helps segregate chromosomes during cell division (Figure 5.46).

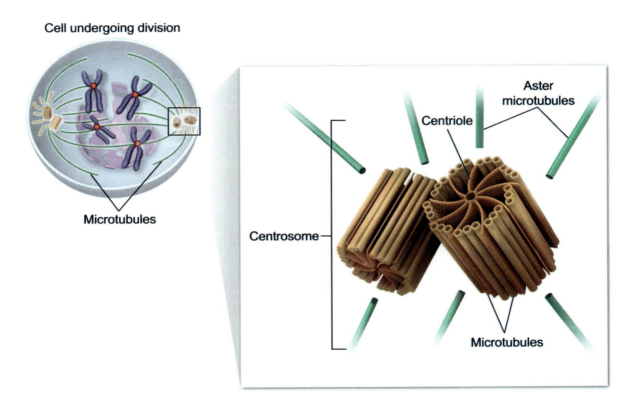

Figure 5.46 A centrosome consists of two centrioles.

Microtubules also facilitate the transport of vesicles and other organelles from one location within the cell to another (Figure 5.47). Movement of intracellular cargo along microtubules is mediated by the two motor proteins **kinesin** and **dynein**. Kinesin moves cargo along microtubules in an anterograde fashion (ie, away from the nucleus), while dynein participates in retrograde transport (ie, toward the nucleus). Kinesin and dynein proteins "walk" along microtubules in a "hand-over-hand" fashion, using ATP hydrolysis as a source of energy.

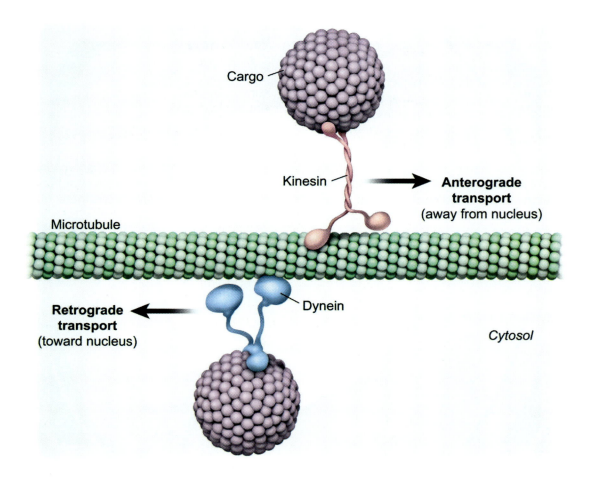

Figure 5.47 Motor proteins move along microtubules.

The types of cytoskeletal fibers and structures covered in this concept are summarized in Table 5.2.

Table 5.2 Types of cytoskeletal fibers.

	Microfilaments	Intermediate filaments	Microtubules
Protein subunits	Actin	Keratin, lamins, vimentin, and desmin	Tubulin
Diameter	7 nm	~10 nm	25 nm
Functions	• Help determine cellular shape • Involved in cellular locomotion • Responsible for muscle contractions • Involved in cytokinesis	• Help determine cellular shape • Make up the nuclear lamina • Help anchor organelles to specific cellular compartments	• Help determine cellular shape • Involved in intracellular transport of vesicles and organelles • Mitotic chromosomal movement • Cellular locomotion (cilia and flagella)
Motor proteins	Myosin	None	Kinesin, dynein

Some eukaryotic cells have microtubule-based cellular extensions known as cilia and flagella (Figure 5.48). **Cilia** are short cellular extensions composed of a specialized arrangement of microtubules. Cilia move in a back-and-forth motion and are usually present in large numbers on the cell surface. Many types of cells utilize cilia; for example, the beating of cilia on respiratory epithelial cells can help move potentially harmful substances away from the lungs (see Concept 14.1.02).

Eukaryotic **flagella** are also composed of a specialized arrangement of microtubules; however, unlike cilia, flagella are usually limited to just one or a few per cell. Flagella are longer than cilia and have a different pattern of movement: a whip-like undulating motion (Figure 5.48). The primary function of flagella is to enable cellular locomotion. For example, during fertilization, the flagellum on a mature sperm cell works to propel sperm towards an oocyte. While prokaryotic and eukaryotic flagella perform similar functions, prokaryotic flagella are not formed from microtubules (see Concept 6.1.03).

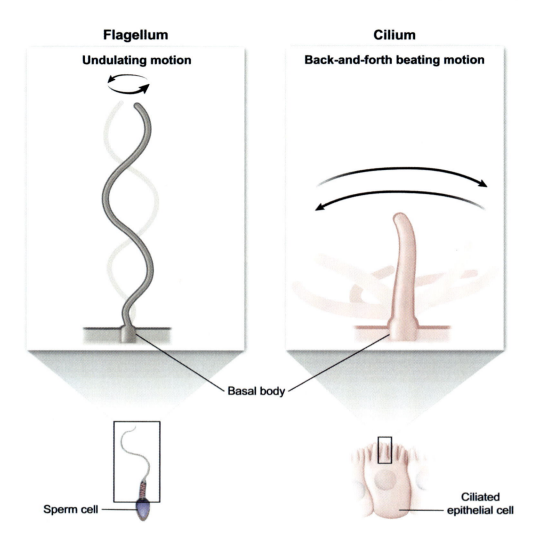

Figure 5.48 Flagella and cilia are involved in cellular motion.

While cilia and flagella have distinct functions, these structures share some common features. Each cilium or flagellum contains a group of microtubules covered by an extension of the plasma membrane. Nine pairs of microtubules are arranged in a ring-like structure, with two single microtubules in the center of this ring, as depicted in Figure 5.49. This 9+2 structure is found in nearly all eukaryotic cilia and flagella. Furthermore, cilia and flagella are both anchored to cells by a structure known as the **basal body**, which is similar to a centriole.

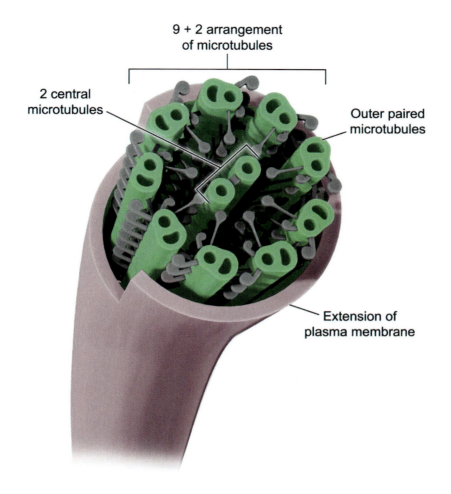

Figure 5.49 Ultrastructure of cilia and flagella.

5.3.07 Extracellular Matrix and Cell Junctions

The plasma membrane is considered the outermost boundary of a living cell, but most cells synthesize and secrete extracellular materials that perform a variety of functions outside of the cell. For example, some eukaryotic and most prokaryotic cells contain a carbohydrate-based extracellular structure known as a cell wall, which is composed of various materials (eg, cellulose, chitin, peptidoglycan) depending on the organism.

Instead of a cell wall, animal cells synthesize a complex structure known as the **extracellular matrix (ECM)**, which is composed mainly of proteins, many of which are glycoproteins (Figure 5.50). The ECM participates in many cellular activities, including attachment and communication among cells, cell growth and migration, mechanical stability, and tissue repair.

The most abundant ECM protein is **collagen**, which forms a strong network of fibers outside of the cell. Collagen fibers are embedded in a web of **proteoglycan** molecules secreted by the cell. Some ECM fibers are attached to the protein **fibronectin**, which is bound to the cell via transmembrane proteins called **integrins**. Inside the cell, integrin proteins are linked to microfilaments via adaptor proteins, thereby connecting the cytoskeleton to the ECM.

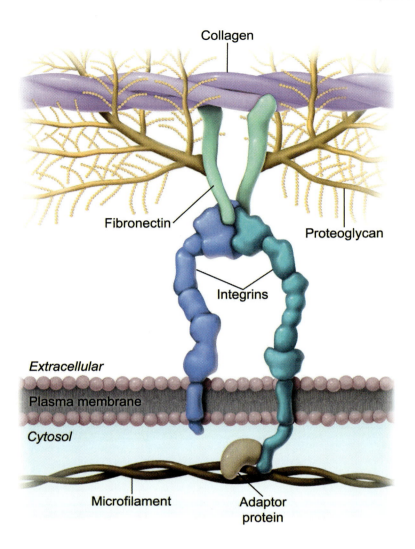

Figure 5.50 Extracellular matrix proteins.

While the ECM connects cells of a multicellular organism indirectly, there are additional physical structures known as **cell-cell junctions**, which form direct attachments between cells. Cell-cell junctions in animal cells include desmosomes, gap junctions, and tight junctions, as depicted in Figure 5.51. These junctions are most prevalent in epithelial tissues (see Concept 5.5.01), which are composed of densely packed cells and very little ECM.

Desmosomes provide tensile strength to tissues by anchoring cytoskeletal intermediate filaments between neighboring cells. These connections create a continuous cytoskeletal network spanning many cells, enabling even distribution of mechanical stress (eg, pulling, stretching, tension). Typically, desmosomes are found in tissues subjected to high levels of mechanical stress (eg, skin, cardiac muscle) and help to prevent tearing in these tissues. Desmosomes are composed of a dense intracellular protein plaque that links intermediate filaments to **cell adhesion molecules (CAMs)** embedded in the plasma membrane.

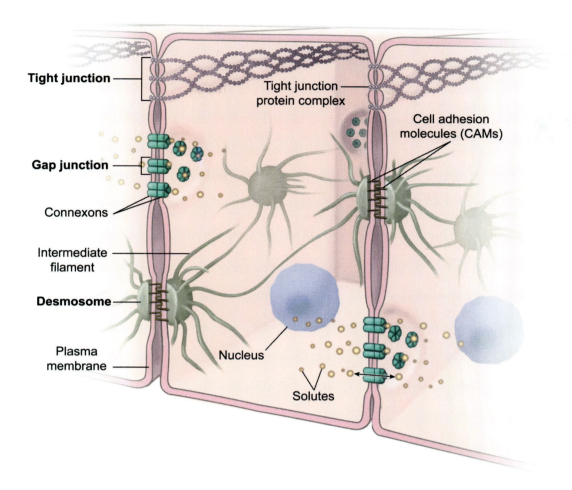

Figure 5.51 Types of cell-cell junctions.

Gap junctions are cell-cell junctions that mediate communication between cells. Protein channels called **connexons** in one cell align with complementary connexon channels in a neighboring cell to form pores that facilitate the passive and bidirectional exchange of ions and small solutes. Gap junctions are found in tissues that depend on coordinated activity, such as smooth and cardiac muscle and neural tissue.

Tight junctions are cell-cell junctions that prevent water and solutes from diffusing between cells and across a layer of epithelial cells. **Tight junction protein complexes** provide a connection preventing the passage of many particles and serve as a barrier to separate tissue space. Tight junctions are found in a number of tissues, including the skin and gastrointestinal tract. For example, tight junctions between cells of the gastrointestinal tract serve as a physical barrier preventing potentially harmful substances in the intestinal lumen (eg, microbes, toxins) from passing between cells.

Concept Check 5.3

A child is found to have an inherited genetic disorder that causes an accumulation of carbohydrates in various organs. A closer look reveals that there are increased numbers of small membrane-bound compartments containing a higher-than-normal concentration of certain carbohydrates. Which organelle does the mutation causing the disorder most likely affect, and why would the impairment of this organelle cause the accumulation of carbohydrates?

Solution

Note: The appendix contains the answer.

Lesson 5.4
Cell Growth and Division

Introduction

Cellular reproduction is at the heart of the modern cell theory, which holds that the cell is the structural and functional unit of all living things and that all cells come from pre-existing cells. A single cell cycle begins at the end of the previous cell's reproductive process (ie, cell division) and ends when that cell completes division and becomes two new cells.

Cell division serves as a method of reproduction for single-celled organisms. For multicellular organisms, cell division serves several purposes: to facilitate the growth and development of the organism and to replace old or damaged cells. This lesson provides a general understanding of what happens during the lifespan of a single cell (the cell cycle), focusing closely on the steps and regulation of cell division.

5.4.01 The Cell Cycle

The continuity of life depends on the successful reproduction of cells and organisms. During its life span, a cell passes through a set of highly regulated phases, known as the **cell cycle**. The most important function of the cell cycle is to facilitate the duplication and segregation of the cell's genetic information to form two identical daughter cells.

The two major phases of the eukaryotic cell cycle are defined as **interphase** and **mitotic (M) phase** (Figure 5.52). The cell spends approximately 90% of its life cycle in interphase, which can be further divided into three distinct parts: G_1 **phase** (first gap phase), **S phase** (synthesis), and G_2 **phase** (second gap phase). Each phase is characterized by distinctive cellular activities. Cell division in prokaryotic organisms occurs via a different mechanism known as binary fission, which is covered in Concept 6.2.01.

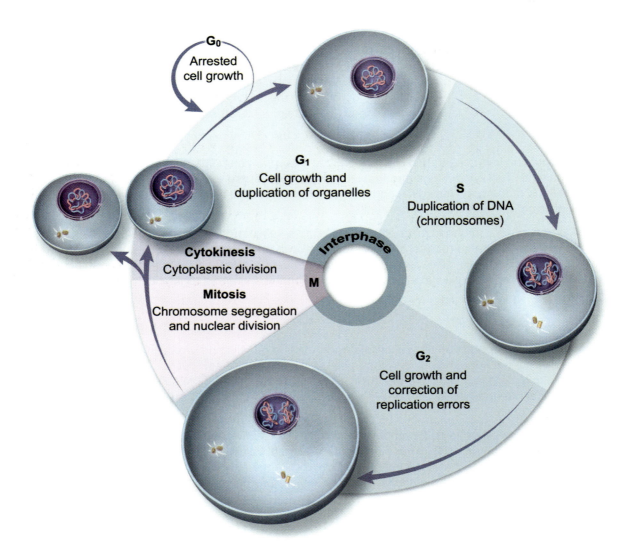

Figure 5.52 Major phases of the cell cycle.

During G_1 phase, the cell enters a growth period in which it produces proteins, stores energy, and assembles new organelles and other molecular machinery. If conditions are favorable, the cell transitions to S phase, during which its DNA is replicated (Figure 5.53). From the end of S phase until chromosomes are separated during mitosis, a duplicate copy of each chromosome is present.

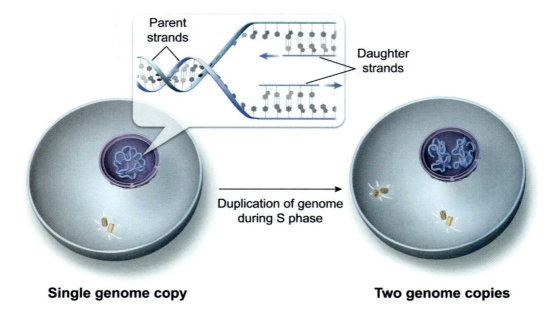

Figure 5.53 The genome is replicated during S phase.

During G_2 phase, the cell replenishes energy stores, synthesizes proteins, and assembles the molecular machinery required to prepare for cell division. Cellular organelles are prepared for partitioning into daughter cells, and any errors in DNA replication are repaired. Once the final preparations for cell division have been made, the transition to M phase can take place.

Two concurrent events are initiated during M phase: mitosis and cytokinesis. **Mitosis** involves the segregation of duplicated chromosomes (nuclear division), while the cell itself is pinched in half during **cytokinesis** (cytoplasmic division) to form two identical daughter cells.

In most cases, after completing G_1 phase the cell enters a specialized resting state known as G_0 **phase**, in which the cell cycle is arrested for a variable period of time depending on cell type and environmental conditions. Some cells may remain in G_0 permanently. For example, during neuronal differentiation, most differentiated neurons enter G_0 and permanently withdraw from the cell cycle.

5.4.02 The Mitotic Process

During the latter part of interphase (S and G_2 phases), the cell is committed to dividing and completes preparations for the final phase of the cell cycle, the **mitotic (M) phase**. At the beginning of mitosis, each duplicated chromosome consists of two identical **sister chromatids**. Each sister chromatid is made up of DNA and its associated proteins and contains a region of repetitive DNA sequences known as a **centromere**, which is where the two sister chromatids are joined, as shown in Figure 5.54.

At the end of mitosis, sister chromatids separate from each other and are pulled to opposite ends of the cell. Once chromatids separate (ie, during anaphase of mitosis), they are no longer considered sister chromatids, but individual (unduplicated) chromosomes, and one chromosome is partitioned into each of the two daughter cells. Therefore, at the beginning of the next cell cycle, each daughter cell contains an identical set of unduplicated chromosomes derived from the parent cell.

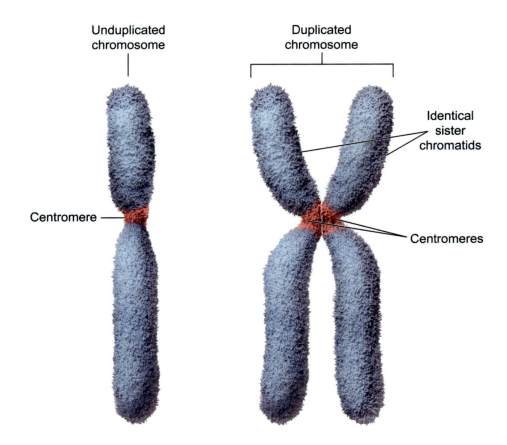

Figure 5.54 Unduplicated versus duplicated chromosomes.

Mitosis is conventionally broken down into four major stages:

1. **Prophase**: During this stage, chromosomes become heavily condensed due to chromatin compaction. At this time, individual chromosomes are observable under a light microscope and nucleoli are no longer visible. The nuclear envelope is broken down and the **mitotic spindle** is assembled. **Spindle fibers** are formed from microtubules originating from each centrosome (ie, microtubule organizing center).

   Spindle fibers on both sides of the duplicated chromosome (ie, one for each sister chromatid) bind to structures called **kinetochores**, which attach to the lateral region of each centromere near the end of prophase (Figure 5.55). During this time, nonkinetochore spindle fibers are also formed, and these fibers assist in elongation of the cell during the later stages of mitosis. As prophase continues, centrosomes migrate to opposite ends of the cell, partially propelled by the growing spindle fibers.

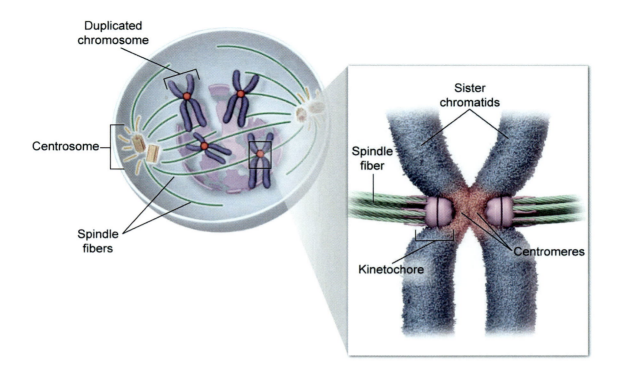

Figure 5.55 Formation of the mitotic spindle begins during prophase.

2. **Metaphase**: By the end of prophase, the centrosomes are located at opposite ends of the cell with chromosomes attached to the spindle fibers. During metaphase, chromosomes are positioned at the "equator" of the cell, along the **metaphase plate**. (Note: Some sources recognize **prometaphase**, an additional stage of mitosis between prophase and metaphase, which combines the events of late prophase and early metaphase.)

3. **Anaphase**: Sister chromatids detach from one another at the centromeres as spindle fibers begin to shorten, pulling the sister chromatids in opposite directions. This results in the separation of sister chromatids, with one identical set of chromosomes segregated to each side of the cell.

4. **Telophase**: Two daughter nuclei form as nuclear envelopes reform around each set of chromosomes. Chromatin condensation relaxes, resulting in less dense chromosomes, and nucleoli reappear. Any remaining spindle fibers are dismantled.

At the conclusion of telophase, the process of mitosis (ie, nuclear division) is complete, and the parent cell contains two genetically identical daughter nuclei (Figure 5.56).

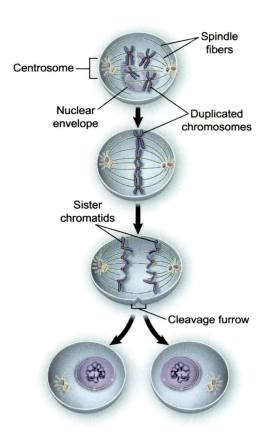

Figure 5.56 Stages of mitosis.

Cytokinesis, or cytoplasmic division, is considered a separate but concurrent event overlapping with mitosis. In animal cells, cytokinesis begins during late anaphase when the cell begins to elongate and a shallow groove called the **cleavage furrow** forms at the cell surface near the original location of the metaphase plate.

On the cytoplasmic side, the cleavage furrow is characterized by a contractile ring of microfilaments (actin) associated with myosin proteins. This association causes the ring to contract, progressively pinching the cell until the two new cells are completely separated from each other, each with its own nucleus, cytosol, and organelles (Figure 5.57). For organisms that contain a cell wall (eg, plants, algae), cytokinesis proceeds via an alternative mechanism.

Chapter 5: Eukaryotic Cells

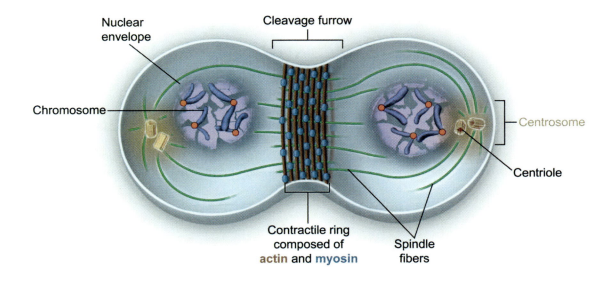

Figure 5.57 Cytokinesis involves the formation of a cleavage furrow and contractile ring.

✓ Concept Check 5.4

During which of the following stages of the cell cycle do cells contain duplicated chromosomes?

	Stage
1	G_1 phase
2	S phase
3	G_2 phase
4	Prophase
5	Metaphase
6	Anaphase
7	Telophase
8	Cytokinesis

Solution

Note: The appendix contains the answer.

5.4.03 Cell Cycle Control

The accurate timing and execution of cell cycle events is critical for organism growth, development, and reproduction. Dysregulation of the cell cycle control system can result in catastrophic effects, such as cell death or cancer development. Therefore, an intrinsic timing mechanism monitoring both internal and external cellular conditions is essential for proper cell cycle control. This mechanism includes regulatory

checkpoints (restriction points) that ensure proper execution of prior steps before entry into the next stage of the cycle.

The three most important cell cycle checkpoints occur at the G_1/S and G_2/M transitions and during M (mitotic) phase, as depicted in Figure 5.58. Checkpoints can be used as mechanisms to promote or inhibit cell proliferation. In healthy cells, entry into the next stage of the cell cycle is prohibited if there are abnormalities in the preceding stages.

Regulation of the G_1/S checkpoint is particularly crucial because it commits the cell to completing a cell cycle. During the G_2/M transition, entry into M phase is blocked if DNA replication is inaccurate or incomplete. If DNA is damaged by radiation or chemicals, progression to the next stage can be delayed until the damage is repaired. Likewise, during M phase, the separation of chromosomes at anaphase is delayed if chromosomes are not properly attached to the mitotic spindle.

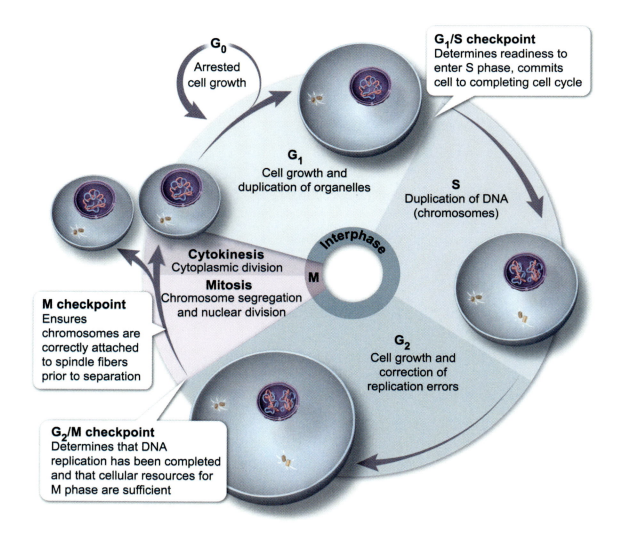

Figure 5.58 Important cell cycle checkpoints.

Cell cycle checkpoints are primarily regulated by proteins known as **cyclins** and **cyclin-dependent kinases (CDKs)**. The relative abundance and/or activity of these proteins fluctuate regularly during the cell cycle. As cyclins accumulate, they are activated by associating with a CDK partner. Activated cyclin-CDK complexes are able to phosphorylate a variety of proteins, advancing the cell through various cell cycle stages.

Cyclin proteins undergo cycles of synthesis and degradation during each cell cycle. Levels of individual CDKs remain constant throughout each cycle, but CDKs become active only when combined with the appropriate cyclin partner. Therefore, the activity of CDKs is dependent on the rise and fall of different cyclins at each cell cycle stage. Cyclin-CDK complexes are inactivated by the rapid proteolysis of the dominant cyclin at each stage, as shown in Figure 5.59.

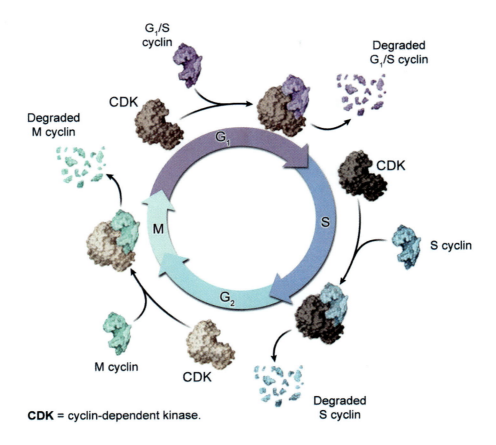

Figure 5.59 Cyclin-CDK complexes regulate cell cycle checkpoints.

Three major types of regulatory cyclins exist, and each is defined by the cell cycle stage in which it functions:

- **G_1/S cyclins** bind to CDKs at the end of G_1 and commit the cell to DNA replication.
- **S cyclins** peak during S phase and are required for DNA replication.
- **M cyclins** peak at the G_2/M transition and promote the events of mitosis.

In multicellular organisms, cell division and cell death must be counterbalanced to keep tissues and organs at an appropriate size. Therefore, regulation of cell death is equally important. **Apoptosis** (ie, **programmed cell death**) is a series of events that occurs as a result of this regulation.

Unlike cell death via necrosis, apoptotic cells shrink and condense without damaging neighboring cells. The apoptotic process is mediated by a cascade of proteolytic enzymes called **caspases**. These enzymes are synthesized as inactive precursors known as procaspases. When activated, caspases cleave and activate other key proteins, leading to DNA fragmentation and disassembly of the cell into subcellular fragments.

As the cell is dismantled, **apoptotic blebs** (ie, irregular bulges in the plasma membrane) are formed. As apoptosis progresses, the apoptotic blebs become separated from the cell undergoing apoptosis, forming

apoptotic bodies. Apoptotic bodies are phagocytosed, either by a neighboring cell or by specialized circulating phagocytic cells of the immune system, such as macrophages (Figure 5.60).

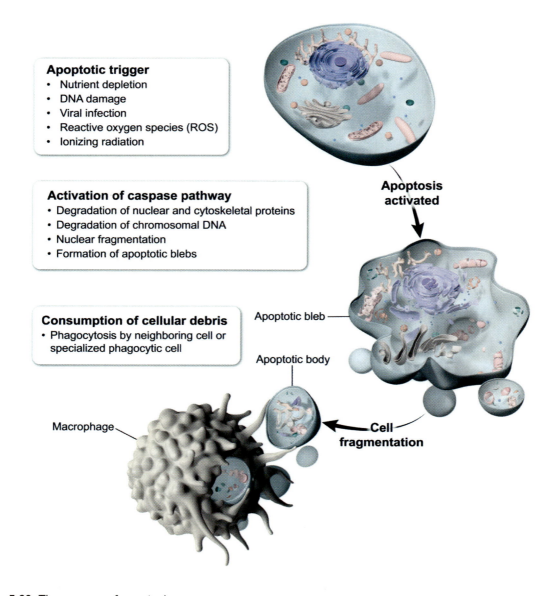

Figure 5.60 The process of apoptosis.

Both chemical and physical external controls also play a role in the regulation of cell division. For example, if essential nutrients are not present, the cell cycle can become arrested. In addition to favorable internal and external conditions, most mammalian cell types require specific growth factors to progress past the G_1/S checkpoint.

Most healthy cells exhibit **density-dependent inhibition**, in which the number of cells present influences the rate of division, and **anchorage dependence**, in which cells must be attached to a surface (eg, extracellular matrix, a culture flask) to divide. These normal limits may be bypassed when the cell cycle is dysregulated, such as in cancer cells, discussed in more detail in Concept 5.4.04.

5.4.04 Cancer

When signals and cell cycle checkpoints are ignored, cells may continue to divide inappropriately, and cancerous cells may develop. **Cancer** can arise when cells begin to make their own growth factors or when cells have abnormalities in the signaling pathways responsible for conveying signals to the cell cycle control system. Alternatively, cells may have defects in the cell cycle control system itself. Most cancers arise from the accumulation of *multiple* genetic defects in one or more of these pathways that allow for uncontrolled cell division without compensation from the normal rate of apoptosis (programmed cell death) (Figure 5.61).

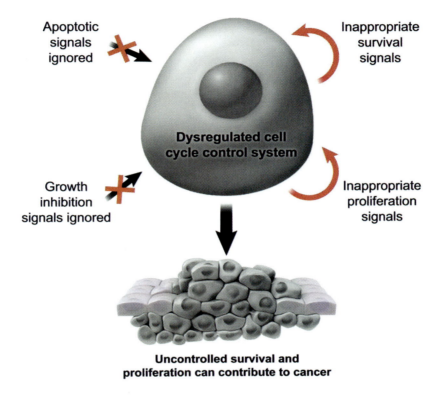

Figure 5.61 Multiple genetic defects can contribute to cell cycle dysregulation and cancer.

A typical cell divides approximately 20–50 times before undergoing programmed cell death. Under typical conditions, the shortening of telomeres (chromosomal ends) during progressive rounds of replication usually leads to a finite cellular life span, as discussed in Concept 1.2.04. Once telomeres shorten past a critical point, the cell cycle control system promotes **senescence**, or growth arrest (see Figure 5.62).

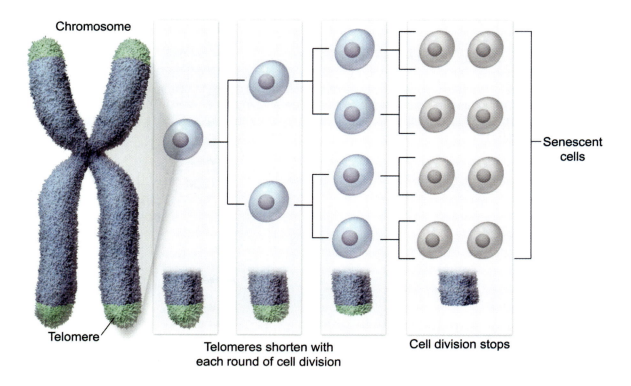

Figure 5.62 When telomeres shorten past a critical point, cells cease to divide.

In contrast to healthy cells, cancerous cells continue to proliferate beyond their expected lifespan. When this proliferation occurs indefinitely, cells are considered "immortal." Inappropriate expression of the enzyme **telomerase** is one reason cells are able to continue dividing beyond their expected lifespan. Telomerase replenishes chromosome ends by adding new telomeric DNA repeats at the ends of chromosomes. Therefore, in cells where telomerase is expressed, chromosome ends are constantly replenished with new telomeric repeats and cell division may occur indefinitely.

Telomerase consists of the protein subunit telomerase reverse transcriptase (TERT) and a type of noncoding RNA called telomerase RNA (TR). Using TR as a template, TERT extends telomeres by repeatedly adding the sequence 5'-TTAGGG-3' to chromosome ends. Once the telomere sequence reaches a sufficient length, DNA polymerase is able to synthesize a complementary DNA strand (Figure 5.63). Therefore, the extension of telomeres allows cells (eg, cancer cells) to avoid senescence. Senescence and aging are covered in more detail in Concept 10.5.02.

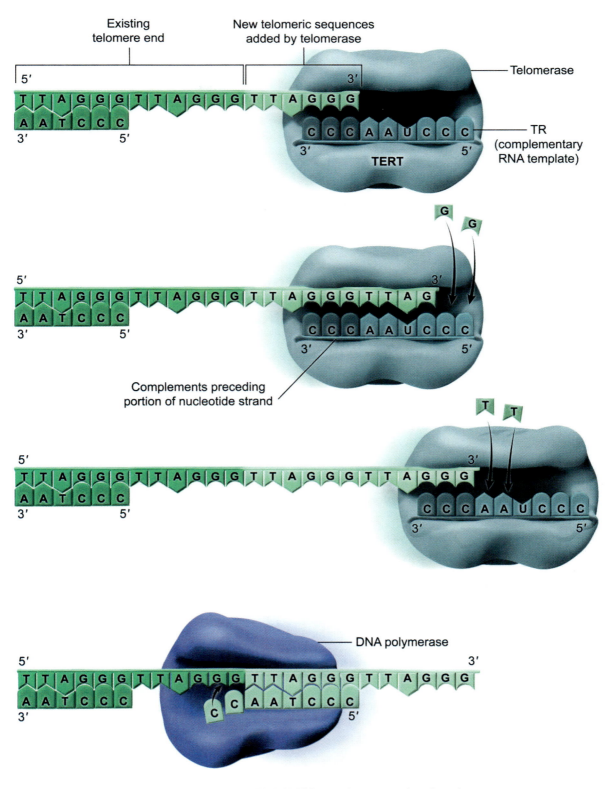

A noncoding telomerase RNA (**TR**) is used as a template by telomerase reverse transcriptase (**TERT**) to extend the ends of chromosomes (**telomeres**)

Figure 5.63 Telomerase extends telomeres.

The regulation of cell growth and division is controlled by many genes, such as genes that encode growth factors and their receptors, intracellular signaling pathways, and molecules responsible for cell cycle regulation. Mutations in any of these genes in somatic cells may lead to unregulated cell growth and ultimately cancer. These mutations may be inherited or can be acquired from environmental exposure to chemical or physical mutagens such as tobacco products, radiation, and certain viruses.

An **oncogene** is a mutated gene that has the potential to cause cancer. **Proto-oncogenes** are the wild-type version of these genes, and they code for proteins that promote regulated cell growth and appropriate division. An activating mutation in at least one allele of a proto-oncogene may convert the proto-oncogene into an oncogene and result in inappropriate activation of the gene or its protein product. Oncogenes promote uncontrolled cell growth and cancer, as shown in Figure 5.64.

Proto-oncogenes may become oncogenes via various mechanisms, including point mutations, epigenetic changes, and chromosomal rearrangements (see Lesson 2.3 and Lesson 7.3). In addition to point mutations in the coding regions of a proto-oncogene, mutations in regulatory regions (eg, promoter, enhancer) can also change gene expression levels (see Concept 2.4.03).

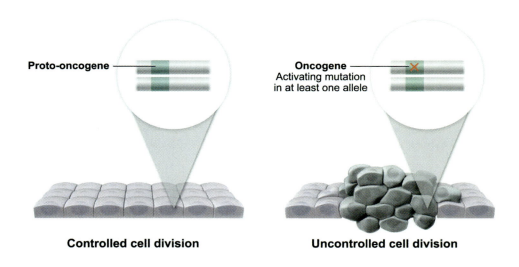

Figure 5.64 Conversion of a proto-oncogene to an oncogene.

Multiple classes of oncogenes exist, including genes that code for growth factors (eg, PDGF), growth factor receptors (eg, HER2), signal transduction proteins (eg, Ras), and transcription factors (eg, Myc). The **c-Myc** gene encodes a transcription factor activated in diverse signaling pathways under typical conditions. When c-Myc expression or activity is upregulated, genes involved in cell growth and metabolism are stimulated, leading to increased cell proliferation. Mutations in c-Myc are also associated with cancer progression from more benign to more invasive forms.

Genes that typically function to repress cell division are called **tumor suppressor genes**. Mutations leading to decreased tumor suppressor gene activity may also contribute to cancer development because cells become able to bypass checkpoint activities (Figure 5.65). Unlike proto-oncogenes, tumor suppressor genes typically require mutations in *both* alleles to have an effect on cell cycle regulation. In many cases, individuals inherit a mutation in one allele, and the second allele undergoes mutation at some point during the individual's lifetime, leading to loss of cell cycle control.

Tumor suppressor genes play various roles in cell cycle regulation. Some are involved in DNA repair and prevent the accumulation of mutations that promote cancer formation. Other tumor suppressor genes are involved in maintaining anchorage dependence and contact inhibition, which are often absent in cancerous cells (see Concept 5.4.03). Some of these genes are directly or indirectly involved in cell signaling pathways that control growth and proliferation.

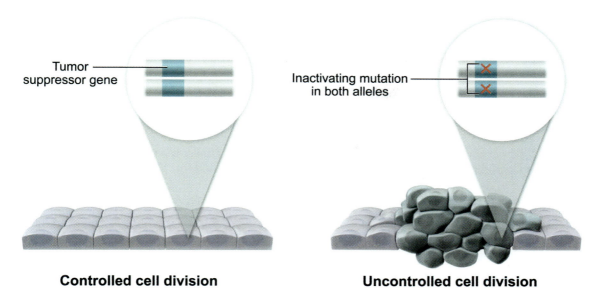

Figure 5.65 Inactivating mutations in both alleles of a tumor suppressor gene lead to uncontrolled cell growth.

The **p53** gene is a tumor suppressor gene that codes for a transcription factor activated in response to various distress signals, including DNA damage, hypoxia (ie, low oxygen levels), and nutrient deprivation. In response to these signals, p53 promotes appropriate response pathways, including activation of DNA repair enzymes and cyclin-dependent kinase (CDK) inhibitors, which pause the cell cycle until the threat has been appropriately addressed.

If cell damage is irreparable or a threat continues, p53 activates apoptotic pathways, resulting in damaged cells undergoing apoptosis so only undamaged cells finish the cell cycle. In the absence of p53 activity, these protective pathways are not activated, and cell cycle progression is uncontrolled.

 Concept Check 5.5

Genetic testing in a cancer patient reveals mutations in a gene that codes for a cell cycle regulation protein that blocks entry into S phase. In one allele, the gene is completely absent, and the other allele contains a mutation that results in loss of regulatory protein activity. Is this gene most likely to act as an oncogene or a tumor suppressor gene?

Solution

Note: The appendix contains the answer.

Lesson 5.5

Eukaryotic Tissues

Introduction

The evolution of multicellular organisms provided an opportunity for cells, the structural and functional units of all living organisms, to associate in specialized groups that perform distinct functions. These functional groups of cells are called **tissues**. This lesson covers the structure and function of the four tissue types found in animal cells: epithelial, connective, muscle, and nervous.

Different types of tissues can be organized to form organs and organ systems. For example, in the intestinal tract, epithelial cells are specialized for the passage of nutrients and water, connective tissue cells provide support, smooth muscle cells contract to aid in the propulsion and absorption of nutrients, and enteric nerves regulate motor functions and gastrointestinal enzyme secretion.

5.5.01 Types of Animal Tissues

In animals, there are four major types of tissues: epithelial, connective, muscle, and nervous (Table 5.3). The organization of each of these tissue types within specific organs and organ systems is discussed in more detail in subsequent chapters. Because nervous and muscle tissue are covered in detail in Chapters 12 and 17, this concept focuses on the characteristics and functions of epithelial and connective tissues.

Table 5.3 Types of animal tissues.

	Epithelial tissue	Connective tissue	Muscle tissue	Nervous tissue
Body tissue type				
Functions	• Lining of surfaces • Protection • Absorption, filtration, and secretion	• Support and protection • Attachment of tissues • Insulation and thermoregulation • Transport	• Voluntary movement (skeletal muscle) • Involuntary movement (cardiac muscle, smooth muscle)	• Transmission and reception of electrical signals • Response to internal and external stimuli (eg, sensory receptors)
Examples	• Skin • Lining of hollow organs (eg, digestive tract, trachea)	• Bone • Blood • Fat (adipose) • Tendons/ligaments • Cartilage	• Skeletal (attached to bone) • Cardiac (heart) • Smooth (hollow organs)	• Brain • Spinal cord • Nerves

Epithelial tissues make up outer and inner body surface linings such as the skin and inner surfaces of the digestive, respiratory, and reproductive tracts (Figure 5.66). The epithelial layer of skin is primarily protective in function, providing a barrier against physical, chemical, and microbial threats.

Epithelial cells that line inner body surfaces are collectively called **mucous membranes** because these cells secrete a lubricating substance called **mucus**. Mucus serves to coat and protect mucous membranes, while also participating in trapping pathogenic organisms and other debris. Epithelial cells lining inner body surfaces also participate in the secretion and absorption of various substances, as well as the excretion of wastes.

Epithelial cells are tightly joined to one another and to an underlying layer of connective tissue (ie, basement membrane). In general, epithelial tissues lack their own blood supply and must gain nutrients from underlying vascularized connective tissue. Rapid proliferation of epithelial cells facilitates replenishment of damaged or senescent cells.

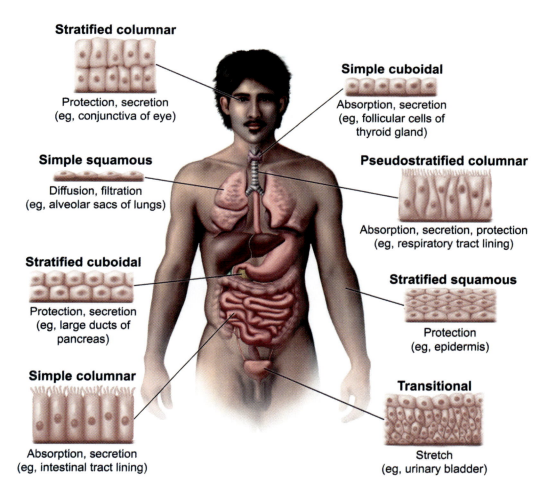

Figure 5.66 Classification and functions of epithelial tissues.

Classification of epithelial tissues is based on cell shape and number of cell layers present (Figure 5.66). While it is unlikely that an exam question would require memorization of epithelial cell classification, the terms used to describe the organization of epithelial tissues have been referenced in exam passages and questions. Tissue containing epithelial cells organized in a single layer is called **simple** epithelium. **Squamous** epithelium consists of thin and flat cells, while cube-shaped cells are called **cuboidal** epithelium. Tall, elongated cells are known as **columnar** epithelium.

Epithelial tissue consisting of two or more layers is called **stratified** epithelium. **Pseudostratified** epithelial cells appear to be organized in multiple layers due to differences in height, but this tissue type is actually composed of a single cell layer. **Transitional** epithelium is a stratified tissue that can appear to consist of cuboidal or squamous cells, depending on the degree to which the tissue is stretched.

Connective tissues are a diverse group of tissues that most often serve to bind and connect structures, thereby providing support and protection. These tissues may also provide structural frameworks, fill spaces, store energy in the form of fat, protect against infections, and help repair tissue damage.

Cells comprising connective tissues are less tightly associated with one another than epithelial cells; in addition, connective tissue cells are typically surrounded by a more extensive extracellular matrix than epithelial cells. Most connective tissues can proliferate and contain a blood supply to varying degrees, depending on the cell types involved.

Connective tissues may be either loosely or densely organized, and some connective tissues have specialized functions. Due to these characteristics, connective tissue is categorized as loose, dense, or specialized (Figure 5.67). **Loose connective tissues** include areolar, adipose, and reticular connective tissues, which support various organs. **Dense connective tissues** may be categorized as regular, irregular, or elastic, and are more fibrous in nature (eg, tendons, ligaments, walls of arteries). **Specialized connective tissues** include cartilage, bone, and blood.

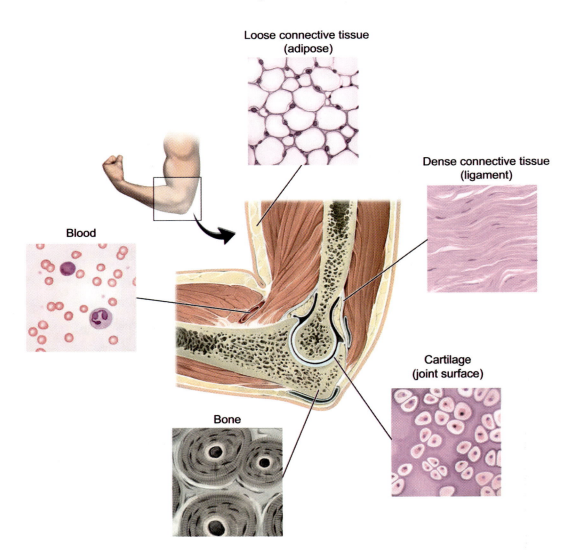

Figure 5.67 Types of connective tissue.

Lesson 6.1

Prokaryotic Cells

Introduction

While early scientists appreciated that plagues and disease could be passed from person to person, the causative agents for these diseases were difficult to identify until the invention of microscope lenses with sufficient **resolving power** (ie, the ability to distinguish two adjacent objects in an image) in the late seventeenth century. It would take another 200 years to develop the framework for what is now known as the germ theory of disease through the work of pioneers such as Louis Pasteur, Joseph Lister, and Robert Koch. The technology necessary to observe viral structures was not developed until well into the twentieth century.

After scientists were able to visualize microorganisms, a classification system based on morphology and cellular structures was devised. This classification system has expanded over time, and the current classification system takes both cellular structures and genetic similarities into account, placing all living organisms within three taxonomic domains (Bacteria, Archaea, and Eukarya).

As presented in Chapter 5, organisms within domains Bacteria and Archaea have a prokaryotic cell structure, whereas organisms within domain Eukarya have a eukaryotic cell structure. This lesson explores the classification and structural features of prokaryotic organisms, with an emphasis on members of the domain Bacteria.

6.1.01 Classification of Prokaryotes

Prokaryotes can be classified as bacteria or archaea based on their ribosomal RNA (rRNA) sequences. Archaea are similar to bacteria in appearance, but at the biochemical level there are notable differences, such as in their cell wall structures and plasma membrane components. While domains Bacteria and Archaea both include organisms that can be considered **extremophiles** (ie, known to thrive in extreme environments), a higher proportion of extremophiles are found within domain Archaea.

Archaea contain some physical features, genes, and metabolic pathways that are more eukaryotic than prokaryotic in nature. For example, some archaea contain chromosomes organized with histone proteins, making their chromosomal organization structure more similar to eukaryotes than prokaryotes.

A summary of the major differences among cell types of the three domains is provided in Table 6.1.

Table 6.1 Major differences between domains Bacteria, Archaea, and Eukarya.

Characteristics	Bacteria	Archaea	Eukarya
Cell walls	Contain peptidoglycan	Peptidoglycan absent, many variations	Carbohydrate-based (eg, cellulose, chitin)
Cell membranes	Contain phospholipids, bilayer	Contain phospholipids (distinct from bacterial), bilayer or monolayer	Contain phospholipids, bilayer
Genome	In general, a single circular chromosome, histones absent	Single circular chromosome, histones may be present	Multiple linear chromosomes, histones present
Ribosomes	70S	70S (distinct from bacterial 70S)	80S
Nucleus	Absent	Absent	Present
Membrane-bound organelles	Absent	Absent	Present
Cellularity	Unicellular	Unicellular	Unicellular or multicellular
Cell division	Binary fission	Binary fission	Mitosis, meiosis
Average cell size	0.1–5 µm	0.1–5 µm	10–100 µm

The primary features that distinguish prokaryotes from eukaryotes include cell wall and ribosome structure, size, and the general absence of a nucleus and other internal membrane-bound organelles. Unlike in eukaryotic cells, DNA in prokaryotic cells is found in a region of the cytosol known as the **nucleoid** and is usually composed of a single circular chromosome (Figure 6.1). In addition, prokaryotic cells may contain **plasmids**, which are small, circular DNA molecules that replicate independently of the chromosome.

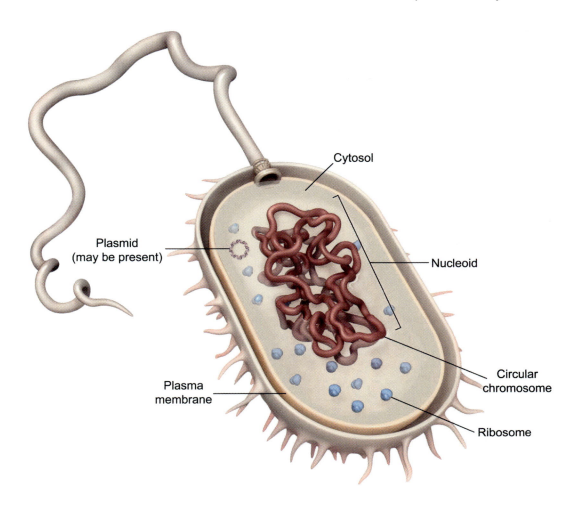

Figure 6.1 Major features of prokaryotic organisms.

The morphological classification of bacteria features three characteristic shapes, as seen in Figure 6.2. A spherical cell is called a **coccus** (plural: **cocci**) and may be observed in various arrangements, such as in chains (streptococci) or clusters (staphylococci). A rod-shaped cell is termed a **bacillus** (plural: **bacilli**) and may also appear in chains (streptobacilli). A spiral-shaped cell is called either a **spirillum** (plural: **spirilla**) or a **spirochete**, depending on factors such as cell thickness, number of twists per cell, and type of motility.

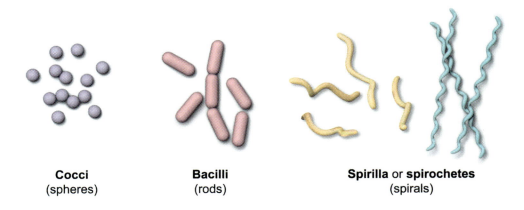

Figure 6.2 Morphological classification of bacteria.

Prokaryotic cells may be further classified based on distinguishing physical and biochemical characteristics such as cell wall type, external structures, and the presence or absence of metabolic pathways. These distinguishing characteristics are discussed throughout this chapter.

6.1.02 Cell Wall

Most prokaryotic organisms are surrounded by an external layer known as the **cell wall**. Unlike the plasma membrane, the cell wall is rigid and provides structure and an added layer of protection to the cell it surrounds. The cell wall contributes to the characteristic morphologies of bacterial cells (see Concept 6.1.01), and it plays a role in bacterial cells' ability to attach to other cells and resist antimicrobial drugs.

One of the most important functions of the prokaryotic cell wall is to prevent cells from rupturing due to osmotic forces. Although some eukaryotes (eg, plants, algae, fungi) also have cell walls, the chemical composition of prokaryotic cell walls is distinct from their eukaryotic counterparts. The two major bacterial cell wall types can be differentiated by performing a staining technique known as a Gram stain. Although not possible for all bacterial organisms, Gram staining is typically one of the first steps in bacterial organism identification and can provide important information regarding diagnosis and treatment of infectious diseases.

Gram-positive cells are surrounded by a cell wall composed of a relatively thick layer of peptidoglycan and appear purple after Gram staining. **Gram-negative** cells are surrounded by a cell wall composed of a thinner peptidoglycan layer along with an additional outer membrane and appear pink after Gram staining (Figure 6.3).

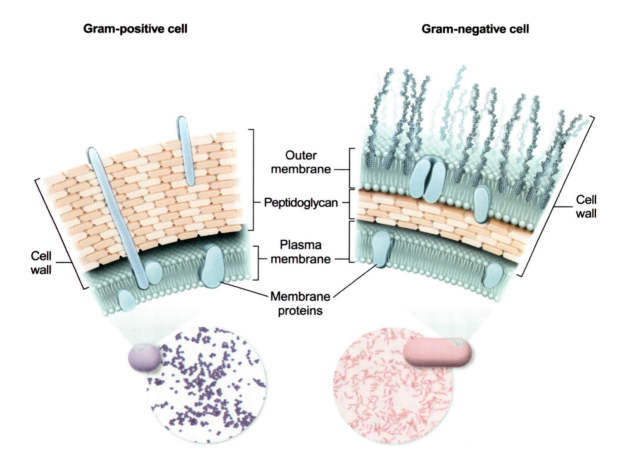

Figure 6.3 The two major bacterial cell wall types can be differentiated via Gram staining.

Chapter 6: Prokaryotes and Viruses

The major component of bacterial cell walls is **peptidoglycan**, a complex polysaccharide connected by short peptide crossbridges. The sugar backbone of peptidoglycan is composed of layers of two alternating sugar molecules, ***N*-acetylglucosamine (NAG)** and ***N*-acetylmuramic acid (NAM)**, with structures similar to glucose. The peptide portion of peptidoglycan is composed of short polypeptides that cross-link with NAMs in adjacent layers (Figure 6.4). These **peptide crossbridges** are composed of D and L amino acids.

As peptidoglycan layers are synthesized, new layers are cross-linked with existing layers to create a continuous, mesh-like cell wall. While peptidoglycan is always composed of the same elements (eg, NAG, NAM, peptide crossbridges), there is considerable diversity in the construction and placement of the connecting peptide crossbridges among species.

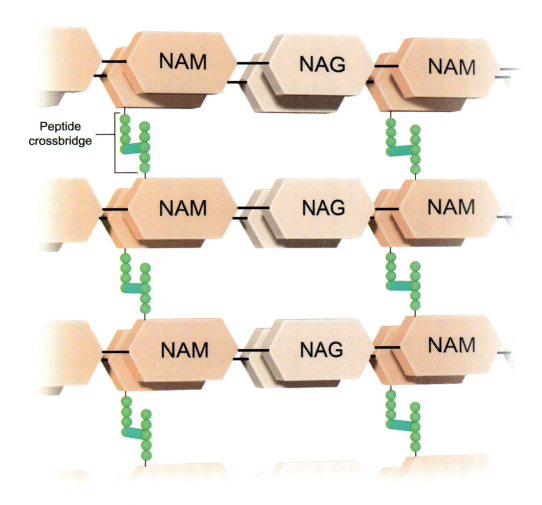

NAM = *N*-acetylmuramic acid; **NAG** = *N*-acetylglucosamine.

Figure 6.4 Possible configuration of peptidoglycan in a gram-positive cell wall.

Peptidoglycan layers in gram-positive and gram-negative cells have different levels of thickness. The peptidoglycan layer in a gram-positive organism is made up of many rows of peptidoglycan and has an average thickness of 30–100 nm. The peptidoglycan layer in a gram-negative organism is much thinner on average (5–10 nm) and contains only a few rows of peptidoglycan.

The peptidoglycan layer of gram-positive organisms is embedded with anionic polymers known as **teichoic acids**, which are anchored to peptidoglycan, and **lipoteichoic acids**, which are anchored to the plasma membrane (Figure 6.5).

Chapter 6: Prokaryotes and Viruses

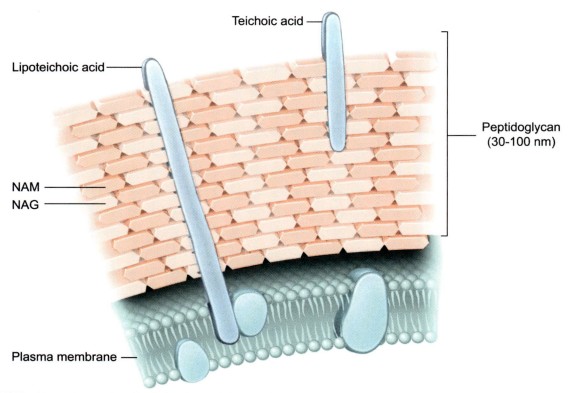

NAM = *N*-acetylmuramic acid; **NAG** = *N*-acetylglucosamine.

Figure 6.5 Gram-positive cell wall.

In addition to differences in peptidoglycan layer thickness, gram-negative cells possess an additional **outer membrane**, external to the peptidoglycan layer, that is not present in gram-positive cells. Notably, the outer leaflet of the outer membrane (ie, the portion facing the extracellular environment) is composed largely of lipopolysaccharides (LPS) rather than phospholipids (Figure 6.6). In gram-negative organisms, the outer membrane provides an additional selectively permeable barrier, mechanical support, and extra protection from lysis due to osmotic forces.

The space between the plasma membrane and the outer membrane (including the peptidoglycan layer) is called the **periplasmic space** and contains a variety of enzymes involved in nutrient binding and transport, peptidoglycan synthesis, and protection from substances harmful to the cell, such as antibiotics.

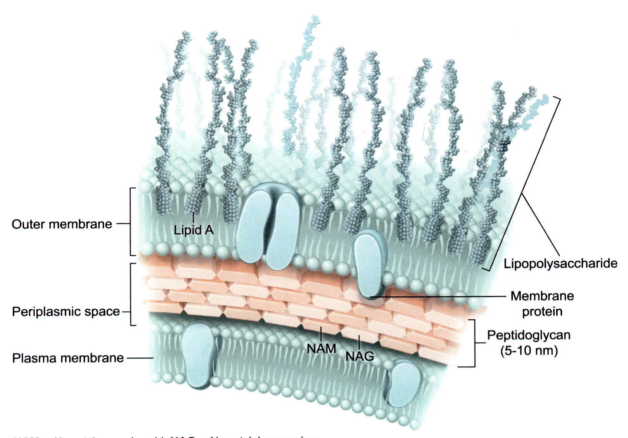

NAM = *N*-acetylmuramic acid; **NAG** = *N*-acetylglucosamine.

Figure 6.6 Gram-negative cell wall.

The lipid portion of LPS, known as **lipid A** (or **endotoxin**), can be toxic under certain conditions. In intact gram-negative organisms, the lipid A portion of the LPS molecule is embedded within the outer membrane. However, if the cell wall is damaged, lipid A may be released. The release of lipid A from dead cells is medically significant because it can trigger a condition known as **endotoxic shock**, which causes fever, inflammation, vasodilation, blood clotting, and circulatory shock. The release of large amounts of lipid A from dead bacterial cells may pose a greater risk than the infection caused by living cells.

Many antibiotics that target the cell wall work by interfering with the linkage of peptide crossbridges between layers of peptidoglycan. For example, β-lactam antibiotics inhibit *new* peptidoglycan synthesis by inhibiting an enzyme (penicillin-binding protein) that catalyzes peptide cross-linking between layers, thereby weakening newly synthesized peptidoglycan, as depicted in Figure 6.7. The weakened peptidoglycan cannot withstand osmotic forces, leading to cell lysis.

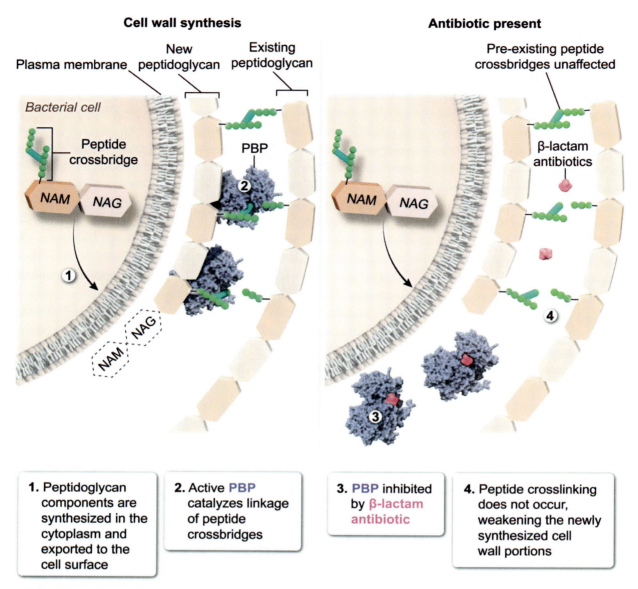

Figure 6.7 β-lactam antibiotics inhibit the formation of peptide cross-links during peptidoglycan synthesis, weakening the cell wall.

One feature that distinguishes archaea from bacteria is the absence of peptidoglycan in archaean cell walls. There are a variety of archaean cell wall types, which often include unique polysaccharide and protein components.

While most prokaryotes possess cell walls, there are some prokaryotic populations that have no cell walls, including members of the genus *Mycoplasma* and related organisms. The lack of rigid cell walls makes mycoplasmas highly **pleomorphic**, meaning these prokaryotes can readily change shape.

6.1.03 Bacterial Structures

In addition to the cell wall, bacterial cells may contain additional structures (both internal and external). For example, some species of bacteria may synthesize an additional carbohydrate coating known as a glycocalyx, contain external appendages such as flagella and fimbriae, and/or form protective endospores under harsh conditions.

Some bacteria secrete a **glycocalyx**, an external coating composed mainly of polysaccharides. A **slime layer** is a glycocalyx loosely attached to the cell wall, and a **capsule** is a more firmly attached glycocalyx. The presence of a capsule can be an important determinant in whether a bacterium can cause disease (ie, a **virulence factor**). For example, the presence of a capsule can prevent bacteria from being phagocytosed by a host organism (Figure 6.8).

Glycocalyces are also an important factor in the formation of **biofilms**, which allow cells to attach to one another and various surfaces in the natural environment (eg, mucosal surfaces), as shown in Figure 6.8. Biofilms confer a survival advantage by providing protection against various threats (eg, antimicrobial substances, detection by the host immune system) and are a factor in most chronic bacterial infections.

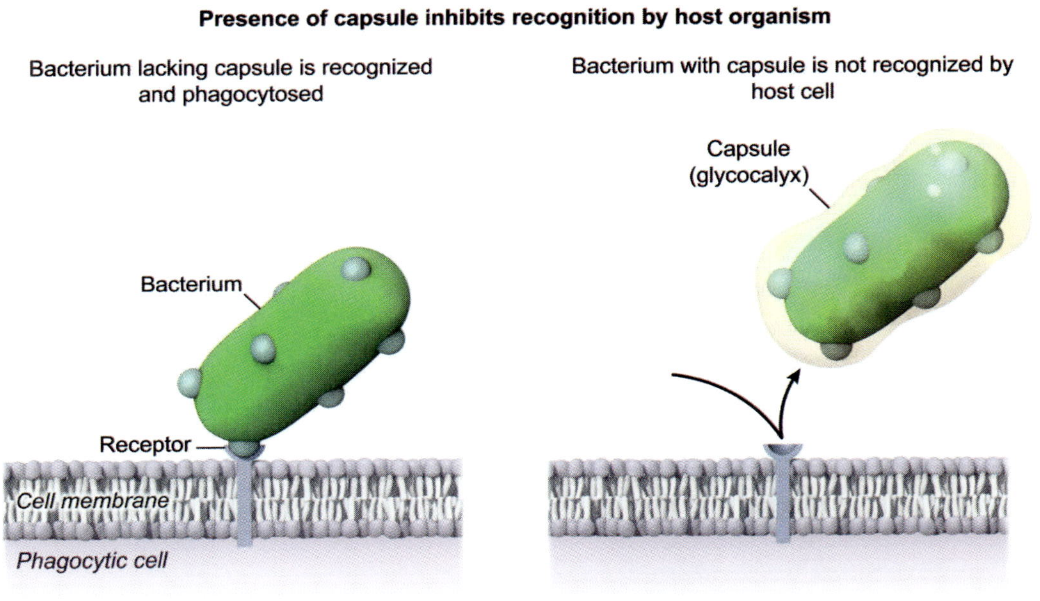

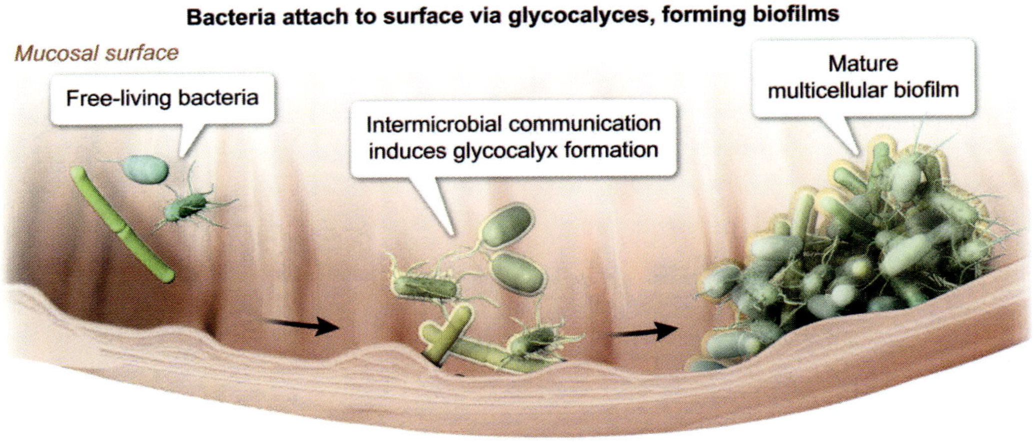

Figure 6.8 Role of glycocalyces in bacterial protection.

Two types of surface appendages can be present in bacterial species: fimbriae and flagella. Common **fimbriae** (singular: **fimbria**) are thin, hairlike appendages composed of the protein pilin, and as many as 200 fimbriae can be distributed over the cell surface (Figure 6.9). The presence of common fimbriae is a virulence factor, mediating attachment of bacteria to host cells/tissues, and common fimbriae are therefore of medical importance.

A specialized form of fimbriae, called **conjugation (sex) pili** (singular: pilus) are distinct from common fimbriae. Conjugation pili are longer and fewer in number (1–6 per cell) than common fimbriae. Conjugation pili are involved in both motility and transfer of genetic material between cells (Figure 6.9). During the process of conjugation, the pilus acts as a tube that mediates DNA transfer between two bacteria, as discussed in Concept 6.3.03.

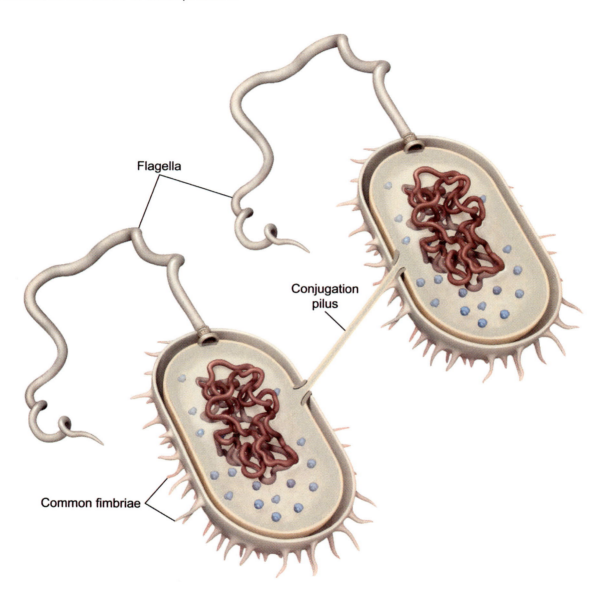

Figure 6.9 Common fimbriae and conjugation pilus.

Bacterial **flagella** (singular: **flagellum**) are long appendages involved in bacterial movement. Not all bacteria have flagella, but those that do are capable of various patterns of motility, allowing flagellated bacteria to move toward favorable environments and away from unfavorable environments. Flagella may be distributed over the entire cell or may be polar (ie, at one or both ends of the cell).

Bacterial flagella are structurally distinct from eukaryotic flagella, which are composed of microtubules (discussed in Concept 5.3.06). A bacterial flagellum consists of three basic parts: a basal body, a hook, and a long tubular filament that extends out of the cell into the extracellular space (Figure 6.10). Unlike eukaryotic flagella, the filament in prokaryotic flagella is not covered by a membrane.

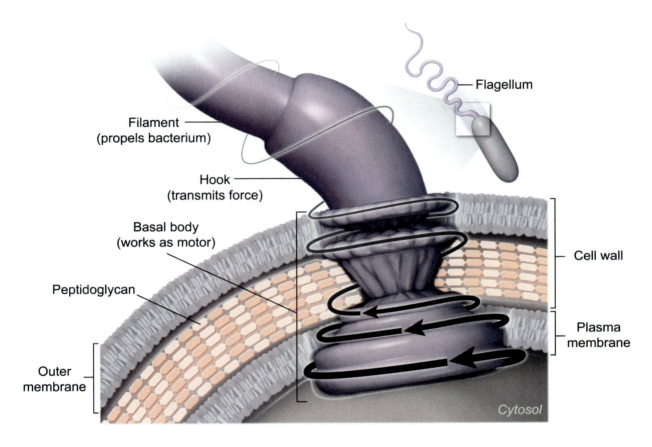

Figure 6.10 Bacterial flagellum in a gram-negative organism.

The **filament** portion of a flagellum is composed of identical subunits of a protein called **flagellin** that surround a hollow core. At its base, the filament is inserted into a curved structure called the **hook**. The **basal body** anchors the hook and filament to the cell wall and plasma membrane (Figure 6.10). Together, the hook and basal body generate the torque required for the filament to rotate 360°. The presence or absence of functional flagella can play a role in an organism's virulence.

Unlike eukaryotic flagella, which propel cells with an undulating motion, prokaryotic flagella propel bacteria with 360° rotation, and the direction of rotation determines the direction of movement. *Counterclockwise* rotation produces a **run**, or continuous movement in a single direction. If there is more than one flagellum, the flagella rotate as a bundle in a coordinated fashion. *Clockwise* rotation results in short, uncoordinated movements of individual flagella, causing **tumbles**, which are abrupt, random changes in direction. When no stimulus is present, organisms alternate between runs and tumbles, producing random movements.

Movement in response to a stimulus is called **taxis**. Specifically, movement in response to a chemical gradient is called **chemotaxis**, and movement in response to light is called **phototaxis**. Movement toward a favorable stimulus is called **positive taxis**, and movement away from an unfavorable stimulus is called **negative taxis**. When an organism senses a positive stimulus (eg, a chemical attractant), the organism increases the duration of runs and decreases the number of tumbles so its net movement is toward the stimulus (Figure 6.11).

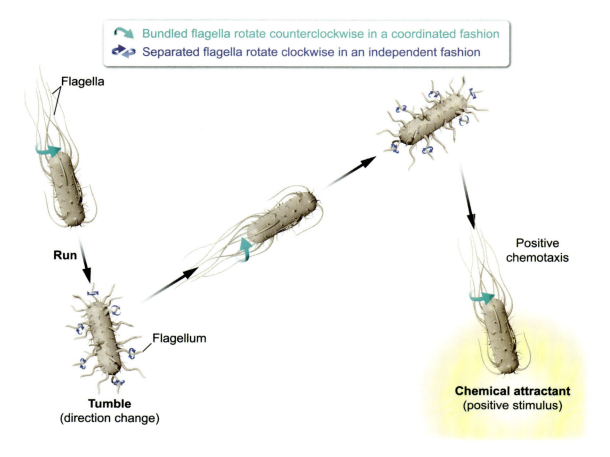

Figure 6.11 Coordinated flagellar rotation toward a chemical attractant.

Some gram-positive bacteria form specialized structures called **endospores** (or simply **spores**) under certain conditions. Bacterial endospores are not reproductive structures but are formed when conditions are unfavorable (eg, nutrient scarcity). As part of an organism's stress response, a bacterium replicates its genome, and a spore coat is synthesized to protect the bacterial genome until environmental conditions improve.

During **sporulation**, a series of coatings are synthesized in layers over a copy of the bacterial genome inside the mother (vegetative) cell. The spore's inner core is formed when two membranes separated by a thick layer of peptidoglycan are deposited around a copy of the bacterial genome. A tough **spore coat** is then deposited on the external surface. This spore coat is composed primarily of proteins, including tough keratin-like proteins, making the spore resistant to a wide variety of physical and chemical conditions. Once sporulation is complete, cell lysis is induced and the spore is released into the environment (Figure 6.12).

Chapter 6: Prokaryotes and Viruses

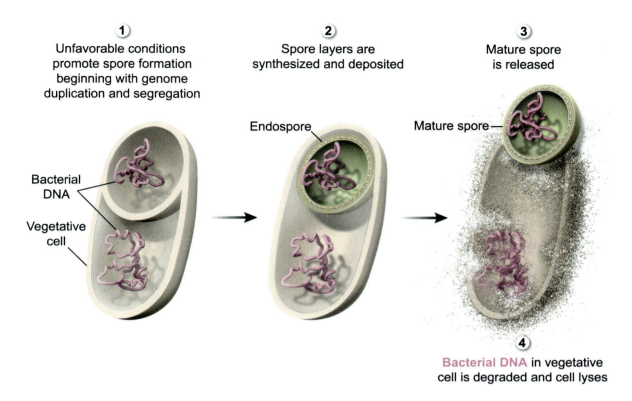

Figure 6.12 Spore formation is induced during unfavorable conditions.

Once released, spores can remain dormant for thousands of years, persisting despite extreme conditions. When environmental signals indicate improved conditions, **germination** of the spore is triggered and the cell is returned to a vegetative (ie, active) state. Disease-causing bacteria capable of forming endospores are medically relevant because these bacteria are resistant to antibiotics, most disinfectants and chemicals, and many physical control methods (eg, heat, dehydration, radiation). The chemical composition of the spore coat, which confers relative impermeability, is a key factor in this resistance.

✓ Concept Check 6.1

Categorize the following cell wall components as belonging to gram-positive and/or gram-negative organisms. Some components may be present in both cell wall types.

- Lipid A
- Lipopolysaccharide (LPS)
- Lipoteichoic acid
- *N*-acetylglucosamine (NAG)
- *N*-acetylmuramic acid (NAM)
- Peptide crossbridge
- Peptidoglycan
- Outer membrane
- Teichoic acid

Solution

Note: The appendix contains the answer.

Lesson 6.2
Growth and Reproduction of Prokaryotes

Introduction

Microbes such as bacteria and archaea can increase in number significantly in a relatively short period of time. To gain control over the growth of microbes that cause disease and food spoilage, an understanding of microbial reproductive mechanisms and the physical and chemical requirements for microbial growth is necessary. The concepts in this lesson provide an overview of bacterial cell division and a discussion of factors that influence bacterial growth.

6.2.01 Binary Fission

In the microbial world, the term *growth* usually refers to reproduction or an increase in cell number, *not* an increase in the size of an individual cell. **Binary fission** is the process by which microbes (eg, bacteria, archaea) reproduce asexually. During binary fission, the parent cell doubles in size and divides into two identical daughter cells. Each daughter cell then divides into two new cells, and the process continues in a similar manner.

Binary fission can be described in four steps, as outlined in Figure 6.13:

1. Replication of the bacterial chromosome (DNA) begins at the origin of replication.

2. As DNA replication occurs, the two bacterial chromosome copies are segregated toward opposite ends of the cell as the cell elongates.

3. The cell begins the process of **septation** by constricting the plasma membrane at the cell midline. Septation is followed by the synthesis of peptidoglycan at the cell's midpoint to form a **septum**.

4. When septum formation is complete, the daughter cells may separate completely or remain attached (ie, in chains or clusters).

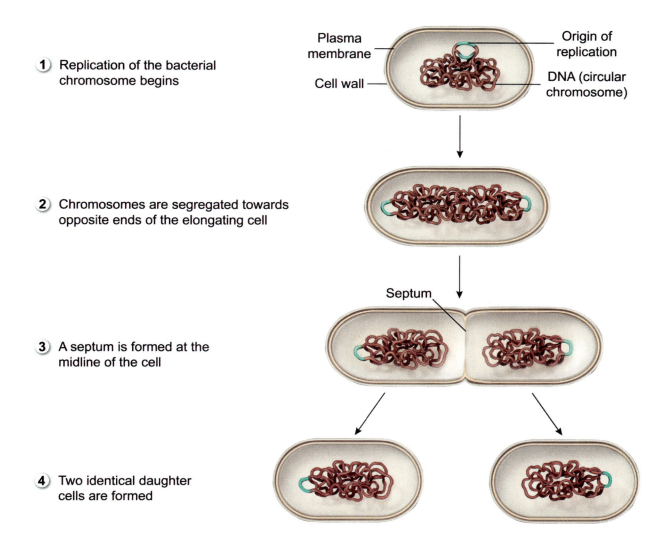

Figure 6.13 Steps of binary fission.

Although both prokaryotic binary fission and eukaryotic mitosis (discussed in Concept 5.4.02) end in the formation of two identical daughter cells, the mechanisms of each process are distinct. In both cases, chromosomal DNA is copied and segregated, and the cell's cytoplasm divides to form two new cells. However, unlike mitotic chromosome segregation in eukaryotic cells, chromosome segregation in prokaryotes is a continuous process in which the chromosomes are separated *as replication occurs*, without the formation of a mitotic spindle.

Prokaryotic reproduction is asexual, so there is no mechanism for genetic recombination between two parent genomes, as there is in eukaryotic sexual reproduction (see Concept 7.1.04). However, genetic information can be transmitted between prokaryotic organisms in a process known as **horizontal gene transfer**, which is discussed in Concept 6.3.03.

6.2.02 Bacterial Growth

The time required for a single prokaryotic cell to complete binary fission (ie, reproduction) is called its **generation time**. Generation time may also refer to the amount of time required for a *population* of cells to double in number and is sometimes referred to as **population doubling time**.

Each type of prokaryotic organism has a unique set of physical and chemical requirements that influence generation time. Under optimal conditions, some organisms (eg, *Escherichia coli*) have a generation time as short as 15–20 minutes. Slow-growing organisms (eg, *Mycobacterium leprae*) may have generation times measured in days rather than in hours or minutes.

When one cell (generation 0) divides by binary fission, the product is two new cells (generation 1). Continued division of each of these cells results in four new cells (generation 2) and so on, as shown in Figure 6.14. This type of growth is **exponential** (ie, growth in which the number of organisms doubles at a constant rate) and is most often expressed on a logarithmic scale rather than on an arithmetic scale.

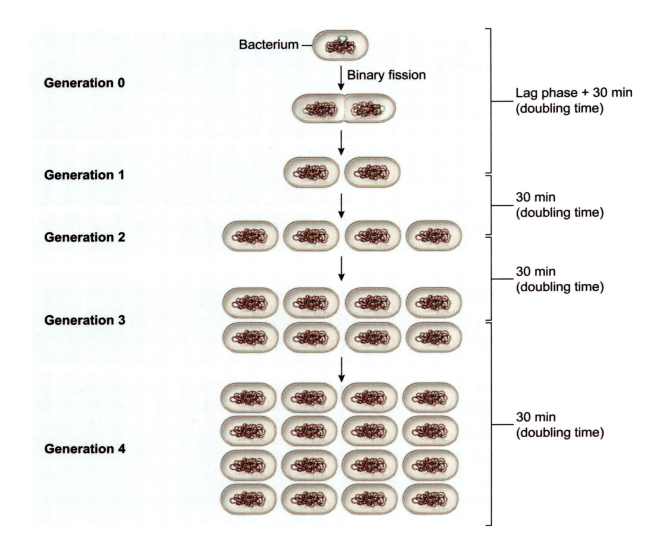

Figure 6.14 Exponential growth.

When a single cell divides by binary fission, the resulting number of cells after a specified amount of time is calculated by multiplying the original number of cells by 2^n, where n equals the number of generations. For example, if a single cell were allowed to multiply for 5 generations, the resulting number of cells would be 32:

$$\text{Number of cells after } n \text{ generations} = \text{Original cell number} \times 2^n$$

$$\text{Number of cells after 5 generations} = 1 \text{ cell} \times 2^5$$

$$\text{Number of cells after 5 generations} = 32 \text{ cells}$$

Due to exponential growth, the number of cells after 20 generations would be 1,048,576 (Figure 6.15):

Number of cells after n generations = Original cell number × 2^n

Number of cells after 20 generations = 1 cell × 2^{20}

Number of cells after 20 generations = 1,048,576 cells

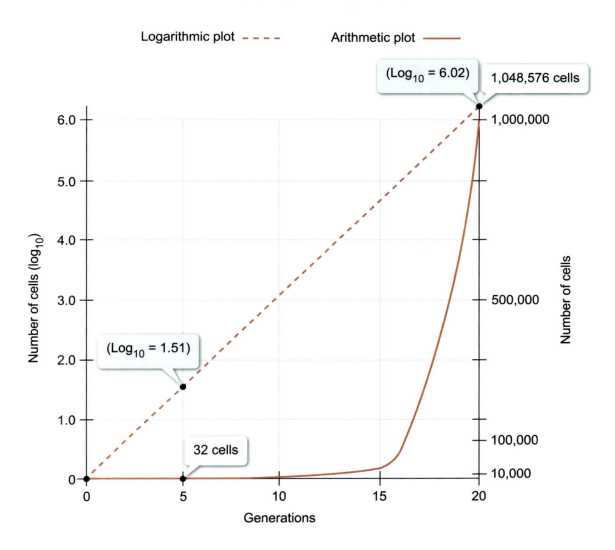

Figure 6.15 Exponential growth plotted on logarithmic and arithmetic scales.

When bacterial growth is plotted on a logarithmic scale, four distinct phases can be identified: the lag, log, stationary, and death phases (Figure 6.16).

Whether in a natural environment or in a laboratory setting, when a bacterial population is moved to new growth conditions, cell division does not begin immediately. The initial period without cell division is called the **lag phase** and the length of the lag phase may vary depending on environmental conditions. Although reproduction does not occur during the lag phase, metabolic activity is high as cells become acclimated to a new environment and synthesize components required for growth (eg, proteins, RNA).

As bacteria become accustomed to the surrounding conditions, binary fission begins, and cells enter a period of exponential (or logarithmic) growth called the **log phase**. During the log phase, generation time is at its shortest, occurring at a constant rate, and metabolic activity is at its peak. Bacteria are typically most sensitive to antibiotics during the log phase. For example, the rate of cell wall synthesis is at its peak when bacteria are growing exponentially, so antibiotics which affect cell wall synthesis (eg, penicillin and its derivatives) are most effective when bacteria are in the log phase.

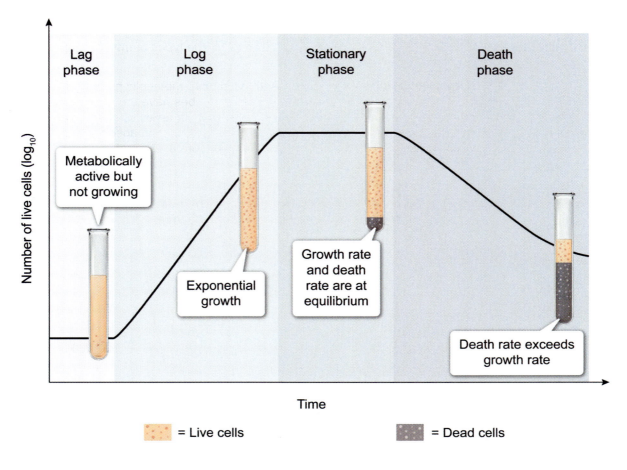

Figure 6.16 Bacterial growth curve.

Eventually environmental nutrients become depleted, metabolic waste products accumulate, and the bacterial growth rate slows. The number of dying cells and the number of new cells being produced reach an equilibrium, and the rate of growth stabilizes. This period is known as the **stationary phase**.

When the death rate exceeds the growth rate, the population enters the **death phase**. The number of viable cells diminishes dramatically unless waste products are removed and new nutrients are added. Although the *growth* of the population declines during the stationary and death phases, metabolically active (ie, live) cells are still present in the population in both phases.

6.2.03 Aerobic and Anaerobic Metabolism

The rate of microbial growth is closely linked to an organism's metabolic capabilities and environmental conditions. Environmental factors such as temperature, pH, and availability of oxygen and nutrients impact metabolism and, therefore, the growth of microbial populations.

Organisms can be described by their mode of nutrition and energy source. Most prokaryotic organisms that infect humans are **heterotrophs**, meaning that they acquire carbon from other organisms to synthesize organic molecules. However, some prokaryotes are **autotrophs**, meaning that they obtain carbon from inorganic sources (eg, carbon dioxide). Prokaryotes can also obtain energy from more diverse sources in either chemical (**chemotrophs**) or light (**phototrophs**) form. Therefore, prokaryotic organisms have a greater diversity of environmental conditions in which they may thrive than do most eukaryotic organisms.

Organisms that require oxygen are known as **obligate aerobes**, and organisms that do not require oxygen are called **anaerobes**. Unlike obligate aerobes, which cannot survive without oxygen, many anaerobes are able to survive and thrive in various oxygen concentrations, including a complete lack of oxygen. Ultimately, an organism's oxygen requirement is the result of the types of metabolic pathways acting in that organism.

Obligate (ie, strict) **anaerobes** and **aerotolerant anaerobes** do not have metabolic pathways capable of utilizing oxygen (ie, aerobic metabolism). However, aerotolerant anaerobes have the necessary enzymes to detoxify reactive oxygen species (ROS) and can survive in aerobic environments, whereas obligate anaerobes cannot. **Facultative anaerobes** can maintain life in the presence and absence of oxygen but prefer the efficiency of aerobic metabolism when oxygen is present (Figure 6.17).

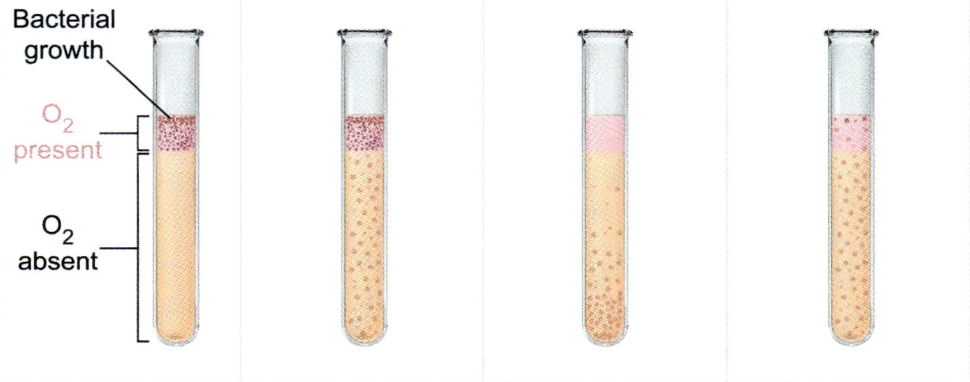

	Obligate aerobe	Facultative anaerobe	Obligate anaerobe	Aerotolerant anaerobe
Conditions for growth	Aerobic conditions required	Aerobic conditions preferred but not required	Anaerobic conditions required	Either aerobic or anaerobic conditions
Final electron acceptor	O_2	O_2 or alternate	Alternate; cannot use O_2	Alternate; cannot use O_2

Figure 6.17 Classification of organisms by oxygen requirement for growth.

Both eukaryotic and prokaryotic organisms catabolize glucose via **cellular respiration** and/or **fermentation**. Both pathways begin with **glycolysis**, during which a molecule of glucose is split into two pyruvate molecules and a small amount of ATP is produced. Glycolysis takes place within the cytosol in both eukaryotes and prokaryotes.

In eukaryotes, pyruvate is decarboxylated in the mitochondria to form acetyl-CoA, which enters the **citric acid (Krebs) cycle**. The citric acid cycle reduces the electron carriers NAD$^+$ and FAD, generating NADH and FADH$_2$, which pass electrons to the **electron transport chain (ETC)** in the inner mitochondrial membrane to oxygen, the final electron acceptor.

As electrons move along the ETC, membrane protein complexes pump protons into the intermembrane space, generating a proton gradient. Protons are then transported down this gradient through the enzyme **ATP synthase** and into the mitochondrial matrix, where ATP synthesis occurs (Figure 6.18).

Prokaryotic cells typically do not contain membrane-bound organelles such as mitochondria. However, prokaryotes carry out aerobic cellular respiration by completing the citric acid cycle in the cytosol and by using the plasma membrane for the ETC, as shown in Figure 6.18. In prokaryotes, ETC protein complexes and ATP synthase are embedded within the plasma membrane, and the proton gradient forms outside the plasma membrane. Protons diffuse back into the cell through ATP synthase, and ATP is synthesized on the cytosolic side of the membrane.

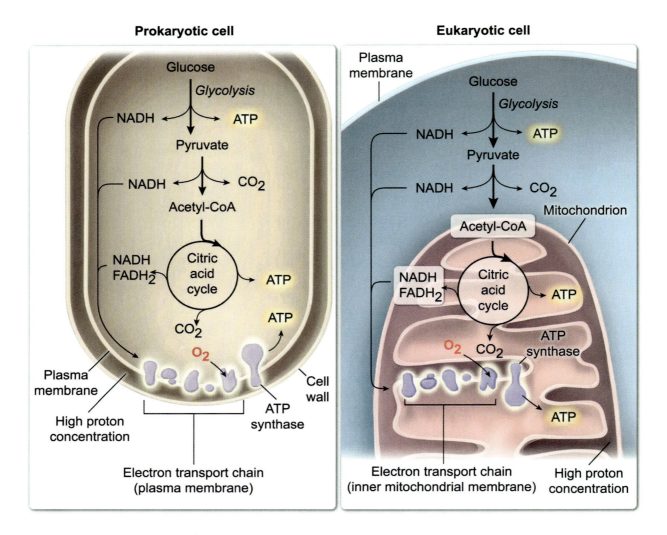

Figure 6.18 Comparison of prokaryotic and eukaryotic cellular respiration.

Aerobic respiration is the preferred metabolic pathway for organisms capable of utilizing oxygen because it yields the highest amount of ATP per glucose molecule. Under anaerobic conditions, obligate aerobes cannot complete aerobic respiration because oxygen is not present to serve as the final electron acceptor.

Without oxygen as the final electron acceptor, cellular respiration cannot continue, so NADH molecules cannot be oxidized to NAD$^+$. In anaerobic conditions, eukaryotes and prokaryotes can also use fermentation to regenerate NAD$^+$ so glycolysis can continue and a small amount of ATP can be produced (Figure 6.19).

Most eukaryotic organisms are limited to either lactic acid fermentation (eg, animals) or ethanol fermentation (eg, yeast). In contrast, prokaryotic organisms have more diverse fermentation pathways and may generate a variety of fermentation products (eg, lactic, acetic, and formic acids, ethanol, acetone), depending on the organism.

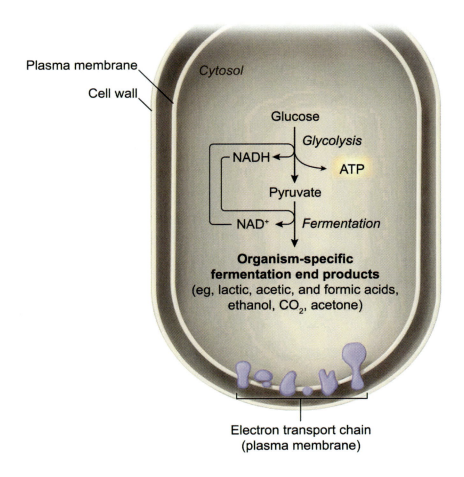

Figure 6.19 Fermentation.

Most eukaryotes are obligate aerobes, so under anaerobic conditions they often utilize fermentation temporarily to continue metabolic activities. In contrast, due to their metabolic diversity, many prokaryotes are not restricted to oxygenated environments and can thrive in anaerobic conditions. In some cases, anaerobic conditions are strictly required (ie, obligate anaerobes).

In addition to fermentation, some anaerobic organisms can perform **anaerobic cellular respiration**. These organisms can utilize the ETC in anaerobic environments by using inorganic compounds other than oxygen as the final electron acceptor for the ETC (Figure 6.20). Some organisms can use molecules containing nitrogen (eg, nitrate, nitrogen gas), sulfur (eg, sulfate, hydrogen sulfide), or carbon dioxide as a final electron acceptor instead of oxygen. Because anaerobic respiration is less efficient than aerobic respiration, ATP yield is not as high, and organisms typically grow more slowly under anaerobic conditions.

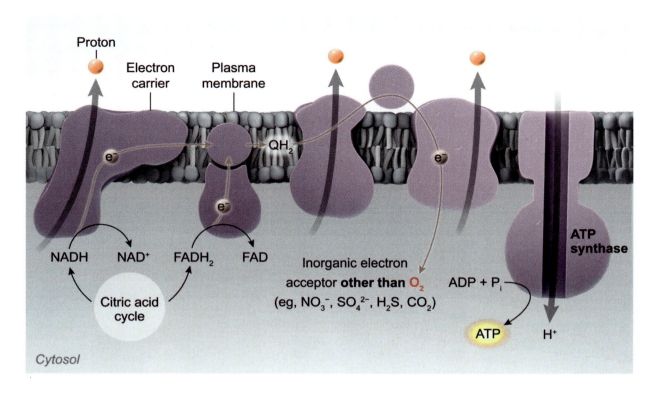

Figure 6.20 In anaerobic cellular respiration, the final electron acceptor is an inorganic molecule other than oxygen.

6.2.04 Symbiotic Relationships

Microbes thrive in a variety of habitats (eg, on land, in water), often in association with other living organisms. A close association between two different organisms can be described as **symbiosis**, and symbiotic interactions can be classified based on the outcome of these relationships (Figure 6.21).

When organisms derive a benefit from their association with one another, the interaction is known as **mutualism**. For example, some bacteria (eg, certain strains of *E. coli*) in the human gut synthesize vitamin K, which is used by the human to support blood clotting and bone health. In return, the intestinal tract of the human provides nutrition and protection to these strains of *E. coli*, helping these organisms survive.

In **commensalism**, one organism derives a benefit (eg, shelter, nutrition) from the association, and the other organism is neither harmed nor helped. For example, there are many bacteria (ie, **gut flora**) that benefit from living in the human gut but neither provide a known measurable benefit nor cause any known harm to the human.

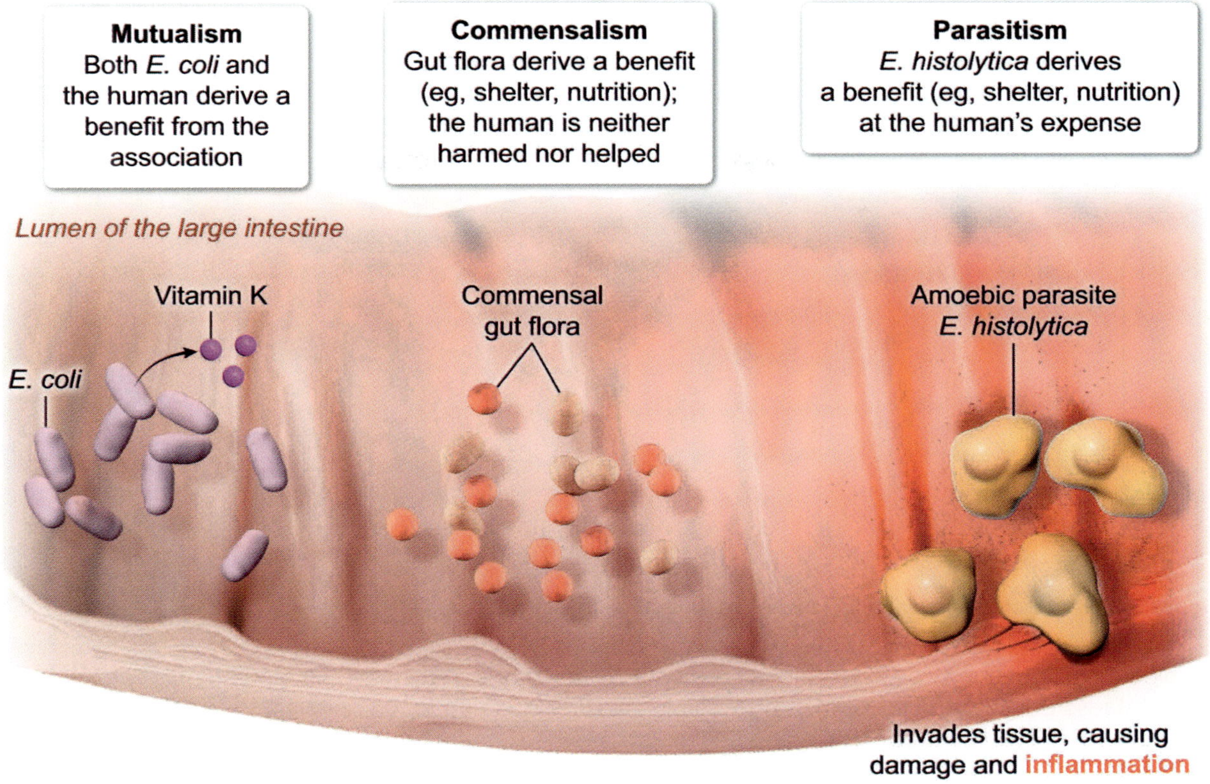

Mutualism
Both *E. coli* and the human derive a benefit from the association

Commensalism
Gut flora derive a benefit (eg, shelter, nutrition); the human is neither harmed nor helped

Parasitism
E. histolytica derives a benefit (eg, shelter, nutrition) at the human's expense

Lumen of the large intestine

Vitamin K
E. coli

Commensal gut flora

Amoebic parasite *E. histolytica*

Invades tissue, causing damage and **inflammation**

Figure 6.21 Examples of symbiotic relationships in the human gut.

In **parasitism** (also known as **antagonism**), one organism derives a benefit at the other organism's expense, harming the other organism. In some cases, the parasite may kill the host organism, but this is often not to the parasite's advantage. An example of a parasitic interaction is infection caused by the organism *Entamoeba histolytica*, which leads to amoebic dysentery. This type of infection causes harm to the human host, but the parasite derives a benefit by gaining shelter within the host and access to host resources.

A parasitic organism that causes disease in the host, such as *E. histolytica*, is considered **pathogenic**. Although some viruses may not cause harm to the host, all viruses are considered **obligate intracellular parasites** because viruses *require* a host to reproduce (see Lesson 6.4).

6.2.05 Antibiotic Resistance

Prior to the development of antibiotics (ie, antibacterial drugs), the death rate from infectious disease was very high. The introduction of antibiotics did not come until well into the twentieth century and is considered one of the highest impact breakthroughs in modern medicine. However, the steep increase in antibiotic-resistant organisms has become a serious public health concern (see Concept 4.5.01).

Some organisms are naturally resistant to certain antibiotics. For example, bacteria lacking a cell wall (eg, mycoplasmas) are naturally resistant to antibiotics that target peptidoglycan synthesis, such as penicillin (see Concept 6.1.02). In other cases, organisms may acquire antibiotic resistance by random mutation of chromosomal genes or by gaining antibiotic resistance genes from other organisms through **horizontal gene transfer** (eg, transformation, transduction, conjugation), which is discussed in more detail in Lesson 6.3.

The use (and overuse) of antibiotics has led to the development of **antibiotic-resistant** strains through the process of natural selection (see Concept 8.1.05). When antibiotics are not present, a bacterial population may consist of a mix of susceptible and resistant cells. After antibiotic introduction, only the susceptible cells are affected, and the resistant cells survive and proliferate (Figure 6.22). Therefore, the presence of the antibiotic does not *cause* resistance; it applies selective pressure to the population, and the subpopulation of cells that were *already resistant* to the antibiotic are able to thrive preferentially.

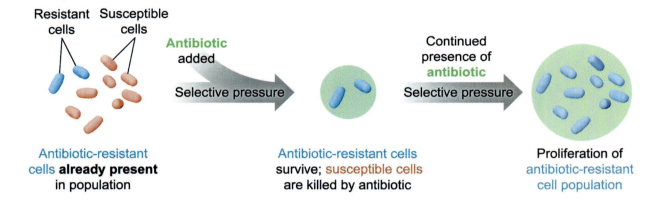

Figure 6.22 Antibiotic-resistant populations are enhanced via natural selection.

Antibiotic resistance may be caused by chromosomal mutations that inhibit the binding of an antibiotic, making it less effective. More commonly, resistance develops due to the acquisition of antibiotic resistance genes from other organisms via horizontal gene transfer of resistance (R) plasmids.

Antibiotic resistance genes typically code for proteins that prevent an antibiotic from working (eg, enzymes that inactivate or degrade an antibiotic, efflux pumps that remove an antibiotic from a cell), as shown in Figure 6.23. For example, organisms that acquire the gene coding for β-lactamase are resistant to antibiotic drugs containing a β-lactam ring, such as penicillin. The β-lactamase enzyme catalyzes degradation of the β-lactam ring, causing the drug to be ineffective at inhibiting cell wall synthesis. R plasmids often contain multiple antibiotic resistance genes, so the acquisition of a single plasmid can lead to **multiple antibiotic resistance**.

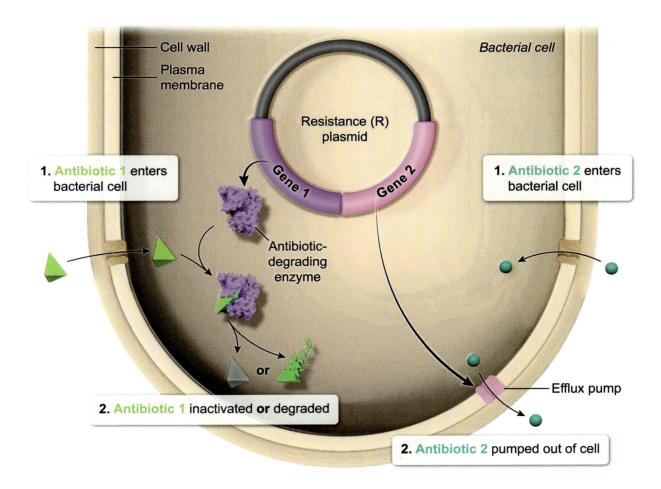

Figure 6.23 Antibiotic resistance mechanisms.

Antibiotic resistance is often investigated using a disk diffusion test, in which disks saturated with different antibiotics are placed on an agar plate inoculated with a bacterial culture. A growth-free halo called a **zone of inhibition** typically forms around disks saturated with an antibiotic to which the bacteria are susceptible due to high local antibiotic concentrations. A zone of inhibition does not form if the bacteria are antibiotic-resistant because the bacteria are not prevented from growing near the antibiotic-saturated disks.

Concept Check 6.2

The graph below depicts a typical bacterial growth curve. Use the following terms to label the blank spaces indicated in the graph: "Death phase," "Lag phase," "Log phase," "Number of live cells ($\log_{10}$)," "Stationary phase," and "Time."

In addition, indicate the phase during which a gram-positive organism would be most susceptible to an antibiotic that inhibits cell wall synthesis.

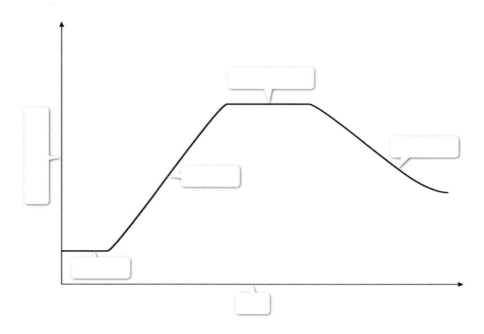

Solution

Note: The appendix contains the answer.

Lesson 6.3
Prokaryotic Genetics

Introduction

Like eukaryotes, prokaryotes store genetic information within double-stranded DNA chromosomes and follow the central dogma of molecular biology to express genetic information. However, prokaryotes exhibit several key differences from eukaryotes in genome organization and gene regulation. This lesson discusses the organization of the prokaryotic genome, mechanisms of horizontal gene transfer, and regulation of prokaryotic gene expression.

6.3.01 Prokaryotic Genome

The genome of most prokaryotes consists of a single chromosome, which is typically circular. Unlike eukaryotic chromosomes, bacterial chromosomes are not packaged with histones, but histones have been found to be associated with some archaean chromosomes. The chromosome in prokaryotic cells is typically located in an area of the cytoplasm known as the **nucleoid**, shown in Figure 6.24.

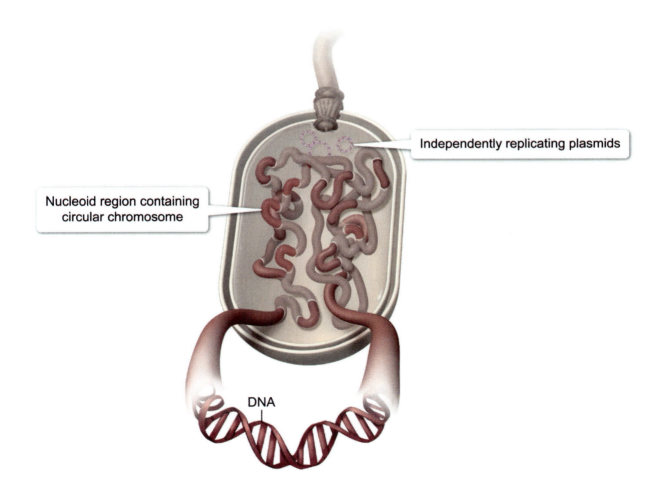

Figure 6.24 Features of prokaryotic genomes.

Each circular chromosome typically has a single origin of replication. Because circular chromosomes lack ends, these chromosomes do not contain telomeres. Prokaryotes typically contain a single copy of each chromosome, so most prokaryotes are considered **haploid** organisms.

6.3.02 Plasmids

Prokaryotic organisms may contain small, extrachromosomal DNA molecules known as **plasmids** (Figure 6.25) in addition to a circular chromosome. Like prokaryotic chromosomes, plasmids are circular and composed of double-stranded DNA; however, plasmid DNA is replicated independently of the prokaryotic chromosome. The number of genes contained in a plasmid can vary greatly, and plasmids may be present in low or high copy numbers. Plasmid-encoded genes are typically not essential for an organism's survival but may provide a survival advantage under certain conditions (eg, presence of antibiotic).

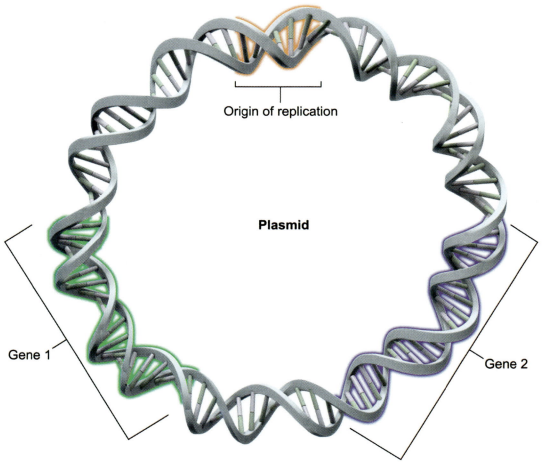

Plasmid features:
- Small, circular molecule composed of double-stranded DNA
- Replicates independently of the genome
- May be present in high or low copy numbers
- Generally contains nonessential genes which may provide a survival advantage

Figure 6.25 Plasmids are small extrachromosomal DNA molecules.

A variety of plasmid types have been identified, including:

- **Resistance (R) plasmids**, which contain genes for resistance to antibiotics or other toxic substances, such as heavy metals (see Concept 6.2.05)
- **Fertility (F) plasmids**, which contain genes needed for conjugation (ie, DNA transfer between two cells, see Concept 6.3.03)
- **Virulence plasmids**, which contain genes that code for enzymes, toxins, or other structures that allow an organism to cause disease (eg, adhesins, which are proteins that allow bacteria to attach to specific tissues).

Plasmids or genes located in plasmids may be transferred through methods of horizontal gene transfer (eg, conjugation, transformation, transduction).

6.3.03 Horizontal Gene Transfer

Prokaryotic organisms are haploid and reproduce asexually, but they can share genetic information through a process known as **horizontal gene transfer**. Horizontal gene transfer may occur between members of the same or different prokaryotic species. The three most common methods of horizontal gene transfer are conjugation, transformation, and transduction.

Conjugation is a process by which DNA is transferred between two bacteria joined temporarily via a connecting tube called a **conjugation (sex) pilus**. The genetic information necessary for the formation of a conjugation pilus is contained in a **fertility (F) plasmid** (also called an **F factor**). Cells that contain an F plasmid are designated as F^+ **cells** and can initiate conjugation, whereas cells that do not contain an F plasmid are designated as F^- **cells**. Conjugation is unidirectional and can be initiated only by an F^+ cell toward an F^- cell.

During conjugation, the F^+ (ie, donor) cell extends its pilus, attaching to the F^- (ie, recipient) cell. One DNA strand of the double-stranded F plasmid is transferred through the pilus from the donor cell to the recipient cell. Once transfer is complete, a complementary DNA strand is synthesized in both the recipient and donor cells, restoring each plasmid to a double-stranded form (Figure 6.26).

1. A donor (F^+) cell initiates conjugation by extending its pilus to the recipient (F^-) cell.

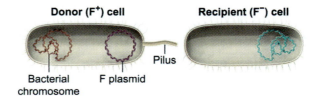

2. One DNA strand from the F plasmid is transferred through the pilus to the recipient cell.

3. After DNA polymerase synthesizes a complementary strand in each cell, both cells are F^+.

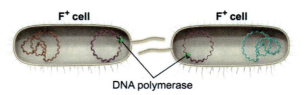

Figure 6.26 Conjugation.

In some cells, the F plasmid may be integrated into the bacterial chromosome. These cells are considered **high frequency of recombination (Hfr) cells** because when conjugation is initiated with an F⁻ cell, Hfr cells often carry portions of the donor chromosome into the recipient. In this way, Hfr cells can transfer chromosomal genes along with the F plasmid genes into a recipient cell and create a **genetically recombinant** organism. In most cases, a full copy of the F plasmid is not transferred from the Hfr cell, and the recipient cell remains F⁻, unable to initiate conjugation with other bacterial cells (Figure 6.27).

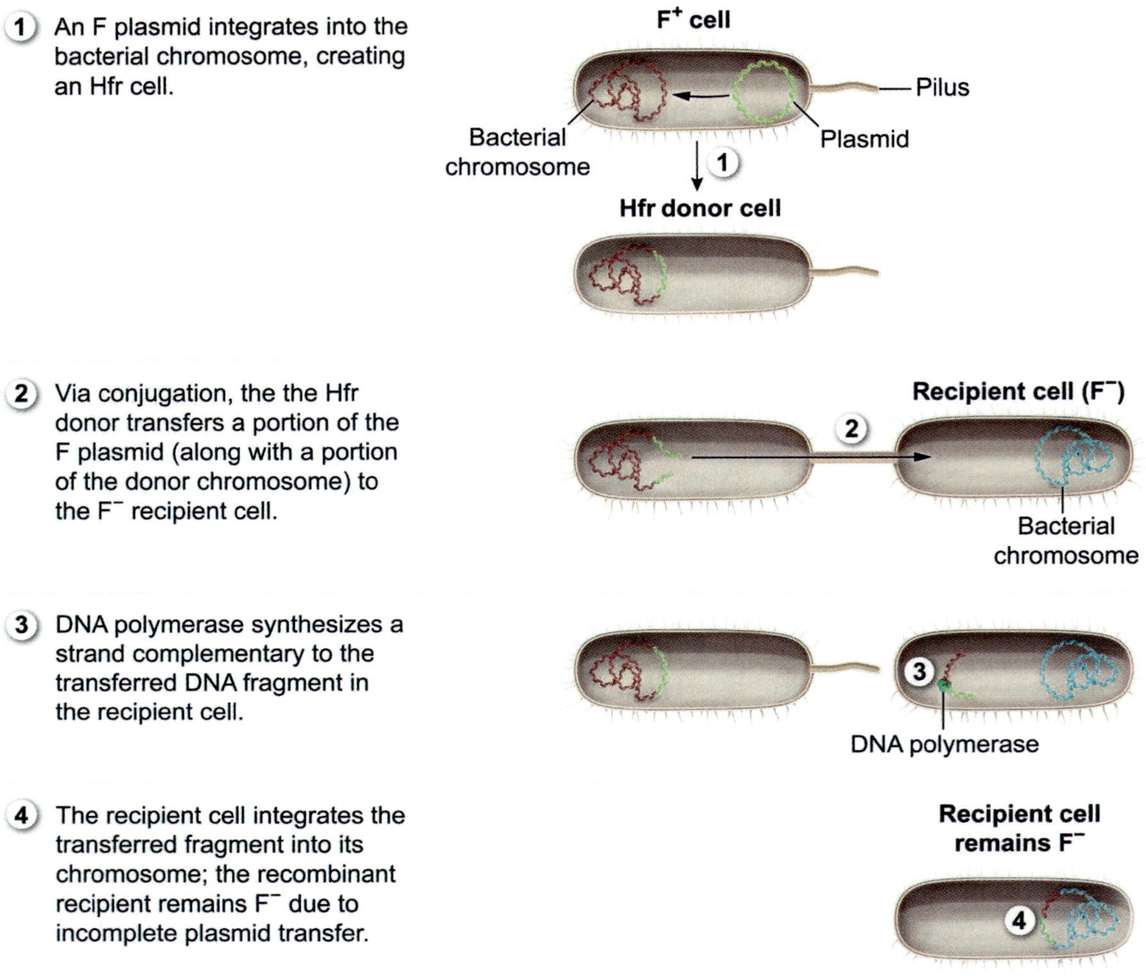

① An F plasmid integrates into the bacterial chromosome, creating an Hfr cell.

② Via conjugation, the the Hfr donor transfers a portion of the F plasmid (along with a portion of the donor chromosome) to the F⁻ recipient cell.

③ DNA polymerase synthesizes a strand complementary to the transferred DNA fragment in the recipient cell.

④ The recipient cell integrates the transferred fragment into its chromosome; the recombinant recipient remains F⁻ due to incomplete plasmid transfer.

Figure 6.27 High frequency of recombination conjugation.

A second method of horizontal gene transfer is **transformation**, in which cells take up DNA (either a linear fragment or a plasmid) from the extracellular environment (eg, via rupture of another cell) through membrane pores or transport proteins. Unlike conjugation, transformation does not require physical joining of donor and recipient cells, and cells able to take up DNA via transformation are called **competent**. Cells may be naturally competent or may be made artificially competent through laboratory manipulation.

If a bacterial cell takes up a linear fragment of DNA from the extracellular environment, the fragment may be integrated into the cell's circular chromosome by genetic recombination (Figure 6.28). This newly integrated DNA is then replicated and maintained in future generations (ie, when the cell undergoes binary fission). If the transferred DNA is a complete plasmid, there is no need for recombination to maintain the plasmid's presence in the bacterial cell because plasmids replicate independently of the bacterial chromosome.

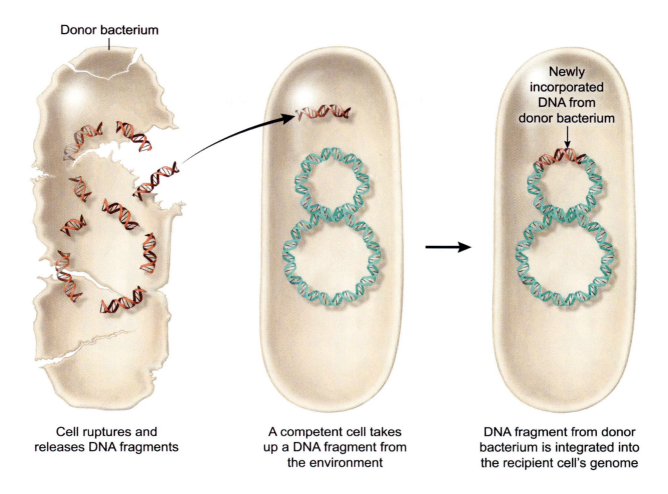

Figure 6.28 Transformation.

Bacterial DNA can also be transferred between two cells with the aid of a viral intermediate via **generalized transduction**, a third method of horizontal gene transfer. **Bacteriophages** (or **phages**) are viruses that infect bacteria. Bacteriophages have a complex protein structure consisting of a capsid head (containing the genetic material), a tail sheath, and tail fibers. The tail sheath allows a phage to inject its viral genome into a bacterial host while the remainder of the viral structures are left outside the bacterium; only the phage DNA enters the host cell.

Once the viral genome enters a host cell, viral replication is initiated. During this process, viral enzymes that direct the replication and assembly of new viruses and the degradation of host bacterial DNA (see Concept 6.5.01) are synthesized. As new viruses are assembled, fragments of bacterial (donor) DNA may be incorporated in place of the viral genome. Phages that take up donor bacterial DNA fragments are called **transducing phages**.

Transducing phages are released by lysis of the infected cell and go on to inject the donor bacterial DNA into new recipient cells. A recipient cell may then integrate bacterial donor cell DNA into the recipient's own chromosome via recombination, as shown in Figure 6.29.

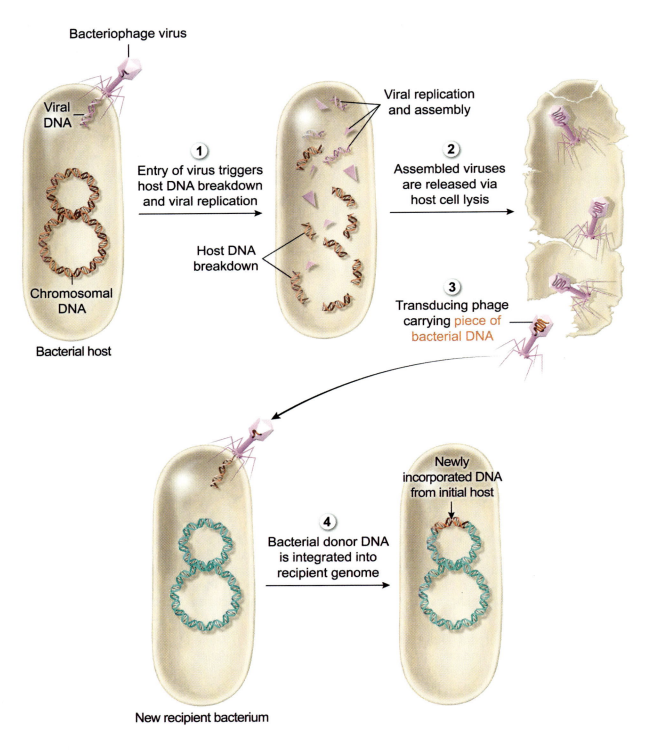

Figure 6.29 Generalized transduction.

> **Concept Check 6.3**
>
> A scientist was studying horizontal gene transfer via conjugation in bacteria. The scientist modified an F plasmid by adding a gene (AmpR) which codes for ampicillin (ie, an antibiotic) resistance to track AmpR movement between cells. After mixing cultures of *E. coli* containing the modified F plasmid with ampicillin-sensitive cultures of *Shigella* (a different species), the scientist did not observe any ampicillin-resistant *Shigella* and concluded that there was no transmission of the modified F plasmid from *E. coli* to *Shigella*. What is a possible reason that the scientist did not observe conjugative F plasmid transfer?
>
> **Solution**
>
> Note: The appendix contains the answer.

6.3.04 Transposons

A **transposon** (or **transposable element**) is a mobile genetic element that can move around the genome, inserting itself in random locations within the genome. Transposons have been observed in most organisms (prokaryotic and eukaryotic), and while transposon origin is undetermined, some transposon types may be linked with certain viruses. In prokaryotes, transposons can be transferred between organisms via horizontal gene transfer (Concept 6.3.03).

Transposons encode the enzyme **transposase**, which must be expressed for the transposon to become active. The mechanism of transfer can be either copy-and-paste (ie, a copy of the transposon is made and inserted at a new location in the genome) or cut-and-paste (ie, the transposon is removed from one location and inserted in a new location). Both mechanisms are illustrated in Figure 6.30.

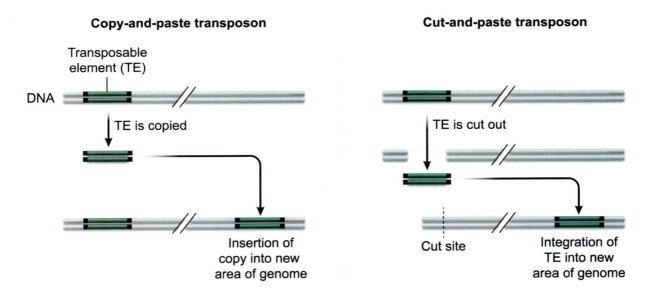

Figure 6.30 DNA transposition mechanisms.

The simplest transposons consist of the transposase gene flanked by **inverted repeat** sequences on either side. The repeats are **palindromes** (ie, reverse complement) of one another, and the repeat on one side is complementary to the repeat at the other side (Figure 6.31). Some transposons, such as retrotransposons (see Concept 6.5.03), are more complex and may contain other genes in addition to the transposase gene between the flanking repeats.

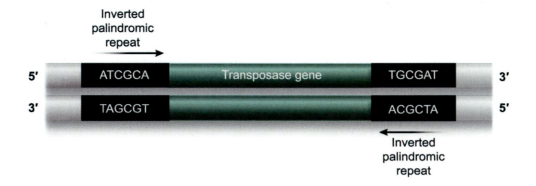

Figure 6.31 Simple transposon with inverted palindromic repeats.

Transposons can have varied effects on coding or regulatory DNA sequences depending on transposon size, number of copies, and insertion location(s). For example, the insertion of a transposon may directly disrupt a gene's coding sequence or disrupt the chromosome in a way that alters the expression or regulation of nearby genes. In addition, nearby genes are sometimes copied and/or moved along with the transposon, which can amplify the copy number of the nearby genes.

6.3.05 Gene Regulation in Prokaryotes

Because prokaryotic genomes are smaller and more streamlined than eukaryotic genomes, most bacterial genes are expressed **constitutively** (ie, constantly) and can be considered **housekeeping genes**. Housekeeping genes are required for essential activities such as gene expression and DNA replication and repair. It is estimated that 60–80% of bacterial genes are expressed constitutively.

Expression of non-housekeeping genes is initiated only when necessary based on environmental conditions. For example, an organism expresses the genes necessary for catabolism of lactose only when lactose is present in the environment.

In prokaryotes, expression of all genes necessary for the same process or function (eg, lactose catabolism) are often regulated together under the control of a switch that coordinately increases or decreases transcription. These co-regulated genes are organized in a unit called an **operon**, which consists of three parts, as shown in Figure 6.32:

- The *regulatory region* consists of the promoter and operator. As in eukaryotes, the **promoter** functions as the starting site for transcription (see Concept 2.2.01). The **operator** is a regulatory element not present in eukaryotes.
- A cluster of *structural genes* is located downstream from the regulatory region and codes for a set of enzymes and proteins needed under specific conditions.
- A constitutively expressed *regulatory gene* codes for a **repressor protein** that binds to the operator under certain conditions.

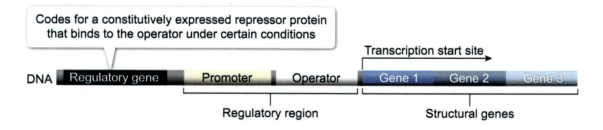

Figure 6.32 General organization of an operon.

When bound by a repressor protein, operator sequences restrict RNA polymerase movement, inhibiting transcription of the downstream structural genes. Therefore, binding of the repressor protein to the operator "flips the switch" to decrease transcription of downstream structural genes. When the repressor protein is not bound to the operator, transcription may proceed under favorable conditions.

When conditions are favorable and a repressor protein is not bound to the operator, transcription of the structural genes within an operon creates one continuous, or **polycistronic**, transcript (ie, mRNA that codes for more than one protein). Each protein is translated independently through internal start codons within the single messenger RNA (mRNA) transcript, as shown in Figure 6.33.

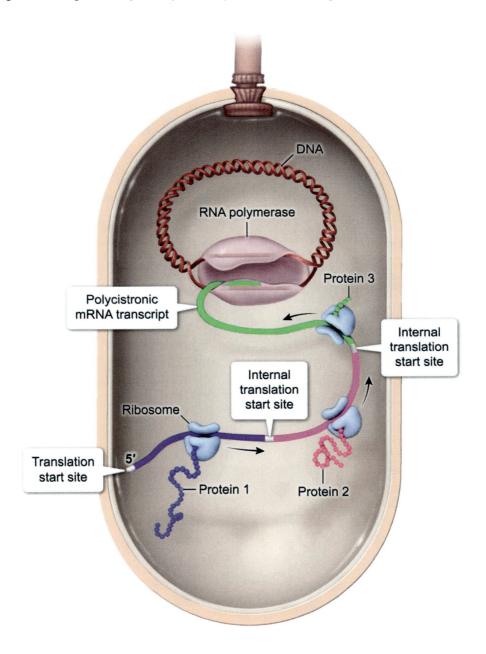

Figure 6.33 Multiple proteins may be translated from a single polycistronic mRNA transcript.

Operons can be either inducible or repressible. Genes controlled by **inducible operons** are typically repressed until conditions are favorable (eg, when a specific nutrient is present). Conversely, genes controlled by **repressible operons** are continually expressed until conditions that inhibit gene expression arise. For example, expression of genes necessary for specific biosynthetic pathways is typically regulated by repressible operons that can be inhibited when sufficient product is present.

The **lactose (*lac*) operon** is a classic example of an inducible operon operating through a negative feedback mechanism. The structural genes in the *lac* operon encode enzymes associated with lactose catabolism. A continuously synthesized repressor protein known as the *lac* repressor blocks transcription when bound to the operator. In the absence of lactose, the bound *lac* repressor blocks RNA polymerase from transcribing the *lac* structural genes, as shown in Figure 6.34.

Lactose present in the environment binds to the *lac* repressor with high affinity, preventing the *lac* repressor from binding to the operator. When the *lac* repressor is not bound to the operator, RNA polymerase can initiate transcription of the *lac* structural genes driving the subsequent catabolism of lactose. When environmental lactose levels are depleted, the *lac* repressor is able to bind to the operator once again, blocking *lac* structural gene transcription.

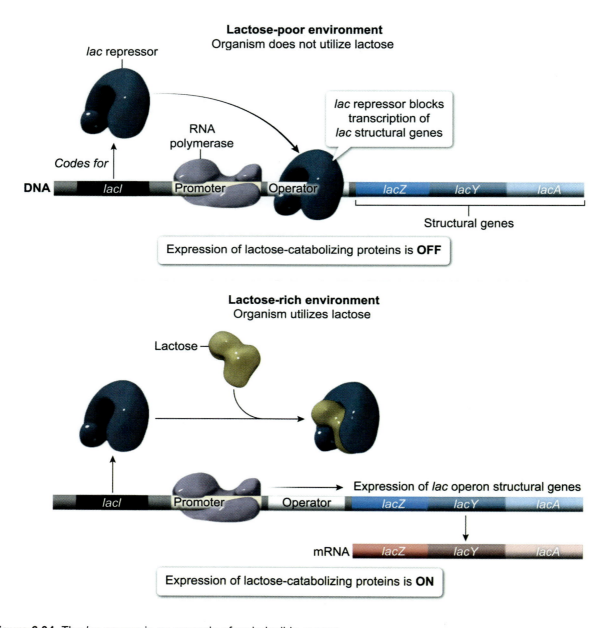

Figure 6.34 The *lac* operon is an example of an inducible operon.

Most organisms prefer to utilize glucose over lactose. To prevent a cell from utilizing lactose when glucose is abundant, a second level of *lac* operon regulation is active. When glucose is not abundant, the *lac* operon is positively regulated by a small molecule known as cyclic adenosine monophosphate (cAMP).

When glucose levels are low, cAMP levels become elevated and cAMP binds to a second regulatory protein known as **catabolite activator protein (CAP)**. When bound to cAMP, activated CAP interacts with RNA polymerase, which allows *lac* structural gene transcription to begin (Figure 6.35). Even if lactose is present, the *lac* operon *also requires* the binding of CAP to activate transcription at a high level, a phenomenon known as **positive regulation**. Therefore, glucose must be low or absent *and* lactose must be present for high-level *lac* gene expression to occur.

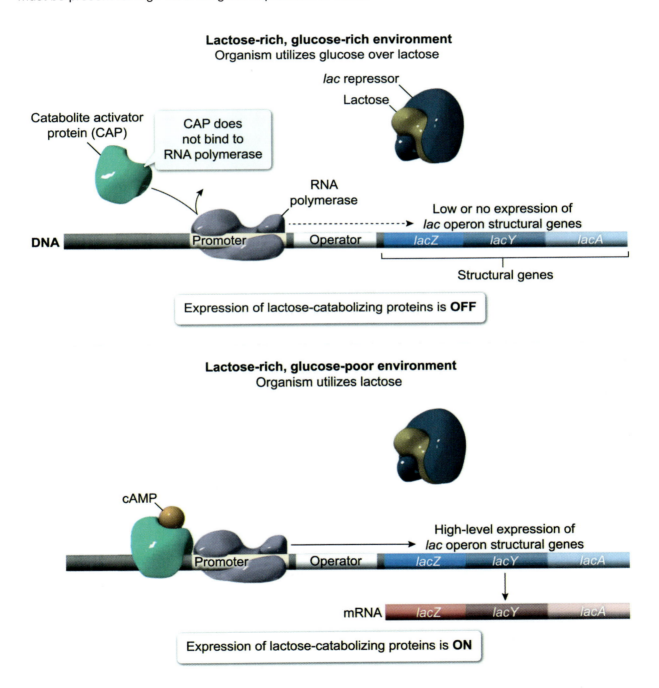

Figure 6.35 Positive regulation of the *lac* operon.

Repressible operons typically contain genes required for anabolic activities, such as amino acid biosynthetic pathways. The **tryptophan (*trp*) operon** is an example of a repressible operon that gives rise to a polycistronic mRNA transcript. The *trp* operon contains five structural genes involved in the biosynthesis of the amino acid tryptophan. When synthesis of tryptophan is required (ie, in a tryptophan-poor environment), the *trp* repressor is inactive and *does not bind* to the operator, permitting expression of the structural genes (Figure 6.36).

When excess tryptophan is present, tryptophan binds to the *trp* repressor. When two tryptophan molecules bind to the *trp* repressor, the resulting complex is able to bind to the operator. When bound to the operator, the *trp* repressor complex blocks RNA polymerase movement and represses transcription of the *trp* structural genes. Therefore, the enzymes required for tryptophan biosynthesis are not expressed when tryptophan is freely available in the cell. When tryptophan levels are depleted, the *trp* repressor cannot bind to the operator, allowing transcription to resume and tryptophan levels to rise.

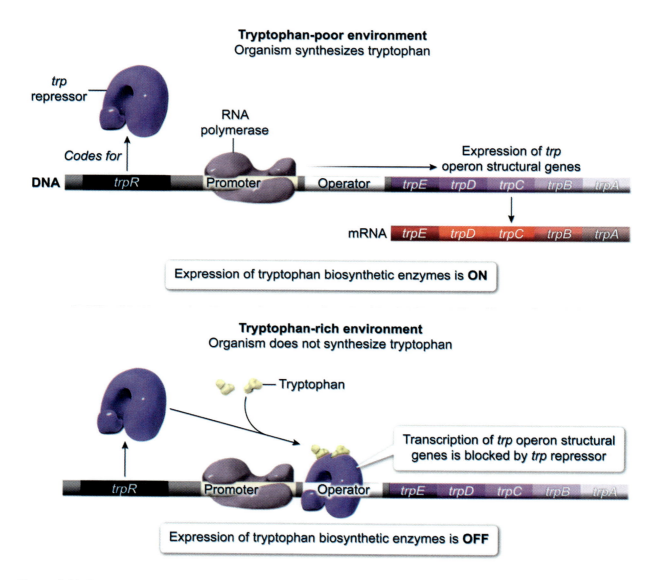

Figure 6.36 The *trp* operon is an example of a repressible operon.

In eukaryotic organisms, gene expression is more complex. In multicellular organisms in particular, differential gene regulation exists in neighboring cells and tissue types (see Lesson 2.4). The prokaryotic use of operons represents a more global type of gene regulation, a strategy not utilized in eukaryotes.

Concept Check 6.4

A mutant bacterial strain in which the *lac* repressor protein binds constitutively (ie, constantly) to the operator of the *lac* operon is discovered. What effect would this mutation have on expression of the *lac* operon structural genes in a lactose-rich, glucose-poor environment?

Solution

Note: The appendix contains the answer.

Lesson 6.4

Viruses

Introduction

By the late 1800s, the germ theory of disease had demonstrated that microscopic organisms such as bacteria were agents of infectious disease. However, it was soon discovered that some diseases could be transmitted by submicroscopic particles not visible via a light microscope (Figure 6.37). It was not until the development of the electron microscope in the 1930s that scientists were able to visualize these particles and to identify them as viruses.

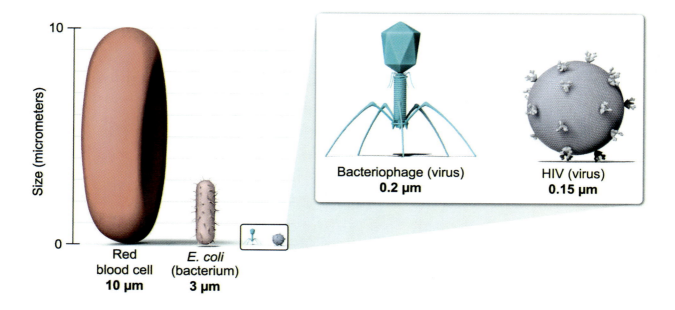

Figure 6.37 Relative sizes of cells, bacteria, and viruses.

Since the development of the electron microscope, many types of viruses and other infectious particles (eg, prions, viroids) have been identified as causative agents for disease in all types of organisms, including bacteria and archaea. Although viruses have some features in common with living organisms, viruses do not conform to the cell theory and therefore are not considered cells or living organisms (Table 6.2).

Like cells, viruses contain genetic material and undergo evolution and natural selection. However, unlike cellular genomes, which consist of DNA, viral genomes are more diverse and may be made up of either DNA or RNA.

Viruses do not carry out metabolism on their own; rather, they require host resources to carry out viral activities. Because viral reproduction is impossible without a host cell, viruses are considered **obligate intracellular parasites**. This lesson provides an overview of the basic features of viruses, including viral structures and genomes.

Table 6.2 A comparison of viral, prokaryotic, and eukaryotic features.

Characteristic	Virus	Prokaryote	Eukaryote
Size	0.02 – 1 µm	0.1 – 5 µm	10 – 100 µm
Reproduction	Dependent on host	Binary fission	Mitosis (cell reproduction) and meiosis (gamete production)
Membrane	Host-derived phospholipid membrane (enveloped viruses only)	Phospholipid cell membrane	Phospholipid cell membrane
Nucleus	Absent	Absent	Present
Genome	Single- or double-stranded DNA or RNA	Double-stranded DNA	Double-stranded DNA
Membrane-bound organelles	Absent	Absent	Present
Ribosomes	Absent	70S ribosomes	80S ribosomes
Cell wall	Absent	Present in most	Present in some
Considered living	No	Yes	Yes

6.4.01 Viral Structures

Viral genetic material is typically protected by a protein coat called a **capsid**. Capsids are composed of protein subunits known as **capsomeres**, to which protein **spikes** may be attached. Most capsids are formed from identical capsomeres, but some viruses contain capsids composed of multiple capsomere types. Viral capsids are a primary determinant of viral morphology; for example, some viral capsids are **polyhedral** (eg, in the shape of an icosahedron), with many faces. Capsids may also be **helical**, with the viral genome covered by capsomeres that create a rod-like or filamentous shape (Figure 6.38).

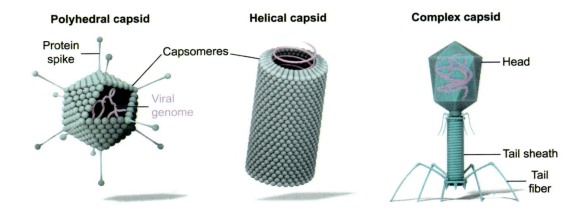

Figure 6.38 Capsid morphologies.

Viruses with more elaborate structures are called **complex viruses**. For example, a bacteriophage (ie, virus that infects only bacteria) has a polyhedral structure known as a **head** with additional structures attached. The head contains the virus's genetic information and is attached to a helical structure called the **tail sheath**. Additional structures (eg, tail fibers) are attached to the tail sheath (Figure 6.38).

Naked viruses are coated solely by a capsid, while **enveloped viruses** are covered by an additional layer called a **viral envelope**, which is derived from the plasma membrane of the host cell (Figure 6.39). The viral envelope is composed of a phospholipid bilayer that contains embedded proteins and glycoproteins (protein **spikes**). Proteins embedded in the viral envelope may be derived from both the host organism and the virus (see Concept 6.5.02).

Whether the virus is enveloped or naked, the proteins present on the outer surface of the virus mediate its entry into the host cell. Although enveloped viruses contain a membrane derived from the host cell, this membrane is not considered a plasma membrane, and enveloped viruses are not considered cells.

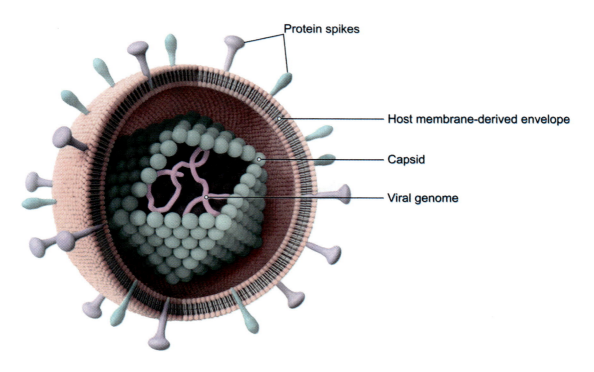

Figure 6.39 An enveloped virus.

6.4.02 Viral Genomes

Like cells, viruses contain nucleic acids; however, unlike cellular organisms, viruses are not limited to double-stranded DNA genomes. Viral genomes are small compared to prokaryotic genomes, with relatively few genes, and may be composed of either single- or double-stranded DNA or RNA. Viral genomes may be either circular or linear, and may be segmented (ie, composed of more than one nucleic acid molecule). Because viruses lack typical cellular structures and metabolic pathways, viruses are most often classified by genome type.

All viruses must use host cell translation machinery (see Concept 2.3.05) for the synthesis and trafficking of viral proteins. However, some viruses contain virally encoded DNA and RNA polymerase enzymes, ensuring their ability to replicate and express viral genes under a variety of conditions. For example, RNA viruses must encode the necessary enzymes for replication and transcription of RNA molecules because host enzymes use double-stranded DNA templates for replication and transcription.

DNA Viruses

Upon infection with a **double-stranded DNA (dsDNA) virus**, viral DNA is typically imported into the host nucleus, where the replication and transcription of viral genes occurs in a process similar to the expression of host genes (Figure 6.40). However, some DNA viruses are known to carry out replication and transcription in the cytoplasm.

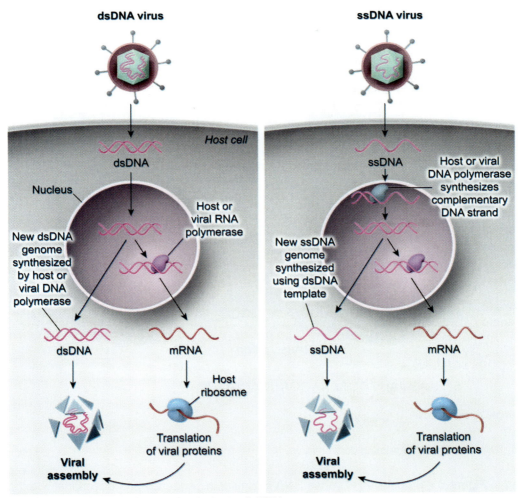

Figure 6.40 Examples of viral replication and gene expression in DNA viruses.

Single-stranded DNA viruses (ssDNA viruses) use a single-stranded DNA genome as a template to synthesize a complementary DNA strand. Subsequently, host or viral RNA polymerases can be used to transcribe viral genes. ssDNA viruses must use the dsDNA template to synthesize a single-stranded DNA copy of the genome before final virus assembly (Figure 6.40).

RNA Viruses

Host RNA polymerase enzymes require a double-stranded DNA template for transcription, so RNA viruses must use virally encoded RNA polymerase enzymes (RNA-dependent RNA polymerases) to transcribe viral RNA. **Single-stranded RNA viruses (ssRNA viruses)** may be classified as either **positive (+)** or **negative (−) sense** (Figure 6.41).

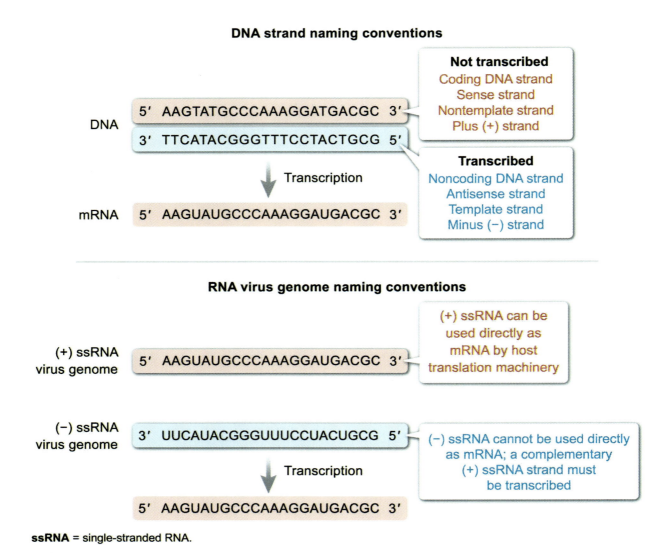

Figure 6.41 RNA virus genome naming conventions.

As depicted in Figure 6.41, positive (+) sense RNA is similar in sequence and directionality to host cell mRNA. Therefore, positive (+) sense RNA can be translated directly by host ribosomes without further modification (see Chapter 2 for a review of gene expression mechanisms). For any type of RNA virus, a positive (+) sense RNA is *required* by the host ribosome to synthesize the correct viral proteins. Therefore, in the case of negative (−) sense RNA, a complementary positive (+) sense copy must be synthesized using the (−) ssRNA strand as a template before translation can take place.

For replication of a (+) ssRNA genome, a complementary (−) ssRNA strand must be synthesized and used as a template to generate more (+) ssRNA to be packaged into new virions (ie, complete, infectious virus particles). Conversely, for replication of a (−) ssRNA genome, the complementary (+) ssRNA synthesized for viral protein translation is also used to generate more copies of the (−) ssRNA genome for final viral assembly (Figure 6.42).

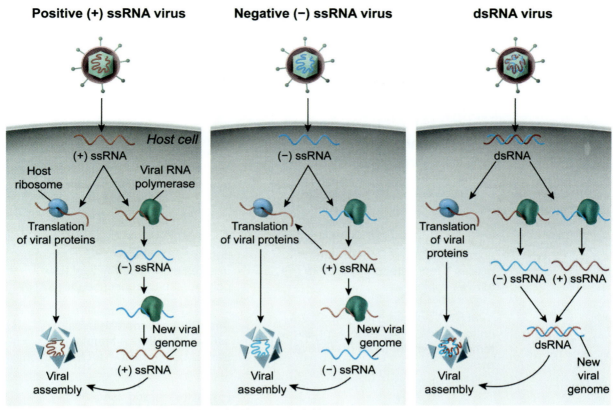

ssRNA = single-stranded RNA; dsRNA = double-stranded RNA.

Figure 6.42 Examples of viral replication and gene expression in RNA viruses.

Double-stranded RNA viruses (dsRNA viruses) are denatured upon host cell entry, and the (+) ssRNA strand can be translated directly by host ribosomes to synthesize viral proteins. Each strand is used as a template to synthesize complementary (+) ssRNA and (−) ssRNA strands that hybridize to create a complete dsRNA genome prior to final virus assembly (Figure 6.42).

Viral replication and gene expression for each viral genome type are summarized in Table 6.3.

Table 6.3 Viral replication and gene expression categorized by genome type.

Type of genome	How is the viral genome replicated?	How are viral genes expressed?
dsDNA	Host or viral DNA polymerase replicates both strands to make more dsDNA copies.	Host or viral RNA polymerase uses dsDNA as a template to transcribe viral mRNA.
ssDNA	Host or viral DNA polymerase synthesizes complementary ssDNA strand to make dsDNA; dsDNA is then used as a template to make more ssDNA copies.	Complementary ssDNA strand is synthesized and used as a template by host or viral RNA polymerase to transcribe viral mRNA.
(+) ssRNA	Viral RNA polymerase synthesizes complementary (−) ssRNA strand to use as a template to make more (+) ssRNA copies.	(+) ssRNA strand used directly during translation; (−) ssRNA can be used as a template by viral RNA polymerase to transcribe (+) ssRNA.
(−) ssRNA	Viral RNA polymerase synthesizes complementary (+) ssRNA for use as a template to make more (−) ssRNA copies.	Viral RNA polymerase uses (−) ssRNA strand as a template to transcribe (+) ssRNA for translation.
dsRNA	Viral RNA polymerase use (+) ssRNA strand as a template to make (−) ssRNA and vice versa.	(+) ssRNA strand used directly during translation; (−) ssRNA can be used as a template by viral RNA polymerase to transcribe more (+) ssRNA copies.

☑ Concept Check 6.5

A scientist has purified viral RNA being translated by host ribosomes in an infected cell. A portion of the viral RNA sequence was determined to be the following: 5'-AAAUGUGUGCCGAAAUGUUGA-3'. If the identified virus has a (−) ssRNA genome, what is its sequence?

Solution

Note: The appendix contains the answer.

Lesson 6.5
Viral Life Cycles

Introduction

As obligate intracellular parasites, viruses are metabolically inactive when outside of a host cell. The viral life cycle begins when a virus enters a host cell. In the host cell, the virus can utilize host cell machinery and resources to replicate and release fully formed viral progeny (**virions**) to infect other host cells. The viral life cycle is comprised of the activities involved from host cell viral entry to exit. This lesson highlights the life cycles of both bacterial and animal viruses, with special consideration paid to retroviruses, a type of animal virus.

6.5.01 Bacteriophages

Bacteriophages (also known simply as **phages**) are viruses that exclusively infect bacteria. Bacteriophages, which typically contain DNA genomes, are capsid coated and, unlike animal viruses, cannot be enveloped due to the rigidity of bacterial cell walls. Some bacteriophages have an elaborate capsid coating (see Concept 6.4.01).

Bacteriophages most often replicate through a cycle known as the **lytic replication cycle**, the steps of which are described as follows and shown in Figure 6.43:

1. **Attachment**: Phage tail fibers attach to the host bacterial cell surface.
2. **Entry**: The phage uses its tail sheath to inject the viral genome into the bacterial cytosol. Only the viral genome enters the cell, and the empty phage remains on the cell exterior.
3. **Synthesis**: Viral enzymes target the bacterial genome for degradation and promote synthesis of viral proteins needed to replicate and assemble new phages.
4. **Assembly**: Viral components (eg, capsid head, tail sheath, tail fibers) are assembled, and viral genomes are packaged inside.
5. **Release**: The host cell lyses (ie, bursts), and the fully assembled virions are released.

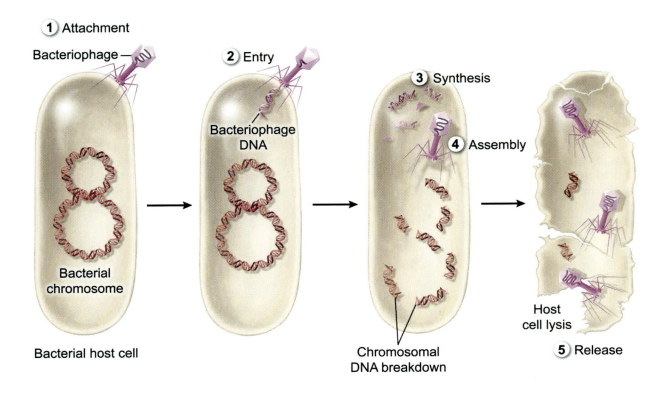

Figure 6.43 Lytic replication cycle of bacteriophages.

Some bacteriophages are able to switch between a lytic replication cycle and an alternative cycle called the **lysogenic replication cycle**. During **lysogeny**, bacteriophages are able to enter a latent (ie, inactive) phase for a variable period of time by incorporating viral DNA into a host chromosome, forming what is known as a **prophage** (ie, viral genetic material recombined into a host chromosome). Expression of most viral genes is repressed, keeping the prophage from stimulating the synthesis and release of new virions.

With each new bacterial division (ie, via binary fission), prophage DNA is replicated along with the bacterial chromosome, and the prophage remains latent within the bacterial population. Under certain conditions, the phage DNA may be excised from the bacterial chromosome, initiating a lytic replication cycle, as shown in Figure 6.44. In some cases, a small piece of the bacterial chromosome may be excised along with the prophage DNA and transferred to the next host via transduction (see Concept 6.3.03).

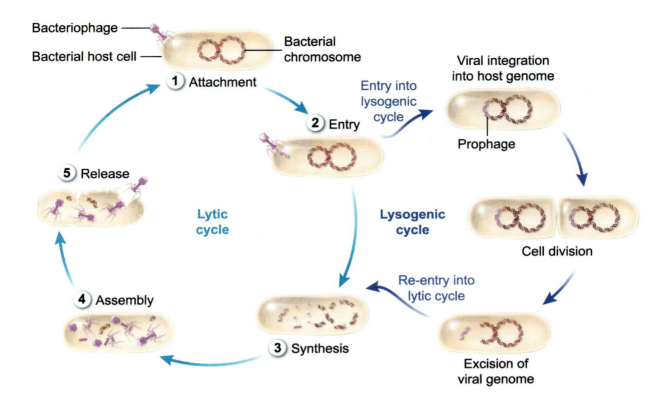

Figure 6.44 Lysogenic replication cycle of bacteriophages.

6.5.02 Animal Viruses

Generally, animal virus life cycles follow the same basic steps as the life cycles of bacteriophages (see Concept 6.5.01), which include attachment, entry, synthesis, assembly, and release. However, due to fundamental differences between prokaryotic and eukaryotic cell biology, there are some significant differences between the life cycles of animal viruses and bacteriophages, as summarized in Table 6.4.

Table 6.4 Similarities and differences between animal viruses and bacteriophages.

	Bacteriophages	**Animal viruses**
Host cell entry	Direct penetration	• Direct penetration • Endocytosis • Membrane fusion
Synthesis and assembly	Host cytoplasm	• Host cytoplasm • Host cellular compartments (eg, nucleus, endoplasmic reticulum, Golgi apparatus)
Host cell release	Host cell lysis	• Host cell lysis • Exocytosis • Budding

Both bacteriophages and animal viruses bind to specific host cell receptor sites complementary to structures found on the outermost virion surface. Unlike bacteriophages, animal viruses may be surrounded by a viral envelope derived from the plasma membrane of the previous host (see Concept 6.4.01). In addition, both capsid-coated and enveloped animal viruses often have glycoprotein **spikes** on the outermost viral surface to mediate host attachment.

Animal virus entry may take place via three mechanisms, as depicted in Figure 6.45:

- Naked virions may enter the cell via **direct penetration**. Insertion of the viral genome into the host cell leaves an empty capsid on the host cell surface.
- The entire virion (naked or enveloped) may enter the cell intact via **endocytosis**.
- Enveloped virions may enter the cell via **membrane fusion**, in which the viral envelope and host plasma membrane fuse, releasing the capsid-coated virus into the host cytoplasm.

Upon entry, viruses that enter the cell through endocytosis or membrane fusion must be **uncoated**, which occurs either within a phagolysosome after endocytosis or within the cytoplasm after membrane fusion.

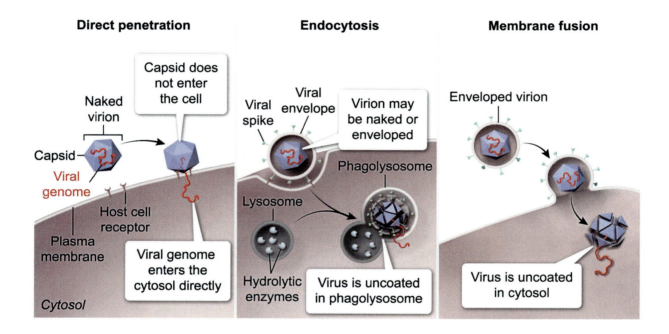

Figure 6.45 Mechanisms of animal virus entry into host cells.

Depending on genome type, the viral genome may be imported into the host cell nucleus to initiate viral replication and gene expression, or these processes may take place within the host cytosol. In either case, DNA replication and transcription may be mediated by host or virally encoded enzymes. Viral proteins may be synthesized by cytosolic ribosomes, or they may be synthesized by ribosomes on the rough endoplasmic reticulum and trafficked through the endomembrane system (Figure 6.46).

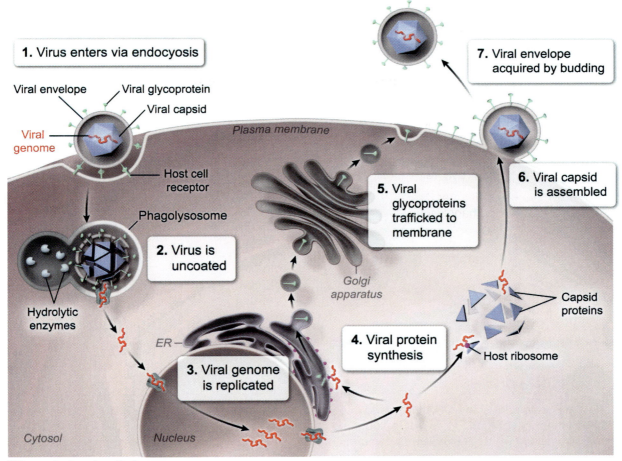

Figure 6.46 Example of viral assembly utilizing host protein synthesis and trafficking machinery.

Naked virions exit the cell via **lysis** (resulting in host cell death) or **exocytosis**, whereas enveloped viruses are released by **budding** (Figure 6.47). In many cases, viral exit via exocytosis or budding does not result in host cell death. During budding, the assembled virus travels to and pushes through the plasma membrane, creating a membrane coating (ie, viral envelope) around the virion. Prior to budding, viral glycoproteins (spikes) are trafficked to the host plasma membrane via the endomembrane system; therefore, the complete viral envelope includes viral glycoproteins, enabling viral attachment to the next host cell.

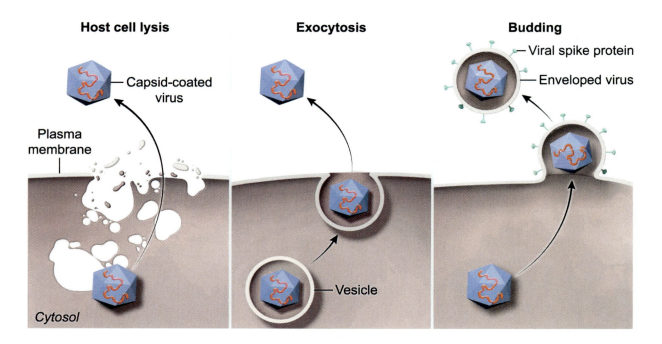

Figure 6.47 Mechanisms of animal virus exit from host cells.

6.5.03 Retroviruses

While it is possible for some DNA viruses to integrate into host cell genomes, most RNA viruses have no mechanism for integration. However, a subgroup of positive sense single-stranded RNA [(+) ssRNA] animal viruses known as **retroviruses** can convert RNA genomes into linear double-stranded DNA (dsDNA), which can then be integrated into host cell DNA. Carrying out this unique viral life cycle requires **reverse transcription** from RNA to DNA (ie, the opposite of traditional transcription) and integration of the viral genome into the host chromosome.

The virally encoded enzyme **reverse transcriptase** (an RNA-dependent DNA polymerase) catalyzes the conversion of retroviral (+) ssRNA to dsDNA in the host cytosol upon viral entry and uncoating. The dsDNA copy is then imported into the nucleus, where the viral DNA integrates into a random region of the host cell's chromosome with the assistance of another virally encoded enzyme known as **integrase**. Subsequently, a (+) ssRNA viral genome can be transcribed by host RNA polymerase and packaged into viral progeny, which then exit the cell via budding, as shown in Figure 6.48.

A virus that integrates into host cell DNA is known as a **provirus** and is replicated along with host DNA during the cell cycle, bearing some resemblance to bacteriophage lysogeny (see Concept 6.5.01). Descendants of the original infected host cell containing the provirus are also infected.

Chapter 6: Prokaryotes and Viruses

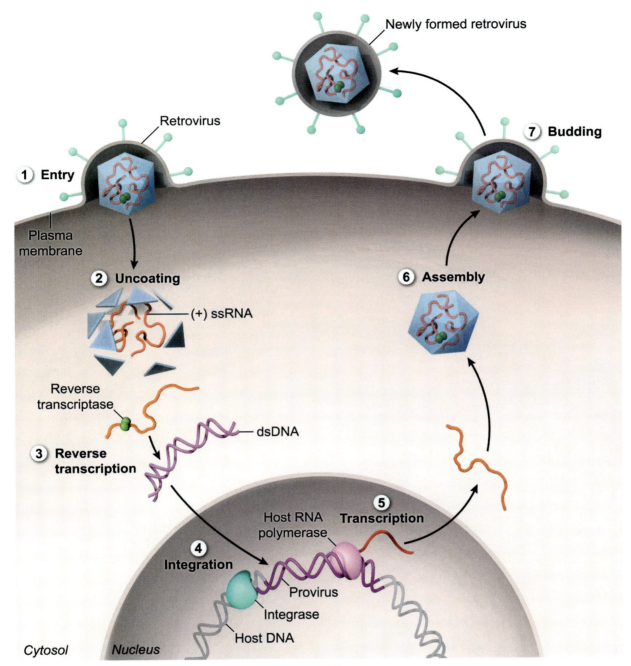

Figure 6.48 Retroviral life cycle.

A subclass of transposable elements (see Concept 6.3.04) known as **retrotransposons** are thought to have originated from retroviruses because retrotransposons move via mRNA intermediates using retroviral mechanisms. For example, transposable elements known as autonomous retrotransposons are able to move from one location to another in the genome via reverse transcription of an mRNA intermediate.

In addition to inverted repeats, autonomous retrotransposons contain genes for the enzymes reverse transcriptase and integrase. Retrotransposon DNA is transcribed in the nucleus to form mRNA, which is transported into the cytosol where translation of retrotransposon encoded genes (eg, reverse transcriptase, integrase) occurs. Retrotransposon mRNA is then converted back into dsDNA by reverse transcriptase and integrated into a new site within the genome by integrase in a process similar to the retroviral life cycle (Figure 6.49).

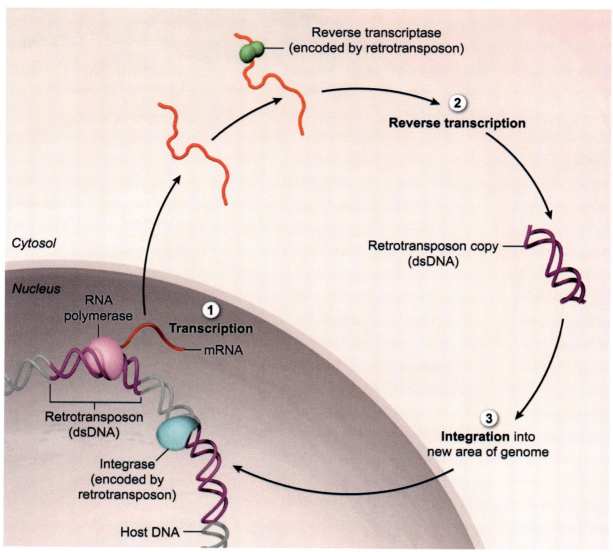

dsDNA = double-stranded DNA.

Figure 6.49 Retrotransposons can move to new locations in the genome using a mechanism similar to the retroviral life cycle.

Lesson 6.6
Sub-Viral Particles

Introduction

In addition to viruses (described in Lessons 6.4 and 6.5), several other types of nonliving infectious agents have been discovered, including **viroids** and **prions**. A viroid is composed of a small, naked RNA molecule with no capsid coat and is not known to code for proteins, while prions are self-replicating proteins that do not contain any genetic material. While the typical features of cells and viruses are not present in these subviral particles, they are still considered infectious and are able to cause disease in some cases. This lesson summarizes the major features of viroids and prions.

6.6.01 Viroids

Viroids are subviral infectious particles consisting of a short, circular single-stranded RNA molecule with no accompanying protein capsid (Figure 6.50). Viroids contain regions that exhibit self-complementarity, resulting in double-stranded regions within their circular RNA genome. Viroid genomes do not typically code for proteins and viroid replication is thought to occur via host RNA polymerases through a mechanism unique to viroids.

When infecting host cells, viroids can bind host RNA sequences via complementary base pairing, resulting in host gene silencing. Most known viroids infect plants; however, hepatitis D virus shows some similarity to known viroids and is an example of a viroid-like virus capable of infecting humans and causing disease.

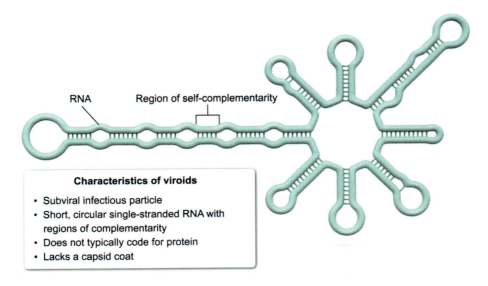

Figure 6.50 Viroids are subviral infectious particles that lack a capsid coat.

6.6.02 Prions

A **prion** (also known as **PrP**^{Sc}) is a misfolded version of a cell surface protein known as **PrP**^{C} (ie, wild-type PrP). Prions can catalyze the misfolding of additional wild-type PrP proteins, resulting in self-propagating protein aggregates that are able to cause disease. Unlike living organisms, viruses, and viroids, prions do not contain genetic material (ie, DNA or RNA).

Because wild-type PrP proteins are highly expressed in the cells of the central nervous system, prion diseases in humans are often neurodegenerative. Misfolded prions can arise by the spontaneous conversion of the wild-type PrP protein structure to a prion form. Some prion diseases affecting humans, such as Creutzfeldt-Jakob disease (CJD) and fatal familial insomnia, may also be heritable (ie, caused by genetic mutations). Humans and animals may also acquire an infectious form of CJD (vCJD) by consuming (ie, becoming inoculated with) products containing prions from cattle with bovine spongiform encephalopathy (BSE), sometimes known as mad cow disease.

Prions act as infectious agents by inducing changes in the secondary structure of other wild-type PrP proteins, resulting in the production of more prions. These structural changes to wild-type PrP proteins occur post-translationally and involve the refolding of α-helices to form β-pleated sheets (Figure 6.51).

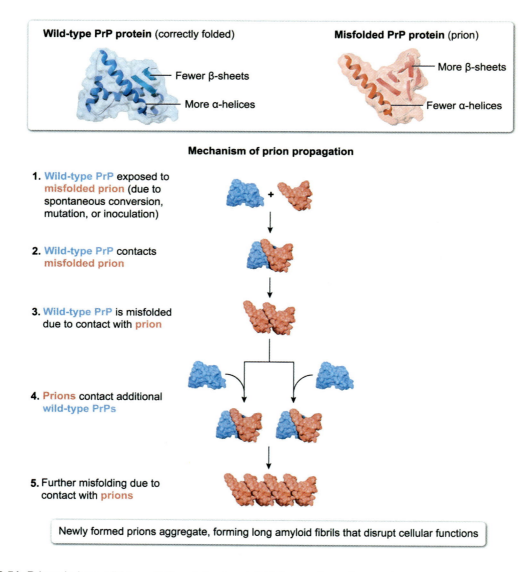

Figure 6.51 Prions induce wild-type PrP proteins to misfold, increasing prion numbers

By inducing wild-type PrP proteins to change conformation, prions perpetuate their own replication, increasing the number of prion proteins without initiating new gene expression. Because the misfolded prion proteins are less soluble than wild-type PrP proteins, the newly formed prions aggregate, and amyloid fibrils are formed. When prion aggregation reaches a critical threshold, cellular functions are disrupted, resulting in disease (Figure 6.51).

END-OF-UNIT MCAT PRACTICE

Congratulations on completing **Unit 3: Cellular Biology**.

Now you are ready to dive into MCAT-level practice tests. At UWorld, we believe students will be fully prepared to ace the MCAT when they practice with high-quality questions in a realistic testing environment.

The UWorld Qbank will test you on questions that are fully representative of the AAMC MCAT syllabus. In addition, our MCAT-like questions are accompanied by in-depth explanations with exceptional visual aids that will help you better retain difficult MCAT concepts.

TO START YOUR MCAT PRACTICE, PROCEED AS FOLLOWS:

1) Sign up to purchase the UWorld MCAT Qbank
   IMPORTANT: You already have access if you purchased a bundled subscription.
2) Log in to your UWorld MCAT account
3) Access the MCAT Qbank section
4) Select this unit in the Qbank
5) Create a custom practice test

Unit 4 Genetics and Evolution

Chapter 7 Genetics

7.1 Meiosis

7.1.01	Cellular Ploidy
7.1.02	Phases of Meiosis
7.1.03	Independent Assortment
7.1.04	Genetic Recombination
7.1.05	Comparing Mitosis and Meiosis

7.2 Mendelian Concepts

7.2.01	Gene and Allele
7.2.02	Genotype and Phenotype
7.2.03	Dominance
7.2.04	Codominance and Incomplete Dominance
7.2.05	Penetrance and Expressivity
7.2.06	Genetic Cross
7.2.07	Punnett Square
7.2.08	Testcross
7.2.09	Pedigree Analysis

7.3 Chromosomes and Inheritance

7.3.01	Determination of Sex
7.3.02	Sex-Linked Traits
7.3.03	Genetic Linkage
7.3.04	Gene Mapping
7.3.05	Biometry
7.3.06	Extranuclear Inheritance Patterns
7.3.07	DNA Mutations
7.3.08	Mutagens
7.3.09	Somatic and Germline Mutations
7.3.10	Outcomes of Mutations

Chapter 8 Evolution

8.1 Factors Affecting Allele Frequency

8.1.01	Gene Pool
8.1.02	Hardy-Weinberg Principle
8.1.03	Genetic Drift
8.1.04	Gene Flow
8.1.05	Natural Selection
8.1.06	Evolutionary Fitness

8.2 Evolution of Species

8.2.01	Species Breeding
8.2.02	Species Hybridization
8.2.03	Species Adaptation
8.2.04	Evolutionary Time

Lesson 7.1

Meiosis

Introduction

Organism reproduction requires **cell division**, which occurs via mitosis and meiosis in eukaryotes. Cell division is preceded by DNA replication, which results in the formation of duplicated chromosomes composed of two identical sister chromatids. These sister chromatids eventually separate to form distinct chromosomes that come to reside in different daughter cells, thereby allowing genetic information to pass from one generation of cells to the next.

The daughter cells produced by **meiosis** contain half the number of chromosomes as the parent cell, a feature that makes meiosis an essential part of sexual reproduction. This lesson covers the concept of cellular ploidy, the events that occur during meiosis, and the effects of meiosis on genetic diversity. In addition, this lesson provides a comparison of meiosis and mitosis.

7.1.01 Cellular Ploidy

Each cell of an organism typically contains a particular number of complete sets of chromosomes, which is referred to as the cell's **ploidy**. As shown in Figure 7.1, cells that possess a single set of chromosomes are **haploid** (1*n*), whereas cells that possess two sets of chromosomes (ie, have two homologous copies of each chromosome) are **diploid** (2*n*).

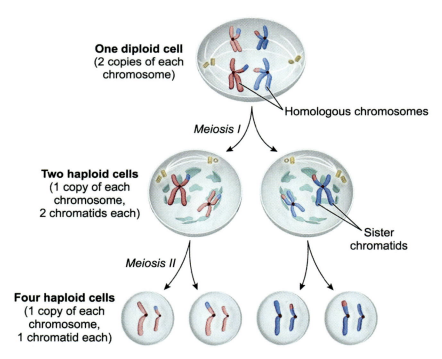

Figure 7.1 Number of chromosome sets in diploid and haploid cells.

In humans, gametes (ie, sperm, ova [eggs]) are haploid, whereas somatic cells (nonreproductive body cells) are diploid. Some organisms are polyploid, which means the cells of these organisms contain more than two complete sets of chromosomes.

7.1.02 Phases of Meiosis

Sexual reproduction involves the joining of haploid **gametes** (ie, egg, sperm) to produce a diploid zygote by **fertilization**. The production of these haploid gametes requires division of diploid parental cells via **meiosis** rather than mitosis. Meiotic cell division consists of two major stages, meiosis I and meiosis II, and like mitosis, each of these stages is divided into four phases: prophase, metaphase, anaphase, and telophase (Figure 7.2).

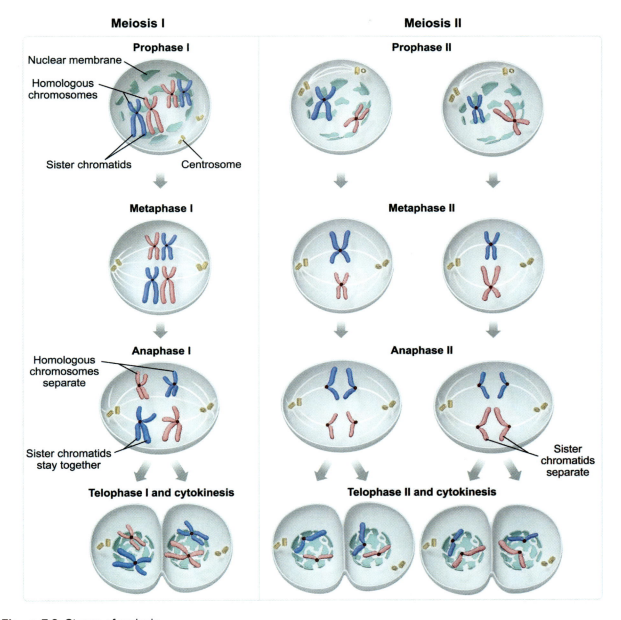

Figure 7.2 Stages of meiosis.

During prophase I of meiosis, homologous chromosomes in the diploid parental cell pair side by side (ie, undergo **synapsis**), and crossing over takes place. The paired homologous chromosomes align at the metaphase plate in metaphase I, and the members of each homologous pair separate to opposite poles of the cell during anaphase I.

Unlike anaphase of mitosis, chromosomal sister chromatids do not separate during meiotic anaphase I. Telophase I and cytokinesis then occur, completing meiosis I and resulting in the formation of two haploid **daughter cells**, the chromosomes of which are still composed of two connected sister chromatids.

Meiosis II then proceeds via a mechanism very similar to mitosis, except that mitosis typically occurs in diploid cells, whereas meiosis II occurs only in haploid cells. During meiosis II, sister chromatids separate to form distinct chromosomes that move to opposite poles of the cell during anaphase II. Following completion of telophase II and cytokinesis, haploid gametes are produced.

7.1.03 Independent Assortment

Offspring that result from sexual reproduction are genetically diverse because the haploid gametes (ie, sex cells) that join to form the offspring are themselves genetically diverse. This diversity arises during formation of the gametes via meiosis. One feature of meiosis that helps generate genetic diversity among gametes is the independent assortment of homologous chromosomes that occurs, as shown in Figure 7.3.

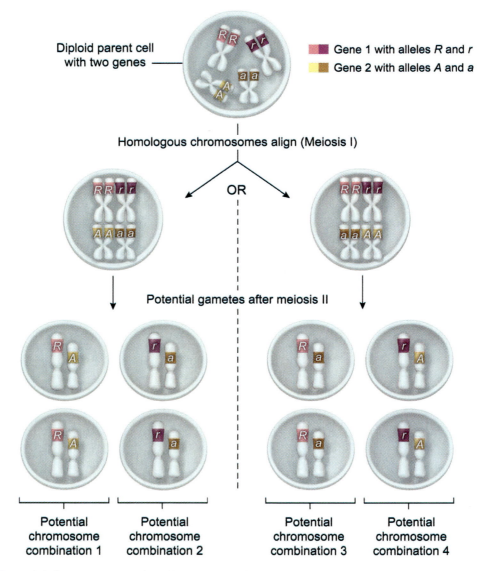

Figure 7.3 Potential chromosome combinations resulting from independent assortment.

Each pair of homologous chromosomes in a diploid cell includes a maternal chromosome (ie, chromosome derived from the mother) and a paternal chromosome (ie, chromosome derived from the father). When pairs of homologous chromosomes align at the metaphase plate during metaphase I of meiosis, the orientation of each pair with respect to which of these chromosomes is closer to a given cellular pole is random. Consequently, at the completion of meiosis I, each daughter cell has an equal probability (ie, 50%) of containing either the maternal or paternal chromosome from a given homologous pair.

Furthermore, because the movement of each pair of homologous chromosomes during meiosis is independent of the movement of every other pair, each haploid gamete produced via meiosis contains a random combination of maternal and paternal chromosomes.

This **independent assortment** of homologous chromosomes into gametes contributes to the gametes' **genetic diversity** because the maternal and paternal chromosomes in a homologous pair typically exhibit genetic differences. Therefore, each different combination of maternal and paternal chromosomes in a gamete creates a unique combination of genetic information.

7.1.04 Genetic Recombination

The process of meiosis allows eukaryotes to produce genetically diverse haploid gametes for sexual reproduction. As depicted in Figure 7.4, meiosis involves a process called **crossing over**, which results in **genetic recombination** between paired homologous chromosomes via the exchange of corresponding DNA segments.

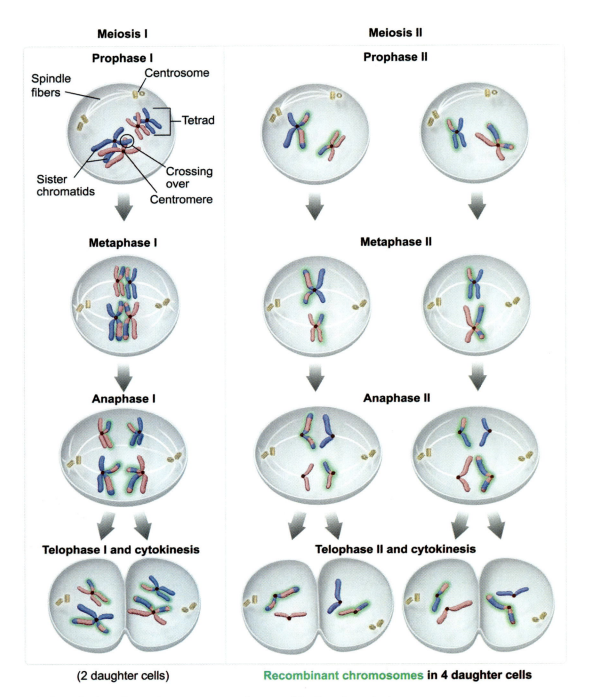

Figure 7.4 Contribution of crossing over to genetic recombination via meiosis.

Crossing over takes place during prophase I. During this stage of meiosis, homologous chromosomes undergo **synapsis**, a process in which homologous chromosomes line up side by side. Because each chromosome consists of two identical (sister) chromatids at this point in meiosis, the adjacent alignment of a pair of homologous chromosomes is called a **tetrad** (because it consists of four chromatids). Tetrads arise when a **synaptonemal complex** (protein structure) forms between homologous chromosomes and holds them tightly together (Figure 7.5).

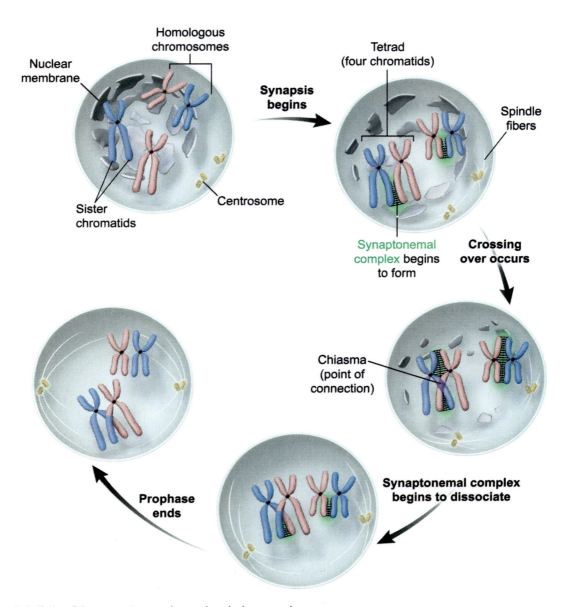

Figure 7.5 Role of the synaptonemal complex during crossing over.

The tetrad structure allows physical contact between the paternal and maternal chromosomes of a homologous pair, which allows equivalent segments of maternal and paternal chromatids to be exchanged between the chromosomes. This exchange of DNA is the hallmark feature of crossing over and results in the formation of **recombinant chromosomes** that consist of DNA derived from two different parents. Crossing over results in the production of these genetically distinct recombinant chromosomes and is therefore an important source of genetic diversity in sexually reproducing species.

7.1.05 Comparing Mitosis and Meiosis

Meiosis and mitosis are both forms of **cell division**, the mechanism by which cells reproduce. Meiosis results in the formation of haploid gametes (ie, egg, sperm), and mitosis occurs in somatic (non-sex) cells for organism growth and repair. A comparison of these two processes is outlined in Figure 7.6.

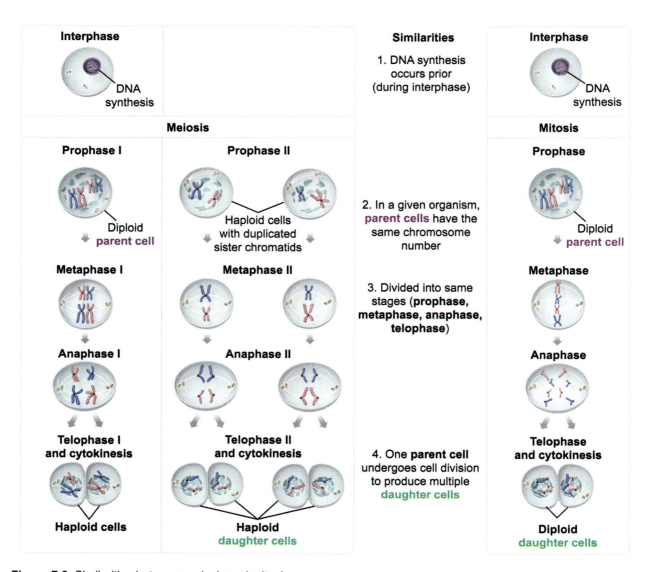

Figure 7.6 Similarities between meiosis and mitosis.

Although meiosis and mitosis are similar in many ways, these processes exhibit important differences, which are summarized in Table 7.1.

Table 7.1 Important differences between meiosis and mitosis.

Characteristic	Meiosis	Mitosis
Number of rounds of cell division	2	1
Number of daughter cells formed	4	2
Ploidy of daughter cells	Haploid	Diploid (same as parent cell)
Occurrence of synapsis and crossing over between homologous chromosomes	Yes	No
Generates genetic diversity	Yes	No
Primary purpose	Gamete production	Organism growth, repair

Ultimately, meiosis functions to produce genetically diverse haploid gametes required for sexual reproduction, whereas mitosis produces genetically identical daughter cells needed for organism growth, renewal, and repair.

Lesson 7.2
Mendelian Concepts

Introduction

Heredity (ie, genetic inheritance) involves mechanisms by which genetically based traits are passed from one generation to the next. Scientific understanding of these mechanisms continues to expand, but some of the most fundamental concepts regarding the study of heredity are associated with experiments performed by Gregor Mendel in the mid-1800s. These fundamental concepts underlie numerous techniques used in the study of genetics today. This lesson covers basic concepts and terminology related to genetics, various mechanisms by which genetic traits are inherited, and graphical tools for investigating inheritance.

7.2.01 Gene and Allele

Genetic information is transmitted across generations from parents to offspring. The basic unit of genetic information is the **gene** (Figure 7.7), which is a sequence of DNA nucleotides that encodes a functional product, such as a protein or RNA molecule. Genes are carried on chromosomes, and the specific site of a gene on a chromosome is referred to as a **locus** (plural: loci). A single chromosome may contain thousands of genetic loci.

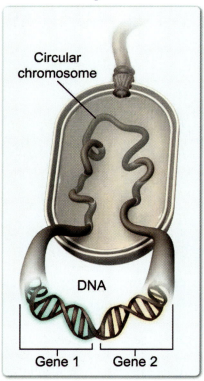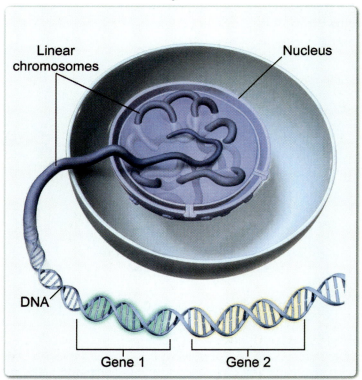

Figure 7.7 Location of genes on chromosomes.

Genes can be present in alternative forms, which are called **alleles** (Figure 7.8). Inheritance of different alleles for a particular gene by different individuals can result in those individuals expressing different forms of the trait controlled by the gene (eg, flower color). The allele that produces the form of a trait most commonly observed in a natural population is called the **wild-type** allele. As opposed to wild-type alleles, **mutant** alleles often produce altered or abnormal forms of a trait.

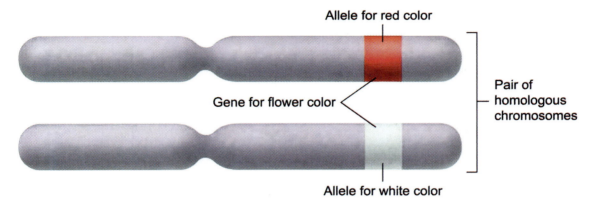

Figure 7.8 Alleles of a gene.

7.2.02 Genotype and Phenotype

Organisms inherit genetic information from their parents. In a general sense, all genetic information that an organism possesses makes up the organism's **genotype**. More specifically, diploid organisms (eg, humans) typically possess two alleles for each gene, and the term genotype refers to the two alleles present at a particular genetic locus. As shown in Figure 7.9, if both alleles at a locus are the same, the genotype of this locus is **homozygous**. Alternatively, if two different alleles are present at a locus, the genotype of the locus is **heterozygous**.

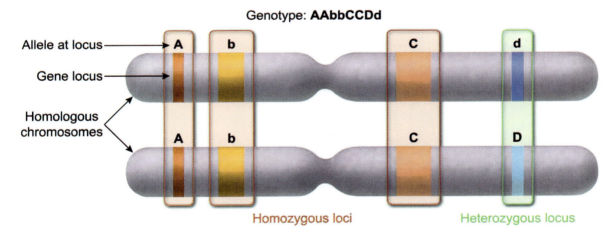

Figure 7.9 Components of a genotype.

An organism's genetic makeup (ie, genotype) is expressed to produce the organism's observable physical characteristics, or **phenotype** (Figure 7.10). An organism's phenotype includes factors such as pattern of development, molecular composition, appearance, and behavior, and phenotype can be affected by the environment.

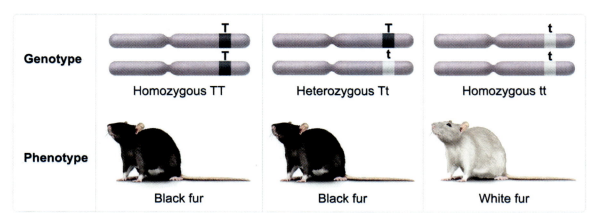

Figure 7.10 Genotype versus phenotype.

7.2.03 Dominance

Alleles, which make up an organism's genotype, can be found on autosomes, sex chromosomes, and mitochondrial DNA. These alleles may be **dominant** or **recessive**, with dominant alleles commonly being represented by capital letters and recessive alleles by lowercase letters. As shown in Figure 7.11, dominant alleles at a particular locus always affect the phenotype, whereas the phenotypic effects of recessive alleles are apparent only when no dominant alleles are present at the locus.

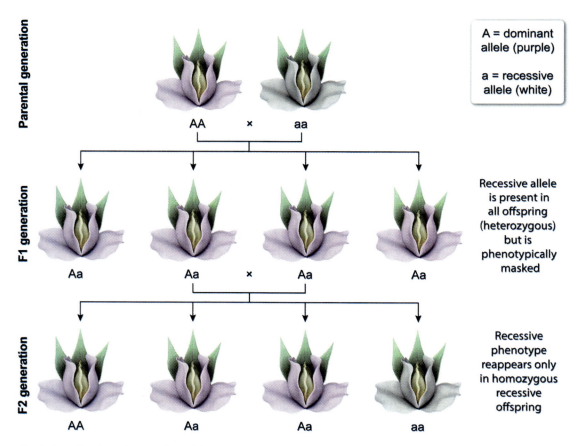

Figure 7.11 Dominant and recessive alleles.

Consequently, **dominant phenotypes** are expressed by organisms that have either homozygous dominant or heterozygous genotypes. Conversely, **recessive phenotypes** are expressed only by organisms that have homozygous recessive genotypes.

7.2.04 Codominance and Incomplete Dominance

Diploid organisms possess two complete sets of chromosomes (ie, have homologous chromosomes). Consequently, the genotype of a diploid organism is composed of two alleles per gene (with the exception of some genes on sex chromosomes). More than one allele exhibiting complete dominance may exist for certain genes in the gene pool, which means that it is possible for a diploid organism to inherit two different dominant alleles for a particular gene (see Figure 7.12).

Alleles exhibiting complete dominance are fully expressed in an organism's phenotype. Therefore, when an organism inherits two different dominant alleles for a particular gene, both alleles can be fully and independently expressed. This situation, in which the phenotype shows the full contribution of two different dominant alleles in a heterozygous genotype, is called **codominance**.

Alternatively, some alleles exhibit a pattern of expression called **incomplete dominance**, in which neither allele in heterozygotes is fully expressed. When incompletely dominant alleles are present, heterozygous individuals exhibit a blended phenotype intermediate to the phenotypes produced by homozygous dominant and homozygous recessive individuals.

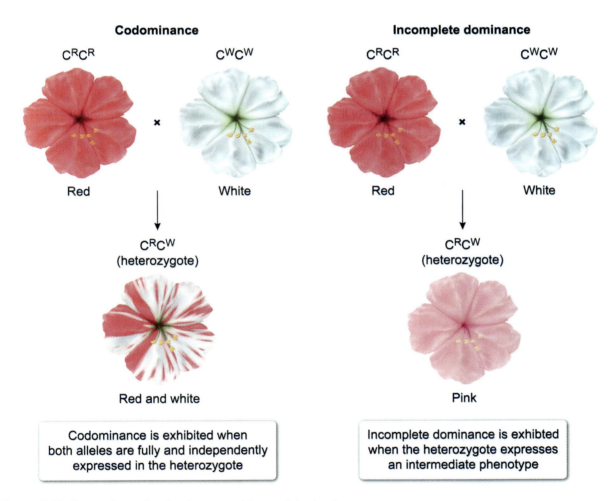

Figure 7.12 Comparison of codominance and incomplete dominance.

7.2.05 Penetrance and Expressivity

The phenotype an organism exhibits is determined by the organism's genotype, environmental conditions, and other factors such as diet. Due to the multiple elements that influence an organism's phenotype, not all organisms that possess the same genetic variant (ie, have the same genotype) exhibit identical phenotypes, as illustrated in Figure 7.13.

The **penetrance** of a genotype is the proportion of individuals with the genotype that express the expected phenotype. If 100% of individuals with a given genotype express the expected phenotype (at least to some degree), the genotype is said to be fully penetrant (ie, have complete penetrance). If less than 100% of individuals with the genotype express the phenotype, the genotype is said to have incomplete or reduced penetrance.

Variable expressivity refers to the ability of a single genotype to produce a range of degrees of expression of the expected phenotype among individuals with the genotype. For example, a disease-causing genotype with variable expressivity results in individuals exhibiting a range of disease severities despite all having the same genotype.

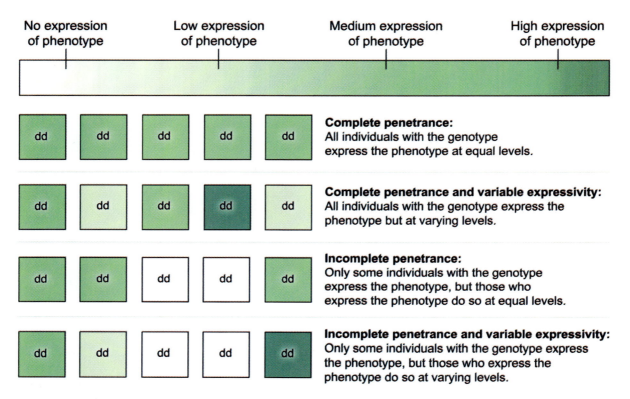

Figure 7.13 Penetrance and expressivity.

7.2.06 Genetic Cross

Experimental **genetic crosses** allow researchers to observe patterns of inheritance across generations. A simple (monohybrid) genetic cross starts by crossing (ie, mating) organisms that have known genotypes and that differ phenotypically in a single trait. The first cross typically involves homozygous parent organisms (ie, true-breeding) for particular alleles. These organisms represent the **parental (P) generation**.

In a simple genetic cross, offspring from the P generation cross are called the **first filial (F1) generation** and are all heterozygous individuals that express the dominant phenotype (ie, all F1 members resemble one parent only). Mating among members of the F1 generation results in offspring that represent the

second filial (F2) generation, with approximately 75% of this generation expressing the dominant trait and approximately 25% expressing the recessive trait. An example of a simple genetic cross is shown in Figure 7.14.

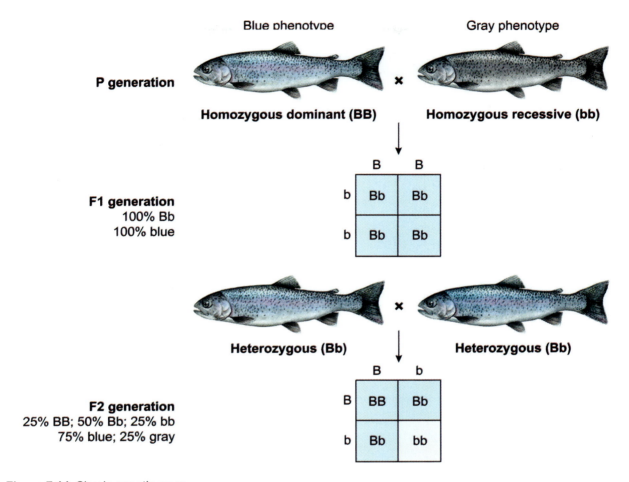

Figure 7.14 Simple genetic cross.

7.2.07 Punnett Square

A **Punnett square** is a graphical tool that can be used to predict the distribution of alleles provided by two individuals with known genotypes involved in a genetic cross (see Figure 7.15). A Punnett square is set up by writing each possible combination of alleles in maternal gametes along the top of a grid and each possible combination of alleles found in paternal gametes down the left side (or vice versa).

The squares of the grid are then filled in with the corresponding maternal and paternal contributions, and each square represents an equally probable outcome (ie, offspring genotype). The probabilities of specific offspring genotypes resulting from the cross can be determined using the Punnett square.

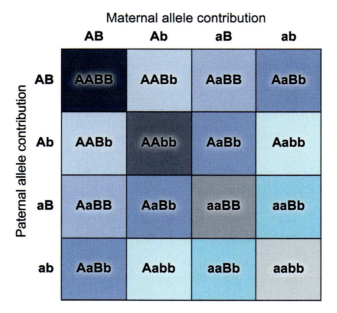

Figure 7.15 A Punnett square.

7.2.08 Testcross

Recessive alleles are phenotypically masked by dominant alleles in heterozygotes. Consequently, organisms with homozygous dominant genotypes and organisms with heterozygous genotypes typically exhibit the same phenotype. To determine whether an organism exhibiting a dominant phenotype is homozygous for the dominant allele or heterozygous, a **testcross** can be performed (Figure 7.16).

In a testcross, an organism exhibiting the dominant phenotype (and unknown genotype) is mated to an organism exhibiting the recessive phenotype (ie, homozygous recessive genotype). If all offspring resulting from the testcross exhibit the dominant phenotype, it can be concluded that the parent exhibiting the dominant phenotype is *homozygous* for the dominant allele. Alternatively, if approximately half the offspring exhibit the recessive phenotype, it can be concluded that the parent exhibiting the dominant phenotype is *heterozygous*.

Chapter 7: Genetics

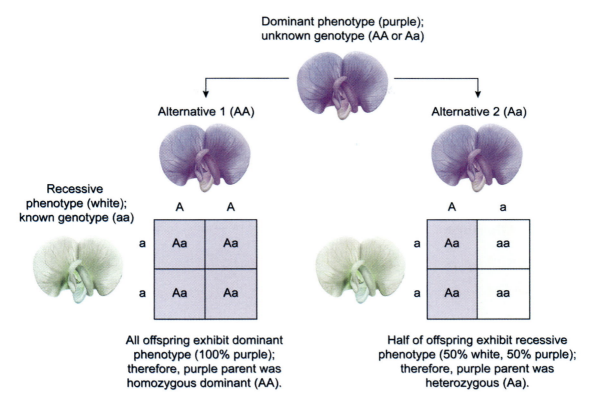

Figure 7.16 A testcross.

7.2.09 Pedigree Analysis

A **pedigree** is a diagram that shows the occurrence of a trait across multiple generations of a family (see Figure 7.17). In the diagram, females (XX) are represented by circles and males (XY) by squares. Direct horizontal lines between individuals represent matings, and offspring resulting from a mating are identified via a vertical line extending down from the horizontal line. Individuals who express the trait being studied are identified by shading.

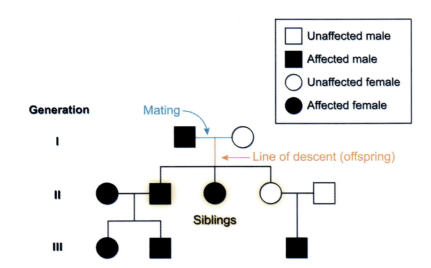

Figure 7.17 Sample pedigree.

Chapter 7: Genetics

Pedigree analysis is often used to determine whether a trait under study is dominant or recessive and whether the allele responsible for the trait is autosomal, X-linked, Y-linked, or mitochondrial. In addition, when two individuals of known genotypes mate, the possible offspring genotypes can be determined from the pedigree. Similarly, if the genotype distribution of the offspring is known, the parental genotypes can often be determined.

Concept Check 7.1

Based on the given pedigree, determine the genotypes of the individuals in Generation I (use H and h to represent the dominant and recessive alleles, respectively).

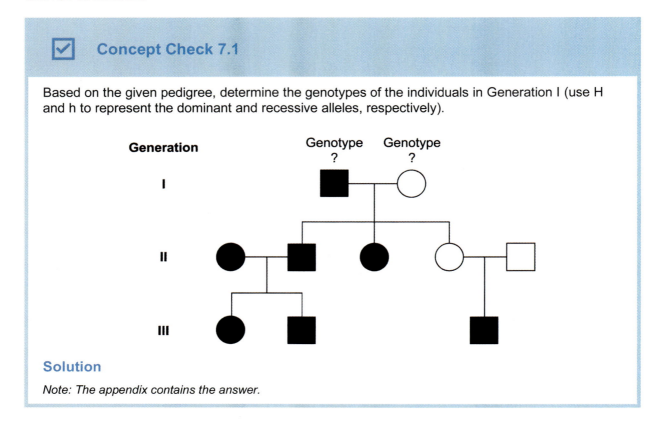

Solution
Note: The appendix contains the answer.

Lesson 7.3

Chromosomes and Inheritance

Introduction

In eukaryotic organisms (eg, humans), genes are found on chromosomes within the nucleus and on DNA present within mitochondria. Therefore, to understand inheritance patterns of genetic traits, chromosome (and mitochondrion) transmission from parents to offspring, as well as factors that affect chromosome structure and function, must be considered.

This lesson deals with sex determination in humans, the inheritance of sex-linked traits, and the concepts of genetic linkage and gene mapping. In addition, this lesson describes the inheritance of extra-nuclear genes (ie, genes present outside the nucleus) and covers the various kinds of mutations that can affect genetic inheritance.

7.3.01 Determination of Sex

In mammals, **genotypic sex** is determined by an inherited combination of sex chromosomes. **Females (XX)** typically inherit one maternal X chromosome and one paternal X chromosome. **Males (XY)** typically inherit one maternal X chromosome and one paternal Y chromosome.

Structurally, the Y chromosome is considerably shorter than the X chromosome and contains fewer genes, some of which are not present on the X chromosome. One of the genes typically present on the Y chromosome but not on the X chromosome is *SRY*, which plays an essential role in mammalian sex determination. Expression of *SRY* induces the development of the male gonads (testes). When the SRY protein is produced, fetuses generally develop a male phenotype, but in the absence of SRY, fetuses generally develop a female phenotype (see Figure 7.18).

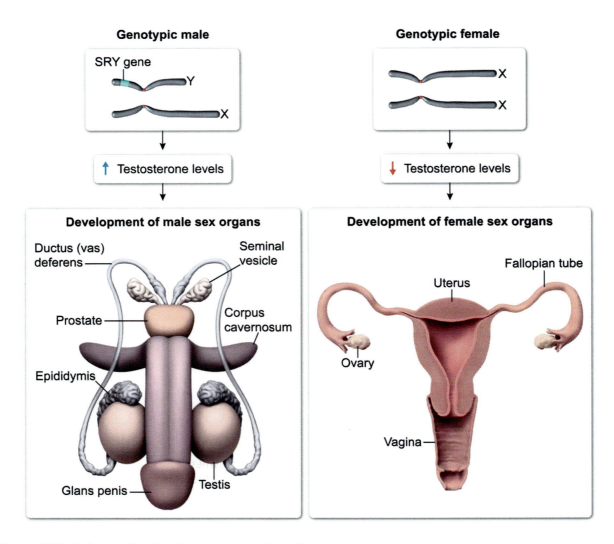

Figure 7.18 Typical mechanism of sex determination in humans.

7.3.02 Sex-Linked Traits

In **sex-linked inheritance**, traits arise from genes on sex chromosomes (ie, X and Y chromosomes). A person who inherits two X chromosomes typically has female characteristics, whereas a person who inherits one X and one Y chromosome typically has male characteristics. Therefore, XX and XY individuals exhibit different inheritance patterns for sex-linked traits.

Recessive X-linked traits are exhibited in XY individuals more often than in XX individuals because in XY individuals, expression of recessive alleles on the X chromosome cannot typically be masked by alleles on the Y chromosome. In contrast, as shown in Figure 7.19, an XX individual who inherits only one copy of a recessive X-linked allele is a **carrier** of the trait (ie, unaffected heterozygote). For an XX individual to *exhibit* a recessive X-linked trait, the person must inherit the recessive alleles on *both* X chromosomes.

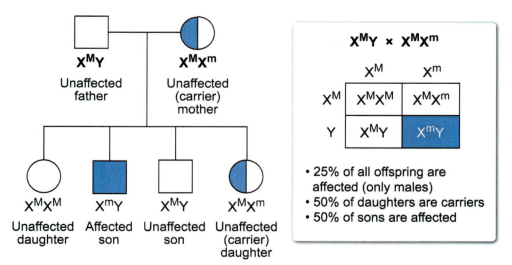

Figure 7.19 Inheritance of an X-linked recessive disorder.

In addition, because the number of genes unique to the Y chromosome is small, Y-linked traits are rare, and those that do occur appear exclusively in individuals with a Y chromosome.

7.3.03 Genetic Linkage

For alleles on different chromosomes, the law of independent assortment states that when the alleles separate during meiosis, they will do so independently of each other (see Concept 7.1.03). However, when genes are close together on the same chromosome, the genes are **linked** and do *not* assort independently into gametes.

Genetic linkage refers to the tendency of alleles on the same chromosome to remain on the same chromosome, as shown in Figure 7.20. Linked alleles do not separate independently during meiosis unless a crossing over event happens between them. As genes become closer in proximity on the same chromosome, the frequency of crossing over between them decreases, and the genes are described as being tightly or closely linked.

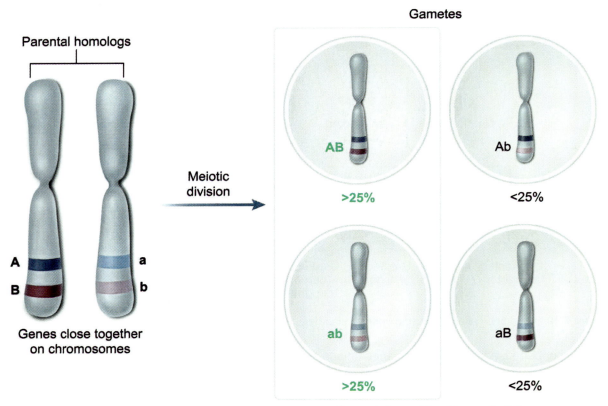

Figure 7.20 Genetic linkage.

7.3.04 Gene Mapping

Homologous chromosomes can exchange genetic information in a process called crossing over, or recombination, during prophase I of meiosis (see Concept 7.1.02). Crossing over can produce new combinations of alleles within a chromosome by moving some alleles of the maternal copy to the paternal copy and vice versa. New combinations of alleles are called **recombinant**, whereas combinations that already existed in a parent are called **parental**.

If two genes are located close together on a chromosome, they are relatively unlikely to be separated by a recombination event because there is little distance between the two genes in which crossing over can occur. Consequently, **recombination frequency** provides an indication of the physical distance between two genes on a chromosome (see Figure 7.21). Analysis of recombination frequencies can be used to construct a **gene map**, which depicts relative positions of genes on a chromosome. Distances between genes are reported in map units, or centimorgans, with a 1% recombination frequency equal to 1 map unit.

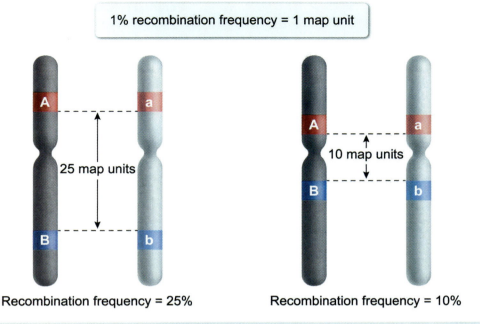

Figure 7.21 Gene map based on recombination frequencies.

7.3.05 Biometry

Analysis of biological data involves the application of statistical and mathematical methods. The use of such methods in biological sciences is referred to as **biometry**, or biostatistics. In addition to its role in data analysis, biometry is essential to researchers during the process of experimental design. Biometry has been particularly important in the development of the field of genetics, beginning with Gregor Mendel's quantitative approach to analyzing the inheritance of traits in pea plants during the mid-1800s (see Lesson 7.2).

The chi-square test is an example of a statistical test used in biological sciences. The chi-square test can be used to evaluate a null hypothesis (H_0), for example, the hypothesis that two genes assort independently, as shown in Figure 7.22.

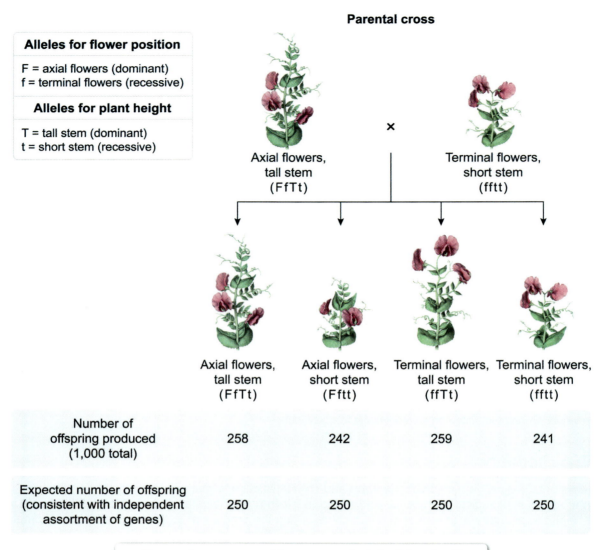

Figure 7.22 Example of the application of biometry.

7.3.06 Extranuclear Inheritance Patterns

Mitochondria are cellular organelles that have their own DNA genome, which contains a very small number of genes compared to the number of genes located on chromosomes within the nucleus. Because mitochondrial genes are **extranuclear** (ie, located outside the nucleus), they are inherited in a distinctly different manner than nuclear genes (ie, genes on autosomes and sex chromosomes).

During fertilization, mitochondria within sperm pass into the ovum but typically are eliminated during early embryonic development (Figure 7.23). Therefore, only *maternal* mitochondria typically persist in offspring, resulting in an inheritance pattern in which transmission of mitochondrial traits is via the mother only.

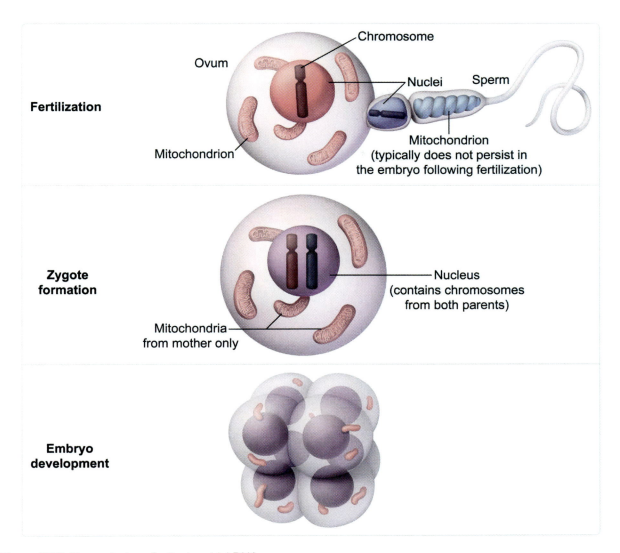

Figure 7.23 Transmission of mitochondrial DNA.

7.3.07 DNA Mutations

Mutations are changes in DNA sequences that result from errors in DNA replication, faulty DNA repair, or exposure to physical or chemical factors (ie, mutagens) that interact with DNA and affect its structure.

Mutations can be broadly characterized as **point mutations**, in which a single nucleotide pair is altered, or **chromosomal mutations**, which involve larger portions of the DNA molecule. In addition, frameshift mutations occur as a result of the insertion or deletion of a nucleotide (or of a small number of nucleotides that is not a multiple of three) that changes the reading frame of the DNA molecule.

As shown in Figure 7.24, point mutations include:

- Missense mutations, in which a nucleotide substitution in the DNA sequence results in the substitution of one amino acid for another in the polypeptide encoded by the DNA.
- Nonsense mutations, in which the nucleotide change results in the formation of a stop codon from a codon that previously encoded an amino acid.
- Silent mutations, in which a nucleotide substitution does not alter the amino acid sequence of the encoded polypeptide in any way.

Chapter 7: Genetics

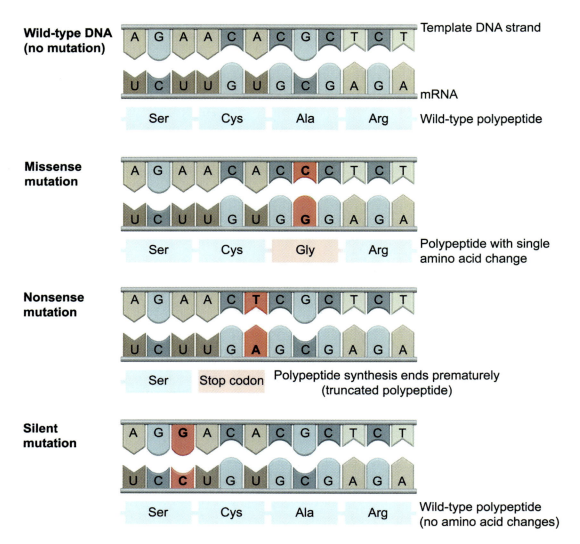

Figure 7.24 Different types of point mutations.

Different types of point mutations, as well as frameshift mutations and possible effects of each mutation type, are discussed in further detail in Lesson 2.3.

As previously mentioned, chromosomal mutations are more extensive than point mutations. Chromosomal mutations include (Figure 7.25):

- **Deletion** mutations, in which a section of DNA is removed from a chromosome.
- **Duplication** mutations, in which a section of DNA is repeated on a chromosome.
- **Inversion** mutations, in which a section of DNA breaks from a chromosome and reattaches in the reverse orientation.
- **Translocation** mutations, in which a section of DNA breaks from one chromosome and attaches to another chromosome.

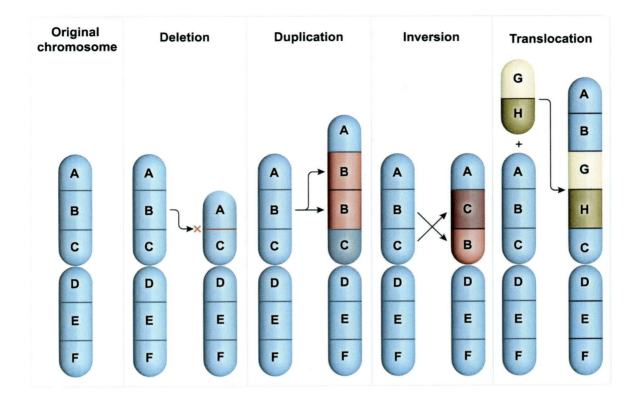

Figure 7.25 Different types of chromosomal mutations.

Unlike point mutations, which typically affect a single gene, chromosomal mutations can potentially disrupt the function of numerous genes (depending on the size of the chromosomal region affected). Therefore, chromosomal mutations are usually harmful.

Some chromosomal mutations, such as inversions and translocations, do not immediately cause a gain or loss of genetic information but can lead to altered expression of the genes involved. Such mutations can cause phenotypic effects because a gene's position on a chromosome relative to other genes and genetic control elements (eg, promotors, enhancers) can affect its expression.

7.3.08 Mutagens

Mutations are changes to the nucleotide sequence that makes up an organism's genome. Mutations can occur spontaneously or be caused by **mutagens**, agents that promote genetic changes or increase their frequency. As shown in Table 7.2, mutagens can be generally categorized as physical, chemical, or biological agents.

Table 7.2 Types of mutagens.

Type of mutagen	Examples
Physical	• Ionizing radiation • Non-ionizing radiation • Heat
Chemical	• Reactive chemicals • Base analogs • Intercalating agents
Biological	• Viruses • Bacteria • Transposable elements

Physical mutagens include heat and various forms of radiation (eg, X-rays, gamma rays, ultraviolet light). Exposure to physical mutagens can result in DNA damage including strand breaks and pyrimidine dimers (ie, distortion of the DNA molecule via formation of covalent bonds between adjacent pyrimidine bases).

Chemical mutagens can cause mutations in a variety of ways. Some chemical mutagens react directly with DNA and cause structural changes. Other chemicals (ie, base analogs) are structurally similar to DNA bases and can be incorporated into DNA during replication, which can result in base mispairing. In addition, some chemical mutagens (ie, intercalating agents) become inserted between DNA bases, which can cause frameshift mutations during DNA replication.

Biological mutagens include certain viruses (eg, human papilloma virus) and bacteria (eg, *Helicobacter pylori*). Viruses can cause mutations via insertion of the viral genome into the host cell genome, while bacterial infections may result in chronic inflammatory conditions that favor mutations. The movement of mobile genetic elements (eg, transposons, retrotransposons) within the genome can also generate mutations.

7.3.09 Somatic and Germline Mutations

The cells that make up an animal can be classified as either **germ cells** (reproductive cells) or **somatic cells** (all other cells). Germ cells are progenitor cells that undergo meiosis to produce gametes (ie, sperm, ova [eggs]). Mutations (ie, heritable changes in DNA sequences) can occur in both germ cells and somatic cells.

Mutations that occur in parental germ cells are called **germline mutations** (Figure 7.26). Such mutations can be passed to offspring via sexual reproduction because gametes, which are derived from germ cells, combine to form the zygote (ie, initial cell of the offspring). In contrast, parental **somatic mutations**, which involve alteration of somatic cell DNA, do *not* pass to offspring because somatic cells are not directly involved in zygote formation (ie, somatic cells do not pass from parents to offspring).

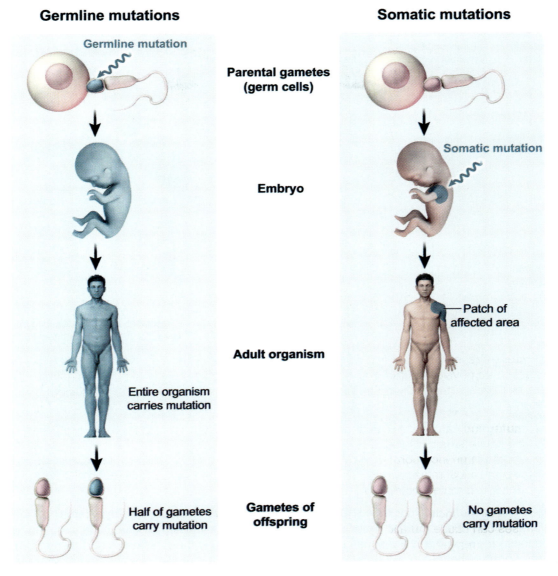

Figure 7.26 Germline mutations versus somatic mutations.

7.3.10 Outcomes of Mutations

Mutations are changes in the nucleotide sequence of DNA molecules. Such changes can produce a wide range of effects in organisms.

Natural selection is an evolutionary mechanism by which adaptive traits that increase the fitness (reproductive success) of an organism are more likely than less favorable traits to be passed to the next generation. This tendency causes **beneficial mutations** (those found in adaptive alleles) to become more common in a population over time and **detrimental mutations** to become less common (Table 7.3).

Table 7.3 Effects of different types of mutations.

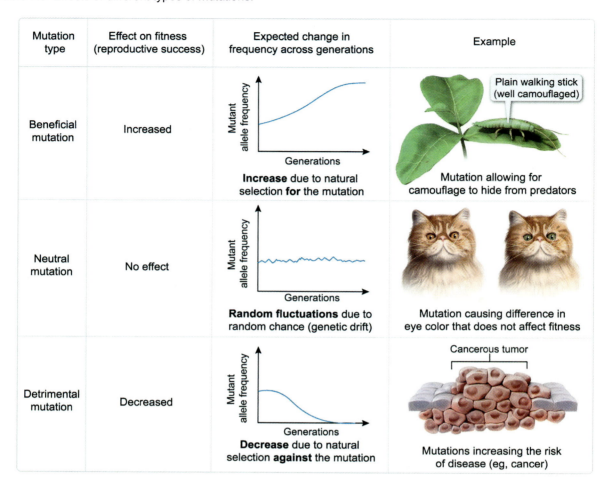

Some types of mutations are more likely than others to cause harmful effects in organisms. In general, frameshift mutations and nonsense mutations are more likely than missense mutations to cause harmful effects because frameshift and nonsense mutations are more likely than missense mutations to cause the production of nonfunctional protein molecules. Conversely, silent mutations, which do not alter the amino acid sequence of encoded proteins, do not affect an organism's fitness (ie, are evolutionarily neutral).

Inborn errors of metabolism are genetic disorders caused by detrimental mutations in genes that code for metabolic enzymes. Because enzyme activity is decreased, these disorders result in the accumulation of metabolites upstream of the affected enzyme and decreased levels of the downstream products in the pathway. The resulting changes in metabolite levels frequently have negative clinical effects such as lethargy, impaired development, and toxicity.

Chapter 8: Evolution

Lesson 8.1

Factors Affecting Allele Frequency

Introduction

The genetic makeup of a given population of organisms can change over time. This change in genetic makeup can be driven by various factors and results in **evolution** of the population. This lesson covers conditions in which a population's genetic makeup (ie, gene pool) would remain unchanged over time as well as factors that would cause a population to evolve. The concept of evolutionary fitness is also addressed.

8.1.01 Gene Pool

A biological population is made up of all members of a given species that live within the same geographical area. For example, all largemouth bass fish (*Micropterus salmoides*) living in a single lake represent a population.

The members of a population have the potential to successfully interbreed; therefore, the population shares a common collection of genetic information, which is referred to as the population's **gene pool**. The gene pool is composed of all alleles of all genes present in the individuals that make up the population. Populations with more genetically diverse gene pools are typically more resilient and can adapt to environmental changes more successfully than populations with less genetic diversity, as shown in Figure 8.1.

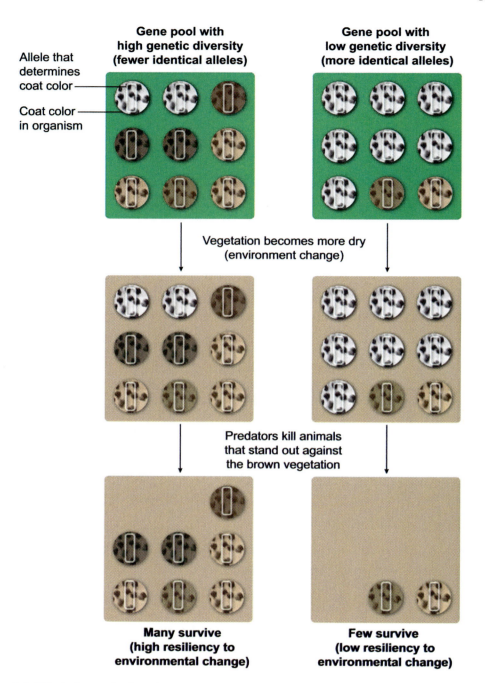

Figure 8.1 Effect of genetic diversity on a population's resiliency to environmental changes.

8.1.02 Hardy-Weinberg Principle

Evolution occurs in a population when the frequencies (ie, relative abundances) of alleles in the population's gene pool change over time. However, according to the **Hardy-Weinberg principle**, allele and genotype frequencies of a population will remain *constant* over time if certain conditions are met (Figure 8.2). Populations that meet these conditions are said to be in **Hardy-Weinberg equilibrium** and are *not* evolving.

For a population to be in Hardy-Weinberg equilibrium, the following conditions must be met:
- Matings within the population must occur randomly.

- No net mutation can occur (ie, mutation cannot create or destroy alleles).
- No migration can occur into or out of the population.
- The population must be very large.
- No natural selection can occur (ie, all genotypes must have equal reproductive success).

Figure 8.2 Conditions for a population to be in Hardy-Weinberg equilibrium.

When a population is in Hardy-Weinberg equilibrium, two equations can be used to predict the population's allele and genotype frequencies:

- $p + q = 1$: Applies to a gene with two alleles (eg, A and a), where p represents the frequency of one allele (A), and q represents the frequency of the other allele (a).
- $p^2 + 2pq + q^2 = 1$: Can be used to calculate genotype frequencies (ie, frequencies of AA, Aa, and aa) from the allele frequencies. Here, p^2 and q^2 represent the frequencies of the two homozygous genotypes AA and aa, respectively. The heterozygous genotype (Aa) frequency is represented by $2pq$.

Concept Check 8.1

In an isolated population of 100,000 largemouth bass (*Micropterus salmoides*), 5,000 are homozygous for a mutant recessive allele. Assuming the population is in Hardy-Weinberg equilibrium, what percentage of largemouth bass in the population are heterozygous for this mutant allele?

Solution

Note: The appendix contains the answer.

8.1.03 Genetic Drift

Genetic drift is the random fluctuation in allele frequencies that occurs in a population due to **chance events** (Figure 8.3). Such events (eg, earthquakes, floods, fires) can result in the death of organisms regardless of their genetic makeup, causing random loss of alleles from the gene pool.

In general, small populations are much more susceptible to genetic drift than large populations because small populations typically contain fewer unique alleles in their gene pools. Genetic drift reduces genetic diversity in small populations by randomly eliminating alleles from their already small collection of alleles. Therefore, rare alleles are much more likely to be completely lost from a gene pool via genetic drift in a small population than in a large population.

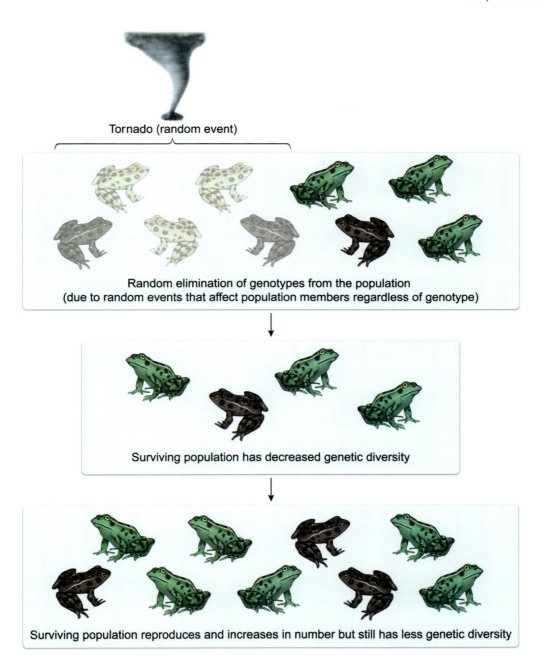

Figure 8.3 Genetic drift.

The **bottleneck effect** (Figure 8.4) is observed in a population when genetic drift occurs following a drastic reduction in the size of the population due to a sudden, unpredictable event (eg, flood, earthquake, human-induced catastrophe). During a bottleneck event, members of a population are eliminated randomly because no particular genotype or phenotype typically confers greater survivability to an organism.

Because the effect of genetic drift on a population is random, a bottleneck event does not necessarily cause a population to become better adapted to its environment. For example, a *beneficial* allele can be completely eliminated from a population's gene pool if no members carrying this allele happen to survive the bottleneck. Furthermore, rare harmful alleles can become *more* prevalent in a population if most members carrying a harmful allele randomly survive the bottleneck and most members carrying a wild-type allele do not.

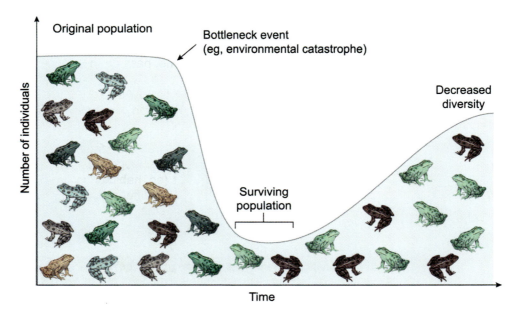

Figure 8.4 Bottleneck effect.

8.1.04 Gene Flow

Movement of individuals (or gametes) among different populations of a species can alter the allele frequencies of these different populations. Such movement of alleles into or out of a population via migration is called **gene flow** (see Figure 8.5).

Individuals that migrate into a population can bring new alleles and therefore increase the genetic diversity of the receiving population. However, if migration occurs in both directions between two populations, gene flow tends to cause the gene pools of the two populations to mix and the genetic makeup of the two populations to become more similar.

Gene flow is caused by migration of individuals (or their gametes) between populations and can result in allele frequency changes

Figure 8.5 Gene flow.

8.1.05 Natural Selection

According to the theory of **natural selection** (see Figure 8.6), organisms with traits that make them well suited to their environment are more likely to survive and reproduce (ie, pass on their alleles) than organisms with less favorable traits. This tendency causes beneficial alleles to become more common and detrimental alleles to become less common in a population over time. As a result, traits that increase an organism's likelihood of survival and reproduction prevail in a population over time, causing the population to become better adapted to its environment.

However, patterns of evolution often vary depending on the selective pressures faced by different species. If two species inhabit similar environments, they often evolve analogous (comparable) characteristics in response to facing comparable selective pressures (eg, habitat, food availability, predation). In contrast, if two species inhabit distinct environments with different selective pressures, they are likely to evolve unique traits best suited to their different environments.

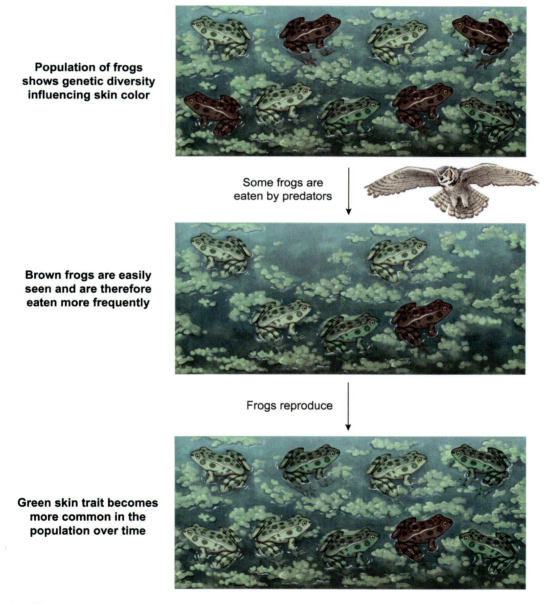

Figure 8.6 Natural selection.

8.1.06 Evolutionary Fitness

Charles Darwin's theory of natural selection is often summarized by the phrase "the survival of the fittest," which can lead to the incorrect conclusion that evolutionary fitness is primarily exhibited in organisms that possess physical strength or superior fighting ability.

Rather than being limited to characteristics such as size and strength, **evolutionary fitness** is measured in terms of an organism's **reproductive success**, as shown in Figure 8.7. *Any* characteristic that promotes an organism's ability to survive and successfully reproduce contributes to its evolutionary fitness, and alleles responsible for such characteristics increase in frequency over time in a gene pool.

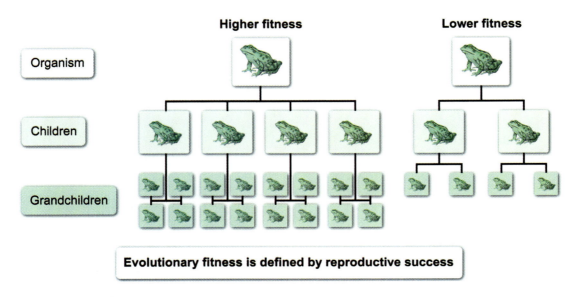

Figure 8.7 Evolutionary fitness.

Lesson 8.2

Evolution of Species

Introduction

A biological species can be defined as a group of organisms whose members have the potential to successfully interbreed (ie, reproduce sexually) in nature to produce fertile offspring. In general, members of different species cannot mate with each other and produce viable, fertile offspring. Consequently, biological species are typically marked by reproductive isolation. This lesson covers factors that influence the extent to which populations are reproductively isolated, factors that influence the adaptation of species, and the concept of evolutionary time.

8.2.01 Species Breeding

Members of all populations of a given species have the potential to interbreed. However, different populations can be more or less isolated from each other, which affects the likelihood of mating between members of those populations.

Inbreeding, which involves mating of individuals with shared ancestry, typically occurs in small, isolated populations. Inbreeding causes an increase in homozygosity of alleles in a population, which can result in inbreeding depression (ie, reduced fitness) due to the expression of harmful recessive alleles in homozygotes. Conversely, **outbreeding** (ie, mating of unrelated individuals) tends to increase genetic diversity in populations and is facilitated by large population size and gene flow between populations, as shown in Figure 8.8.

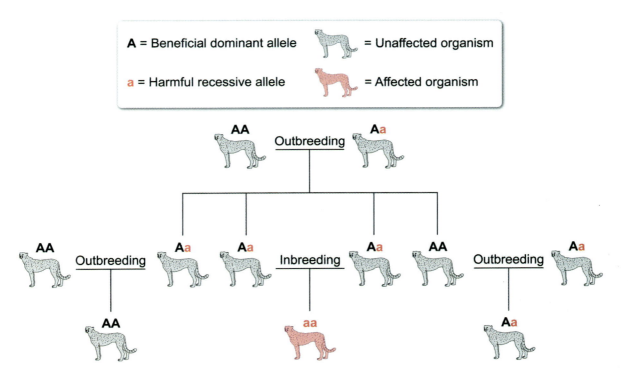

Figure 8.8 Effects of inbreeding and outbreeding.

8.2.02 Species Hybridization

Mating between two different species may result in **hybrid offspring**, some of which may be inviable (ie, unable to survive) or infertile. When two different species cannot produce viable or fertile offspring by interbreeding, they are considered reproductively isolated. Reproductive isolation typically results from barriers that fall into two main categories: **prezygotic barriers** and **postzygotic barriers**.

Prezygotic barriers prevent mating or, if mating does occur, interfere with successful fertilization, thus preventing the formation of hybrid zygotes. Table 8.1 summarizes the prezygotic mechanisms that contribute to species' reproductive isolation.

Table 8.1 Prezygotic isolating mechanisms.

Mechanism	Description
Temporal isolation	Species mate during different years, in different seasons, or at different times of the day
Habitat isolation	Species live in the same geographical area but occupy different habitats and do not interact
Behavioral isolation	Species engage in particular courtship rituals or produce specific signals to attract mates
Mechanical isolation	Anatomical differences between species prevent successful mating
Gametic isolation	Sperm and eggs of different species cannot successfully fuse to accomplish fertilization

Postzygotic barriers do not function until *after* hybrid zygotes have been successfully produced. Postzygotic reproductive isolation occurs when species produce hybrid offspring that are inviable or infertile. Table 8.2 summarizes the postzygotic mechanisms that contribute to reproductive isolation.

Table 8.2 Postzygotic isolating mechanisms.

Mechanism	Description
Hybrid inviability	The genetic makeup of hybrid offspring causes impaired development and death
Hybrid sterility	Hybrid offspring may develop normally but are unable to reproduce successfully
Hybrid breakdown	First-generation hybrids are viable and fertile, but second-generation hybrids are inviable or sterile

If hybrid offspring successfully mate with members of either parental species, **genetic leakage** may occur (see Figure 8.9). Genetic leakage is the transfer of genetic information (ie, genes) between different species.

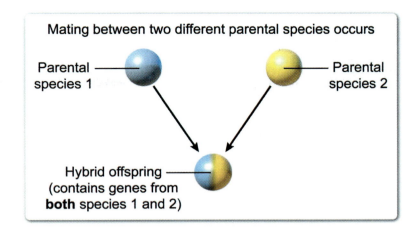

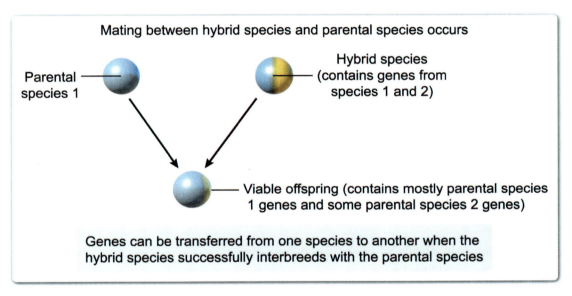

Figure 8.9 Genetic leakage.

8.2.03 Species Adaptation

Natural selection is an evolutionary mechanism in which **adaptive traits** that increase the fitness (ie, reproductive success) of an organism are more likely to be passed to the next generation than less favorable traits. This tendency causes beneficial traits (those encoded by adaptive alleles) to become more common and detrimental traits to become less common over time.

There are multiple types of natural selection. One type, disruptive selection, favors phenotypic extremes and tends to eliminate intermediate phenotypes from a population. Disruptive selection can result in **polymorphism** within a species, or when two or more different phenotypic forms (ie, morphs) are maintained in a population, as depicted in Figure 8.10.

Figure 8.10 Different phenotypic forms of *Neochromis omnicaeruleus*.

Adaptive radiation occurs when natural selection results in rapid diversification of a single species into multiple forms adapted to exploit different environmental resources. As shown in Figure 8.11, adaptive radiation reduces intraspecific competition (competition for resources by members of a single species).

Reduced intraspecific competition improves fitness for the entire species because each subgroup has a role within an ecological community (ie, niche) that is different from the rest of the subgroups. Adaptive radiation can eventually lead to speciation, or the formation of new species, if the subgroups diverge enough to become reproductively isolated from each other.

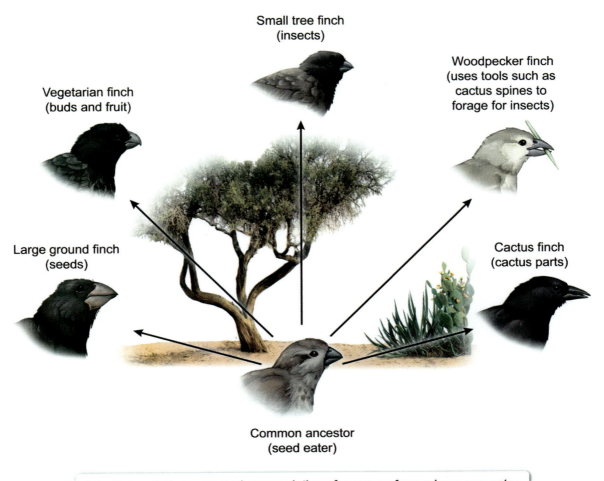

Adaptive radiations occur when speciation of a group of organisms generates multiple new species that can utilize different environmental resources

Figure 8.11 Adaptative radiation.

8.2.04 Evolutionary Time

The **neutral theory of molecular evolution** states that most genetic mutations are neutral, which means they do not affect the fitness of an organism. Within species, neutral mutations are randomly fixed or lost due to genetic drift and occur at fairly constant rates over time. After one species diverges to form two separate species, the accumulation rate of neutral mutations in a region of homologous DNA is mostly similar in both species. Because the number of neutral mutations tends to increase at a constant rate, accumulation of these mutations can serve as a molecular clock of evolution.

The **molecular clock model** (see Figure 8.12) allows researchers to measure **evolutionary time** by analyzing random changes in the genome over time. The molecular clock model assumes that species that diverged more recently from a common ancestor have accumulated fewer mutations over a shorter time and are more genetically similar to one another (ie, more closely related). In contrast, species that diverged less recently (ie, longer ago) from a common ancestor have accumulated more mutations across a longer period and are less genetically similar to one another (ie, less closely related).

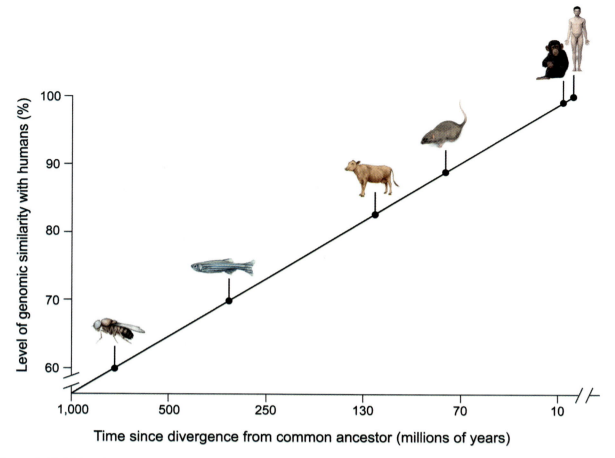

Figure 8.12 Molecular clock model.

END-OF-UNIT MCAT PRACTICE

Congratulations on completing **Unit 4: Genetics and Evolution**.

Now you are ready to dive into MCAT-level practice tests. At UWorld, we believe students will be fully prepared to ace the MCAT when they practice with high-quality questions in a realistic testing environment.

The UWorld Qbank will test you on questions that are fully representative of the AAMC MCAT syllabus. In addition, our MCAT-like questions are accompanied by in-depth explanations with exceptional visual aids that will help you better retain difficult MCAT concepts.

TO START YOUR MCAT PRACTICE, PROCEED AS FOLLOWS:

1) Sign up to purchase the UWorld MCAT Qbank
   IMPORTANT: You already have access if you purchased a bundled subscription.
2) Log in to your UWorld MCAT account
3) Access the MCAT Qbank section
4) Select this unit in the Qbank
5) Create a custom practice test

Unit 5 Reproduction

Chapter 9 Reproductive Systems

9.1 Biological Sex and Reproduction

- 9.1.01 Biological Sex Differences
- 9.1.02 Human Reproduction

9.2 Male Reproductive System

- 9.2.01 Male Reproductive Anatomy
- 9.2.02 Spermatogenesis
- 9.2.03 Hormonal Control of the Male Reproductive System

9.3 Female Reproductive System

- 9.3.01 Female Reproductive Anatomy
- 9.3.02 Oogenesis
- 9.3.03 Hormonal Control of the Female Reproductive System

Chapter 10 Pregnancy, Development, and Aging

10.1 Pre-Implantation Development

- 10.1.01 Fertilization and Blastulation
- 10.1.02 Implantation and Placental Development

10.2 Post-Implantation Development

- 10.2.01 Gastrulation
- 10.2.02 Organogenesis and Neurulation

10.3 Cellular Mechanisms of Development

- 10.3.01 Cell Potency and Specialization
- 10.3.02 Gene Regulation in Development
- 10.3.03 Cell Signaling and Migration
- 10.3.04 Programmed Cell Death

10.4 Gestation

- 10.4.01 Pregnancy
- 10.4.02 Fetal Circulation
- 10.4.03 Parturition
- 10.4.04 Lactation

10.5 Cellular Regeneration and Senescence

- 10.5.01 Regeneration
- 10.5.02 Cell and Organism Aging

Lesson 9.1

Biological Sex and Reproduction

Introduction

In humans, **biological sex** is defined by physical characteristics such as:

- Chromosomal and genetic features (eg, sex chromosomes, genes involved in sex determination)
- A set of anatomical and physiological conditions (**sex traits**) that include internal and external reproductive organs, secondary sex characteristics (eg, body morphology, hair growth pattern, breast tissue distribution), and hormone levels and function

The concept of biological sex is distinct from the concepts of gender and gender identity. **Gender** is a social construct based on norms, behaviors, and cultural roles associated with a particular biological sex. **Gender identity** refers to a person's inner sense of self as male, female, or another identity (eg, gender-fluid) and may or may not correspond with an individual's biological sex traits. Further information on the distinctions between biological sex, gender, and gender identity can be found in Behavioral Sciences Concept 44.2.01.

9.1.01 Biological Sex Differences

Primary sex determination in humans is defined by the inheritance pattern of sex chromosomes from each parent. The inheritance of particular sex chromosomes typically initiates the differentiation of reproductive organs during embryonic development (Figure 9.1). A person who inherits two X chromosomes (ie, XX individual) usually develops female reproductive organs and sex traits, whereas a person who inherits one X and one Y chromosome (ie, XY individual) usually develops male reproductive organs and sex traits (see Concepts 9.2.03 and 9.3.03).

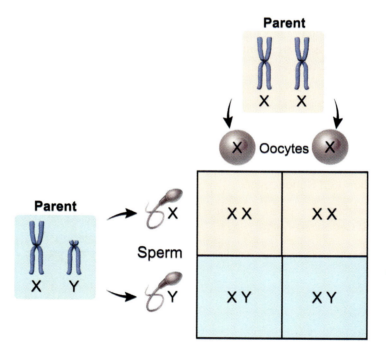

Figure 9.1 Primary sex determination in humans.

Although the composition of sex chromosomes is determined at fertilization, there is variation in human sexual development. Individuals may be born with or develop differences in sex traits (eg, reproductive anatomy, sex hormones, secondary sex characteristics) that vary from sex traits in individuals typically characterized as "male" or "female."

In some cases, individuals may display variations in sex chromosome numbers (eg, XXY). Alternatively, there may be variations in hormonal signaling pathways involved in differentiation of reproductive organs. For example, some individuals who inherit a Y chromosome may develop both male and female sex traits due to an inability to respond to androgenic hormones (eg, testosterone).

Biologically, sex differences affect the functions of various body systems, such as susceptibility to certain diseases (immune system), pain processing (nervous system), and heart health (cardiovascular system). Taken together, these differences in biology can affect how an individual experiences disease and responds to treatment.

9.1.02 Human Reproduction

Human offspring are produced via sexual reproduction, in which haploid (1n) gametes (ie, **sperm** and **oocytes**) are united to form a diploid (2n) cell. This diploid cell is known as a **zygote** and contains one set of chromosomes from each parent (Figure 9.2).

To facilitate the process of sexual reproduction, female and male reproductive systems are involved in the production of sex hormones and gametes, and the facilitation of a delivery mechanism (ie, sexual intercourse) that brings gametes together. In addition, female reproductive systems are involved in the maintenance of a supportive environment in the female reproductive tract, where **fertilization** (ie, zygote formation) and **pregnancy** (ie, fetal development) can occur (see Chapter 10).

Anatomical features of the male and female reproductive systems, gamete production, and hormonal control of reproduction are covered in subsequent lessons in this chapter.

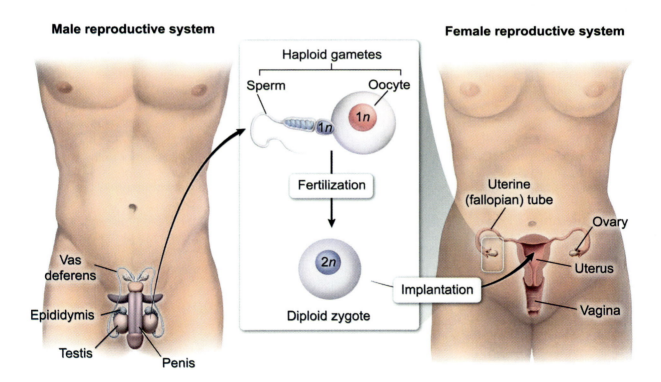

Figure 9.2 Male and female reproductive systems facilitate human reproduction.

Lesson 9.2
Male Reproductive System

Introduction

The male reproductive system participates in human reproduction by generating haploid gametes known as **sperm** and facilitating the delivery of these gametes to a female reproductive tract. The concepts in this lesson provide an overview of male reproductive anatomy, the process of gamete formation via **spermatogenesis**, and hormonal regulation of the male reproductive system.

9.2.01 Male Reproductive Anatomy

The male reproductive system consists of internal and external structures that participate in producing sperm, producing and regulating sex hormones, and facilitating reproduction. As depicted in Figure 9.3, the following structures are associated with the male reproductive system:

- **Testes** (singular, **testis**) are a pair of reproductive structures (ie, **gonads**) where spermatogenesis and sex hormone production occur.
- The **scrotum** is an external sac that hangs below the penis and contains the testes. The scrotum maintains the temperature of the testes at 2–4 °C below body temperature, as required for spermatogenesis.
- The **epididymides** (singular, **epididymis**) are a pair of long, tightly coiled tubes on the posterior surface of each testis where sperm undergo maturation and become motile. Mature sperm are stored in the epididymides until release.
- The **ductus (vas) deferentia** (singular, **ductus [vas] deferens**) are a pair of long muscular ducts that transfer mature sperm from the epididymides to the urethra.
- The **seminal glands** (also known as **seminal vesicles**) are a pair of accessory glands responsible for the production of seminal fluid, which contains chemicals (eg, fructose, prostaglandins) that provide energy and stimulate contractions to propel sperm through the male and female reproductive tracts.
- The **prostate gland** is an unpaired accessory gland surrounding the urethra between the bladder and penis. The prostate gland secretes prostatic fluid, which contains the enzymes necessary to prevent coagulation of sperm in the vagina.
- The **bulbourethral (Cowper's) glands** are paired accessory glands located inferior to the prostate gland. These glands secrete thick, alkaline mucus, which lubricates the tip of the penis. The alkalinity of the mucus neutralizes acids to protect sperm from the acidic environment of the urethra.
- The **urethra** is a long duct that originates from the bladder and runs through the prostate gland and the shaft of the penis. The urethra is used to transport both semen and urine out of the body (at different times), functioning in both the male reproductive and excretory systems.
- The **penis** is an external erectile structure that contains the urethra, through which ejaculation of **semen** (ie, a combination of sperm and seminal fluid) occurs.

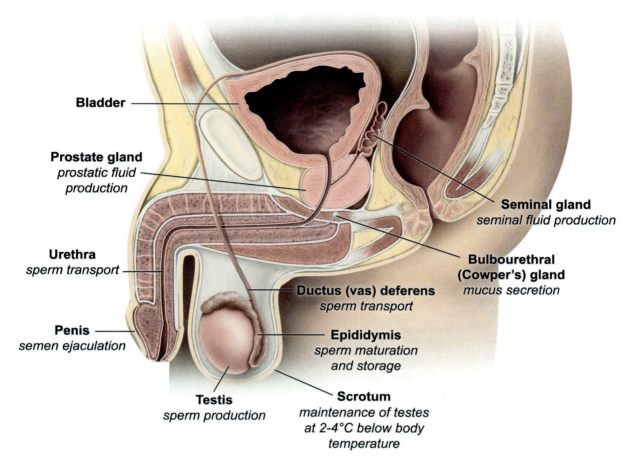

Figure 9.3 Anatomical structures of the male reproductive system.

The testes, surrounded by an outer fibrous capsule, contain numerous compartments filled with hundreds of tightly coiled **seminiferous tubules**. A cross section of these tubules (Figure 9.4) shows developing sperm cells and supporting cells known as **Sertoli (nurse) cells**, which play a critical supportive and regulatory role (eg, providing nutrients, fluids, hormones) in sperm production.

Leydig cells are interstitial endocrine cells found in spaces between adjacent seminiferous tubules (Figure 9.4). Leydig cells stimulate sperm cell development by secreting the steroid hormone testosterone in response to hormones that are released from the anterior pituitary gland (see Concept 9.2.03).

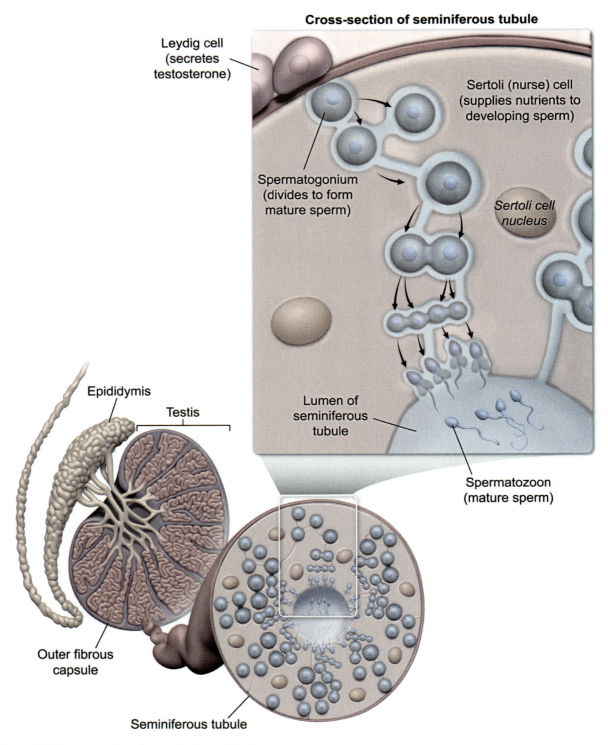

Figure 9.4 Cross section of a seminiferous tubule.

Semen (ie, ejaculatory fluid) is composed of sperm cells suspended in seminal fluid. The highest proportion of seminal fluid comes from the seminal glands, followed by the prostate gland and bulbourethral glands. Seminal fluid has various functions, including serving as a transport medium, supplying nutrients, providing chemicals that protect and activate sperm, and facilitating propulsion of sperm through the male and female reproductive tracts.

A summary of the major components of semen and the origin of each component is provided in Table 9.1.

Table 9.1 Major components of semen.

Source	Major products	Function	Percentage of ejaculate
Testes	Sperm, sex hormones	Fertilization of oocyte (ie, female gamete)	1–5%
Seminal glands	Slightly alkaline seminal fluid containing fructose, prostaglandins	Nourishment of sperm, contraction of male and female reproductive tracts	60–65%
Prostate gland	Prostatic fluid containing anticoagulant enzymes	Prevention of sperm coagulation in vagina	30–35%
Bulbourethral (Cowper's) glands	Thick, alkaline mucus	Neutralization of urinary acids in urethra prior to ejaculation, lubrication of the tip of the penis	<5%

9.2.02 Spermatogenesis

Generation of gametes (ie, spermatozoa) in the male reproductive tract occurs in the seminiferous tubules of the testes. During the fetal period, a population of stem cells known as **spermatogonia** undergo continuous rounds of mitotic division to yield identical diploid (2*n*) daughter cells. After the fetal period, spermatogonia are growth arrested (ie, dormant) until puberty.

At puberty, spermatogonial cell division resumes, yielding identical diploid daughter cells. After each division, one daughter cell differentiates, giving rise to a spermatocyte that enters meiosis to generate four haploid (1*n*) spermatozoa. The other daughter cell remains in reserve as an undifferentiated stem cell which maintains the spermatogonial population.

The differentiation of spermatogonia to form mature, active sperm is called **spermatogenesis** (Figure 9.5). This process begins at puberty and typically continues throughout the lifetime of an individual. During spermatogenesis, a **primary spermatocyte** (a diploid daughter cell produced from a mitotic division of a spermatogonium) undergoes meiosis I to form two haploid daughter cells known as **secondary spermatocytes**. Both secondary spermatocytes undergo meiosis II to form four haploid **early spermatids**, which are nonmotile and do not yet have the characteristic morphology of mature sperm.

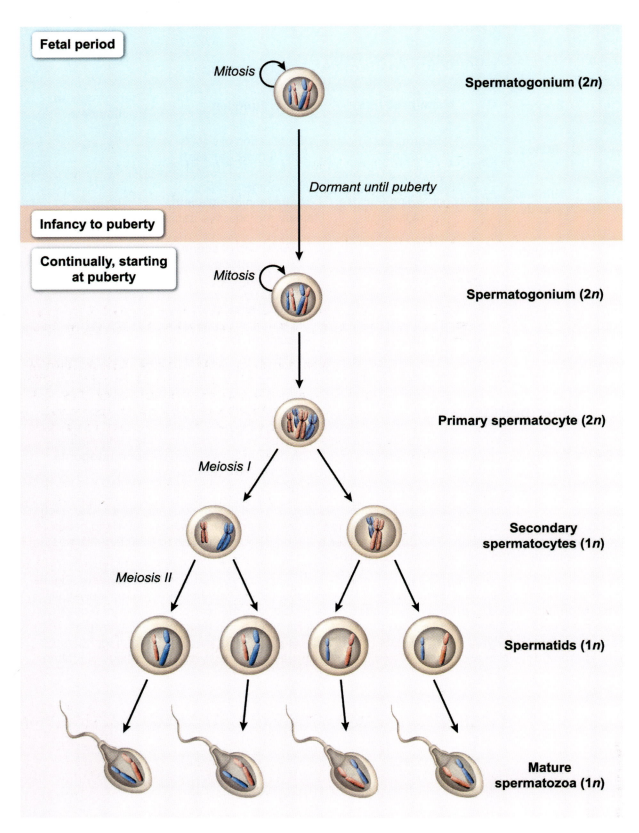

Figure 9.5 Spermatogenesis.

The early spermatids proceed into a maturation phase in which they are transformed into mature sperm cells (ie, **spermatozoa**), as shown in Figure 9.6. Each early spermatid differentiates into an individual elongated sperm cell known as a **late spermatid**. During this process, additional structures (eg, acrosome, flagella, spiral arrangement of mitochondria) are formed and excess cytoplasm is removed, but no additional cell division takes place. Finally, the nonmotile spermatozoa are released into the lumen of the seminiferous tubules and pushed toward the epididymis.

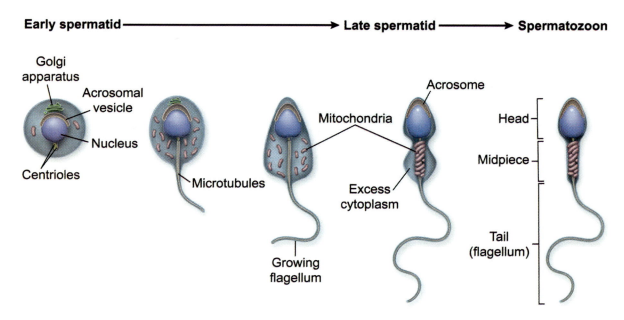

Figure 9.6 Sperm maturation.

The final steps of maturation occur in the epididymis. Over a period of 14 days, spermatozoa gain motility and become capable of fertilizing an oocyte. Mature sperm may be stored in the epididymis for several months. Over time, any unejaculated sperm are eventually reabsorbed at the epididymis lining. In humans, the entire process of spermatogenesis, from spermatogonium to spermatozoon, happens over 64–75 days and occurs continually beginning at puberty. Once spermatogenesis has been initiated (ie, in post-pubertal individuals), developing sperm in all stages of spermatogenesis may be observed in the seminiferous tubules.

A mature spermatozoon is divided into three segments, as depicted in Figure 9.7:

- The **head** contains the nucleus and the **acrosome**, which encapsulates the tip of the nucleus. The acrosome is a flattened vesicular structure that originates from a specialized acrosomal vesicle in the Golgi apparatus. Specialized enzymes (eg, hyaluronidase, proteases) within the acrosome allow sperm to penetrate the zona pellucida (ie, thick extracellular matrix) of an oocyte during fertilization.
- The **midpiece** is packed with mitochondria arranged in a spiral around the microtubules that form the tail. These mitochondria produce the ATP required for flagellum-driven sperm motility. The midpiece also contains a pair of centrioles anchored to microtubules, from which the tail originates.
- The **tail** (ie, flagellum) is a singular elongated structure composed of microtubules that originate from the centrioles of the midpiece (see Concept 5.3.06 for more information on eukaryotic flagella). Sperm motility in the female reproductive tract is powered by the action of the flagellum.

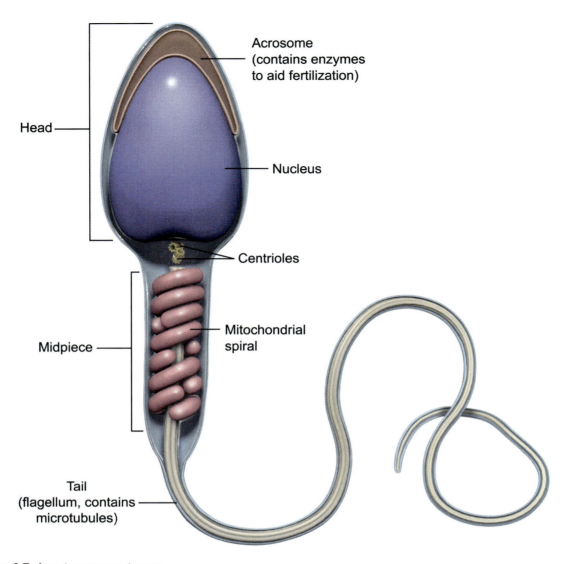

Figure 9.7 A mature spermatozoon.

9.2.03 Hormonal Control of the Male Reproductive System

During the early weeks of embryogenesis, an XY embryo begins to develop male sexual characteristics based on the balanced expression of genes and hormonal signals that activate testis development and repress ovarian development. Internal and external male reproductive structures are typically formed in the presence of these specific signals.

In an XY individual, expression of *SRY* and other genes involved in sex determination typically initiate a cascade of events that result in the development of male reproductive organs. Developing Sertoli (nurse) cells begin to secrete **anti-Müllerian hormone (AMH)**, repressing the development of **Müllerian (paramesonephric) ducts**, from which female reproductive structures are derived. This repression allows androgens (eg, testosterone) secreted by the developing testes to promote development of male reproductive structures derived from **Wolffian (mesonephric) ducts** (Figure 9.8).

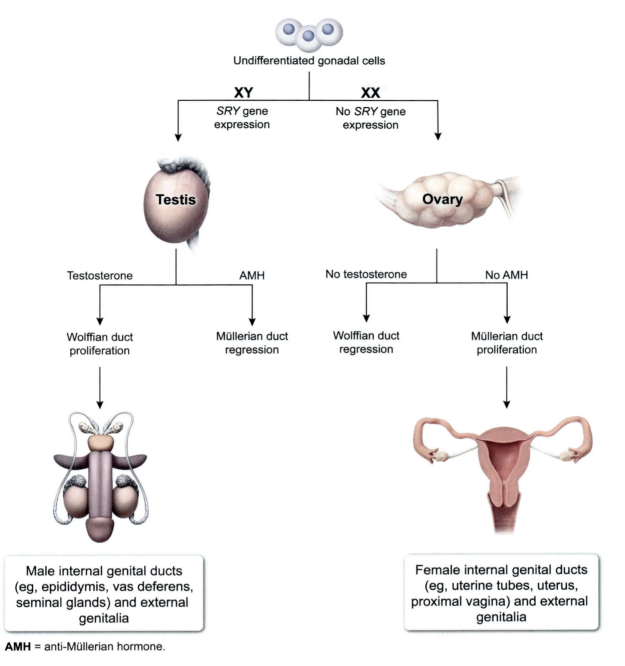

Figure 9.8 Differentiation of male sex organs during development.

During infancy and childhood, sex hormone levels are low. At the onset of puberty, gametogenesis and secondary sexual development are regulated by the coordinated action of the hypothalamus, anterior pituitary gland, and testes, collectively known as the **hypothalamic-pituitary-gonadal (HPG) axis** (Figure 9.9). The hypothalamus secretes **gonadotropin-releasing hormone (GnRH)**, which stimulates the secretion of two gonadotropin hormones, **luteinizing hormone (LH)** and **follicle-stimulating hormone (FSH)**, from the anterior pituitary gland.

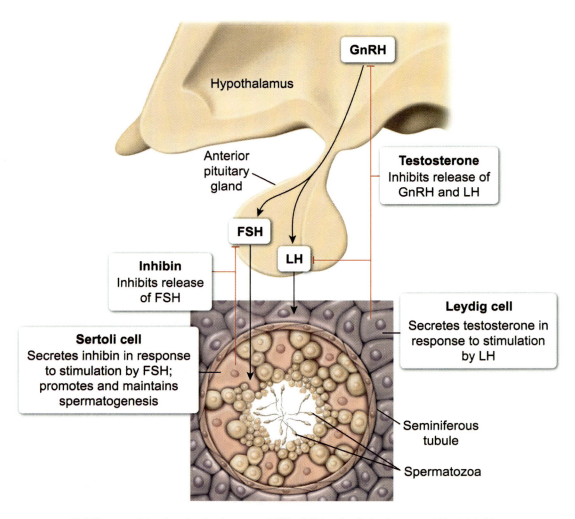

Figure 9.9 Hormonal control of the male reproductive system.

In **Leydig cells**, LH stimulates the production of androgens such as **testosterone**, the dominant male sex steroid hormone essential for spermatogenesis and development of male sex traits. FSH stimulates **Sertoli cells**, thereby promoting and supporting the process of spermatogenesis. Both testosterone and FSH are required to initiate and maintain spermatogenesis.

The male HPG axis is regulated via two negative feedback systems (Figure 9.9). In the first system, testosterone levels regulate secretion of GnRH and LH. High testosterone levels inhibit GnRH and gonadotropin secretion, repressing the production of additional testosterone. If testosterone levels fall too low, this repression is relieved and GnRH is again released by the hypothalamus, stimulating secretion of LH and FSH by the anterior pituitary. In turn, Leydig cells are stimulated to increase testosterone production.

In a second negative feedback mechanism acting on the HPG axis, Sertoli cells regulate the rate of spermatogenesis by secreting the peptide hormone **inhibin**. Inhibin acts on the anterior pituitary to inhibit FSH secretion, thereby inhibiting Sertoli cell stimulation. A decrease in inhibin secretion allows FSH levels to rise, and Sertoli cells are again stimulated to resume the promotion of spermatogenesis.

Concept Check 9.1

Using the following terms, correctly label the structures and hormones that control the male reproductive system: "anterior pituitary gland," "FSH," "GnRH," "hypothalamus," "inhibin," "LH," and "testosterone."

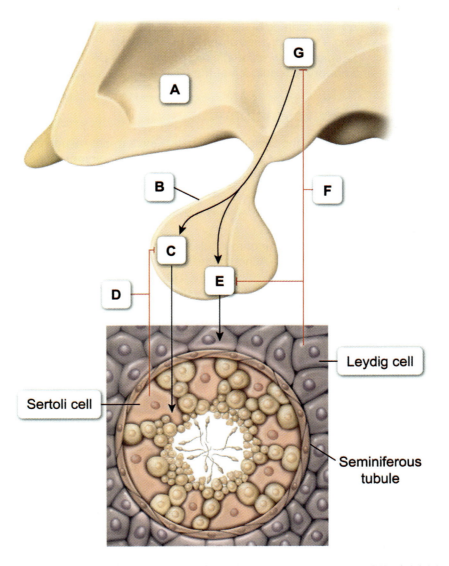

GnRH = gonadotropin-releasing hormone; **FSH** = follicle-stimulating hormone; **LH** = luteinizing hormone.

Solution

Note: The appendix contains the answer.

Lesson 9.3

Female Reproductive System

Introduction

The female reproductive system participates in human reproduction by generating haploid ($1n$) gametes known as **oocytes** and providing a supportive environment for fertilization and growth of offspring. The concepts in this lesson provide an overview of female reproductive anatomy, the process of gamete formation via **oogenesis**, and hormonal regulation of the female reproductive system.

9.3.01 Female Reproductive Anatomy

The internal and external structures of the female reproductive system are responsible for producing gametes (ie, oocytes), receiving sperm, and providing a supportive environment for fertilization, gestation, and nourishment of offspring. As depicted in Figure 9.10, the following structures are associated with the female reproductive system:

- **Ovaries** are reproductive gonads that secrete sex hormones (eg, estrogens, progesterone) and serve as the sites of oogenesis.
- The **uterine (fallopian) tubes** are a pair of muscular tubes originating from the uterus and extending laterally toward each ovary. Although the open ends of the uterine tubes are near the ovaries, there is not a direct connection between the ovary and uterine tube. The open ends of uterine tubes are surrounded by **fimbriae**, fingerlike projections that facilitate direction of the oocyte from the abdominal cavity into the uterine tubes. Ciliated cells line the interior of the uterine tubes and help propel the oocyte toward the uterus. Fertilization typically takes place in the uterine tubes.
- The **uterus** is a muscular organ responsible for protecting and nourishing the embryo and fetus. The inner lining of the uterus, called the **endometrium**, undergoes cyclical changes in thickness at different points in the menstrual cycle. The uterus also contains a thick layer of smooth muscle called the **myometrium**, which contracts during menstruation and childbirth.
- The **cervix** is the most inferior portion of the uterus and serves as the opening into the vagina.
- The **vagina** is a muscular tube that functions in elimination of menstrual fluids during the menstrual cycle, in reception of the penis during sexual intercourse, and as the final segment of the birth canal during childbirth.
- **Female external genitalia (vulva)** include the folds of the **labia majora** (singular: **labium majus**) and **labia minora** (singular: **labium minus**), which protect the external opening of the vagina and the **clitoris**. The external portion of the clitoris lies at the junction of the labia minora folds. Stimulation of the densely innervated clitoris triggers genital changes (eg, lubrication, pH changes) that facilitate reproductive success.
- **Mammary glands** are accessory glands located within the chest wall that become fully developed during pregnancy to facilitate **lactation** (ie, the synthesis and secretion of breast milk) and nursing.

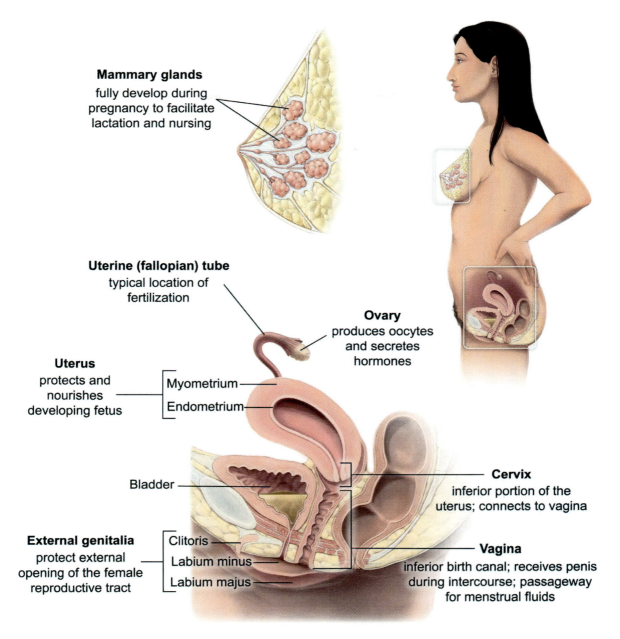

Figure 9.10 Structures of the female reproductive system.

The ovaries are a pair of female gonads homologous (ie, possess the same embryonic origin) to testes. Each ovary is surrounded by an outer fibrous capsule covered by epithelial cells. Within an ovary, the cortex contains many **ovarian follicles** in various stages of maturation. Each follicle contains an immature primary oocyte surrounded by supporting epithelial cells (Figure 9.11).

In the early stages of follicle maturation, the supporting follicular cells proliferate and mature into **granulosa cells**. Granulosa cells support developing oocytes by secreting sex steroid hormones such as estrogens (eg, estradiol) during the ovarian cycle.

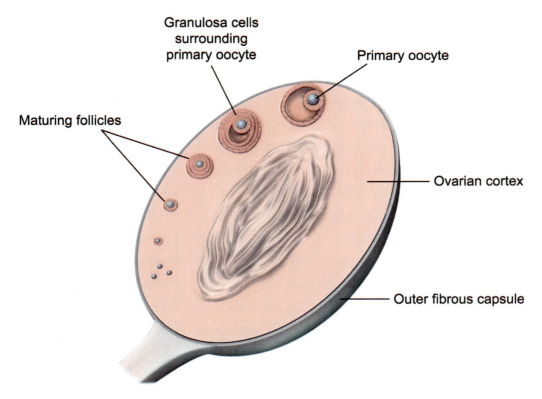

Figure 9.11 Ovary structure.

9.3.02 Oogenesis

Oogenesis begins during embryogenesis, as **oogonia** (stem cells) within developing ovaries undergo continuous rounds of mitotic division to yield identical diploid (2n) daughter cells. Unlike spermatogonia, which initiate meiosis I beginning at puberty, oogonia initiate meiosis I to form **primary oocytes** during the fetal period.

Primary oocytes are arrested (ie, paused) in late prophase I and do not resume the stages of meiosis I again until the beginning of puberty, in response to hormonal changes (see Concept 9.3.03). At the onset of puberty, these primary oocytes are surrounded by follicular support cells in the ovary, forming **ovarian follicles**. Hormonal changes during each menstrual cycle typically result in a single ovarian follicle being selected to continue meiosis and proceed with **ovulation**.

At the completion of meiosis I, unequal cell division of the chosen primary oocyte results in cells of two different sizes: a larger haploid (1n) **secondary oocyte** and a smaller haploid cell called the **first polar body** (Figure 9.12). In humans, the first polar body typically undergoes apoptosis and degenerates but may complete meiosis in some cases. The secondary oocyte initiates meiosis II but arrests at metaphase II until fertilization.

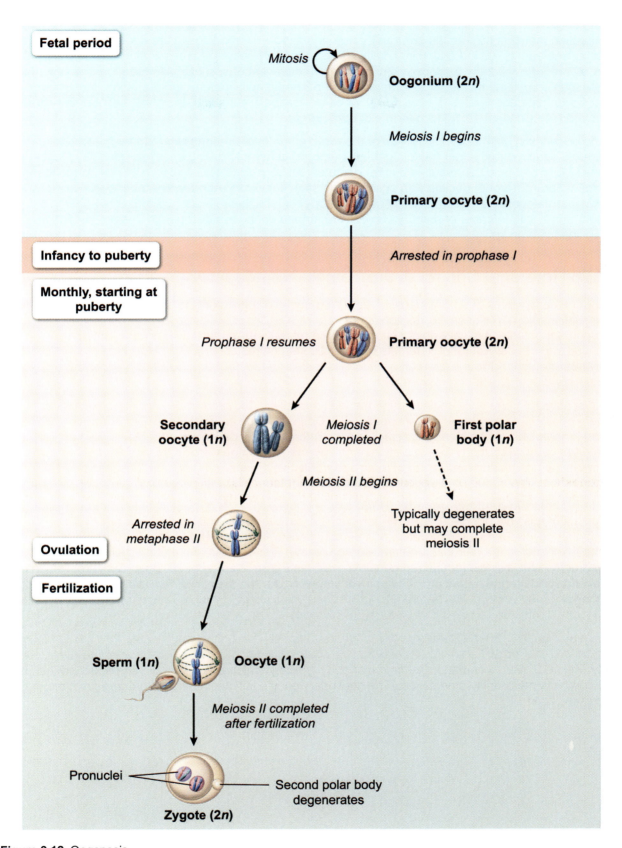

Figure 9.12 Oogenesis.

During ovulation, the ovarian follicle ruptures and the secondary oocyte, surrounded by a thick glycoprotein matrix known as the **zona pellucida** and a group of granulosa cells known as the **corona radiata**, is released into the abdominal cavity. The secondary oocyte is drawn into a **uterine (fallopian) tube**, where fertilization by a sperm cell can occur. In the event of successful fertilization, the secondary oocyte then completes meiosis II to form one large **ovum** (ie, fully mature gamete) and a small **second polar body** that degenerates (Figure 9.12). After fertilization is complete, the male and female **pronuclei** are fused (see Concept 10.1.01) to form a diploid **zygote**.

During oogenesis, each oogonium yields one haploid gamete (ovum) and two to three haploid polar bodies, as opposed to the four haploid gametes (sperm) generated during spermatogenesis. Oogenesis begins before birth and can result in mature gametes 13–50 years later. As levels of sex hormones (eg, estrogens, progesterone) decline with age during a state known as **menopause**, oogenesis ceases. Unlike spermatogenesis, which typically occurs throughout the lifespan after the onset of puberty, it is thought that a finite number of primary oocytes are formed during the fetal period that are not typically replenished later in life.

9.3.03 Hormonal Control of the Female Reproductive System

During the early weeks of embryogenesis, an XY embryo typically begins to develop male reproductive organs initiated by *SRY* gene expression (see Concept 9.2.03). In XX embryos, *SRY* is absent and development of testes typically does not occur. As a result, **anti-Müllerian hormone (AMH)**, is not produced and the development of female reproductive structures derived from **Müllerian (paramesonephric) ducts** is promoted. Because testes are not formed, **Wolffian (mesonephric) ducts** degenerate due to a lack of testosterone.

The first stage of oogenesis begins during the fetal period with the production of follicles containing primary oocytes, as discussed in Concept 9.3.02. During infancy and childhood, sex hormone levels are low and progression of oogenesis is repressed; therefore, sexual development remains dormant until the onset of puberty. At puberty, the secretion of **estrogens** and **progesterone** (ie, sex steroid hormones) promotes the growth and maturation of female reproductive organs, development of secondary sex characteristics, resumption of oogenesis, and initiation of the menstrual cycle.

The **hypothalamic-pituitary-gonadal (HPG) axis** regulates a series of cyclical changes that make reproduction possible (Figure 9.13). The female reproductive cycle includes the **ovarian cycle** (ie, events that occur during the maturation of a primary oocyte) and the **uterine (menstrual) cycle** (ie, events that occur in the uterus to prepare for pregnancy). The ovarian and uterine cycles occur concurrently, with a single female reproductive cycle lasting an average of 28 days. Synchronization of the ovarian and uterine cycles provides optimal conditions for the support of fertilization and early pregnancy.

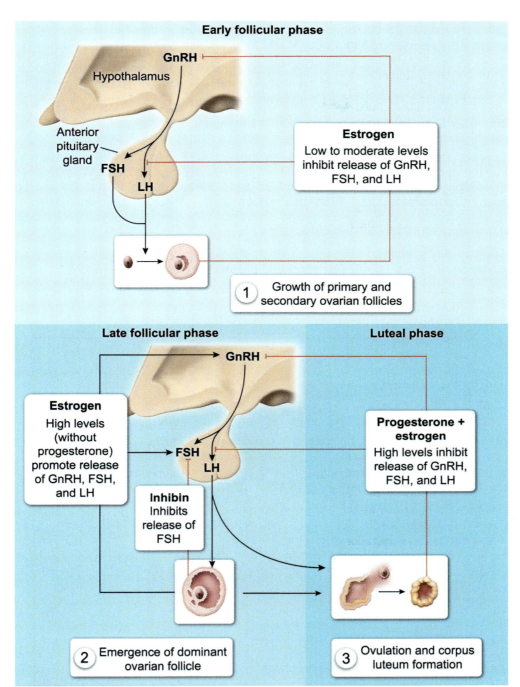

Figure 9.13 Regulation of the female reproductive system.

The ovarian cycle can be divided into three phases (Figure 9.14):

- **Follicular phase** (days 1–13): Stimulation of the HPG axis results in the release of **gonadotropin-releasing hormone (GnRH)** from the hypothalamus, stimulating the anterior pituitary gland to release small amounts of **follicle-stimulating hormone (FSH)** and **luteinizing hormone (LH)**.

FSH and LH stimulate developing ovarian follicles to release estrogens. During the early follicular phase, low to moderate estrogen levels exert negative feedback on the HPG axis, and only one developing follicle typically survives. The surviving (dominant) follicle secretes **inhibin**, which acts on the anterior pituitary to inhibit FSH, repressing maturation of additional follicles during the same ovarian cycle.

Although estrogen initially inhibits the HPG axis, the emergence of a dominant follicle stimulates the secretion of high levels of estrogen, exerting a *stimulatory* effect on the HPG axis during the late follicular phase. This effect results in a surge of LH and, to a lesser extent, FSH.

- **Ovulation** (day 14): Soon after the LH surge, the mature ovarian follicle ruptures, and a secondary oocyte is released into the abdominal cavity, entering a nearby uterine tube.
- **Luteal phase** (days 15–28): LH stimulates the conversion of the ruptured follicle into a structure known as the **corpus luteum**, which secretes high levels of progesterone and estrogens, exerting negative feedback on the HPG axis. Lower FSH and LH levels during the luteal phase prevent maturation of additional follicles. If fertilization does not occur, the corpus luteum degenerates, causing a sharp decline in progesterone and estrogens. This decline relieves the inhibition of FSH and LH, allowing the next ovarian cycle to begin.

If the secondary oocyte is fertilized following ovulation, the corpus luteum persists and continues to secrete progesterone and estrogens. The continued presence of progesterone and estrogens keeps FSH and LH levels low, promoting the pregnancy and preventing a new ovarian cycle from being initiated during pregnancy.

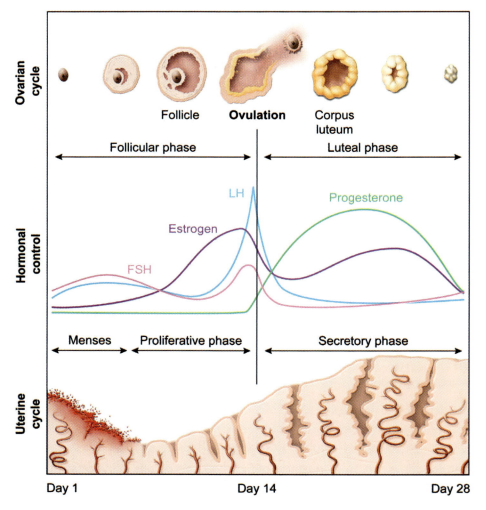

FSH = follicle-stimulating hormone; **LH** = luteinizing hormone.

Figure 9.14 The events of the ovarian and uterine cycles.

The uterine (menstrual) cycle occurs concurrently with the ovarian cycle and is also divided into three phases (Figure 9.14):

- **Menstrual phase** (days 1–4): If fertilization of a secondary oocyte does not occur in the prior uterine cycle, **menstruation** begins. During **menses**, the uterus sheds the majority of its endometrial (ie, inner) layer. The detached tissue and blood pass out of the body through the vagina. Toward the end of this phase, ovarian follicles are stimulated to grow and produce estrogens, resulting in the cessation of blood flow.
- **Proliferative phase** (days 5–14): The endometrial layer proliferates, doubling in thickness. Endometrial glands develop and vascularization of the endometrium is increased to prepare for embryo implantation.
- **Secretory phase** (days 15–28): Increased progesterone and estrogen secretion from the corpus luteum triggers the further thickening and development of the endometrium. These changes result in secretion of nutrients from endometrial glands and the creation of an environment that can sustain a developing embryo in the event of fertilization and implantation.

 Concept Check 9.2

Ovarian function declines naturally with age, resulting in the cessation of the ovarian and menstrual cycles (ie, menopause) around age 45–55 for most individuals. As menopause advances, the number of functional ovarian follicles decreases, and less estrogen is produced by the ovaries. What effect would this decrease in estrogen levels have on the regulation of the HPG axis in a menopausal individual?

Solution

Note: The appendix contains the answer.

Lesson 10.1
Pre-Implantation Development

Introduction

Embryonic development begins when male and female gametes join to form a zygote (ie, fertilization). Although fertilization typically takes place in a uterine tube within 12–24 hours after ovulation, the developing embryo must then travel to the uterine cavity and implant for a viable pregnancy to occur. The developmental steps between fertilization and implantation take place within 12 days after ovulation. This lesson covers **pre-implantation development** from the time of fertilization until the completion of implantation, as well as development of the **placenta** (ie, transient organ which mediates fetal-maternal gas and nutrient exchange).

10.1.01 Fertilization and Blastulation

Fertilization (ie, joining of male and female gametes to form a zygote) becomes possible with the release of the secondary oocyte from the ovary (ie, **ovulation**). Upon ovulation, the oocyte travels into the abdominal cavity and is drawn into the **uterine (Fallopian) tube** by densely ciliated projections known as **fimbriae**. The ciliated lining of the uterine tube helps propel the oocyte into the uterine cavity. Sperm typically make contact with the oocyte within the uterine tube.

Successful fertilization requires the following steps, as outlined in Figure 10.1:

1. **Capacitation:** To reach the oocyte, sperm must undergo a final maturation step in the uterine tube; this step is known as capacitation. Female reproductive tract secretions induce changes in plasma membrane permeability at the sperm head and trigger increased flagellar movement (ie, motility).

2. **Contact with oocyte:** The sperm moves past follicular cells of the oocyte's **corona radiata**, toward the **zona pellucida** (ie, thick matrix of glycoproteins between the corona radiata and plasma membrane) to reach the oocyte. Receptors located in the sperm head bind to glycoproteins in the zona pellucida.

3. **Acrosome reaction:** Located within the sperm head, the acrosome is a specialized vesicle filled with hydrolytic enzymes (see Concept 9.2.02). These enzymes are released near the oocyte, leading to zona pellucida degradation, thereby enabling the sperm to reach the oocyte's plasma membrane.

4. **Fusion:** Oocyte and sperm plasma membranes fuse, triggering depolarization of the oocyte's plasma membrane.

5. **Sperm contents entering the oocyte:** The sperm nucleus, mitochondria, and a pair of centrioles enter the oocyte. However, most sperm mitochondria are destroyed following oocyte entry (see Concept 7.3.06).

6. **Cortical reaction:** Upon sperm fusion, oocyte plasma membrane depolarization leads to increased intracellular calcium (Ca^{2+}) levels, which triggers **cortical granules** (ie, secretory vesicles containing hydrolytic enzymes) to fuse with the oocyte's plasma membrane. Hydrolytic enzymes are released into the space between the plasma membrane and the zona pellucida, causing the zona pellucida to lift away from the oocyte and harden. The cortical reaction results in the formation of a protective envelope that blocks additional sperm from entering (ie, **polyspermy**).

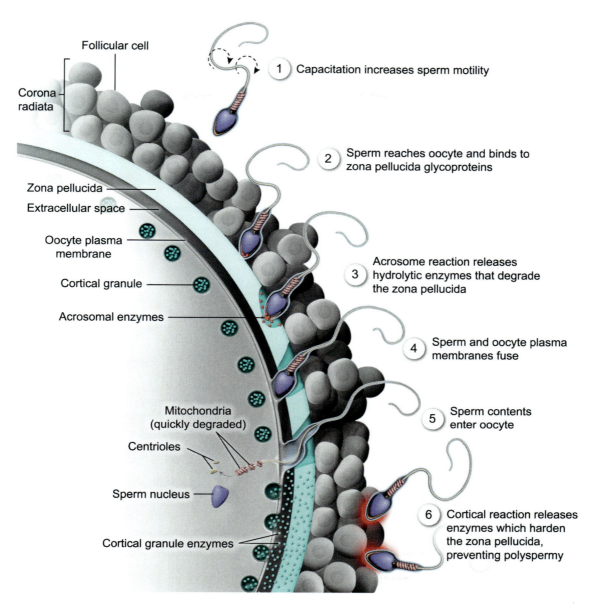

Figure 10.1 Fertilization sequence.

Following sperm and oocyte plasma membrane fusion, sperm contents can enter the oocyte. The secondary oocyte must then complete meiosis II, dividing to form a mature ovum and second polar body, which degenerates (see Concept 9.3.02). The nucleus of the mature ovum develops into a **female pronucleus**. Simultaneously, the **male pronucleus** (ie, sperm nucleus) is propelled toward the female pronucleus. Fusion of haploid (1n) pronuclei produces a diploid (2n) cell known as a **zygote**.

Figure 10.2 provides an overview of the events of spermatogenesis and oogenesis, which culminate in fertilization and zygote production.

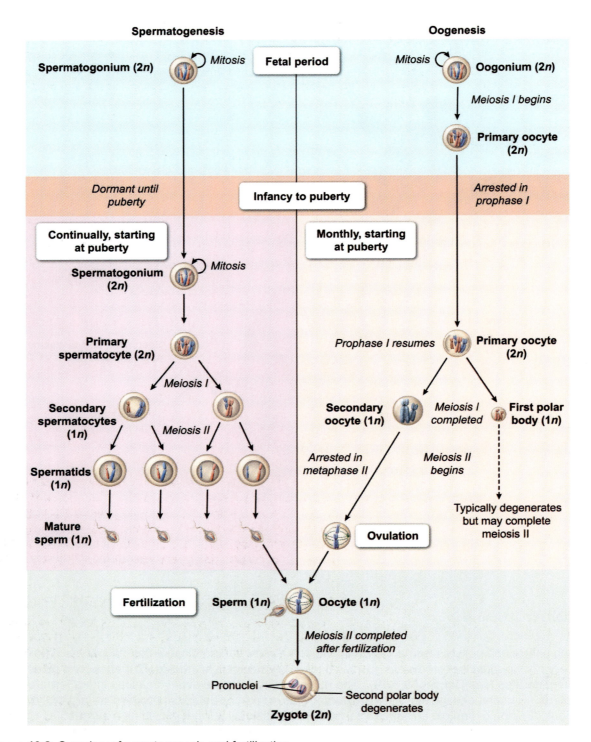

Figure 10.2 Overview of gametogenesis and fertilization.

Immediately following zygote formation, epigenetic modifications (eg, chromatin remodeling, DNA methylation changes) occur and allow for developmental totipotency, the ability to give rise to all embryonic cell types. Notably, some parental epigenetic modifications (eg, imprinted genes) are maintained (see Concept 10.3.02).

Within 24 hours after fertilization, the zygote begins a period of rapid mitotic cell division known as embryonic **cleavage** (Figure 10.3). During cleavage, cell division proceeds without concurrent cell growth (ie, daughter cells become progressively smaller compared to the zygote), forming a mass of identical cells known as **blastomeres**. After 3–4 days, the embryo enters the **morula** stage. The morula, roughly the same size as the original zygote, consists of 16–32 blastomeres. During the transition from zygote to morula, the developing embryo travels from the uterine tube towards the uterine cavity.

By days 4–5, the rapidly dividing embryonic cells reorganize into a **blastocyst**, which contains a hollow, fluid-filled cavity known as the **blastocoel**. The blastocyst contains two distinct cell populations: an **inner cell mass**, which contains cells that develop into the embryo, and a **trophoblast**, the outer cell layer that develops into extraembryonic tissues (eg, placenta).

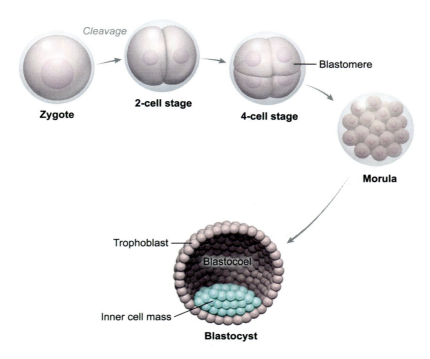

Figure 10.3 Early embryonic development from zygote to blastocyst.

Concept Check 10.1

For successful fertilization of the oocyte by the sperm, a series of steps must occur in sequence. Using the following list, provide a rationale for why unsuccessful completion of each step could lead to a failed fertilization sequence:

1. Capacitation
2. Acrosome reaction
3. Cortical reaction

Solution

Note: The appendix contains the answer.

10.1.02 Implantation and Placental Development

For development to continue beyond the blastocyst stage, the blastocyst must implant into the uterine lining (Figure 10.4). By 6–7 days after fertilization, surges in progesterone and estrogen trigger the secretory phase of the uterine cycle, and the endometrial lining is primed to support **implantation** of the blastocyst (see Concept 9.3.03).

Developing trophoblastic cells (ie, cells from the outer portion of the blastocyst) adhere to the endometrial surface, causing the blastocyst to burrow into the endometrial lining. Endometrial cells proliferate and surround the blastocyst, now known as the **embryo**, sealing it off from the uterine cavity. Until the placenta is completely formed, the endometrium supports and nourishes the developing embryo.

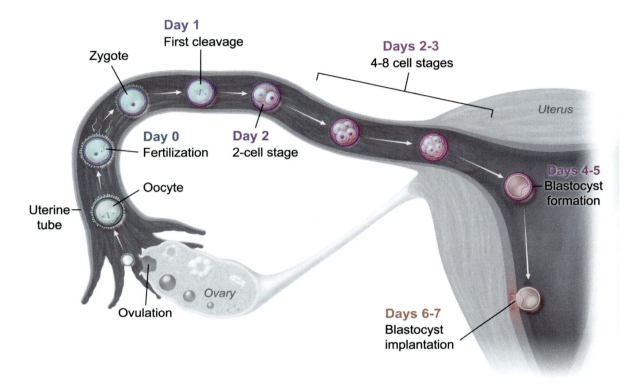

Figure 10.4 Blastocyst implantation.

The process of implantation is usually complete by day 12. If fertilization does not occur, the corpus luteum typically degenerates, and declining progesterone and estrogen levels trigger menstruation. However, when successful implantation occurs, the developing trophoblastic cells begin to secrete the hormone **human chorionic gonadotropin (hCG)**, and the corpus luteum is maintained while the placenta develops.

The **placenta** forms over the first 3 months of pregnancy to facilitate nutrient, gas, and waste exchange between the maternal circulation and the developing embryo. An extraembryonic membrane known as the **chorion** is derived from the trophoblast following implantation. **Chorionic villi**, fingerlike projections that invade the endometrium, form from the chorion and become vascularized. The chorion and chorionic villi form the bulk of the placenta (Figure 10.5).

The corpus luteum continues to secrete estrogen and progesterone for approximately the first 9 weeks of pregnancy, effectively inhibiting the initiation of another menstrual cycle. However, the developing placenta gradually takes over secretion of hCG, estrogens, and progesterone to support the pregnancy, leading to corpus luteum degeneration.

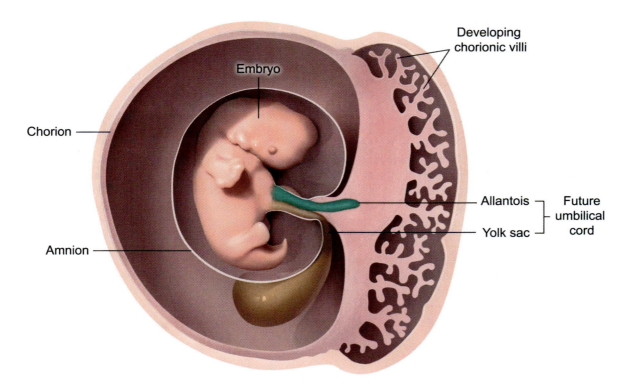

Figure 10.5 Developing placenta and extraembryonic structures.

The embryo is supported by several extraembryonic structures derived largely from the blastocyst inner cell mass before and during placenta formation (Figure 10.5):

- **Yolk sac:** a region where maternal-fetal nutrient and gas exchange occurs prior to placenta formation and the main site of embryonic blood cell production prior to the fetal period.
- **Allantois:** formed from the yolk sac, a structure that mediates fluid and waste exchange between the embryo and yolk sac. The umbilical cord is eventually formed from the yolk sac and allantois.
- **Amnion:** a tough membrane filled with amniotic fluid that surrounds the embryo.

Once fully developed, the placenta receives blood from both the maternal and the fetal circulatory systems, although these two systems do not typically intermix (see Concept 10.4.02).

Lesson 10.2
Post-Implantation Development

Introduction

Successful implantation of the blastocyst signals the beginning of the embryonic period of development. **Organogenesis** (ie, the development of embryonic tissues and organs) begins with **gastrulation**, a period of rapid growth and reorganization of embryonic cells. By the end of gastrulation, the three primary germ layers are established, from which all embryonic tissues and organs are formed. Shortly thereafter, mesodermal tissue induces the overlying ectoderm to become neural tissue, in a process known as **neurulation**.

By the end of the embryonic period, the development of all major body systems is well underway and continues throughout the remainder of the pregnancy. This lesson covers the major events of post-implantation development during the embryonic period. More information about the fetal period, along with other aspects of pregnancy and parturition, can be found in Lesson 10.4.

10.2.01 Gastrulation

At the time of implantation, the blastocyst consists of two major cell types: trophoblast cells and the inner cell mass (ICM). Before gastrulation, the ICM cells are organized into two cell types: the **epiblast** (upper layer) and **hypoblast** (lower layer). A bilaminar (ie, two-layered) disk, consisting of a portion of the epiblast and the hypoblast, eventually becomes the embryo proper. The cells of the epiblast form the embryo and amnion, and the cells of the hypoblast form a portion of the chorion and the yolk sac, structures that are discussed in more detail in Concept 10.1.02.

During week three of development, ICM cells undergo a period of rapid growth, differentiation, and cellular rearrangement known as **gastrulation**, establishing the three **primary germ layers** from which all embryonic tissues and organs arise: ectoderm, mesoderm, and endoderm.

Gastrulation begins with the formation of a groove known as the **primitive streak** within the epiblast which establishes the anterior-posterior axis of the embryo. Along the primitive streak, cells that will form the internal organs migrate into the interior space of the embryo, and cells that will form the skin and nervous system migrate along the outer surface. The process of gastrulation is illustrated in Figure 10.6.

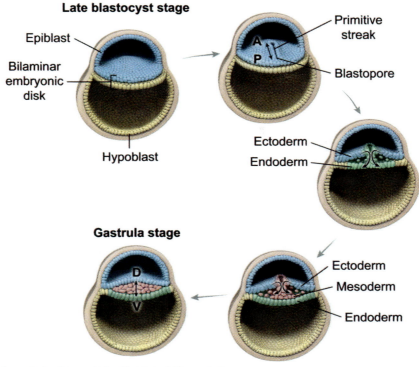

Figure 10.6 Gastrulation.

The **blastopore**, or first point of invagination during gastrulation, forms as endodermal cells invaginate at the posterior end of the primitive streak. In mammals, the blastopore eventually becomes the anal opening of the embryo. As development continues, further migration of endodermal cells creates a primitive gut cavity called the **archenteron**.

As the three primary germ layers are formed, the relatively flat embryonic disk also undergoes significant changes in size and shape. Because of asynchronous rates of growth and cell movement, the embryo folds into a three-dimensional form in which ectodermal cells are largely found on the exterior of the embryo, endodermal cells are found on the interior, and mesodermal cells are found in between (Figure 10.6).

Gastrulation results in a three-layered embryo called a **gastrula**. Following gastrulation, the three primary germ layers have different fates, as depicted in Figure 10.7:

- The **ectoderm** develops into the skin and nervous system.
- The **mesoderm** gives rise to muscle, bone, and other connective tissues.
- The **endoderm** forms the lining of the digestive tract and other internal organs.

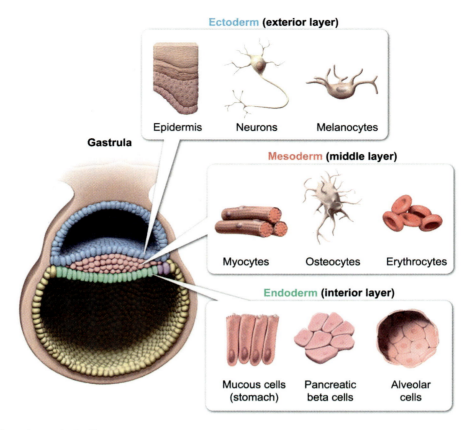

Figure 10.7 Germ layer derivatives.

A more comprehensive list of germ layer derivatives is included in Table 10.1. Note: Although it is unlikely that every germ layer derivative must be memorized, past exams have included highly specific questions regarding germ layer derivatives, with particular attention to counterintuitive examples (eg, adrenal medulla is derived from ectoderm, whereas adrenal cortex is derived from endoderm).

Table 10.1 Detailed derivatives of the three primary germ layers.

Ectoderm	Mesoderm	Endoderm
• Brain and spinal cord • Neuroepithelia of sense organs (eg, retina, inner ear, olfactory) • Neural crest derivatives (eg, cranial, spinal, and sympathetic ganglia, adrenal medulla, melanocytes, calcitonin-producing thyroid cells, some facial connective tissues) • Anterior and posterior pituitary, pineal gland • Epidermis, epithelia of hair follicles, skin glands, and mammary glands • Lens and cornea of eye • Internal and external ear • Epithelia of oral cavity, nasal cavity, salivary glands, and inferior portion of anal canal • Skeletal and connective tissue of head	• All skeletal and cardiac muscle tissue, most smooth muscle tissue • Most cartilage, bone, and other connective tissues • Cardiovascular and lymphatic systems • Spleen and hematopoietic cells • Dermis • Adrenal cortex • Kidneys and ureters • Gonads and genital ducts (except germ cells, which are specified at the epiblast stage)	• Epithelial lining of digestive canal (except oral and anal cavities) • Epithelial lining of urinary bladder, gallbladder, and liver • Epithelial lining of respiratory tract • Epithelia of thyroid, parathyroid glands, pancreas, and thymus • Epithelial lining of middle ear • Epithelial lining of reproductive tract

10.2.02 Organogenesis and Neurulation

The formation of specific tissues and organ systems begins after primary germ layer establishment. During **organogenesis**, progressive cell differentiation turns the relatively undifferentiated primary germ layers into functional organs and organ systems. Exposure to signaling molecules from an underlying mesodermal structure called the **notochord** induces overlying ectodermal cells to form neural tissues in a process called **neurulation**.

Three ectodermal cell types arise during neurulation: **epidermal** cells (eg, the outer layer of the skin), **neural crest** cells (eg, precursor cells of the peripheral nervous system [PNS], melanocytes) and **neural plate** cells (eg, precursor cells of the central nervous system [CNS]). At this stage of embryogenesis, the embryo is called a **neurula**.

Once ectodermal cell types are specified, neurulation continues in the neural plate via the following steps (Figure 10.8):

1. Thickening and elongation of the neural plate in response to additional signals from the underlying mesoderm.

2. Upward movement of the edges of the neural plate to form **neural folds** on either side of a central **neural groove**.

3. Migration of lateral edges of the neural folds toward the midline of the embryo, fusing to form a **neural tube** that sits inferior to an overlying ectodermal layer. Neural crest cells **delaminate** (ie, detach) from the neural tube as the tube closes and these cells migrate to various regions throughout the embryo.

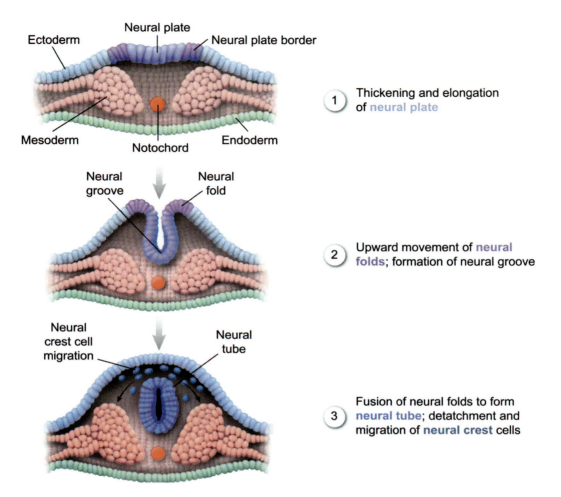

Figure 10.8 Neurulation.

Neural crest cells differentiate into a number of diverse cell types, most notably the neurons and glia of the PNS (eg, sensory and autonomic ganglia, adrenal medulla, Schwann cells), melanocytes (pigment cells), calcitonin-producing thyroid cells, and facial connective tissue.

After neural tube formation, organogenesis continues with the segmentation of mesodermal tissues adjacent to the neural tube into **somites**, clusters of muscle and connective tissue precursor cells found in the back (eg, precursors to vertebrae, ribs, dermis). Concurrently, tissues derived from nearby mesoderm and endoderm give rise to the remaining organs and organ systems. See Concept 10.2.01 for more details about the structures derived from each germ layer.

A progression of the stages of pre- and post-implantation development is shown in Figure 10.9.

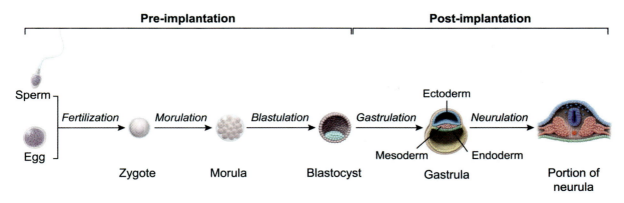

Figure 10.9 Summary of pre- and post-implantation development from fertilization to neurulation.

> ☑ **Concept Check 10.2**
>
> Sort the following tissues by the germ layer of origin (ectoderm, mesoderm, or endoderm):
>
> | | | |
> |---|---|---|
> | Adrenal medulla | Dermis | Rib cartilage |
> | Alveoli of lung | Oral epithelium | Schwann cells |
> | Anterior pituitary gland | Red blood cells | Sympathetic ganglia |
> | Capillary endothelium | Reproductive tract epithelium | Urinary bladder epithelium |
>
> **Solution**
>
> *Note: The appendix contains the answer.*

Lesson 10.3
Cellular Mechanisms of Development

Introduction

Mammalian development begins with the zygote, a single cell that gives rise to all tissue types and organs in the mature organism. The organization and transformation of cells during the developmental period are carried out by a complex array of cellular mechanisms that depend on the precise coordination of signal and response (ie, development is highly regulated in both space and time).

This lesson provides an overview of some of the cellular and molecular mechanisms that can be used to execute a developmental plan. Mechanisms such as asymmetrical cell division, cell signaling and migration, differential gene expression, and programmed cell death act synergistically to drive the proper patterning and growth of the embryo.

10.3.01 Cell Potency and Specialization

A single-celled zygote has the potential to give rise to every cell type (ie, both embryonic and extraembryonic) and is thus considered **totipotent** (Figure 10.10). Because of asymmetrical cell division during embryonic cleavage, the fate of specific embryonic blastomeres becomes progressively restricted beyond the 8-cell stage.

As an embryo progresses toward blastulation, individual blastomeres are restricted to either embryonic or extraembryonic cell types; therefore, the blastomeres are no longer totipotent. Instead, these cells are considered **pluripotent** (ie, able to give rise to either embryonic or extraembryonic cell types but not both). For example, as the embryo approaches the blastocyst stage, trophoblast cells are already restricted to extraembryonic cell fates (eg, precursors to the placenta) and are no longer able to contribute to the embryo itself (Figure 10.10).

As development continues, the fate of individual embryonic cells becomes even more restricted. At neural tube closure, the cells of the neural tube are **multipotent**, or able to generate multiple cell types within a restricted population or lineage (eg, ectodermal cells are restricted to neural, epidermal, and neural crest cell fates). Continued development may yield further restriction of multipotent cells or **unipotent** cells (ie, generate only one cell type). Through differential gene expression, **terminal differentiation** of these cells leads to unique cell types with characteristic biochemical signatures, structures, and functions.

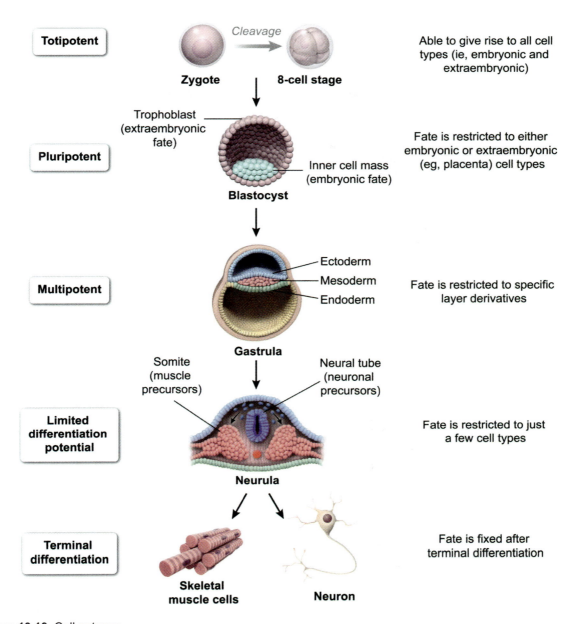

Figure 10.10 Cell potency.

Stem cells retain their potency and ability to divide but may generate daughter cells that terminally differentiate into specific cell types. Depending on the source, stem cells can give rise to many cell types (eg, pluripotent embryonic stem cells of the inner cell mass) or only a few (eg, multipotent hematopoietic stem cells). Although stem cells are usually discussed in the context of embryonic development, it is important to note that stem cell populations also exist in many adult tissues (eg, bone marrow, brain).

Some stem cell populations can **self-renew**. As a result of asymmetrical cell division, one daughter cell may terminally differentiate into a specific cell type, whereas the other retains stem cell properties that maintain the progenitor cell pool (Figure 10.11). For example, during early spermatogenesis, unipotent spermatogonial stem cells generate daughter cells that either differentiate into primary spermatocytes or maintain the spermatogonial stem cell population so that additional spermatozoa can be generated continually throughout life (see Concept 9.2.02).

If stem cell division is symmetrical, both daughter cells may retain stem cell properties in some instances, thereby enlarging the stem cell population. In other instances, both daughter cells derived from the stem cell may terminally differentiate, effectively depleting the stem cell population over time (Figure 10.11).

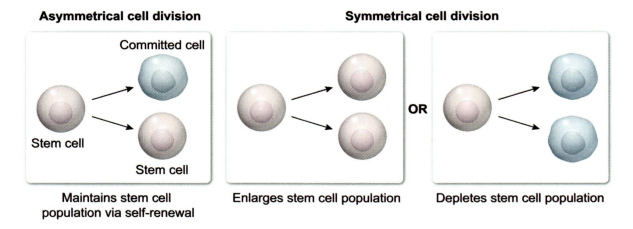

Figure 10.11 Stem cell division and self-renewal.

The process by which cells become committed to a specific cell fate occurs in stages. Totipotent cells are said to be unspecified. **Cell specification** begins very early in development when a cell becomes capable of differentiating into a particular cell type on its own in a neutral environment. However, the fate of a specified cell is not fixed and may be changed by external cues. For example, a cell specified as mesodermal may retain the potential to become a neuron if placed in conditions which induce neuronal development.

A cell is said to be **determined** when it is tied to a specific fate, even when placed in an alternate environment (ie, the cell's developmental fate is *not* altered by external cues). For example, once a mesodermal cell is determined as a muscle cell precursor, that mesodermal cell cannot become a neuron, even when placed in conditions which induce neuronal development.

Once determined, progenitor cells are induced to undergo terminal differentiation, developing specialized structures and functions. In some cases (eg, neuronal progenitor cells), terminally differentiating cells may leave the cell cycle and do not divide again.

10.3.02 Gene Regulation in Development

A fundamental question in development centers on how differences arise in seemingly equivalent (ie, totipotent) embryonic cells. Although the full mechanism is unknown, the progressive determination and differentiation of unique cell types are directed by differential gene expression resulting from a combination of cell **intrinsic** (internal) and **extrinsic** (external) **factors**. Early embryogenesis is driven largely by intrinsic factors, with later patterning involving extrinsic factors, resulting in progressive cell fate restriction, as summarized in Table 10.2.

Table 10.2 Cell intrinsic and extrinsic factors in early embryogenesis.

Cell intrinsic factors	Cell extrinsic factors
Epigenetic • DNA methylation • Chromatin modification	Paracrine/juxtacrine signaling factors • Growth factors • Morphogen gradients • Lateral inhibition
Asymmetric cell division • Unequal distribution of cellular factors (eg, mRNA transcripts, transcription factors, proteins)	Environmental conditions • Oxygen concentration • Nutrient availability • Temperature • Physical forces (eg, flow, tension) • Exposure to exogenous chemicals
Mutations • Mistakes in DNA replication/repair	Extracellular matrix (ECM) • Three-dimensional organization • ECM-bound factors

In the early cleavage stage of embryogenesis, cell division is rapid and the embryo is dependent on oocyte stores of mRNA and proteins. The embryonic genome becomes active beginning at the 4- to 8-cell stage, but oocyte contributions persist through the blastocyst stage. The embryonic genome must undergo extensive epigenetic remodeling before genome activation. This remodeling allows transcription factors better access to chromatin, which is necessary for the proper regulation of embryonic gene expression.

During remodeling, cytosine nucleotides may be methylated, which generally has a silencing effect on gene expression. Most genomic DNA is demethylated shortly after fertilization; however, for certain genes, the maternal and paternal methylation patterns remain intact, and alleles are not functionally equivalent. Therefore, methyl group retention on specific alleles leads to parent-specific gene expression in offspring, a phenomenon known as **genomic imprinting** (Figure 10.12). When the paternal allele is imprinted (ie, methylated), gene expression occurs only from the maternal allele and vice versa.

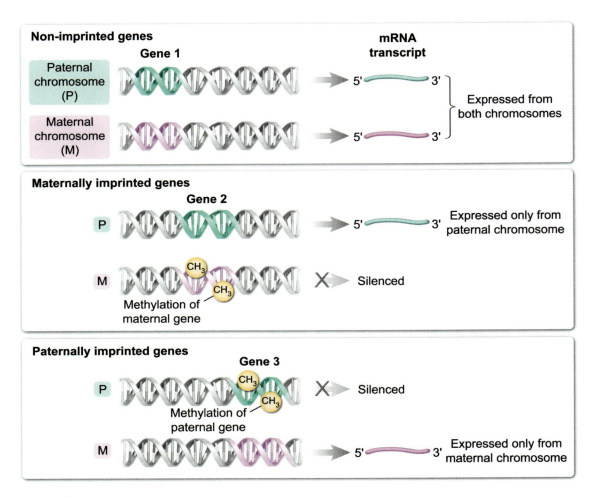

Figure 10.12 Genomic imprinting.

Because of differences in primary sex determination, individuals with two X chromosomes have twice as many copies of X chromosome genes than XY individuals. This phenomenon has important implications in terms of X-linked inheritance (discussed in Concept 7.3.02).

XX individuals compensate for this disparity via **X chromosome inactivation (X inactivation)** in early embryonic development. During X inactivation, one X chromosome in each cell becomes inactivated due to expression of the *XIST* (X-inactive specific transcript) gene. *XIST* expression initiates epigenetic X chromosome modification, resulting in extensive chromatin remodeling, DNA methylation, and gene silencing (see Concept 2.4.02). This causes the X chromosome expressing *XIST* to condense, forming a compact structure known as a **Barr body**.

X chromosome inactivation occurs in each embryonic cell on a random basis. Therefore, in any given cell from an XX individual, the active X chromosome may be of paternal or maternal origin. If an XX individual is heterozygous for an X-linked trait, roughly half of the individual's cells express the maternal allele, and the other half express the paternal allele (Figure 10.13).

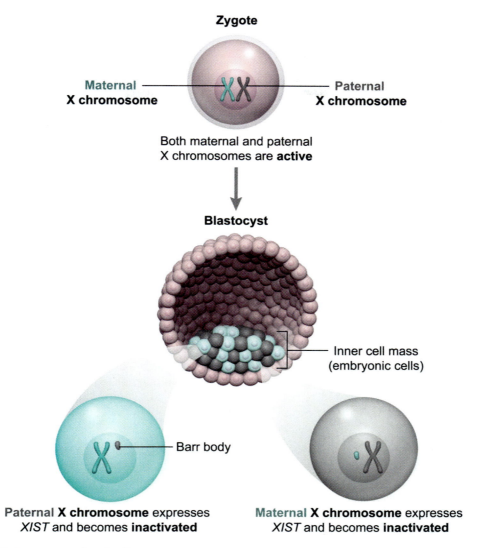

Figure 10.13 X chromosome inactivation.

As discussed in Concept 2.4.03, regulatory elements found outside of DNA coding regions contain enhancer sequences that, when bound by transcription factors, modulate gene expression levels. Tissue-specific gene expression during development arises due to unique combinations of transcription factors expressed in individual cells.

Enhancer sequences are often modular, meaning that distinct enhancers associated with a single gene can regulate that gene's expression in different tissues and at different times during development. Only specific combinations of transcription factors bound to an enhancer allow a gene to be expressed in a particular cell type. This same gene remains unexpressed in cells that do not have the correct combination of transcription factors.

For example, the transcription factor PTF1A is known to be expressed in both pancreatic and cerebellar progenitor cells during development (Figure 10.14). Two distinct regulatory enhancers controlling *PTF1A* expression exist: one that controls expression in the developing cerebellum and one that controls expression in the developing pancreas. Mutations within individual enhancers disrupt development in a tissue-specific fashion, affecting either the pancreas or cerebellum but not both (ie, a mutation in the pancreatic enhancer affects the pancreas, but not the cerebellum).

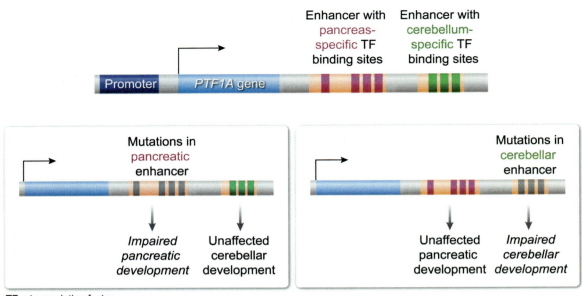

Fig 10.14 Example of enhancer sequence modularity.

In addition to an organism's genetic makeup, environmental conditions also play a role in development. Environmental cues may influence organism development by altering cell signaling and gene expression at crucial times during development. Epigenetic changes due to diet, temperature, and other environmental exposures may alter the progression of development in both small and highly impactful ways.

One well-known example of environmental effects during human development involves deficiencies in dietary folate. Low maternal folate intake can affect neural tube closure in the developing embryo and is one of the leading causes of neural tube defects in humans. Although the exact mechanism is unknown, it is hypothesized that disruption of folic acid metabolism may affect DNA methylation in the developing nervous system. It is estimated that 25%-30% of neural tube defects in humans can be prevented by taking supplemental folate during pregnancy.

10.3.03 Cell Signaling and Migration

Cell-cell signaling drives many developmental cellular activities (eg, division, adhesion, migration, differentiation). For example, ectodermal cells are induced to become neural plate cells in response to signal secretion by the underlying notochord (see Concept 10.2.02). **Induction** involves the cells or tissues that produce signaling molecules (ie, **inducers**) and the responding cells or tissues, which must be **competent** (ie, able to receive and respond to inductive signals). For example, a secreting cell induces differentiation only in neighboring cells that express the correct receptors and signal transduction machinery.

Often, inductive interactions are reciprocal: The inducing cells trigger the secretion of new inductive signals in the neighboring cells, which then act on the original inducing cells to further refine developmental fates. For example, in vertebrate eye development, the ectoderm overlying the optic vesicle (ie, developing eye) induces lens formation, and the newly formed lens cells reciprocate, instructing the optic vesicle to form the retina.

Inducers are most often paracrine factors but may also be autocrine or juxtacrine signals (Figure 10.15). Secreted **paracrine** signals exert their effects on nearby cells via diffusion, whereas **autocrine** signals exert effects on the same cell from which they are secreted. **Juxtacrine** signals are cell surface ligands from one cell that interact with receptors on adjacent cells. Before circulatory system development, endocrine signaling is not prominent.

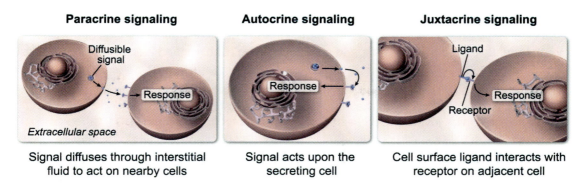

Figure 10.15 Types of inductive cell signaling.

Signaling gradients are a key feature of early embryonic development. **Morphogens** are inductive paracrine factors that diffuse from a signaling cell to form a concentration gradient within nearby tissues. The fates of receiving (ie, competent) cells within a diffusible distance of the morphogen are influenced by the concentration of morphogen to which the receiving cells are exposed (Figure 10.16).

In addition to diffusion, morphogen gradients are influenced by time-dependent morphogen destruction and/or uptake into cells. Overlapping morphogen gradients may be used to establish precise patterns of gene expression and cell differentiation during development.

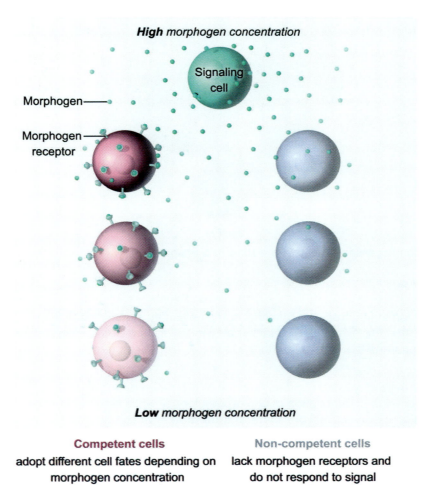

Figure 10.16 Morphogens influence the fates of competent cells based on morphogen concentration.

For example, morphogen concentration in the neural tube influences the differentiation of spinal neuron populations along the dorsal-ventral axis of the neural tube. Opposing concentrations of the morphogens bone morphogenetic protein (BMP), Wnt, and sonic hedgehog (Shh) in the neural tube specify dorsal and ventral cell fates (Figure 10.17).

High levels of BMP and Wnt are secreted from the overlying ectoderm and most dorsal regions of the neural tube, inducing gene expression that leads to dorsal cell fates (ie, differentiation into sensory neurons). In contrast, high levels of Shh secretion from the notochord and most ventral regions of the neural tube induce the expression of genes that lead to ventral cell fates (ie, differentiation into motor neurons).

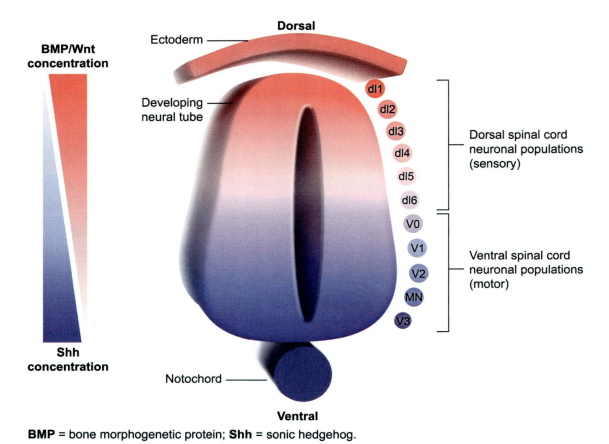

BMP = bone morphogenetic protein; Shh = sonic hedgehog.

Figure 10.17 Opposing morphogen gradients in neural tube development.

Although less common than paracrine signaling, some juxtacrine signaling pathways are crucial in developmental biology. For example, the Notch pathway involves a Notch receptor protein embedded in an inducing cell membrane that interacts with a ligand (eg, Delta) embedded in the receiving cell membrane.

Adjacent progenitor cells may often express both the Notch receptor and its ligand. This type of signaling may be used to induce the differentiation of one progenitor cell while keeping the other in an undifferentiated state, a process known as **lateral inhibition**. For example, in two adjacent cells, the cell in which the Notch pathway is first engaged leads to the inhibition of the Delta ligand within the same cell, thereby preventing the ligand from activating Notch in the adjacent cell. Therefore, the first cell to activate the Notch pathway undergoes differentiation, and the adjacent cell remains undifferentiated, as shown in Figure 10.18.

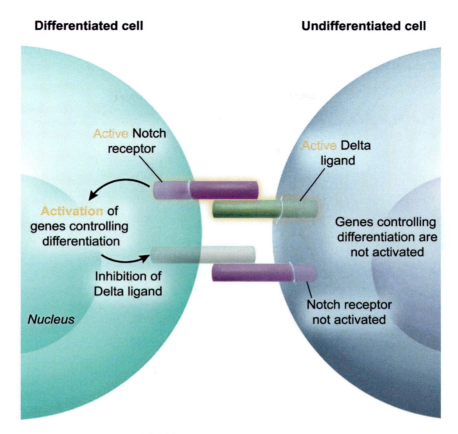

Figure 10.18 Notch signaling and lateral inhibition.

As development proceeds, some cells must migrate short and long distances to arrive at the appropriate locations. For example, during gastrulation, many complex cell movements and rearrangements occur to transform the embryo from a flat, two-layered disk into a complex tubular structure. Cells migrate by responding to environmental cues and surrounding cells. Because of embryonic patterning, a migrating cell encounters different environments as it moves within the embryo. Cells migrate (either individually or in groups) in response to short- and long-range signals to reach their final destination.

The migration of neural crest cells is an example of cell migration during development (see Concept 10.2.02). Following neural tube closure, some cells (ie, neural crest cells) lose their adhesive junctions and separate from the epithelium in a process called **delamination**. Once delaminated, neural crest cells migrate extensively along the anterior-posterior axis, giving rise to a wide variety of different cell types, including cells of the peripheral nervous system, adrenal medulla, melanocytes, and head/neck connective tissues.

Both external (eg, environmental signals) and internal factors (eg, changing patterns of adhesion proteins) drive neural crest migration. These factors guide cells via both attractive and repulsive interactions through different environments and over long distances.

10.3.04 Programmed Cell Death

Just as cell proliferation is crucial during embryonic development and morphogenesis, so too is restricting the number of cells produced. Normal development often generates an excess number of cells, with surplus cells later removed via apoptosis (ie, programmed cell death). During mammalian development, apoptosis occurs at several points, including during blastocyst formation, shaping of tubular structures, and separation of digits (Figure 10.19).

Correctly timed apoptosis is critical during vertebrate limb development. During development, each digit is specified by exposure to sequential signaling gradients (eg, morphogens) from the posterior interdigital tissue (ie, mesodermal tissue between each digit). After digit formation, apoptosis of the interdigital webbing is induced to separate the limb bud into distinct digits.

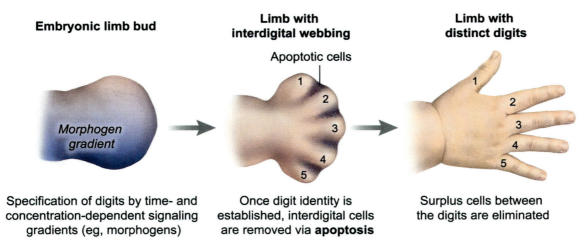

Figure 10.19 Apoptosis in the separation of digits.

 Concept Check 10.3

Indicate whether each of the following statements about cell-cell signaling during development are true or false and why.

1. Cells in a developing tissue can respond to multiple morphogens, leading to the establishment of more complex differentiation patterns.
2. Lateral inhibition is a mechanism by which cells in a developing tissue enhance each other's differentiation, leading to uniform cellular identities.
3. During embryonic development, the timing of induction is critical because cells can respond to certain inductive signals only during specific developmental stages.

Solution

Note: The appendix contains the answer.

Lesson 10.4

Gestation

Introduction

Gestation (ie, pregnancy) generally lasts approximately 38 weeks, from ovulation until birth. However, the gestation period is often expressed as 40 weeks when considered from the beginning of the last menstrual cycle. During gestation, many maternal changes occur to support the growing fetus and to prepare the body for delivery (ie, parturition) and lactation (ie, synthesis and secretion of milk). This lesson provides an overview of gestation, fetal circulation, parturition, and lactation.

10.4.01 Pregnancy

As discussed in Lesson 10.2, the first stage of gestation (the **embryonic period**) occurs within the first eight weeks following fertilization. The embryonic period begins with zygote formation (week two of pregnancy), and by the end of this period, the embryo has initiated the development of all major organ systems. The **fetal period** begins at the ninth week after fertilization (week 11 of pregnancy) and lasts until birth. Relatively few new structures are formed during the fetal period, but fetal growth is significant during this time.

Gestation is generally divided into three equal parts known as **trimesters**. The first trimester (weeks 1–12) includes fertilization, implantation, the embryonic period, and the first stages of the fetal period. By the second trimester (weeks 13–26), the placenta is fully formed and functions to sustain the fetus while secreting pregnancy hormones (eg, human chorionic gonadotropin, estrogen, progesterone). The third trimester begins at week 27 and lasts until the end of pregnancy (weeks 38–40).

A timeline of the development of select embryonic and fetal structures during gestation is shown in Figure 10.20.

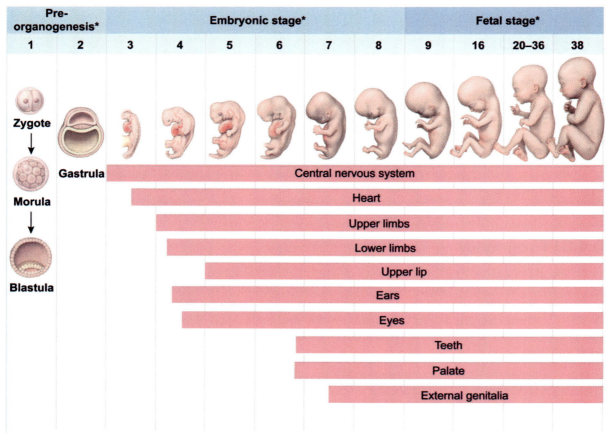

Figure 10.20 Timeline of the development of select embryonic and fetal structures during gestation.

10.4.02 Fetal Circulation

The fetus acquires oxygen (O_2) and nutrients and eliminates carbon dioxide (CO_2) and waste through the maternal circulatory system. Once the fetal circulatory system is functional, fetal-maternal materials are exchanged through the **placenta**. The placenta is attached to the fetus at the **umbilicus** (ie, navel) via the **umbilical cord**, which is derived from the embryonic yolk sac and allantois (see Concept 10.1.02).

Typically, there is no mixing of maternal and fetal blood because the exchange of materials occurs via diffusion through capillary walls. Fetal red blood cells contain a specialized type of hemoglobin (**fetal hemoglobin [hemoglobin F]**), which has a greater O_2 affinity than adult hemoglobin (**hemoglobin A**), thereby favoring O_2 transfer from the maternal blood supply to fetal blood (see Concept 13.1.03).

Embedded in the uterine wall, the placenta contains many small blood vessels emerging from the umbilical cord and branching into capillaries in the **chorionic villi** of the placenta (Figure 10.21). O_2 and nutrients are transported from a uterine artery into the **intervillous space** (ie, space around the chorionic villi), and eventually diffuse into fetal capillaries connected to the unpaired **umbilical vein**. CO_2 and wastes from fetal blood are transported via paired **umbilical arteries** into fetal capillaries at the placenta and diffuse into the intervillous space and subsequently into a uterine vein.

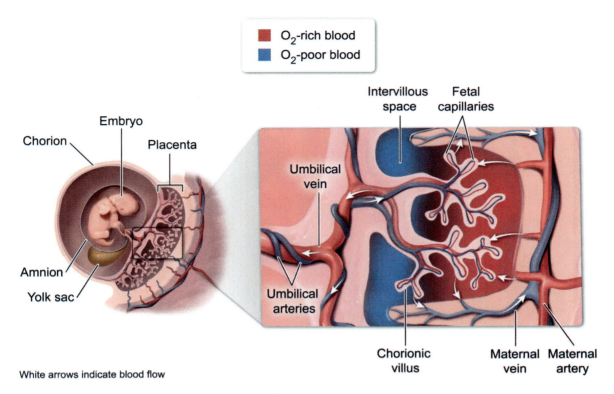

Figure 10.21 Fetal and maternal circulation at the placenta.

The placenta also functions as a protective barrier. Although most bacteria and viruses cannot cross the placenta, certain pathogens (eg, HIV, *Treponema pallidum* [causes syphilis], rubella virus [causes German measles]) can travel through the placenta, potentially leading to birth defects and/or fetal disease. Certain maternal immunoglobulins (ie, antibodies) can cross the placenta, conferring partial pathogen immunity to the fetus, and are thought to function in fetal immune system "training."

Oxygenation of fetal blood occurs in the placenta, not the fetal lungs. Therefore, fetal circulation to and from the placenta is reminiscent of the pulmonary circulation, in which pulmonary veins carry oxygenated blood from the lungs to the heart and pulmonary arteries carry deoxygenated blood from the systemic circulation to the lungs (see Concept 13.1.07).

After acquiring O_2 and nutrients from the maternal circulation at the placenta, oxygenated fetal blood returns to the fetus via a single umbilical vein in the umbilical cord, and deoxygenated blood is transported into the placenta from the fetus via two umbilical arteries in the umbilical cord (Figure 10.22).

Because some fetal organs (eg, lungs, liver) are not functional during gestation, there are several key differences between the fetal and postnatal circulatory systems. Oxygenated blood (ie, high O_2 saturation [SpO_2]) enters the fetus via the umbilical vein, which divides into two branches at the fetal liver. Most of the blood flows into one branch, the **ductus venosus**, which bypasses the liver and connects with the inferior vena cava. Deoxygenated blood (low SpO_2) returning from the fetal systemic circuit enters the venae cavae, mixes with the oxygenated blood from the ductus venosus, and passes into the right atrium (Figure 10.22).

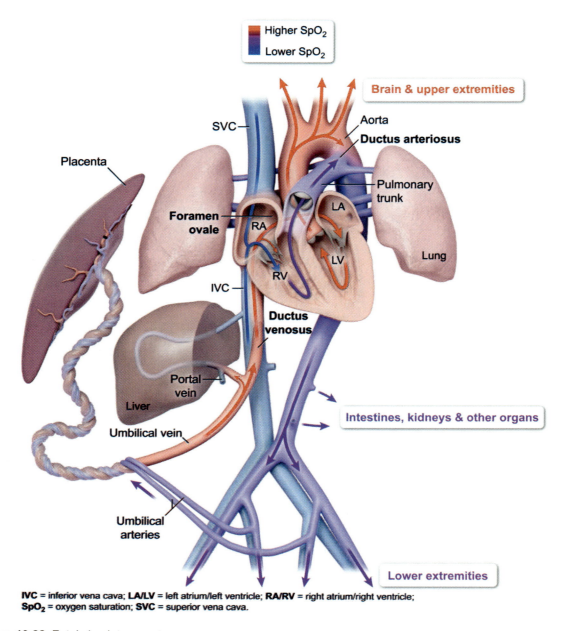

Figure 10.22 Fetal circulatory system.

An opening in the fetal heart called the **foramen ovale** acts as a one-way valve that connects the right and left atria. Most of the oxygenated blood entering the right atrium (ie, from the ductus venosus) passes through the foramen ovale into the left atrium and joins the systemic circulation via the aorta.

The small amount of blood that passes into the right ventricle (rather than moving via the foramen ovale) is pumped into the pulmonary trunk and directly into the **ductus arteriosus**, a vessel that bypasses the lungs, connecting the pulmonary trunk directly to the descending aorta. This deoxygenated blood mixes with the O_2-rich blood in the descending aorta, sending moderately oxygenated blood to the lower body of the fetus. Deoxygenated blood returning from the fetal systemic circulation is transported back to the placenta via the umbilical arteries (Figure 10.22).

A summary of fetal circulatory structures and their functions can be found in Table 10.3.

Table 10.3 Fetal circulatory structures.

Fetal circulatory structures	Function
Umbilical vein	Transports oxygenated blood from the placenta toward the fetus
Ductus venosus	Connects the umbilical vein with the inferior vena cava, shunting oxygenated blood away from the liver and toward the heart
Foramen ovale	Allows oxygenated blood to travel directly from the right atrium to the left atrium
Ductus arteriosus	Connects the pulmonary artery with the aorta, shunting blood away from lungs and into aorta
Umbilical arteries	Transport deoxygenated blood from the fetus toward the placenta

Directly following birth, the infant takes its first breaths, and O_2 in the now-functional lungs signals the cessation of placental blood flow. In addition, breathing via the lungs leads to pressure changes that cause the ductus arteriosus, ductus venosus, and foramen ovale to close, a process that typically begins 12-24 hours after birth.

10.4.03 Parturition

Typically, when a pregnancy has reached full term (38–40 weeks), the fetus exits the uterus through the vagina, a process known as **parturition** (ie, childbirth). **Labor** is the process by which childbirth occurs and is triggered by hormones from both the placenta and the fetus, along with mechanical cues. The initiation of labor is not well understood, but the process likely depends on a combination of maternal and fetal signals.

When labor begins, increased pressure on the cervix causes impulses to be sent to neurosecretory glands in the hypothalamus, causing the hormone oxytocin to be secreted via the posterior pituitary gland. Oxytocin secretion stimulates the release of **prostaglandins** in the uterine myometrium, causing uterine contractions. A **positive feedback** loop is thus initiated: Each contraction causes more pressure on the cervix, sending increased signals for oxytocin release. When the infant is born, the positive feedback loop is broken because pressure on the cervix is relieved, as illustrated in Figure 10.23.

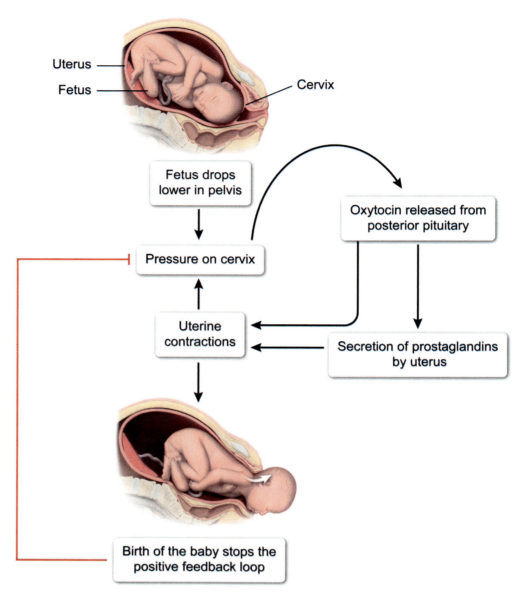

Figure 10.23 Hormonal control of parturition.

Labor can be divided into three stages, as shown in Figure 10.24:

1. **Dilation of the cervix:** From the onset of regular uterine contractions, cervical dilation typically lasts 6–12 hours. During this time, the amniotic sac (ie, fluid-filled sac surrounding the fetus) is usually ruptured. Dilation of the cervix is considered complete at 10 cm.

2. **Expulsion:** Delivery of the infant through the vagina typically occurs within 10 minutes to several hours after complete cervical dilation.

3. **Delivery of the placenta:** The placenta separates from the uterine wall and is typically delivered via the birth canal within minutes to an hour after birth of the infant.

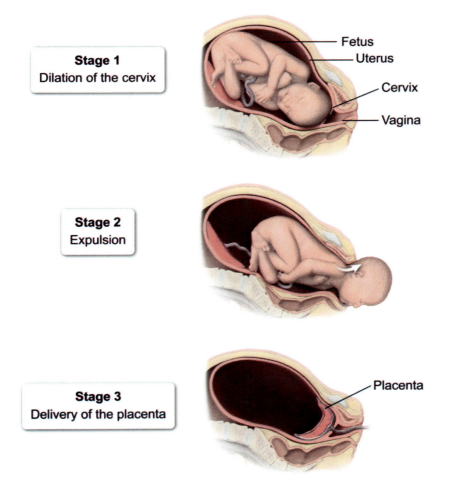

Figure 10.24 Stages of labor.

10.4.04 Lactation

Following childbirth, an infant can no longer rely on the placenta for nutrition and must depend instead on external sources, including breast milk. During puberty, the breasts develop under the influence of estrogen, but do not **lactate** (ie, produce milk). Near the end of pregnancy, the milk-producing **mammary glands** of the breasts develop further and are converted into secretory structures. Fully developed mammary glands are composed of 15–20 lobes, which secrete milk when stimulated.

Before delivery, when estrogen and progesterone are high, the mammary glands produce small amounts of a thin, low-fat secretion called colostrum. Within days of delivery, when estrogen and progesterone are decreased, the mammary glands produce milk with a higher fat and calcium concentration than colostrum. Mammary glands also secrete immunoglobulins (ie, antibodies), providing passive immunity to the infant.

When the infant latches to the breast, stimulation of mechanoreceptors in the nipples sends signals to the hypothalamus, which stimulates the anterior pituitary gland to secrete **prolactin** and the posterior pituitary gland to secrete **oxytocin**. Prolactin stimulates milk production, whereas oxytocin stimulates the **let-down reflex**, a positive feedback loop resulting in the ejection of milk through milk ducts in the nipples. When the infant stops suckling, the stimulus for hormone release ends, and milk is no longer ejected (Figure 10.25).

Chapter 10: Pregnancy, Development, and Aging

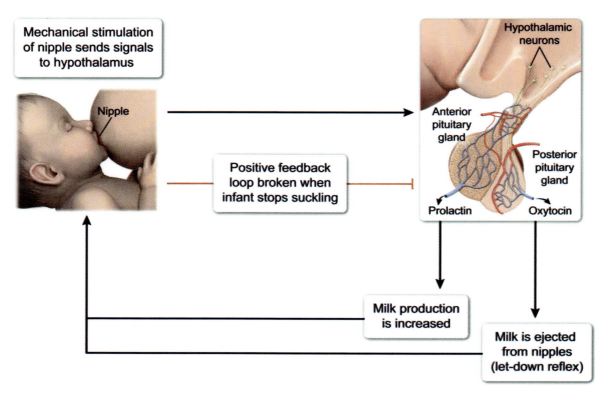

Figure 10.25 Hormonal control of milk secretion.

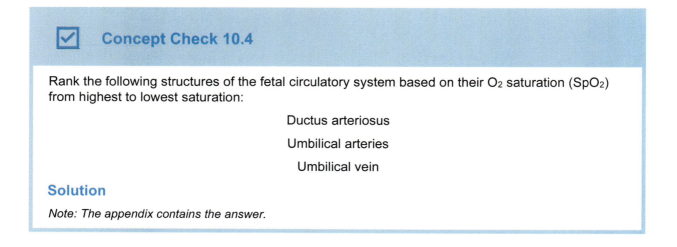

Concept Check 10.4

Rank the following structures of the fetal circulatory system based on their O_2 saturation (SpO_2) from highest to lowest saturation:

Ductus arteriosus

Umbilical arteries

Umbilical vein

Solution

Note: The appendix contains the answer.

Lesson 10.5
Cellular Regeneration and Senescence

Introduction

Fetal development ends at birth, but growth and development continue throughout the lifespan of the organism. The ability of a fully developed organism to regenerate tissues varies throughout life, declining with age. In response to cellular injury or damage, an organism may initiate a program of **tissue regeneration** or repair. Alternately, older or damaged cells may enter a **senescent** (ie, growth-arrested) state, with senescent cells accumulating as the organism ages. This lesson discusses the regenerative capabilities of mammalian tissues, as well as the effects of aging and senescence at both the cell and organism levels.

10.5.01 Regeneration

With the assistance of adult stem cells, many tissues possess self-renewing and self-repairing capabilities. **Tissue regeneration** is a program of cell proliferation and growth that renews some structures and tissues throughout life (eg, endometrial lining, red blood cells, epidermis) or restores damaged tissues with the same functional tissue present before the injury.

For example, when the epidermis is injured, a series of regenerative steps leads to the ultimate replacement of the damaged area with functional (ie, identical) epidermal tissue. However, some tissues (eg, nerve cells) cannot be replaced. Likewise, when certain tissues (eg, cardiac muscle) are injured, the injured tissue is replaced with scar (ie, fibrotic) tissue instead of functional tissue during **repair**. Figure 10.26 illustrates the differences between tissue regeneration and repair.

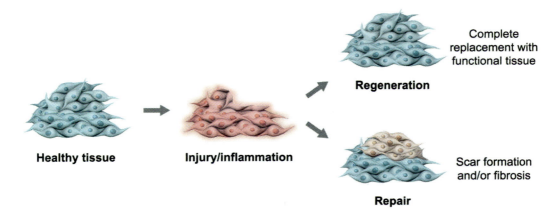

Figure 10.26 Tissue regeneration and repair.

In general, mammalian tissues have limited regenerative capabilities. For example, it is impossible for humans to regenerate an amputated limb, but healing near the area of amputation via thick scar tissue is possible. In addition to healing bone fractures and replacing lost blood, humans can also regenerate several other tissues in the body (eg, liver).

Although tissue renewal and repair involve the differentiation of stem cells, regenerating tissues are formed from local stem cell populations in the context of a fully developed organism, rather than in the context of embryonic development. The potency of adult stem cells is much more restricted than the potency of embryonic stem cells. Mammalian tissues are limited in their regenerative potential due to differences between the embryonic and adult cellular environments.

10.5.02 Cell and Organism Aging

Senescence is a cellular response that limits the proliferation of aged or damaged cells (Figure 10.27). Although senescence plays a physiological role in tissue homeostasis during normal development, it is also a stress response triggered by events associated with aging (eg, telomere shortening, genome instability, mitochondrial dysfunction). For example, cells are limited to a finite number of divisions due to the shortening of telomeres with each round of cell division and when telomere lengths are shortened past a critical point, a program of cellular senescence may be initiated.

Cellular senescence involves a series of programmed events including chromatin remodeling, metabolic changes, and increased autophagy, culminating in stable growth arrest of the senescent cell. Senescent cells often secrete pro-inflammatory cytokines, leading to both short- and long-term consequences (Figure 10.27).

Senescence has numerous physiological roles. For example, a cellular senescence program can be implemented during normal embryonic development, in response to cellular injury, or in transformed (ie, cancerous) cells to prevent potential detrimental effects.

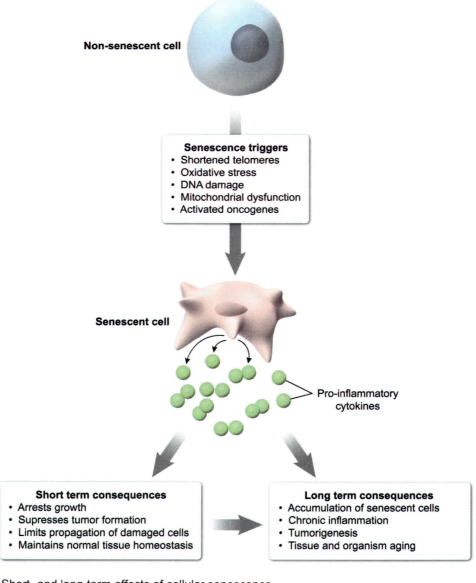

Figure 10.27 Short- and long-term effects of cellular senescence.

The role of senescence in cancer development is paradoxical. Senescence can be considered a mechanism of tumor suppression, *limiting* the development of cancerous cells by removing the cells from the cell cycle. However, accumulation of senescent cells may also play a role in *promoting* tumor progression through the secretion of pro-inflammatory cytokines, increasing cancer incidence with age. In addition, cancer cells may begin expressing the enzyme telomerase, which can replenish telomere ends. Telomerase expression allows cells to divide indefinitely, thereby preventing a protective senescent state.

Senescent cells are sometimes eliminated (ie, via apoptosis), but, in many cases, senescent cells can accumulate over time, leading to decreased organism function. The accumulation of senescent cells is thought to be a contributing factor in organism aging, likely due to the secretion of pro-inflammatory factors by senescent cells (Figure 10.27). As more and more cells enter a senescent state, stem cell numbers decline, leading to the tissue deterioration associated with aging.

 Concept Check 10.5

A cell that has entered a state of senescence due to telomere shortening is able to re-enter a non-senescent (ie, active) state. Propose a possible mechanism by which this could occur.

Solution

Note: The appendix contains the answer.

END-OF-UNIT MCAT PRACTICE

Congratulations on completing **Unit 5: Reproduction**.

Now you are ready to dive into MCAT-level practice tests. At UWorld, we believe students will be fully prepared to ace the MCAT when they practice with high-quality questions in a realistic testing environment.

The UWorld Qbank will test you on questions that are fully representative of the AAMC MCAT syllabus. In addition, our MCAT-like questions are accompanied by in-depth explanations with exceptional visual aids that will help you better retain difficult MCAT concepts.

TO START YOUR MCAT PRACTICE, PROCEED AS FOLLOWS:

1) Sign up to purchase the UWorld MCAT Qbank
   IMPORTANT: You already have access if you purchased a bundled subscription.
2) Log in to your UWorld MCAT account
3) Access the MCAT Qbank section
4) Select this unit in the Qbank
5) Create a custom practice test

Unit 6 Endocrine and Nervous Systems

Chapter 11 Endocrine System

11.1 Endocrinology

- 11.1.01 Hormone Structure and Function
- 11.1.02 Endocrine Glands
- 11.1.03 Hormone Transport
- 11.1.04 Endocrine Signaling
- 11.1.05 Regulation of Endocrine System

11.2 Hormones

- 11.2.01 Hypothalamic Hormones
- 11.2.02 Posterior Pituitary Hormones
- 11.2.03 Anterior Pituitary Hormones
- 11.2.04 Adrenal Hormones
- 11.2.05 Renal Hormones
- 11.2.06 Pancreatic Hormones
- 11.2.07 Thyroid Hormones
- 11.2.08 Parathyroid Hormones
- 11.2.09 Sex Hormones
- 11.2.10 Appetite Hormones
- 11.2.11 Other Hormones

Chapter 12 Nervous System

12.1 Cells of the Nervous System

- 12.1.01 The Neuron
- 12.1.02 Myelin Sheath
- 12.1.03 Glial Cells

12.2 Neural Communication

- 12.2.01 Resting Membrane Potential
- 12.2.02 The Action Potential
- 12.2.03 Synaptic Transmission
- 12.2.04 Neurotransmitters
- 12.2.05 Summation of Postsynaptic Potentials

12.3 Nervous System Structure and Function

- 12.3.01 Organization of the Nervous System
- 12.3.02 The Sympathetic and Parasympathetic Nervous Systems
- 12.3.03 Afferent and Efferent Pathways
- 12.3.04 Reflexes
- 12.3.05 Feedback Control

Lesson 11.1
Endocrinology

Introduction

The **endocrine system** participates in cellular communication and regulation of responses in the body. Unlike the nervous system (see Chapter 12), which provides responses on a short-term scale (eg, reflexes, sensory perception), the endocrine system typically modulates body responses over longer periods of time (eg, growth, development).

The endocrine system uses chemical messengers known as **hormones**, which are carried in the bloodstream for long-distance communication among the body's cells. For example, cortisol, produced by cortical cells of the adrenal glands, is carried in the bloodstream to distant target cells, including liver and skeletal muscle cells, to mediate the stress response (Figure 11.1).

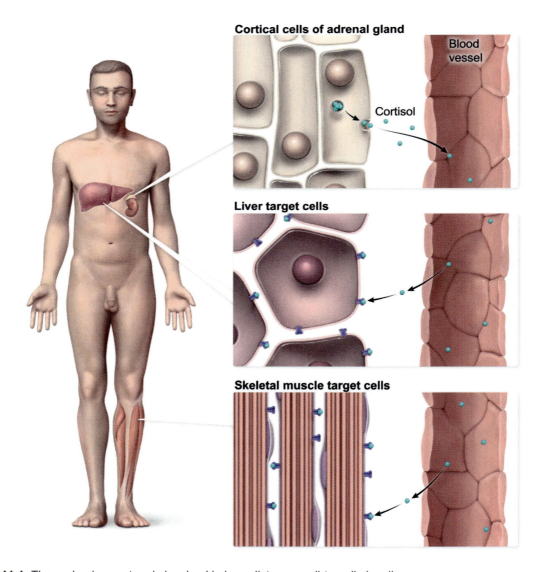

Figure 11.1 The endocrine system is involved in long-distance, cell-to-cell signaling.

This lesson explores types of hormones, endocrine glands, endocrine signaling, and how the endocrine system is regulated.

11.1.01 Hormone Structure and Function

Hormones are signaling molecules that are secreted from endocrine tissues into the circulation. Although hormones are typically present at low concentrations in the bloodstream, binding of hormones to target cell hormone receptors can cause profound responses. Hormones can be classified using several schemes.

One way to classify hormones is based on structure and divides hormones into three broad categories, as shown in Figure 11.2:

- **Peptide hormones** are small proteins; therefore, peptide hormones consist of amino acids linked by peptide bonds.
- **Steroid hormones** are lipids with a carbon skeleton derived from cholesterol molecules.
- **Amino acid–derived hormones** are small molecules that are modified versions of the amino acids tyrosine or tryptophan. These hormones are sometimes grouped together with peptide hormones.

Chapter 11: Endocrine System

Peptide hormones

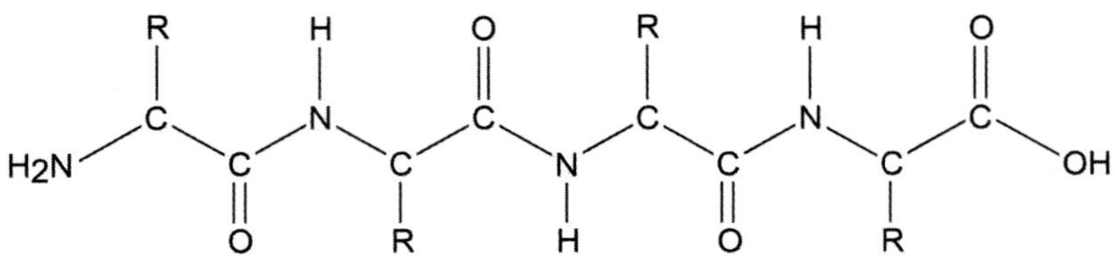

Composed of a chain of amino acids

Steroid hormones

Cholesterol

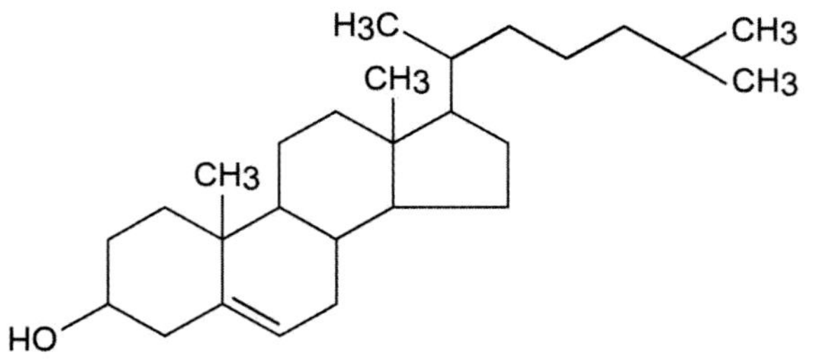

Derived from cholesterol (precursor)

Amino acid–derived hormones

Tyrosine

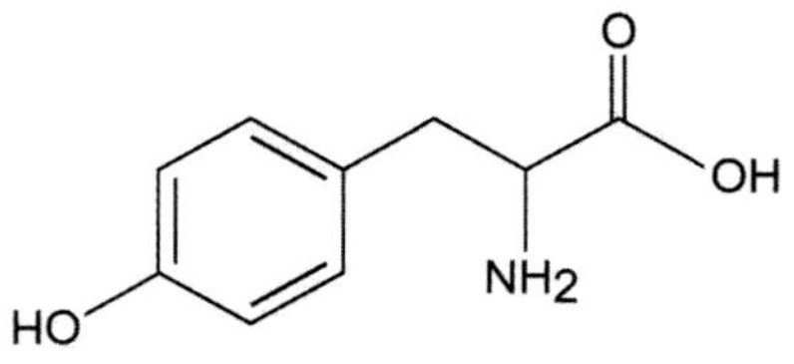

Tryptophan

Derived from tyrosine (precursor)

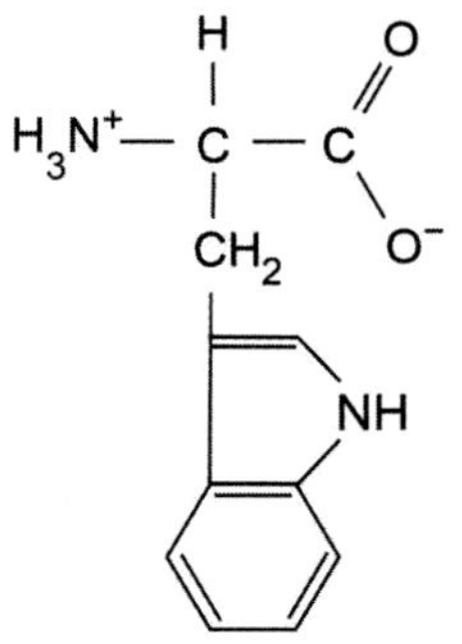

Derived from tryptophan (precursor)

Figure 11.2 Classifying hormones based on structure.

Another way to classify hormones is based on their influence on other hormones. **Tropic hormones** are hormones that target endocrine tissues to influence secretion of other hormones. Through actions on other hormones, tropic hormones influence and contribute to a specific response. For example, thyroid-stimulating hormone (TSH), a tropic hormone produced by the anterior pituitary gland, causes the release of thyroid hormone from the thyroid gland, which stimulates increased metabolism in tissues.

In contrast to tropic hormones, **direct hormones** act directly on target cells to elicit nonendocrine responses (ie, responses other than hormone secretion). For example, the anterior pituitary hormone prolactin, a direct hormone, causes a nonendocrine response (ie, promotion of milk production and secretion by exocrine mammary glands).

Some hormones can act as both a tropic hormone and a direct hormone. For example, growth hormone (GH) released by the anterior pituitary gland causes tropic effects by stimulating the liver to release other hormones that mediate growth in bone, cartilage, and soft tissue. GH also causes direct tissue effects by stimulating glucose uptake, fat utilization, and protein synthesis. Therefore, GH can be classified as both a tropic and a direct hormone (Figure 11.3).

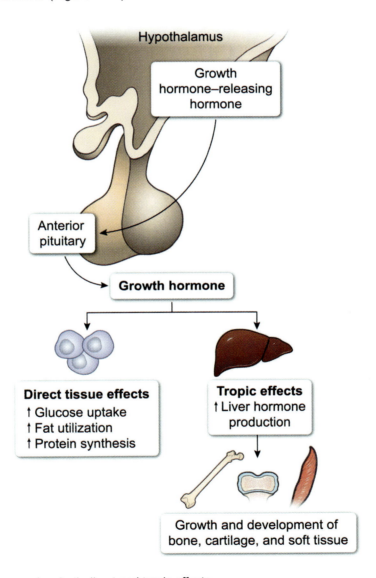

Figure 11.3 Growth hormone has both direct and tropic effects.

11.1.02 Endocrine Glands

Unlike **exocrine glands**, which contain ducts and secrete products (eg, sweat, tears, milk) onto epithelial surfaces, **endocrine glands** lack ducts and secrete products (ie, hormones) into the bloodstream. The major endocrine glands in the body are shown in Figure 11.4. Other organs and some tissues can also produce hormones (eg, renal erythropoietin production, adipose tissue leptin production). Specific hormones produced by endocrine glands and the effects of these hormones are detailed in Lesson 11.2.

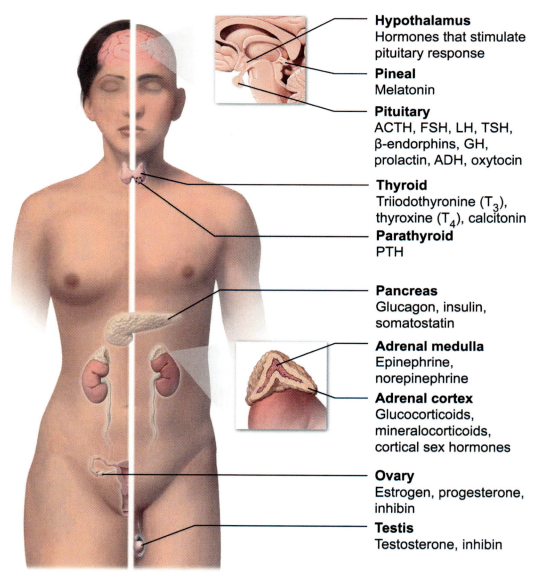

ACTH = adrenocorticotropic hormone; FSH = follicle-stimulating hormone; LH = luteinizing hormone; TSH = thyroid-stimulating hormone; GH = growth hormone; ADH = antidiuretic hormone; PTH = parathyroid hormone.

Figure 11.4 Major endocrine glands and their products.

11.1.03 Hormone Transport

Hormones travel to target cells via the blood, which is composed primarily of water. Therefore, the solubility of a hormone in water affects how a particular hormone is transported in the blood and how that hormone exerts a physiological effect on target cells.

Peptide hormones are produced in the rough endoplasmic reticulum (ER), and like most proteins in the body, they typically possess an overall charge at physiological pH. The net charges on peptide hormones make them water soluble (ie, hydrophilic) and allow peptide hormones to circulate largely in a free form (ie, unbound to proteins) in the bloodstream. However, peptide hormones cannot cross the hydrophobic lipid bilayer of the target cell plasma membrane and therefore rely on intracellular molecules to transmit a signal to the interior of a cell, as depicted in Figure 11.5.

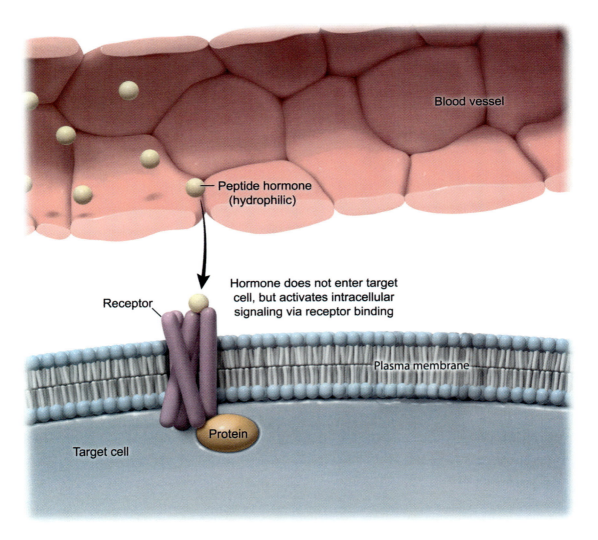

Figure 11.5 Peptide hormone transport.

After production in the rough ER and modification in the Golgi apparatus, peptide hormones are stored within vesicles in the secreting cell until needed. When the appropriate signal is received, the vesicles travel to the plasma membrane and undergo exocytosis. Released peptide hormones are then carried in the bloodstream to target cells (ie, cells possessing specific peptide hormone receptors) upon which peptide hormones exert their effects, most of which involve changes to the activity of existing proteins. Due to this mechanism of production and action, peptide hormones are relatively fast acting with relatively short-lived results.

Steroid hormones are produced in the smooth ER, and most cells that secrete steroid hormones have greater amounts of smooth ER than cells that do not secrete steroid hormones. Because steroid hormones are derived from cholesterol, the resulting hormones are lipophilic (ie, hydrophobic) and can diffuse readily through the secreting cell's membrane. Unlike peptide hormones, steroid hormones are not stored in vesicles for later use; rather, steroid hormones diffuse through the secreting cell's membrane and enter into the circulation once produced.

Because they are lipophilic, steroid hormones are predominantly bound to protein carriers to increase solubility during circulation in the aqueous bloodstream. Steroid hormones are typically inactive when bound to a protein carrier; therefore, carrier dissociation is typically required for hormone activation. Protein carrier–hormone pairs serve as a ready reserve of steroid hormones in the blood. The hydrophobicity of steroid hormones facilitates diffusion through target cell plasma membranes, after which the hormones bind receptors in the cytoplasm or nucleus to regulate gene expression (Figure 11.6).

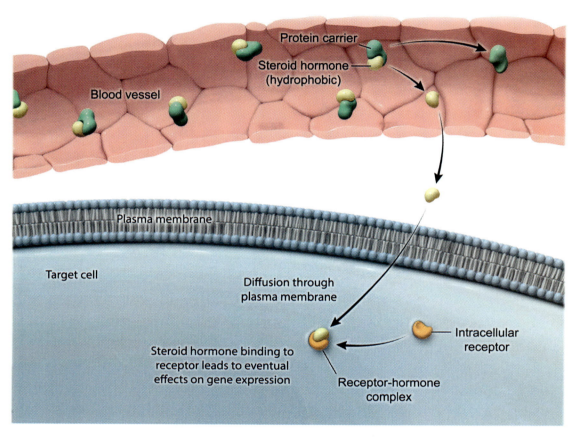

Figure 11.6 Steroid hormone transport.

Because steroid hormones act upon DNA to influence transcription and gene expression (ie, synthesis of new proteins), these hormones are slower acting than peptide hormones, the effects of which typically involve protein modification (eg, phosphorylation) rather than protein synthesis. However, the effects of steroid hormones are typically longer lasting than those of peptide hormones.

Amino acid–derived hormones can function similarly to peptide hormones and steroid hormones based on their structure and regulation (Figure 11.7). For example, the catecholamines epinephrine and norepinephrine behave like peptide hormones because they circulate unbound in the bloodstream and bind to surface receptors on target cells. Thyroid hormones are similar to steroid hormones because they are lipophilic, circulate in protein-bound form, and activate intracellular hormone receptors, but thyroid hormones are also similar to peptide hormones because they do not simply *diffuse* through the target cell membrane.

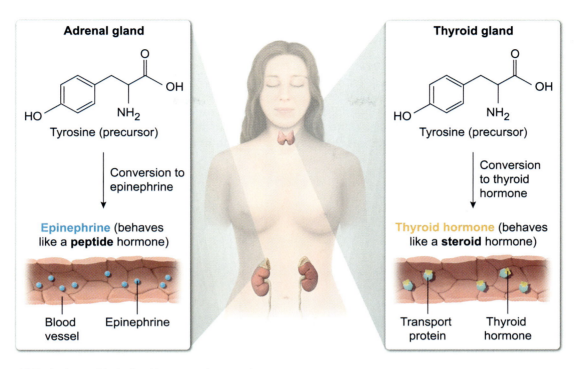

Figure 11.7 Amino acid–derived hormone transport.

The effects in the target cell once a hormone binds to its receptor (ie, endocrine signaling) are discussed in detail in Concept 11.1.04.

11.1.04 Endocrine Signaling

Cell signaling occurs when a signal acts upon a target cell and causes a response. There are three broad ways in which cell signaling occurs (Figure 11.8). In **autocrine signaling**, the signal acts upon the same cell that releases the signal. During **paracrine signaling**, the signal (ie, paracrine factor) diffuses through the interstitial fluid to act on a nearby cell. Finally, in **endocrine signaling**, signals (ie, hormones) released by secreting cells travel through the bloodstream to act on a more distant target cell.

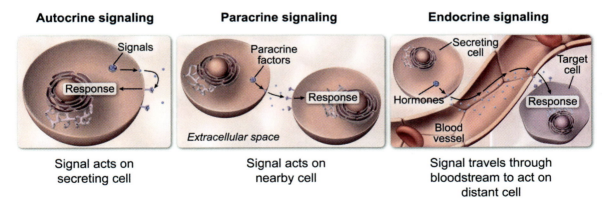

Figure 11.8 Types of cell signaling.

During endocrine signaling, hormones affect the function of many diverse cells and tissues throughout the body. This pattern occurs because hormone receptors can be expressed in a variety of cell types, and any cell that expresses the specific receptor for a particular hormone can respond to the receptor's ligand (ie, hormone). The reception of a signal by a target cell is the first step in endocrine signaling.

Receptors for peptide hormones are located in the plasma membrane because peptide hormones are charged and cannot readily cross the phospholipid bilayer. The receptor is often coupled with a G protein, the function of which is detailed in Concept 5.2.02. The peptide hormone is described as the **first messenger**, which binds its receptor and triggers a **second messenger** (eg, cyclic adenosine monophosphate), another molecule inside the target cell that transmits the signal, often through phosphorylation of products. This sequence is known as a **signaling cascade** (Figure 11.9).

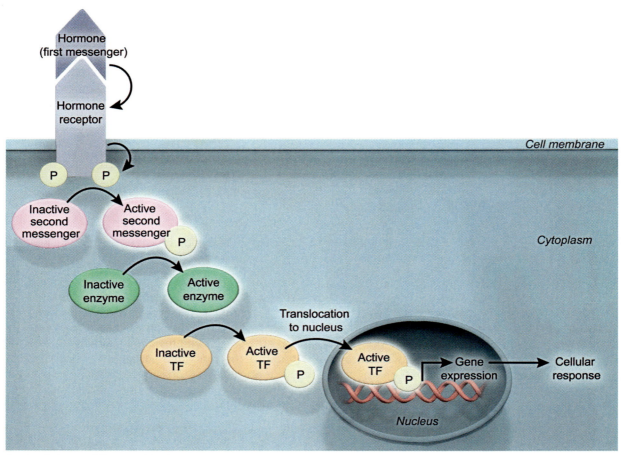

P = phosphate group; **TF** = transcription factor.

Figure 11.9 Example of a signaling cascade.

The responses elicited by peptide hormones can occur quickly, but the effects are relatively short lived. However, a small concentration of peptide hormones can have a great effect due to the potential for signal amplification along the signaling cascade. **Signal amplification** occurs when a small number of molecules produce many more activated products. For example, a single enzyme activated by a membrane receptor can catalyze the production of hundreds of second messengers, greatly amplifying the eventual response in the cell.

Because steroid hormones are lipophilic, these hormones are able to diffuse directly across the plasma membrane and bind intracellular receptors in the cytosol or nucleus, and they typically do not make use of second messengers. Rather, the steroid hormone–receptor complex typically acts as a transcription factor and binds DNA, resulting in increased or decreased transcription of target genes. Often, two steroid hormone–receptor complexes bind together (ie, dimerize) prior to DNA binding, as illustrated in Figure 11.10 for glucocorticoids (ie, steroid hormones produced by the adrenal cortex).

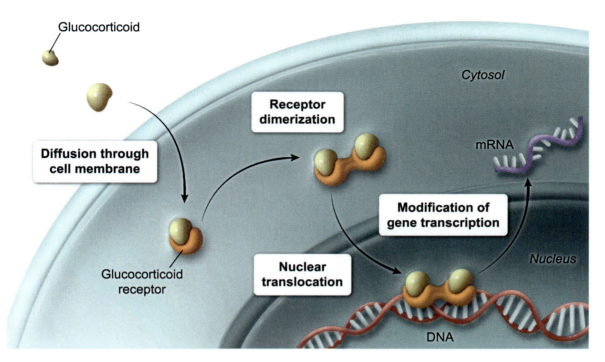

Figure 11.10 Steroid hormone signaling.

Because steroid hormones influence gene expression, the cellular effects of steroid hormones take longer to produce but last longer than the cellular effects of peptide hormones.

As noted in Concept 11.1.03, amino acid–derived hormones can exhibit characteristics of peptide and steroid hormones depending on the specific hormone and context. For example, like peptide hormones, binding of the amino acid–derived hormone epinephrine to its receptor activates a cascade of events in which second messengers are rapidly activated to influence glycogen metabolism as well as transcription of certain genes. In addition, like steroid hormones, the amino acid–derived thyroid hormones influence transcription of target genes upon binding thyroid hormone receptors that are members of the nuclear hormone receptor family.

11.1.05 Regulation of Endocrine System

As a regulator for the rest of the body, the endocrine system *itself* must be regulated. Regulation of the endocrine system allows a return to homeostasis when the body's normal balance is disturbed. Hormones are constantly regulated in response to the concentrations of molecules governed by those hormones. When the concentration of a particular product molecule increases, less of the hormone that caused the increased concentration is released. This process is known as **negative feedback**, as illustrated in Figure 11.11.

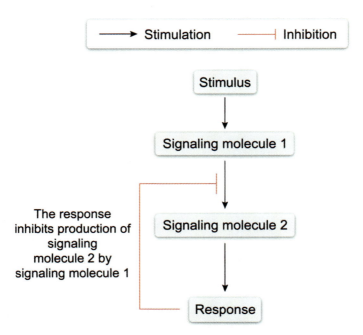

Figure 11.11 Negative feedback regulates the endocrine system.

In addition to negative feedback regulation, the nervous system influences much of the endocrine system's actions. This influence is accomplished via the hypothalamus, a brain structure that serves as a link between the nervous and endocrine systems. The hypothalamus regulates other endocrine structures (notably, the anterior pituitary gland) by secreting releasing and inhibiting factors. Lesson 11.2 goes into further detail about the hypothalamic-pituitary connection and how specific hormones are regulated.

Lesson 11.2
Hormones

Introduction

The endocrine system is comprised of various endocrine glands and organs throughout the body, which, in response to specific signals, produce and release hormones to maintain homeostasis. This lesson explores the production and release of important hormones in the body as well as the function and regulation of these hormones.

Together with the nervous system (see Chapter 12), the endocrine system is responsible for signaling and maintaining homeostasis of other body systems. The endocrine system presents a unique way to synthesize and review information about other body systems. Therefore, some of the information in this lesson is a review or deeper exploration of information presented in other chapters, which are referenced throughout.

Due to its role in integrating the functions of multiple body systems, the endocrine system is a particularly high-yield topic on the exam. Gaining a thorough understanding of the endocrine system facilitates an understanding of the required knowledge of many other systems as well.

11.2.01 Hypothalamic Hormones

The **hypothalamus**, a forebrain structure located inferior to the thalamus, regulates the synthesis and secretion of multiple hormone classes. The hypothalamus both processes inputs from the brain's cerebral cortex and senses the plasma concentration of numerous hormones; therefore, the hypothalamus serves as the critical interface between the nervous and endocrine systems. In response to various inputs, the hypothalamus controls body-wide endocrine function by influencing activity within the pituitary gland (also known as the hypophysis), an endocrine organ located inferior to the hypothalamus.

The interface of the nervous and endocrine systems creates what are known as neuroendocrine pathways in the body. Neuroendocrine pathways are involved in the release of either hormones (from endocrine cells) or neurotransmitters and neurohormones (from neurons and neuroendocrine cells). An example of a neuroendocrine pathway is the stress response, depicted in Figure 11.12:

1. When stress is perceived, the hypothalamus sends electrical signals (ie, neural signals) via sympathetic neurons to the adrenal medulla.
2. The adrenal medulla releases epinephrine and norepinephrine (ie, neurohormones of the endocrine system) from neuroendocrine cells.
3. Epinephrine and norepinephrine are transported in the blood and bind receptors in specific body cells to cause responses to stress; for example, epinephrine binding specific receptors in the smooth muscle of blood vessels causes vasoconstriction and a subsequent increase in blood pressure.

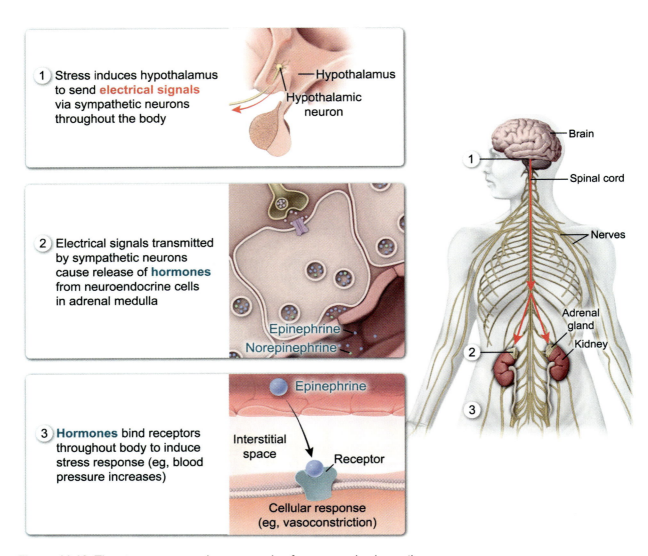

Figure 11.12 The stress response is an example of a neuroendocrine pathway.

The hypothalamus is able to exert control over the endocrine system through peptide hormones known as releasing and inhibiting factors. These factors are examples of tropic hormones, or hormones that act to influence an endocrine response (ie, alter secretion of other hormones, see Concept 11.1.01). Hypothalamic tropic hormones act on the anterior pituitary as part of the hypothalamic-pituitary (hypothalamic-hypophyseal) control axis.

The interaction of the hypothalamus and anterior pituitary is an example of a **portal system** in the body. In a portal system, blood circulates through two capillary beds connected in series by small portal veins, rather than through only one capillary bed (see Concept 13.1.10). The hypothalamic-pituitary portal system allows hypothalamic hormones to remain concentrated as they travel a short distance to anterior pituitary target cells, rather than becoming diluted in the entire bloodstream. In this way, the hypothalamus and nearby anterior pituitary have a direct line of chemical communication.

An example of communication via the hypothalamic-pituitary portal system is seen in the regulation of growth hormone (GH) release from the anterior pituitary. Growth hormone–releasing hormone (GHRH), a tropic hormone produced in the hypothalamus, travels via portal veins to the anterior pituitary, causing the anterior pituitary to release growth hormone, as shown in Figure 11.13.

Chapter 11: Endocrine System

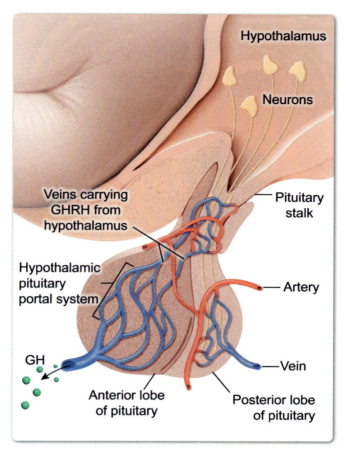

GHRH = growth hormone-releasing hormone; GH = growth hormone.

Figure 11.13 The hypothalamic-pituitary portal system.

Through the endocrine actions of releasing and inhibiting factors on the anterior pituitary, the hypothalamus promotes homeostasis in the body. As the serum level of a given hormone rises, that hormone inhibits the signaling pathway that promotes the synthesis and secretion of the hormone itself in a control process known as a negative feedback loop. Many negative feedback loops exist to ensure a constant appropriate level of hormones in the body.

Thermoregulation, appetite and satiety, the stress response, body water content and blood pressure, metabolism, growth, and reproduction are all examples of processes governed by the hypothalamus. In addition to secreting tropic hormones to regulate the anterior pituitary (discussed in more detail in Concept 11.2.03), the hypothalamus produces hormones for storage in, and release from, the posterior pituitary (discussed further in Concept 11.2.02).

11.2.02 Posterior Pituitary Hormones

The **pituitary gland** consists of the **anterior** and **posterior lobes**, both of which are derived from embryonic ectodermal tissue. The anterior pituitary is derived from epithelial cells of the developing roof of the mouth and contains typical glandular endocrine cells, whereas the posterior pituitary (neurohypophysis) is derived from ectodermal neural tissue in the developing brain and does *not* contain typical glandular endocrine cells. Therefore, hormones are not *produced* in the posterior pituitary; instead, the lobe is a storage site for hormones released from hypothalamic neurons (ie, **neurohormones**) rather than from typical glandular cells.

Stimulation of these hypothalamic neurons results in exocytosis of neurosecretory vesicles containing the stored neurohormones from axon terminals located in the posterior pituitary. The exocytosed hormones are secreted into small vessels that carry blood away from the posterior pituitary, as illustrated in Figure 11.14. The two neurohormones stored and released from the posterior pituitary are the peptide hormones **antidiuretic hormone (ADH, vasopressin)** and **oxytocin**.

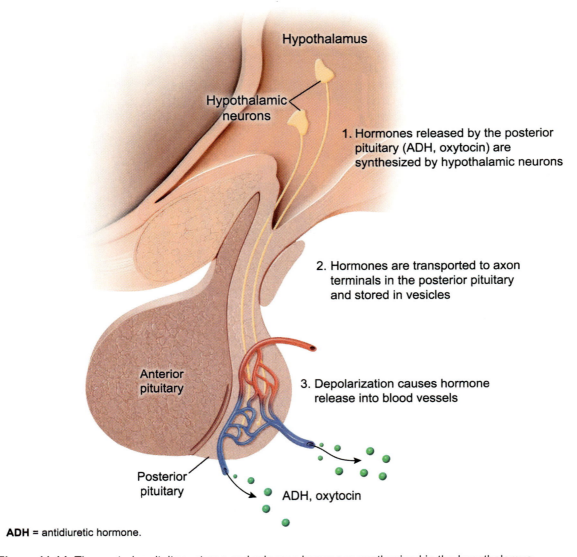

ADH = antidiuretic hormone.

Figure 11.14 The posterior pituitary stores and releases hormones synthesized in the hypothalamus.

ADH is involved in the regulation of body water content (see Lesson 16.2). Special stretch receptors monitor the degree of atrial stretching in the heart (ie, volume of blood entering the atria) and transmit this information to the hypothalamus. In addition, the hypothalamus monitors changes in blood osmolarity. In response to changes in atrial stretching and blood osmolarity, the hypothalamus signals the posterior pituitary to adjust ADH secretion accordingly.

When blood volume is reduced or blood osmolarity is increased, ADH causes the insertion of aquaporins (ie, water channels) into kidney nephron tubules. Aquaporin insertion increases the permeability of the nephrons to water, thereby increasing the reabsorption of water into the bloodstream, which increases blood volume and decreases blood osmolarity to homeostatic levels. When blood volume increases or blood osmolarity decreases, less ADH is released to maintain homeostasis. The effect of ADH on body water content is summarized in Figure 11.15.

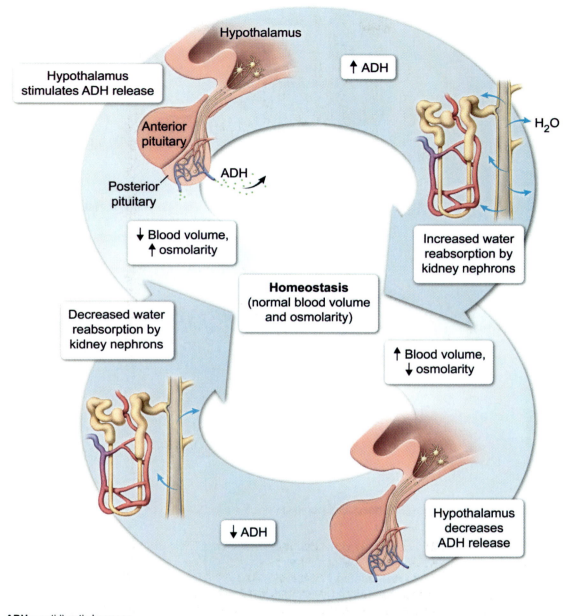

ADH = antidiuretic hormone.

Figure 11.15 Actions of antidiuretic hormone.

Oxytocin is released by the posterior pituitary in response to pressure against the cervix (ie, from the head of the fetus) during childbirth. Oxytocin release increases uterine contractions, further stimulating the release of oxytocin until parturition (Figure 11.16). This situation is one of the rare examples of positive feedback in the body, in which a stimulus causes a response that further *increases* (rather than *decreases*, as in negative feedback) the initial stimulus. Oxytocin also targets mammary glands to stimulate milk ejection during breast feeding.

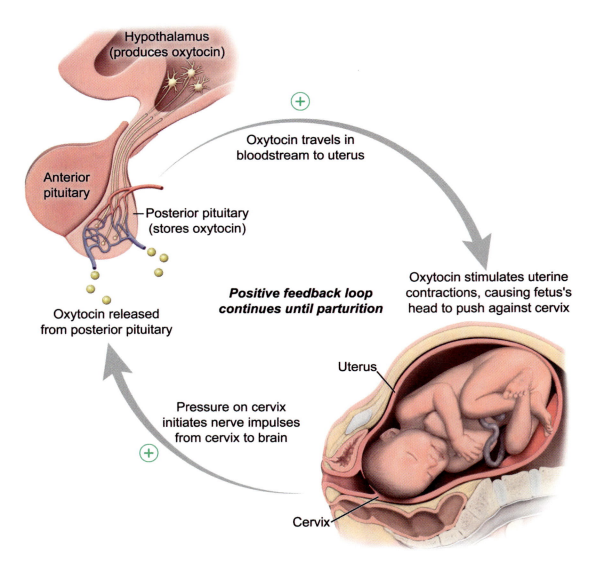

Figure 11.16 Oxytocin works in a positive feedback loop just prior to parturition.

11.2.03 Anterior Pituitary Hormones

The pituitary gland is a small structure located inside a bony cavity inferior to the hypothalamus at the brain's base. Two separate lobes of the pituitary gland exist: the anterior and posterior pituitary. The **anterior pituitary** (also known as the adenohypophysis) is made up of glandular endocrine tissue containing different cell types that synthesize and secrete several tropic peptide hormones (ie, hormones that produce endocrine effects). These tropic peptide hormones include adrenocorticotropic hormone (ACTH), follicle-stimulating hormone (FSH), luteinizing hormone (LH), and thyroid-stimulating hormone (TSH).

The anterior pituitary also secretes several direct peptide hormones: β-endorphins, growth hormone (GH), and prolactin. More information about the posterior pituitary, which does not contain typical endocrine glandular cells, can be found in Concept 11.2.02.

The synthesis and secretion of many anterior pituitary hormones are controlled by tropic neurohormones released by hypothalamic neurons (Figure 11.17). These neurohormones are secreted into the hypophyseal portal system, which allows direct chemical communication between the hypothalamus and the anterior pituitary (see Concept 11.2.01).

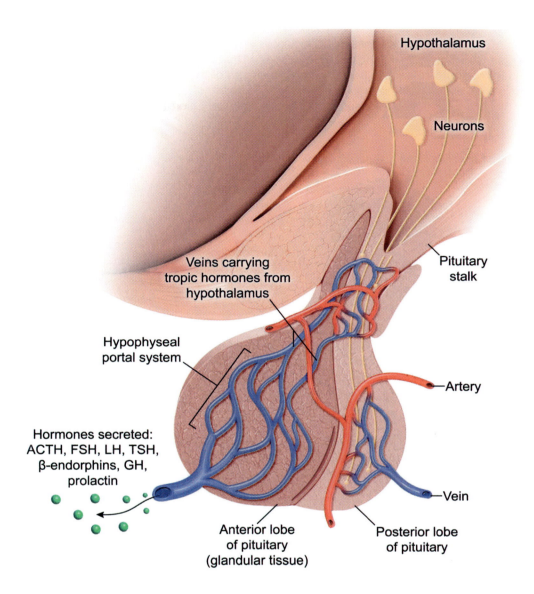

ACTH = adrenocorticotropic hormone; FSH = follicle-stimulating hormone; LH = luteinizing hormone; TSH = thyroid-stimulating hormone; GH = growth hormone.

Figure 11.17 Anterior pituitary hormones.

ACTH acts on the adrenal cortex to promote synthesis and secretion of glucocorticoids, steroid hormones that function to increase blood glucose and mediate the stress response (see Concept 11.2.04). Secretion of **corticotropin-releasing hormone (CRH)** by the hypothalamus promotes ACTH release as part of the hypothalamic-pituitary-adrenal pathway (HPA axis). Glucocorticoids inhibit secretion of CRH and ACTH.

FSH and **LH** regulate reproductive function (see Concept 11.2.09) by serving as gonadotropic hormones (ie, hormones that regulate the endocrine activity of gonads [ovaries, testes]). FSH stimulates follicle maturation in ovaries and spermatogenesis in testes. LH stimulates estrogen production and triggers ovulation in ovaries and promotes spermatogenesis and testosterone synthesis in testes. **Gonadotropin-releasing hormone (GnRH)** produced by the hypothalamus promotes synthesis and secretion of FSH and LH by the anterior pituitary. Estrogens and testosterone suppress the release of GnRH, FSH, and LH.

TSH acts on the thyroid gland to stimulate production of the thyroid hormones triiodothyronine (T_3) and thyroxine (T_4). Thyroid hormones affect most body cells and function to increase the rate of cellular metabolism (see Concept 11.2.07) and other processes. The secretion of TSH is controlled by hypothalamic **thyrotropin-releasing hormone (TRH)** as part of the hypothalamic-pituitary-thyroid pathway (HPT axis), and thyroid hormones suppress the release of TRH and TSH. Table 11.1 summarizes the tropic hormones released by the anterior pituitary gland and the regulation of these hormones.

Table 11.1 Tropic hormones of the anterior pituitary.

Hormone	Secretion stimulated by	Secretion inhibited by	Action
Adrenocorticotropic hormone (ACTH)	Corticotropin-releasing hormone (CRH)	Glucocorticoids	Promotes glucocorticoid release by adrenal glands
Follicle-stimulating hormone (FSH)	Gonadotropin-releasing hormone (GnRH)	Estrogens (low to moderate levels), testosterone	Stimulates ovarian follicle maturation, testosterone synthesis, and spermatogenesis
Luteinizing hormone (LH)			Promotes estrogen production, ovulation, and testosterone synthesis
Thyroid-stimulating hormone (TSH)	Thyrotropin-releasing hormone (TRH)	Thyroid hormone	Stimulates thyroid gland to release thyroid hormones (T_3, T_4)

T_3 = triiodothyronine; T_4 = thyroxine.

β-endorphins act as neurotransmitters, binding opioid receptors in the brain to decrease pain perception and act as natural opiates (in contrast to exogenous opiate drugs [eg, morphine, heroin], which also bind opioid receptors). In addition, β-endorphins can cause feelings of euphoria; for example, these chemicals are thought to be partially responsible for the so-called runner's high experienced by athletes. β-endorphins may have additional roles, including functions in memory, learning, and thermoregulation. β-endorphins are co-secreted with ACTH and as such, are regulated in the same manner.

GH regulates growth by promoting protein synthesis, fat utilization, and glucose uptake in many tissues. Secretion of GH is highest in childhood and remains elevated until mature height is reached in early adulthood. Even so, GH remains an important hormone in adults, with effects on processes other than growth (eg, metabolism, cell turnover). Hypothalamic **growth hormone–releasing hormone (GHRH)** stimulates GH release, whereas hypothalamic **somatostatin** inhibits GH release. Proper GH levels are particularly important in childhood: a GH deficit can lead to dwarfism, and excessive GH can lead to gigantism.

Prolactin is released after parturition in response to nipple stimulation (ie, infant suckling) and promotes the mammary glands to produce and secrete milk (see Concept 10.4.04). However, milk ejection requires the posterior pituitary hormone oxytocin. Prior to parturition (ie, during pregnancy), high levels of estrogen and progesterone inhibit prolactin release.

Hypothalamic **prolactin-inhibiting factor (PIF, dopamine)** decreases anterior pituitary prolactin secretion. Note that hypothalamic prolactin regulation is the opposite of other anterior pituitary hormones, whose secretion requires *release* of stimulatory hypothalamic hormones. In contrast, the hypothalamus must *not* release PIF (ie, an inhibitory hormone) for prolactin secretion to occur. Table 11.2 summarizes the direct hormones produced by the anterior pituitary and their regulation.

Table 11.2 Direct hormones produced by the anterior pituitary.

Hormone	Secretion stimulated by	Secretion inhibited by	Action
β-endorphins	Corticotropin-releasing hormone (CRH)	Glucocorticoids	Decrease pain perception, cause feelings of euphoria
Growth hormone (GH)	Growth hormone–releasing hormone (GHRH)	Somatostatin	Promotes growth in children and adults
Prolactin	Infant suckling	Prolactin-inhibiting factor (PIF, dopamine), high estrogen and progesterone levels	Stimulates lactation

11.2.04 Adrenal Hormones

The capsule-covered **adrenal glands** are located atop each kidney. Each adrenal gland can be subdivided into two portions: an outer portion called the **adrenal cortex** and a central inner portion called the **adrenal medulla**. In addition to being separated anatomically, the two portions of the adrenal glands secrete different hormones (Figure 11.18).

Adrenal hormones mediate the **stress response**, the physiological changes that allow the body to cope with and react to physical, social, and emotional stressors. Specifically, adrenal hormones promote the fight-or-flight response, influencing both cardiovascular function and energy metabolism (ie, cellular processes by which energy is stored or utilized), the latter by promoting energy utilization and inhibiting energy storage.

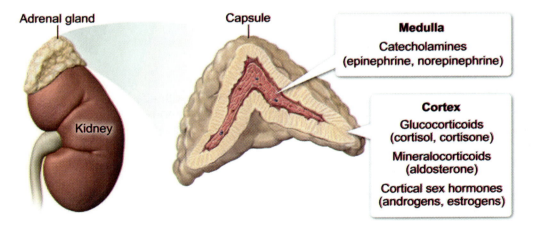

Figure 11.18 Adrenal gland anatomy and hormones.

Adrenal Cortex

The adrenal cortex is responsible for producing three classes of **corticosteroids** (ie, steroid hormones produced by the adrenal cortex). The first corticosteroid class consists of **glucocorticoids**, including **cortisol** and **cortisone**. Glucocorticoids function to increase blood glucose by acting on the liver to increase the synthesis of glucose from other molecules (ie, gluconeogenesis) and promoting the breakdown of fats into fatty acids (ie, lipolysis) to provide energy for gluconeogenesis. Both glucose and free fatty acids are forms of cellular energy that can be readily utilized.

In addition to increasing blood glucose levels, glucocorticoids increase the breakdown of proteins to amino acids, which can be used to create new proteins or for ATP production. Glucocorticoids also have anti-inflammatory effects. Glucocorticoid secretion is regulated by the hypothalamic-pituitary-adrenal pathway (HPA axis), as discussed in Concept 11.2.03.

The second corticosteroid class consists of **mineralocorticoids**, primarily **aldosterone**. Mineralocorticoids act to regulate blood pressure and blood volume. Blood filtration by kidney nephrons yields a fluid (ie, filtrate) that undergoes significant changes in volume and composition prior to excretion as urine (see Lesson 16.2). Most filtered sodium ions (Na^+) and water are reabsorbed; however, Na^+ and water reabsorption is subject to some degree of adjustment based on physiological conditions such as blood pressure and blood volume.

Aldosterone secretion is regulated by the **renin-angiotensin system (RAS)**. Low blood pressure promotes the renal release of the enzyme renin, which catalyzes cleavage of the plasma protein angiotensinogen to form angiotensin I. The lungs release another enzyme, angiotensin-converting enzyme, which catalyzes the cleavage of angiotensin I to form angiotensin II. Angiotensin II increases blood pressure by inducing vasoconstriction and by promoting aldosterone release, as shown in Figure 11.19.

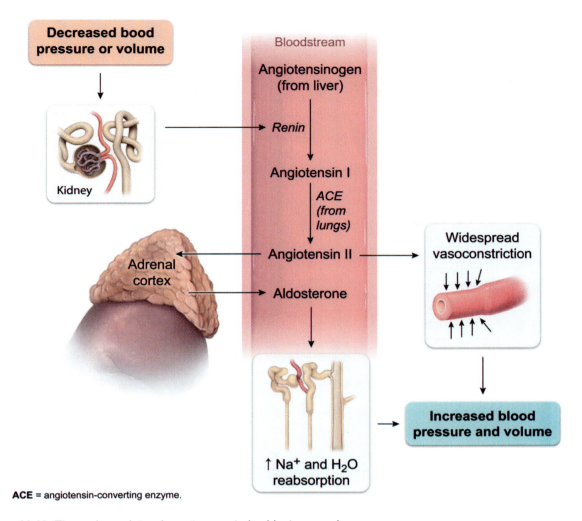

Figure 11.19 The renin-angiotensin system controls aldosterone release.

Aldosterone acts on nephron distal convoluted tubules and collecting ducts to promote Na^+ reabsorption and potassium (K^+) secretion. However, more Na^+ is reabsorbed than K^+ is secreted, generating an ion gradient between the filtrate and the interstitial fluid (ie, increases interstitial fluid osmolarity), causing water to be reabsorbed into the interstitial fluid (and eventually the bloodstream) via osmosis. As a result, blood volume and blood pressure increase, restoring homeostasis.

The third corticosteroid class consists of **cortical sex hormones** known as **androgens** (sex steroids promoting male characteristics) and **estrogens** (sex steroids promoting female characteristics). Androgens stimulate sperm cell differentiation and development of masculine sexual traits, whereas estrogens stimulate endometrial growth and development of feminine sexual traits. Note that both androgens and estrogens are secreted by sex organs as well (see Concept 11.2.09).

Adrenal Medulla

The adrenal medulla releases the catecholamines **epinephrine** and **norepinephrine**, amino acid–derived hormones that enable increased body mobility and promote rapid information processing under extreme stress. Catecholamine release is controlled by the autonomic nervous system (ANS), which is composed of two branches with broadly antagonistic effects: the sympathetic nervous system (SNS) and the parasympathetic nervous system (PNS). Many tissues are innervated by both ANS branches; however, the adrenal medulla is innervated only by the SNS. Acetylcholine release by sympathetic neurons stimulates catecholamine release (Figure 11.20).

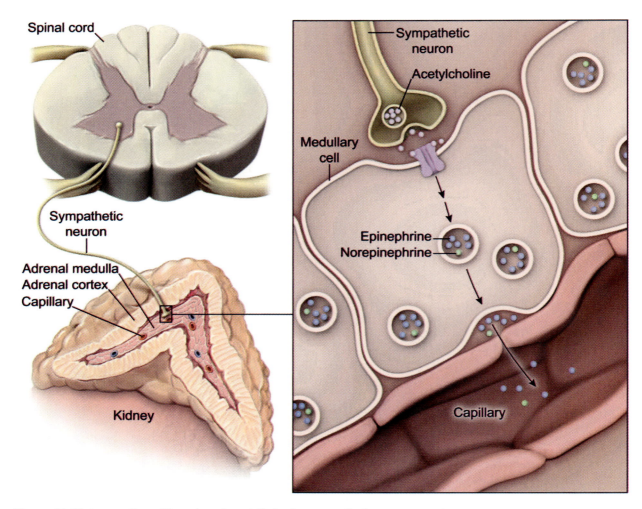

Figure 11.20 Innervation of the adrenal medulla by the sympathetic nervous system.

Catecholamines modulate the metabolism of glycogen, the polymer storage form of glucose. Low blood glucose levels (ie, hypoglycemia) promote catecholamine release. Along with pancreatic glucagon, these hormones promote glycogenolysis (ie, breakdown of glycogen into glucose monomers). At the same time, catecholamines inhibit enzymes that mediate glycogen synthesis from glucose (ie, glycogenesis), thereby maintaining higher levels of free glucose in the blood. The response to hypoglycemia is depicted in Figure 11.21.

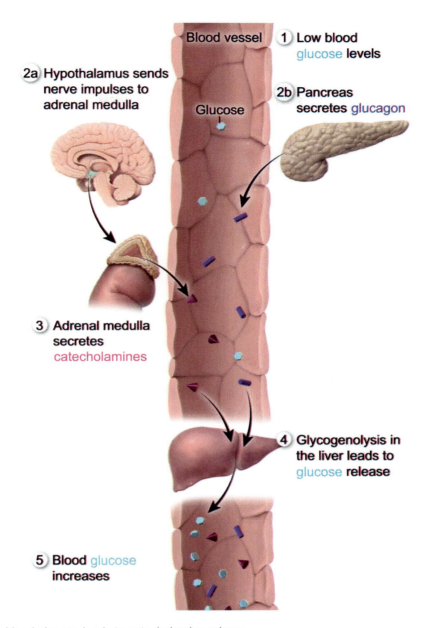

Figure 11.21 Low blood glucose leads to catecholamine release.

Catecholamine release redirects blood flow via vasoconstriction (ie, narrowing of blood vessels) and vasodilation (ie, widening of blood vessels). During times of stress, vasoconstriction reduces blood supply to organs carrying out nonessential functions (eg, stomach, intestines, kidneys) to conserve oxygen and nutrients for organs necessary for immediate survival (eg, heart, skeletal muscle). Vasodilation increases blood flow to these organs necessary for immediate survival.

Catecholamines can have opposing effects (ie, vasoconstriction or vasodilation) on blood vessels in different locations due to catecholamine ability to bind different receptor types. Binding of catecholamines to α-adrenergic receptors during times of stress leads to reduced blood flow to nonessential organs, and catecholamine binding to β-adrenergic receptors results in increased blood flow to essential organs, as shown in Figure 11.22.

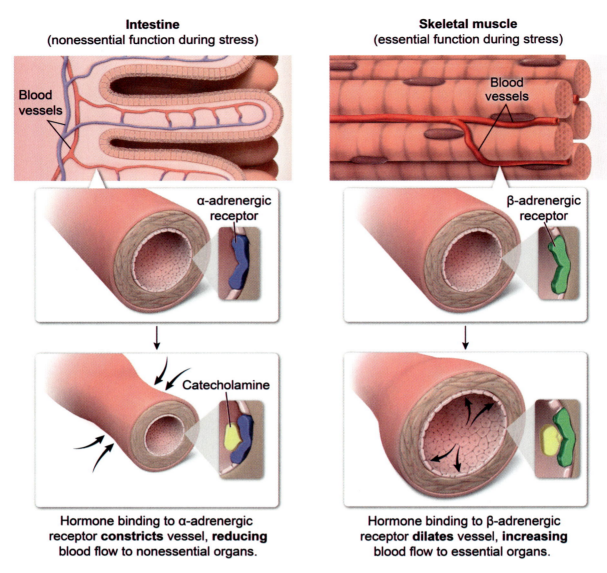

Figure 11.22 Catecholamine effects on blood vessels.

In addition to effects on blood vessel diameter, catecholamines increase heart rate and cardiac muscle contractility, promoting increased blood flow to the brain, lungs, and skeletal muscles, and thereby allowing the processing and execution of responses to stressful stimuli. At the same time, catecholamines dilate airways (bronchioles) to increase respiratory function and oxygen delivery to necessary tissues.

The complete stress response involves both catecholamines released by the adrenal medulla and glucocorticoids released by the adrenal cortex (Figure 11.23). Catecholamines mediate very quick responses to immediate stressors, whereas glucocorticoids mediate slower responses to long-term stressors. The presence of high glucocorticoid levels for extended periods of time can lead to negative effects (eg, immune system suppression).

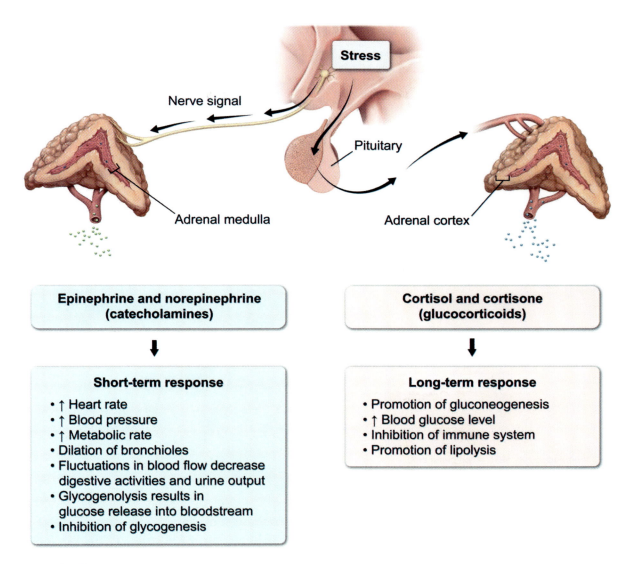

Figure 11.23 The stress response.

11.2.05 Renal Hormones

In addition to its involvement in several other important bodily functions, the kidney possesses endocrine function via the production of the peptide hormone **erythropoietin (EPO)**. Decreased oxygen levels in the blood delivered to the kidneys stimulate EPO production. Via the bloodstream, EPO reaches target cells in the red bone marrow, signaling an increase in red blood cell (ie, erythrocyte) production. More erythrocytes in circulation increases the oxygen-carrying capacity of the blood, thereby raising blood oxygen content back to homeostatic levels, as shown in Figure 11.24.

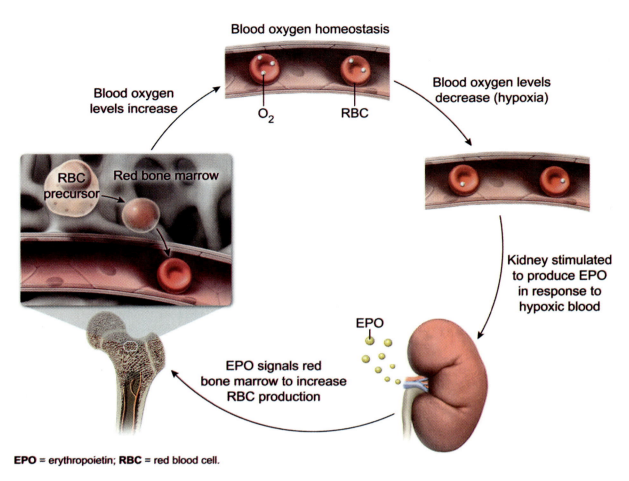

Figure 11.24 Erythropoietin maintains blood oxygen homeostasis.

In addition to EPO production, the kidney is involved in the activation of vitamin D to vitamin D_3, also known as **calcitriol**. Calcitriol acts as a steroid hormone that promotes calcium absorption from ingested food in the small intestine. Calcitriol also acts on the kidney itself to promote reabsorption of calcium in the nephron tubules. The overall effect of calcitriol activation is a net increase in serum calcium.

11.2.06 Pancreatic Hormones

The **pancreas** is an organ with both exocrine and endocrine functions. The exocrine pancreas is composed of glands known as pancreatic acini and secretes products (eg, pancreatic lipase, pancreatic amylase) as a solution known as pancreatic juice into the duodenum of the digestive system. The role of the exocrine pancreas in facilitating digestion is discussed in more detail in Concept 15.2.03.

The endocrine pancreas consists of functional units known as islets of Langerhans, which are made up of three types of cells, as shown in Figure 11.25:

- **Alpha cells** produce the peptide hormone **glucagon**. Alpha cells can also exhibit paracrine function through the inhibition of beta cell function in certain settings.
- **Beta cells** produce the peptide hormone **insulin**, which inhibits neighboring alpha cell function; therefore, beta cells also exhibit paracrine function.
- **Delta cells** produce **somatostatin**, a peptide hormone that has a generalized inhibitory effect on digestive function and has been shown to suppress insulin and glucagon release.

The products of alpha and beta cells (ie, glucagon and insulin) are critical in maintaining glucose homeostasis. Glucose is an essential molecule in the body, serving as a source of necessary metabolic intermediates (eg, oxaloacetate for the oxidation of macronutrient metabolites in the Krebs cycle) and as an energy substrate for ATP generation.

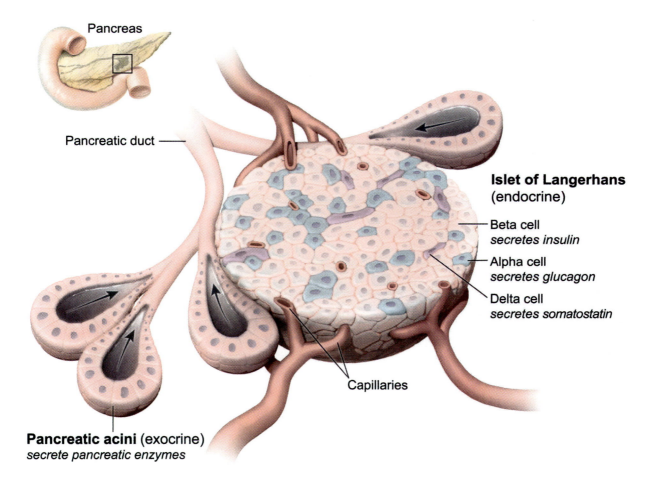

Figure 11.25 The pancreas has exocrine and endocrine functions.

In the fasted state, circulating blood glucose levels are maintained due to the breakdown of glycogen stores in the liver (ie, glycogenolysis) and through gluconeogenesis (ie, the synthesis of glucose from other molecules). Conversely, in the fed state (ie, after consumption of a carbohydrate-containing meal), blood glucose levels initially increase because glucose is absorbed from the digestive system lumen into the bloodstream, from which glucose can be taken up by tissues and stored as glycogen (ie, glycogenesis).

When blood glucose levels are low (ie, hypoglycemia), glucagon acts on target cells (eg, in the liver) by binding a G protein–coupled receptor on the plasma membrane and inducing the adenylate cyclase/cAMP second messenger cascade (see Concept 5.2.02). In response, gluconeogenesis and glycogenolysis are promoted in target cells, raising blood glucose levels. In addition, higher glucagon levels are often accompanied by higher availability of amino acids and glycerol for gluconeogenesis.

Glucagon limits glucose use in insulin-sensitive tissues and decreases glucose uptake by peripheral tissues. Catecholamines from the adrenal medulla stimulate glucagon release and directly promote glycogenolysis (see Concept 11.2.04). Figure 11.26 summarizes the responses to hypoglycemia.

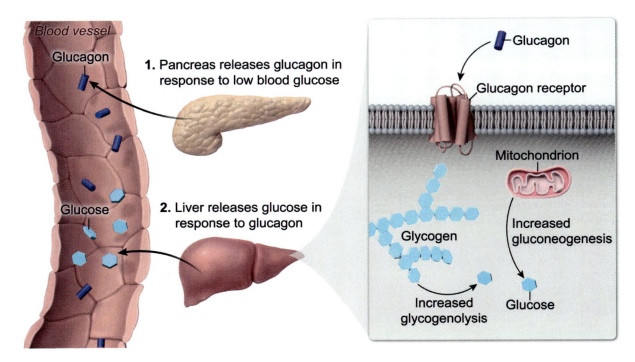

Figure 11.26 Cellular responses to low blood glucose.

When blood glucose levels are high (ie, hyperglycemia), insulin binds insulin receptors in target tissues, causing translocation of glucose transporters to the plasma membrane of these cells. Therefore, binding of insulin allows markedly increased rates of glucose uptake, promoting glycogenesis in the liver and muscle.

Insulin also stimulates fat synthesis (via glucose transport into adipocytes) and protein synthesis (via entry of plasma amino acids into tissues), increasing fat and protein stores for use during the fasted state. Insulin inhibits glucagon release, decreases liver gluconeogenesis, and decreases liver and muscle glycogenolysis. Figure 11.27 summarizes the responses to hyperglycemia.

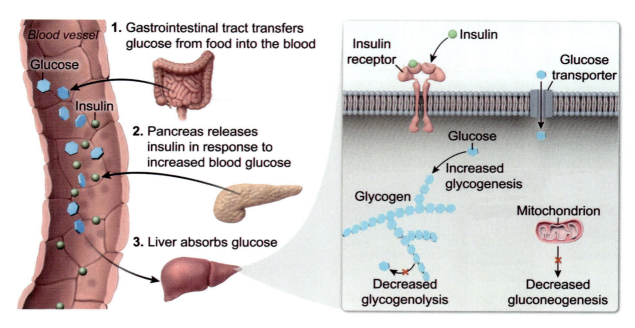

Figure 11.27 Cellular responses to high blood glucose.

Diabetes mellitus is a condition characterized by hyperglycemia and is divided into two subtypes. In type 1 (insulin-dependent) diabetes mellitus, insulin production is low or absent due to autoimmune destruction of pancreatic beta cells. Individuals with type 1 diabetes mellitus benefit from supplemental insulin. Type 2 (non-insulin-dependent) diabetes mellitus is characterized by insulin resistance and increased glucose production by the liver. Due to insulin resistance, supplemental insulin is not as helpful in individuals with type 2 diabetes mellitus.

The final hormone produced by the pancreas, somatostatin, is also produced in the hypothalamus, where it inhibits growth hormone secretion by the anterior pituitary (see Concept 11.2.03). Somatostatin is secreted by pancreatic delta cells in response to high glucose and amino acid concentrations in the blood and has a general inhibitory effect on digestive processes.

11.2.07 Thyroid Hormones

The **thyroid gland** is located in the neck, anterior to the trachea, as depicted in Figure 11.28. Two endocrine cell types compose the thyroid: follicular cells and C (parafollicular) cells.

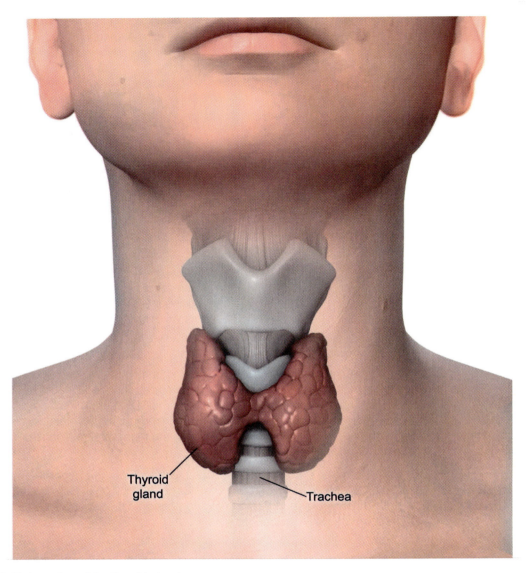

Figure 11.28 Location of the thyroid gland.

The follicular cells of the thyroid produce **triiodothyronine (T_3)** and the less potent **tetraiodothyronine (thyroxine, T_4)**, collectively known as **thyroid hormones**. Both thyroid hormones are tyrosine derivative hormones that contain iodine atoms. Thyroid hormones are lipophilic (ie, water-insoluble); therefore, these hormones travel in the bloodstream bound to plasma proteins. Most circulating thyroid hormone is T_4, but target tissues increase circulating T_3 levels by enzymatically converting T_4 to T_3.

In adults, T_3 and T_4 influence the function of most cells in the body and help to set the basal rate of metabolism. Activation of nuclear thyroid hormone receptors has a general stimulatory effect on many processes and increases cellular metabolism and body temperature via regulation of transcription in target cells. In children, thyroid hormones are necessary for proper development. Thyroid hormone production is regulated through the linked activities of the hypothalamus, pituitary, and thyroid gland (ie, the hypothalamic-pituitary-thyroid [HPT] axis), as discussed in Concept 11.2.03.

The C cells of the thyroid gland produce **calcitonin**, a peptide hormone involved in calcium homeostasis. Calcitonin is secreted in response to increased plasma calcium concentration and is thought to reduce plasma calcium in several ways. First, calcitonin decreases bone resorption by osteoclasts and promotes bone calcium storage. Calcitonin also increases renal excretion of calcium and decreases intestinal absorption of calcium. The effects of calcitonin are illustrated in Figure 11.29.

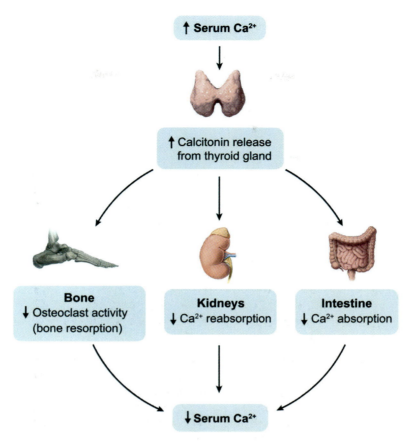

Figure 11.29 Physiological effects of calcitonin.

 Concept Check 11.1

Hypothyroidism is a condition in which thyroid hormones are secreted at lower levels than normal. Predict the effects of hypothyroidism on hypothalamic thyrotropin-releasing hormone (TRH) and anterior pituitary thyroid-stimulating hormone (TSH) secretion. In addition, predict some clinical features of patients with hypothyroidism.

Solution

Note: The appendix contains the answer.

11.2.08 Parathyroid Hormones

The vast majority of the body's calcium is stored in the bone; however, free calcium ions play an integral role in processes throughout the body. Extracellular calcium concentrations are much higher than cytosolic calcium concentrations, and cytosolic calcium entry can affect the membrane potential of a cell. In addition, calcium entry into the cytosol from outside the cell or from intracellular compartments (eg, sarcoplasmic reticulum) can act as a signal to stimulate intracellular processes (eg, muscle contraction).

The **parathyroid glands** are four small endocrine glands located in the neck on the posterior surface of the thyroid gland (Figure 11.30). Through the production of the peptide hormone **parathyroid hormone (PTH)**, the parathyroid glands are important in regulating calcium homeostasis in the body, and to a lesser extent, phosphate homeostasis. The actions of PTH are antagonistic to those of calcitonin produced by the thyroid gland (see Concept 11.2.07).

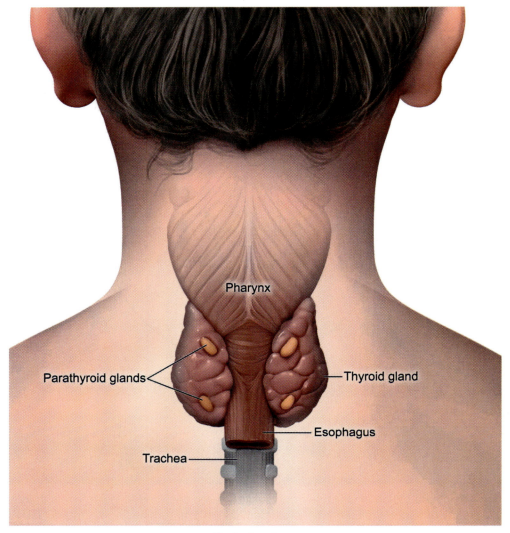

Figure 11.30 The parathyroid glands.

PTH is released in response to decreased plasma calcium and acts in three primary ways to produce a net calcium increase. First, PTH stimulates bone resorption (ie, breakdown) through indirect actions on osteoclasts, the bone cells responsible for resorption. The majority of calcium in the body is stored as hydroxyapatite, a mineral found in the bone matrix that primarily contributes to bone strength and hardness (see Concept 18.1.03).

When bone is resorbed through the actions of osteoclasts, calcium stored in the bone matrix is released into the blood, dissolving mineralized bone and increasing plasma calcium levels. However, osteoclasts do not possess PTH receptors. Therefore, the impact of PTH on osteoclasts is indirect and is achieved through PTH actions on osteoblasts, bone cells with PTH receptors.

In response to increased PTH, osteoblasts are stimulated to differentiate from osteoblast precursor cells. At the same time, PTH promotes the release of ligands by osteoblasts to promote differentiation of osteoclast precursor cells into osteoclasts. In this way, prolonged PTH exposure leads to increased osteoclast numbers and an increase in blood calcium levels, as shown in Figure 11.31.

Before PTH release

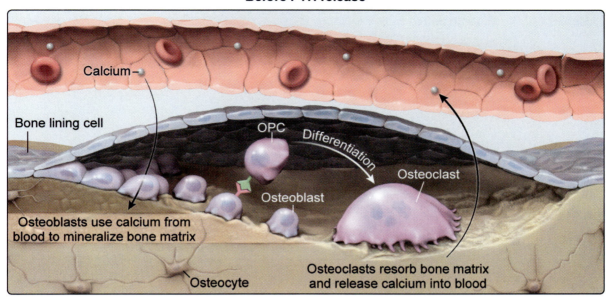

After PTH release

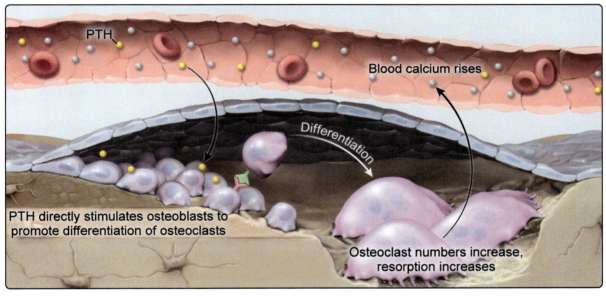

OPC = osteoclast precursor cell; PTH = parathyroid hormone.

Figure 11.31 Parathyroid hormone promotes bone resorption.

The second way PTH increases blood calcium occurs in the kidneys. When PTH levels rise, the kidneys increase calcium reabsorption into peritubular capillaries, increasing blood calcium levels. As a result of increased calcium reabsorption, less calcium is excreted in the urine. PTH also inhibits phosphate reabsorption in the kidneys, increasing excretion of phosphate in the urine (Figure 11.32).

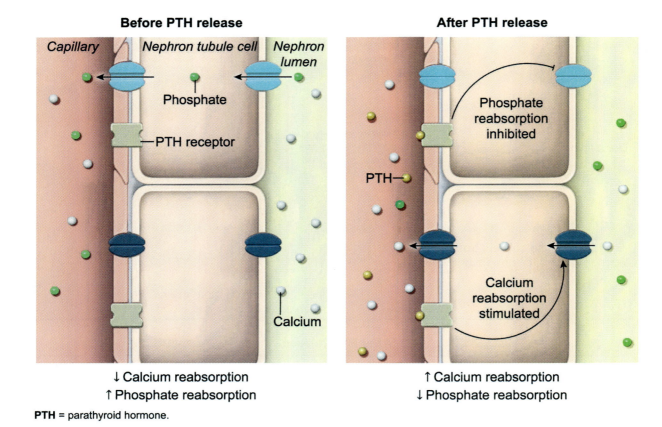

Figure 11.32 Parathyroid hormone promotes calcium reabsorption and phosphate excretion in the kidney.

The final way PTH affects blood calcium is through actions on intestinal cells. PTH increases the activity of the enzyme that catalyzes the final step in the conversion of inactive circulating vitamin D into its active form, **calcitriol**, in the kidneys. Calcitriol promotes absorption of calcium and phosphate in the intestinal lumen. However, because PTH also promotes phosphate excretion in the kidneys, the net effect of PTH on phosphate homeostasis is limited. In addition to promoting intestinal calcium absorption, calcitriol promotes reabsorption of calcium in the kidneys.

Figure 11.33 summarizes the effects of PTH on calcium homeostasis.

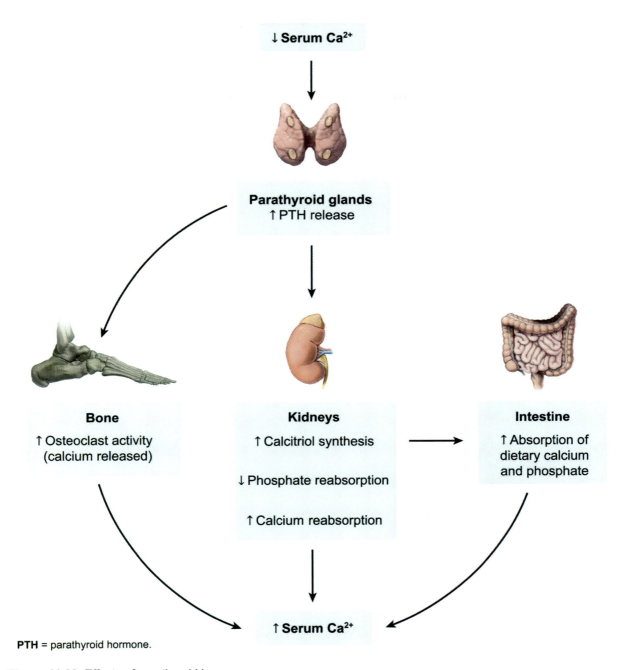

Figure 11.33 Effects of parathyroid hormone.

When blood calcium levels are high, PTH release from the parathyroid glands is prevented via a negative feedback mechanism so that blood calcium levels remain in homeostasis.

11.2.09 Sex Hormones

Sex hormones are steroid hormones with a role in sexual development and reproduction. The term **androgen** refers to the hormones involved in developing and maintaining masculine sexual characteristics, and the term **estrogen** refers to the hormones involved in developing and maintaining feminine sexual characteristics.

As discussed in Lessons 9.1 and 9.2, in an XY embryo, expression of the SRY gene on the Y chromosome promotes sex determination and the development of the testes, the sex organs where most testosterone (the primary androgen) is produced (note: the adrenal cortex secretes small amounts of testosterone, see Concept 11.2.04). The production of testosterone in a fetus typically promotes the development of male sex organs. Because an XX embryo does not express the SRY gene, XX fetuses are exposed to less testosterone, typically resulting in the development of female sex organs (Figure 11.34).

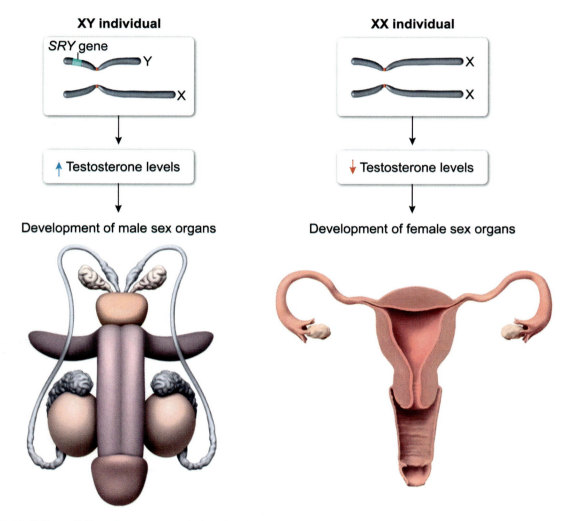

Figure 11.34 Differentiation of sex organs during development.

The production of gametes (ie, sperm and egg cells) via meiosis and the development of secondary sexual characteristics are under the control of various sex hormones.

Male Reproductive Hormones

Beginning at puberty, spermatogenesis (ie, production of sperm) and development of secondary sexual characteristics are regulated by the **hypothalamic-pituitary-gonadal (HPG) axis**. Gonadotropin-releasing hormone (GnRH) from the hypothalamus stimulates the release of luteinizing hormone (LH) and follicle-stimulating hormone (FSH) from the anterior pituitary (see Concept 11.2.03). In the testes, LH promotes the secretion of testosterone by Leydig cells, and FSH promotes and maintains spermatogenesis by stimulating Sertoli (nurse) cells.

Two negative feedback systems operate to maintain appropriate sex hormone levels. First, high levels of testosterone inhibit both GnRH and LH, limiting excessive testosterone production. Second, Sertoli cells secrete a hormone known as **inhibin**, which functions to inhibit FSH, controlling the process of spermatogenesis. The effects of male reproductive hormones are summarized in Figure 11.35.

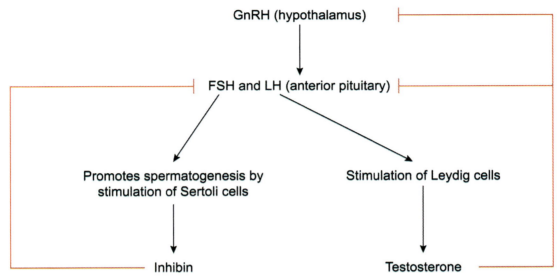

GnRH = gonadotropin-releasing hormone; FSH = follicle-stimulating hormone; LH = luteinizing hormone.

Figure 11.35 Summary of male reproductive hormones.

Female Reproductive Hormones

Oogenesis (ie, production of ova) begins prior to birth, at the fetal stage, but the process is arrested during infancy and childhood due to low sex hormone levels. At puberty, hormonal stimulation from estrogens (eg, estradiol) and progesterone promotes the uterine (menstrual) cycle, the monthly resumption of oogenesis during the ovarian cycle, and the development of secondary sexual characteristics. The ovarian and uterine cycles occur concurrently, and the cycles last approximately 28 days. These cycles provide optimal conditions for fertilization and pregnancy.

The HPG axis promotes the ovarian and uterine cycles through the production of FSH and LH. FSH promotes oogenesis by stimulating the release of estrogens from ovarian follicles, and a surge in LH secretion stimulates ovulation. Early in the ovarian cycle, low to moderate levels of estrogen released by ovarian follicles inhibit the release of GnRH, FSH, and LH in a negative feedback loop. However, later in the ovarian cycle (but prior to ovulation and high levels of progesterone), high levels of estrogen released from the dominant ovarian follicle promote GnRH, FSH, and LH release.

Progesterone, released later in the ovarian cycle by the corpus luteum (ie, a structure that forms from the follicle after ovulation), promotes thickening of the endometrium to prepare an environment conducive to implantation. Progesterone also inhibits GnRH, FSH, and LH release. In yet another negative feedback loop, inhibin, produced by the dominant ovarian follicle, inhibits the release of FSH, which would otherwise cause additional follicles to mature during the same ovarian cycle.

If fertilization does not occur, estrogen and progesterone levels drop, and another ovarian cycle and menstrual cycle begin. However, if fertilization occurs, estrogen and progesterone levels remain high to promote the proper uterine environment for embryonic and fetal development. A summary of female reproductive hormones is illustrated in Figure 11.36.

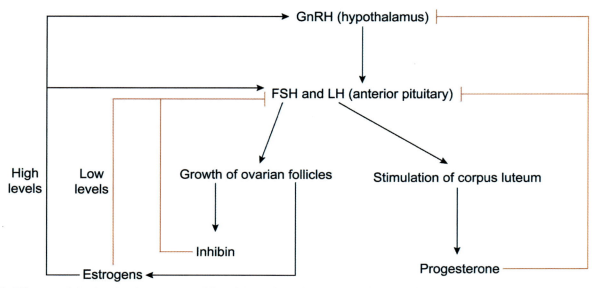

Figure 11.36 Summary of female reproductive hormones.

The complex actions and interactions of various sex hormones are discussed in further detail in Concepts 9.2.03 and 9.3.03.

11.2.10 Appetite Hormones

Sufficient energy and essential nutrients must be taken in via the diet to supply the body's many energy- and nutrient-demanding processes. If dietary intake is chronically too low to meet metabolic demands, body mass is lost via the breakdown of endogenous molecules to compensate for deficient intake of energy and essential molecules, potentially leading to compromised function. Chronic excessive dietary intake can also compromise function by increasing metabolic costs and susceptibility to certain metabolic or body composition-related disorders (eg, insulin resistance, type 2 diabetes mellitus).

The balance between intake and utilization of dietary nutrients is regulated by numerous factors, including hormones released from cells throughout the body. Some of these hormones promote desire for food intake (ie, increased appetite), whereas others promote a feeling of satiety (ie, fullness or dietary satisfaction). **Ghrelin** is a hormone released by cells in the stomach and transported via the bloodstream to the brain, where it acts on the hypothalamus to stimulate appetite prior to a meal.

After a meal, the body is in an energy-rich state (ie, high concentrations of glucose and lipids) and the hormone **leptin** is released by adipose tissue. Leptin triggers feelings of satiety by communicating to the hypothalamus that the stomach is full, thereby suppressing appetite. In general, the greater the adipose tissue stores, the higher the leptin levels in the serum. In addition, delta cells of the pancreas produce **somatostatin**, a hormone that has a generalized inhibitory effect on digestive function (see Concept 11.2.06). Figure 11.37 summarizes the effects of hormones on appetite.

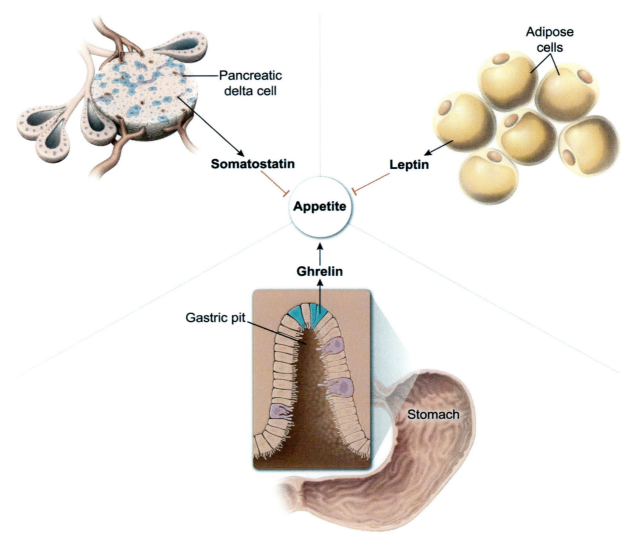

Figure 11.37 Hormonal influences on appetite.

Concept 15.3.01 explores endocrine control of digestion further.

11.2.11 Other Hormones

There are several hormones that are important to know for the exam in addition to those covered in the previous concepts of this lesson.

The amino acid–derived hormone **melatonin** is produced by the pineal gland, a brain structure. Melatonin is thought to influence circadian rhythms, promoting drowsiness and sleep. The pineal gland is stimulated to secrete melatonin when retinal photoreceptors detect low light levels (Figure 11.38).

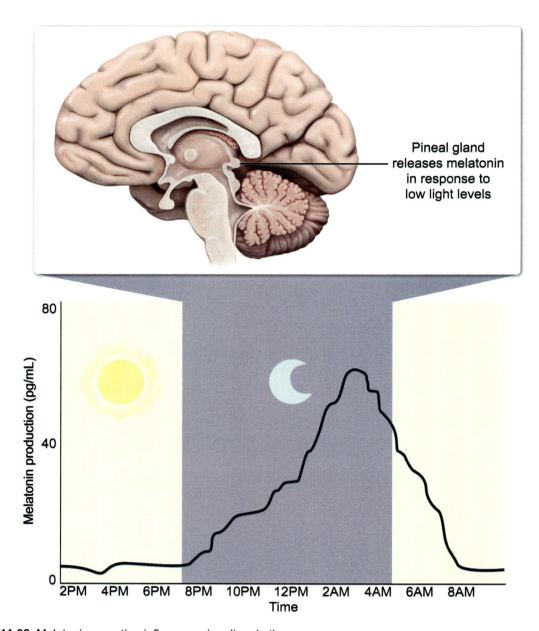

Figure 11.38 Melatonin secretion influences circadian rhythms.

The thymus, a lymphoid organ located anterior to the heart, releases the peptide hormone **thymosin**. Thymosin promotes T cell development and differentiation, which are discussed in more detail in Concept 20.1.03. The thymus shrinks in size after puberty; accordingly, thymosin is most active in children.

Atrial myocytes (ie, cardiac muscle cells of the atria) produce the peptide hormone **atrial natriuretic factor (ANF)** in response to excessive atrial stretch, a characteristic of high blood pressure. ANF targets the kidneys to increase glomerular filtration rate via afferent arteriole vasodilation and efferent arteriole vasoconstriction. At the same time, ANF causes less sodium to be reabsorbed in the nephrons, allowing more water to remain in the filtrate to be excreted in the urine, thereby lowering blood volume. ANF also inhibits the secretion of renin, ultimately inhibiting aldosterone release. The actions of ANF are summarized in Figure 11.39.

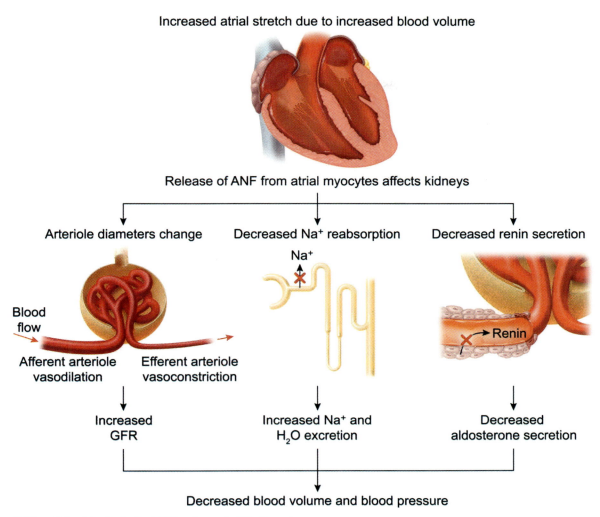

Figure 11.39 Actions of atrial natriuretic factor on the kidneys.

Several peptide hormones are important in digestion, including gastrin, cholecystokinin (CCK), and secretin. **Gastrin** is released by the G cells of the stomach wall in response to food ingestion and targets parietal cells of the stomach wall (see Concept 15.1.02). Secretion of gastrin promotes stomach motility and the release of hydrochloric acid (HCl) by parietal cells to promote digestion.

CCK is produced by epithelial cells of the duodenum (ie, small intestine) and promotes secretion of digestive enzymes from the pancreas and bile from the gallbladder to facilitate digestion. In addition, CCK decreases stomach motility and promotes the feeling of satiety (ie, fullness). **Secretin**, also produced by duodenal epithelial cells, promotes digestive enzyme and bicarbonate release from the pancreas into the duodenum to promote digestion. Secretin also inhibits HCl secretion by stomach parietal cells as chyme moves from the stomach to the small intestine. The actions of CCK and secretin are covered in further detail in Concept 15.1.03.

Tables 11.3, 11.4, and 11.5 provide an overview of important hypothalamic and pituitary hormones, other hormones produced by endocrine glands, and hormones produced by organs and tissues outside the endocrine system, respectively.

Table 11.3 Important hypothalamic and pituitary hormones.

Hormone name	Hormone type	Hormone source	Hormone action
Antidiuretic hormone (ADH)	Peptide	Posterior pituitary (produced by hypothalamus)	Increases blood volume and pressure by increasing nephron water reabsorption
Oxytocin	Peptide		Stimulates uterine contractions and milk ejection from mammary glands
Adrenocorticotropic hormone (ACTH)	Peptide	Anterior pituitary	Promotes glucocorticoid synthesis and secretion in adrenal cortex
Follicle-stimulating hormone (FSH)	Peptide		Stimulates ovarian follicle maturation and spermatogenesis
Luteinizing hormone (LH)	Peptide		Stimulates estrogen production, triggers ovulation, promotes testosterone synthesis
Thyroid-stimulating hormone (TSH)	Peptide		Promotes thyroid hormone synthesis and secretion in thyroid gland
β-endorphins	Peptide		Decreases pain perception and causes feelings of euphoria
Growth hormone (GH)	Peptide		Regulates growth by promoting protein synthesis and fat utilization
Prolactin	Peptide		Promotes production and secretion of milk by mammary glands
Corticotropin-releasing hormone (CRH)	Peptide	Hypothalamus	Stimulates ACTH production in the anterior pituitary
Gonadotropin-releasing hormone (GnRH)	Peptide		Stimulates FSH and LH production in the anterior pituitary
Thyrotropin-releasing hormone (TRH)	Peptide		Stimulates TSH production in the anterior pituitary
Prolactin-inhibiting factor (PIF, dopamine)	Amino acid–derived		Prevents prolactin production in the anterior pituitary

Table 11.4 Important hormones produced by endocrine glands.

Hormone name	Hormone type	Hormone source	Hormone action
Glucocorticoids (cortisol, cortisone)	Steroid	Adrenal cortex	Increase blood glucose during stress response
Mineralocorticoids (aldosterone)	Steroid		Increase blood volume and pressure via increasing reabsorption of water and salt in the kidney
Cortical sex hormones	Steroid		Stimulate sperm cell differentiation, development of male/female sexual traits
Catecholamines (epinephrine, norepinephrine)	Amino acid–derived	Adrenal medulla	Increase blood glucose levels, heart rate, blood flow to critical organs during stress response
Glucagon	Peptide	Pancreas (alpha cells)	Increases blood glucose levels
Insulin	Peptide	Pancreas (beta cells)	Decreases blood glucose levels
Somatostatin	Peptide	Pancreas (delta cells), hypothalamus	General inhibitory effect on digestion, inhibits GH secretion by anterior pituitary
Thyroid hormone (T_3, T_4)	Amino acid–derived	Thyroid gland (follicular cells)	Stimulatory effects, increases cellular metabolism, increases body temperature
Calcitonin	Peptide	Thyroid gland (C [parafollicular] cells)	Reduces blood calcium levels
Parathyroid hormone (PTH)	Peptide	Parathyroid glands	Increases blood calcium levels
Testosterone	Steroid	Testes	Promotes spermatogenesis and male sex trait development
Estrogens	Steroid	Ovaries and placenta	Promotes oogenesis and female sex trait development
Progesterone	Steroid		Promotes thickening of endometrium

| Inhibin | Peptide | Testes and ovaries | Inhibits FSH |
| Melatonin | Amino acid–derived | Pineal gland | Influences circadian rhythms, promotes drowsiness and sleep |

T_3 = triiodothyronine; T_4 = thyroxine.

Table 11.5 Important hormones produced by organs and tissues outside the endocrine system.

Hormone name	Hormone type	Hormone source	Hormone action
Erythropoietin (EPO)	Peptide	Kidney	Increases blood oxygen levels through promoting red blood cell production
Calcitriol	Steroid	Kidney	Promotes intestinal calcium absorption and nephron calcium reabsorption
Ghrelin	Peptide	Stomach	Acts on hypothalamus to stimulate appetite
Gastrin	Peptide	Stomach	Promotes stomach motility and HCl release in stomach parietal cells
Leptin	Peptide	Adipose tissue	Acts on hypothalamus to suppress appetite
Thymosin	Peptide	Thymus	Promotes T cell development and differentiation
Atrial natriuretic factor (ANF)	Peptide	Heart	Acts on kidneys to reduce blood volume and pressure
Cholecystokinin (CCK)	Peptide	Duodenum	Promotes secretion of pancreatic enzymes and bile, decreases stomach motility
Secretin	Peptide	Duodenum	Promotes secretion of pancreatic enzymes and bicarbonate, inhibits stomach HCl secretion

HCl = hydrochloric acid.

Concept Check 11.2

Three primary hormones that influence water balance in the body are discussed in this lesson. Complete the given table with the names and properties of these three hormones.

Hormone	Location of hormone production	Stimulus for hormone release	Hormone target	Hormone function
	Hypothalamus (released from posterior pituitary)			
		Low blood pressure		Promotes Na$^+$ reabsorption and K$^+$ secretion to increase H$_2$O reabsorption, leading to increased blood volume and blood pressure
	Atrial myocytes			

Solution

Note: The appendix contains the answer.

Lesson 12.1

Cells of the Nervous System

Introduction

The **nervous system** includes both neurons, cells responsible for communicating information throughout the body via electrochemical signaling, and glial cells (sometimes called neuroglia or neuroglial cells), which provide various means of support to neurons. Some glial cells modify neurons by providing an electrically insulating layer called myelin. This lesson presents the basic characteristics of neurons and the ways that the various types of glial cells interact with neurons.

12.1.01 The Neuron

Neurons (ie, nerve cells) are the cells of the nervous system specialized to receive, integrate, and transmit information via electrochemical signaling. The sites of information transmission, called **synapses**, are areas of neuron plasma membrane either in close proximity to or in actual contact with other cells, including other neurons.

A typical neuron (Figure 12.1) consists of a **soma** (cell body), which contains the nucleus and participates in information reception and integration, and specialized branching extensions. **Dendrites** are neuron extensions generally specialized for signal *reception* and contain many receptors for molecules involved in information transfer. **Axons** are neuron extensions typically specialized for signal *transmission*. The **axon hillock** is the region where an axon originates from a neuron cell body, and the **axon terminal** is the end of the axon, where signal transmission occurs.

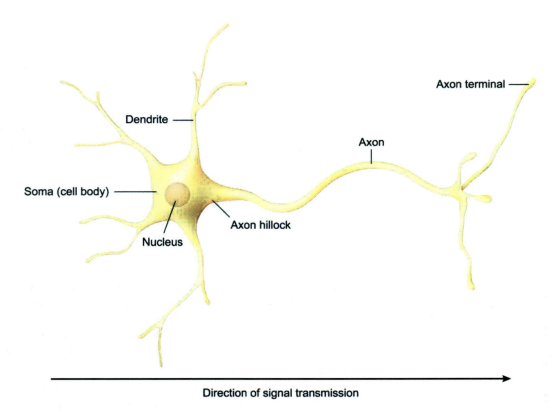

Figure 12.1 The neuron.

Although all neurons have at least one synapse, neuron morphology varies considerably. **Bipolar neurons** have a single dendrite and a single axon, whereas **multipolar neurons** have more than one dendrite and a single axon (Figure 12.1 illustrates a multipolar neuron). **Unipolar neurons** have cell bodies that lay to the side of a single extension formed from the fusion of a dendrite and an axon. Signals in unipolar neurons travel from the "dendritic" (ie, receptive) end to the "axonal" (ie, transmissive) end; however, the entire extension is called an axon.

Neurons can also be classified according to their function and location. Neurons in the brain and spinal cord are part of the central nervous system (CNS), and neurons in the periphery are part of the peripheral nervous system (PNS). **Sensory (afferent)** neurons transmit information from the periphery to the CNS and are often linked to a specific type of sensory receptor. **Motor (efferent** or **somatic)** neurons transmit signals from the CNS to the periphery (eg, skeletal muscle fibers) to cause an action (eg, movement). Together, sensory and motor neurons account for roughly 1% of all neurons; the other 99% are **interneurons**, which are CNS neurons connecting two neurons.

The flow of ions across the plasma membrane plays an essential role in neuron function, as discussed in Lesson 12.2, and, to regulate such flow, neurons invest a large amount of energy in pumping ions across the plasma membrane. In addition, the structure of neurons necessitates the transport of molecules to and from distant cellular locations (eg, from the cell body to the axon terminal). Neurons use cellular transport systems with ATP-dependent motor proteins to move molecules throughout the cell. To help meet the cell's energy demands, neurons have an abundance of mitochondria.

Concept Check 12.1

Identify each region in the diagram of a bipolar neuron as a region that primarily receives information at synapses or a region that primarily transmits information at synapses.

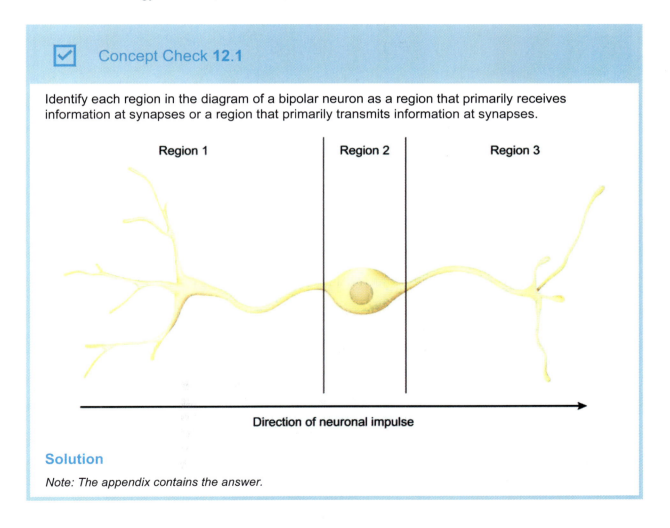

Region 1 Region 2 Region 3

Direction of neuronal impulse

Solution

Note: The appendix contains the answer.

12.1.02 Myelin Sheath

As part of their specialization for transmitting information via electrical impulses, certain neurons display different structural characteristics. **Myelinated neurons** have an electrically insulating sheath known as **myelin** around their axons, whereas **unmyelinated neurons** lack this layer. The myelin sheath consists of multiple layers of a glial cell phospholipid membrane wrapped around the axon, as shown in Figure 12.2. Myelination alters ion and nutrient transport along axons, and myelinating cells provide nutrients to myelinated regions via gap junctions and transporters. Cell bodies and dendrites are not myelinated.

The myelin layer is not continuous along the length of a myelinated axon; it is interrupted at small, regularly spaced sites called the **nodes of Ranvier**. These nodes are rich in ion channels (ie, membrane proteins through which ions can pass). By focusing ion movement to the nodes of Ranvier, myelination causes electrical impulses to "jump" along the axon. This form of electrical impulse transmission is called **saltatory conduction** and greatly augments the rate of signal transmission along myelinated axons (discussed in detail in Lesson 12.2).

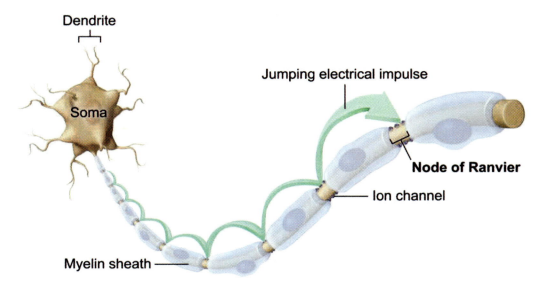

Figure 12.2 A myelinated axon.

Of the various types of glial cells (discussed in Concept 12.1.03), two participate in myelination: **Oligodendrocytes** myelinate central nervous system (CNS) neurons and **Schwann cells** myelinate peripheral nervous system (PNS) neurons.

In demyelinating diseases (eg, multiple sclerosis, Guillain-Barré syndrome) or when nerves are otherwise damaged (eg, through mechanical trauma such as cutting or compression), the myelin sheath is damaged or removed (Figure 12.3). Demyelination can profoundly slow or block signal transmission along neurons that are myelinated when healthy.

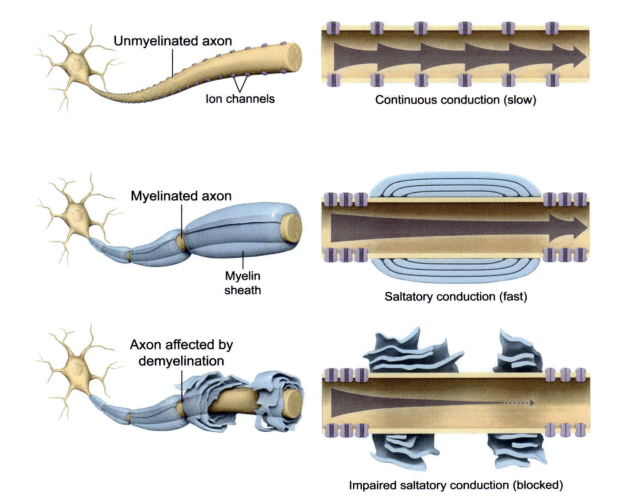

Figure 12.3 Axon myelination states.

12.1.03 Glial Cells

Like a surgeon requires a supporting group of individuals carrying out specific tasks (eg, monitoring vital signs, administering anesthesia, assisting with surgical procedures) to safely perform surgery, neurons require a group of support cells for the protection and optimization of neuronal impulse transmission. **Glial cells** (sometimes called glia or neuroglia) are nervous system cells that help fuel, protect, and structurally support neurons.

There are multiple types of glial cells. Some of the central nervous system (CNS) glial cells are shown in Figure 12.4:

- **Ependymal cells** line compartments in the CNS, thereby forming a selectively permeable barrier between the compartments, and produce cerebrospinal fluid, which helps cushion and support structures, thereby protecting neurons from mechanical trauma.
- **Oligodendrocytes** are myelinating cells, forming myelin sheaths around multiple axons to reduce ion leakage, decrease capacitance, and increase action potential propagation.
- **Microglia** are immune cells that can transform into specialized macrophages in response to neuronal damage or infection and can phagocytose pathogens, damaged cells, and waste materials. Microglia functions are essential because other immune cells (eg, B cells, cells of the innate immune system) are largely prevented from entering the CNS.
- **Astrocytes** are a network of cells connected by gap junctions and perform diverse roles. Some contact with blood vessels and regulate blood flow or form the blood-brain barrier; others are found

near synapses, where they help regulate chemical composition by taking up K⁺ ions and signaling molecules that might interfere with signal transmission. Yet others associate with neurons to exchange metabolic substrates.

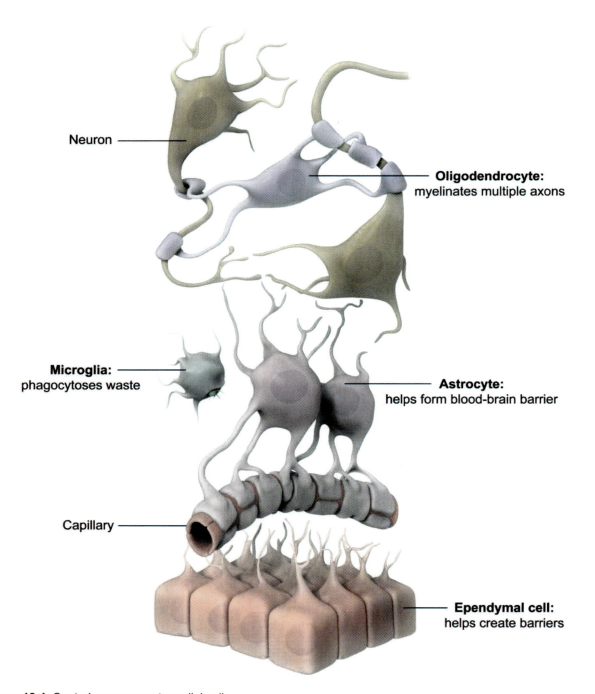

Figure 12.4 Central nervous system glial cells.

Two of the glial cell types found in the peripheral nervous system (PNS) are illustrated in Figure 12.5:

- **Schwann cells** each associate with a single axon to form a myelin sheath that increases the speed of electrical impulse conduction.
- **Satellite cells** are nonmyelinating Schwann cells that provide structural support and supply nutrients to neuron cell bodies, similar to the role that some astrocytes play in the CNS.

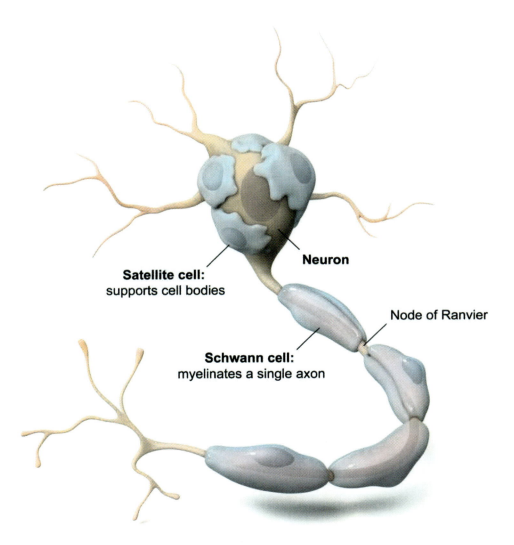

Figure 12.5 Peripheral nervous system glial cells.

Lesson 12.2
Neural Communication

Introduction

Neural communication with other cells (eg, other neurons, myocytes) is enabled by special neuronal impulses called **action potentials** that cause a temporary reversal of the charge gradient at their axon terminals. This lesson first discusses the electrical gradient present in neurons under unstimulated conditions, called the resting membrane potential, before presenting the mechanisms by which action potentials are generated and elicit responses in target cells.

12.2.01 Resting Membrane Potential

Phospholipid bilayers are not very permeable to charged molecules, so ion transport in and out of cells occurs predominantly via proteins embedded in the cell membrane (see Lesson 5.2). For example, channels allow ions to diffuse across the plasma membrane, whereas pumps use energy to move ions across the membrane. Because of the various types, numbers, and activities of the membrane transport proteins, cell plasma membranes are selectively permeable.

As a result of this selective permeability, concentrations of charged molecules in the intracellular and extracellular fluid differ, and electrical and concentration gradients exist. The concentration gradients of the various ions favor the movement of the ions from areas of higher concentration to areas of lower concentration, whereas the electrical gradient across the plasma membrane attracts ions to oppositely charged regions, as depicted in Figure 12.6.

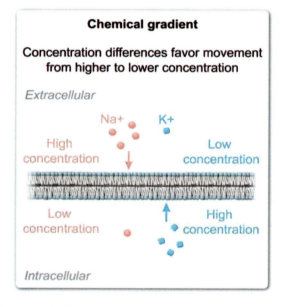
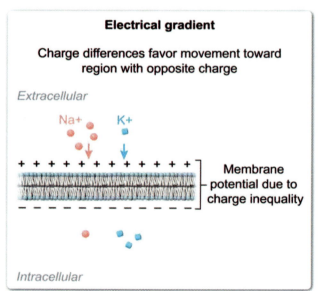

Figure 12.6 The effect of chemical and electrical gradients on ions.

Because opposite charges attract, the separation of charge across the cell membrane represents a form of potential energy that can be used to drive cellular processes (eg, secondary active transport), and the magnitude of the charge difference is accordingly called the **membrane potential**. By convention, the membrane potential is measured using the extracellular fluid as a reference (assigned a value of 0 mV).

In a cell at rest, the membrane electrical polarization is such that the intracellular fluid is typically more negative than the extracellular fluid.

If only a single ion type could cross a plasma membrane, that ion's movement would be influenced by both its concentration gradient (favoring movement from higher to lower concentration) and the membrane's electrical gradient (favoring movement away from like charges and toward opposite charges). For a given concentration gradient, there is an electrical gradient that would exactly oppose the concentration gradient and prevent net ion movement across the membrane. This membrane potential, at which the ion's concentration and electrical gradients cancel each other, is called the ion's **equilibrium potential** (Figure 12.7).

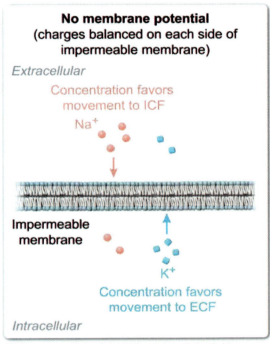
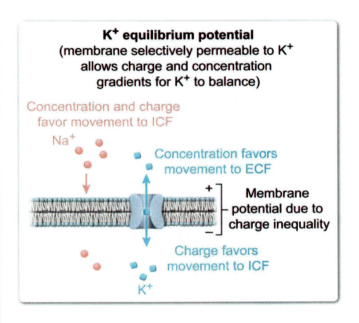

ECF = extracellular fluid, ICF = intracellular fluid.

Figure 12.7 Equilibrium potential.

Because both electrical and concentration (ie, chemical) gradients influence ionic movement across the plasma membrane, the combined influence is referred to as the **electrochemical gradient**. Ions diffuse across cellular membranes down the ion's electrochemical gradient.

In an actual cell, the plasma membrane is permeable to more than one ion, so the electrochemical gradients of multiple ions simultaneously influence the cell's membrane potential. Ions with greater permeability exert a greater influence on the membrane potential, pushing the membrane potential in the direction of the equilibrium potential of these more permeable ions. The **resting membrane potential (RMP)** is the electrical gradient across a cell's membrane under baseline (unstimulated) conditions.

Although numerous ions can cross the plasma membrane, Na^+ and K^+ play a major role in establishing the RMP (Figure 12.8). Leak channels for each ion allow a small, continual stream of the two ions across the plasma membrane via diffusion, and Na^+/K^+ ATPase pumps the ions back across the plasma membrane against their concentration gradients. The net result of transport through these and other membrane proteins is that the plasma membrane is ~40 times more permeable to K^+ than to Na^+. Accordingly, the RMP of approximately −70 mV is closer to the equilibrium potential of K^+ (−90 mV) than that of Na^+ (+60 mV).

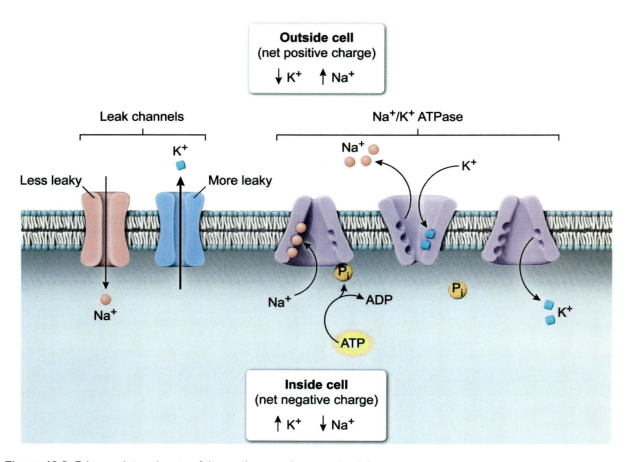

Figure 12.8 Primary determinants of the resting membrane potential.

12.2.02 The Action Potential

Just as the combination of open and closed channels determines the resting membrane potential (RMP), a change in the complement of open and closed channels in a cell's membrane can alter the cell's membrane potential. An **action potential** is a brief, regenerative wave of membrane potential fluctuation that travels away from the site of initiation in an excitable cell (eg, neuron, muscle). In neurons, action potentials originate from a region called the **trigger zone**, consisting of the axon hillock and the initial part of the axon, when the trigger zone membrane potential becomes more positive than a certain threshold (around −55 mV).

In addition to its use to describe these electrical impulses that travel long distances, the term *action potential* is used to denote the stereotypic pattern of membrane potential changes that occur in a single location during one of these electrical impulses. These changes include a period of **depolarization** (ie, in which the membrane potential becomes more positive) followed by a period of **repolarization** (ie, in which the membrane potential returns to its baseline level). Figure 12.9 depicts these membrane potential changes in a neuron and a muscle cell membrane.

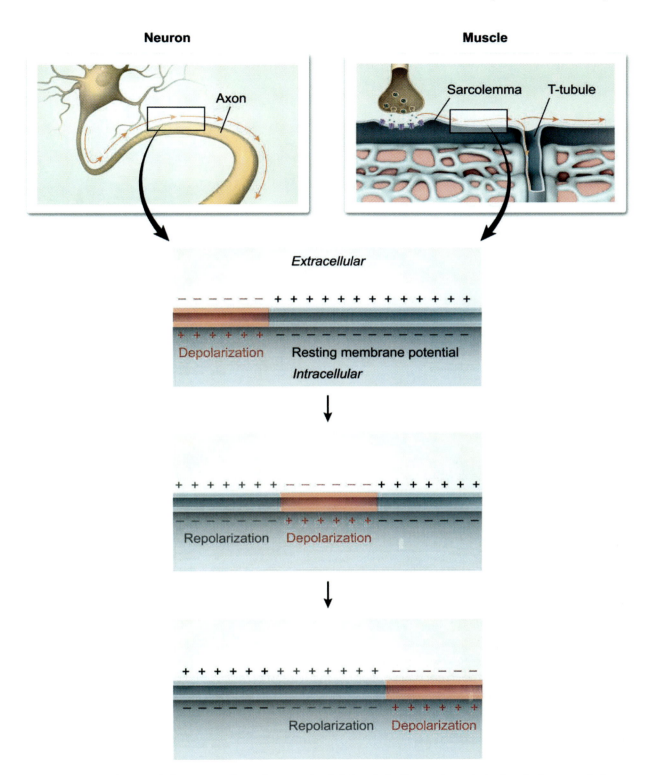

Figure 12.9 The characteristic pattern of membrane potential depolarization and repolarization that occurs during an action potential.

The use of the term *action potential* to describe both the electrical impulse moving along a cell's membrane, as well as the characteristic local membrane potential changes accompanying such a signal, alludes to an important point: The moving electrical impulse is composed of multiple isolated events. By analogy, a "wave" in a stadium consists of multiple sections of people standing and sitting in sequence (not one set of people running around the stadium). Similarly, the regenerating depolarization wave that travels along the membrane during an action potential consists of ion channels in multiple membrane regions being activated sequentially (Figure 12.10).

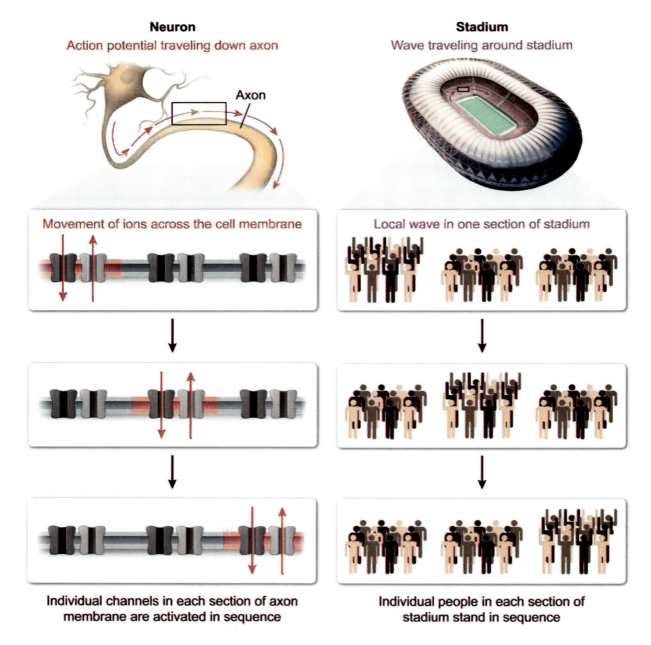

Figure 12.10 Like a wave in a stadium, the large-scale action potential that travels down an axon consists of multiple, local events in sequence.

As with the RMP, the movement of sodium (Na^+) and potassium (K^+) ions is responsible primarily for the fluctuations in membrane potential that occur during an action potential. The changes in ion concentrations underlying a single action potential are very small and do not affect the overall Na^+ and K^+ concentration gradients across the cell membrane. However, this ion movement does affect the electrical gradient, as manifested in membrane potential changes.

During an action potential, **voltage-gated channels** for Na^+ and K^+ (Figure 12.11) play essential roles, as do the Na^+ and K^+ leak channels that contribute to the RMP (see Concept 12.2.01). Voltage-gated Na^+

and K⁺ channels both have a structural component called an **activation gate** that opens to allow ion flow through the channel in response to depolarization. The voltage-gated Na⁺ channel also possesses a second, **inactivation gate** that closes in response to depolarization. Note that the naming of these gates varies across sources, but this is how they will be referred to in this book.

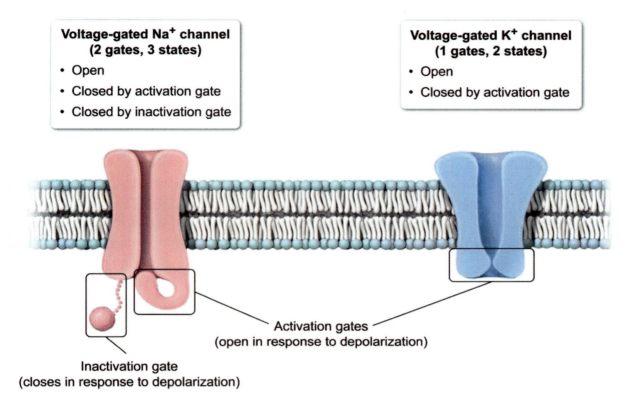

Figure 12.11 Voltage-gated Na⁺ and K⁺ channels.

The rates at which these gates open and close are very important, and Na⁺ activation gates open before the movements of the other gates are completed.

An action potential encompasses three phases:

1. A **rising phase** in which the membrane potential becomes progressively more positive

2. A **falling phase** in which the membrane potential becomes progressively more negative, eventually becoming hyperpolarized

3. A **restoring phase** in which the membrane potential returns to the resting level from a hyperpolarized state

In the rising phase (Figure 12.12), the membrane potential becomes more positive due to increased Na⁺ permeability caused by the opening of Na⁺ channel activation gates. Depolarization to the **threshold potential** (around −55 mV) is pivotal because this initiates a positive feedback cycle of depolarization-induced Na⁺ channel opening, leading to further depolarization. Eventually, this feedback cycle leads to the activation of all the voltage-gated Na⁺ channels in the vicinity. Therefore, if the threshold potential is reached, an action potential occurs, and, if the threshold is not reached, an action potential does not occur (ie, action potentials are **"all-or-none"** events).

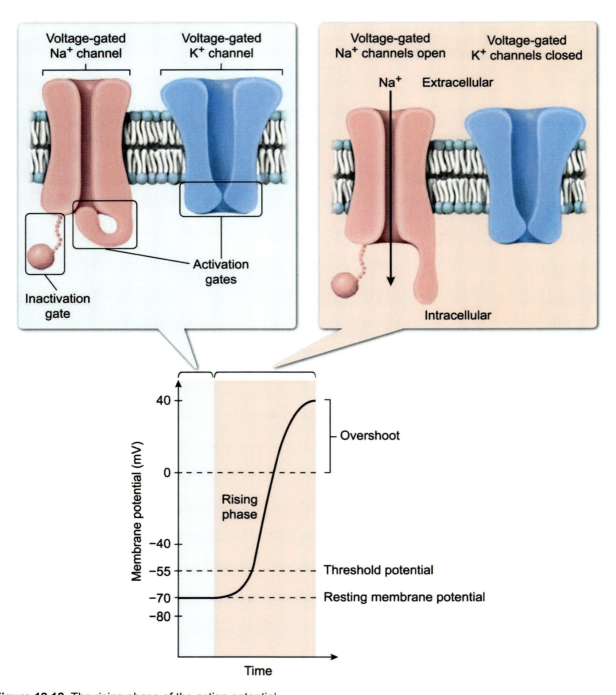

Figure 12.12 The rising phase of the action potential.

In response to the influx of positively charged Na⁺ into the intracellular fluid, the inside of the cell becomes more positive and the extracellular fluid more negative. Because of this Na⁺ influx, the membrane potential eventually reverses sign, becoming positive, and peaks at a membrane potential of around +40 mV. The part of the action potential above 0 mV is sometimes called **overshoot**.

After the peak of the rising phase, the membrane potential begins the sharp decline of the falling phase (Figure 12.13). To start this phase, voltage-gated Na⁺ channel inactivation gates close, thereby ending the inward flow of Na⁺. At the same time, voltage-gated K⁺ channels open, reversing and eventually hyperpolarizing the membrane (ie, becoming more negative than the RMP). As membrane potential falls below the threshold potential, Na⁺ channels begin resetting to the resting state. The portion of the action potential more negative than the RMP is sometimes called **undershoot**.

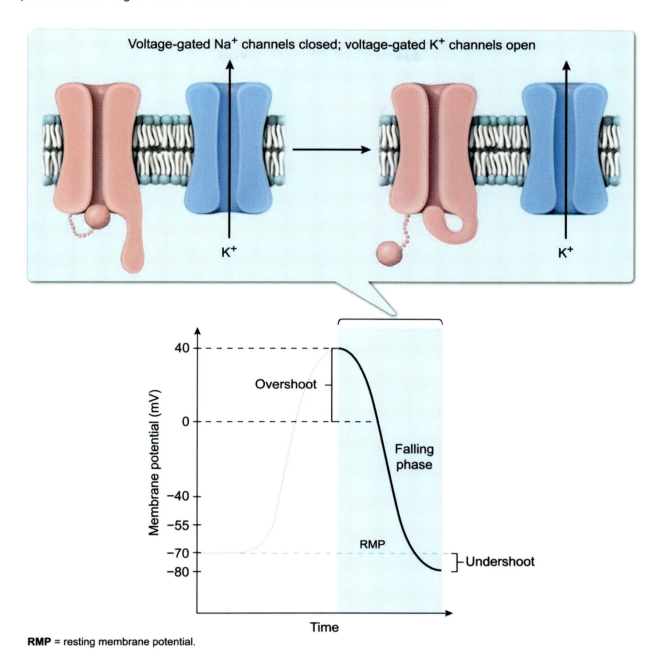

RMP = resting membrane potential.

Figure 12.13 The falling phase of the action potential.

For most of the falling phase, the voltage-gated Na⁺ channels are closed, primarily via the inactivation gates, and the voltage-gated K⁺ channels are open, neither of which occur during the resting state.

The restoring phase (Figure 12.14) of the action potential occurs when voltage-gated Na⁺ and K⁺ channels are finished being reset to the resting state and the RMP is restored. In this phase, the Na⁺ channel activation gates return to the closed position while the inactivation gates return to the open position. At the same time, voltage-gated K⁺ channels close. With closure of the voltage-gated Na⁺ and K⁺ channels, Na⁺ and K⁺ leak channels, along with the Na⁺/K⁺ pump, are of primary importance in determining membrane potential, and membrane potential rises back to the RMP (approximately −70 mV).

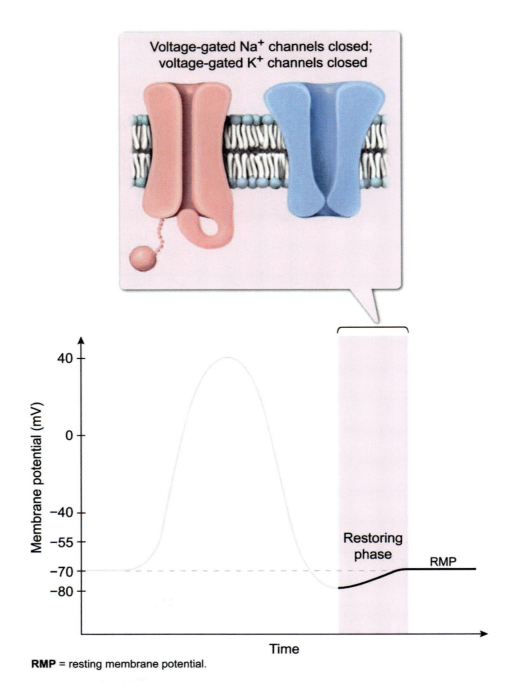

Figure 12.14 The restoring phase of the action potential.

The complete action potential is shown in Figure 12.15.

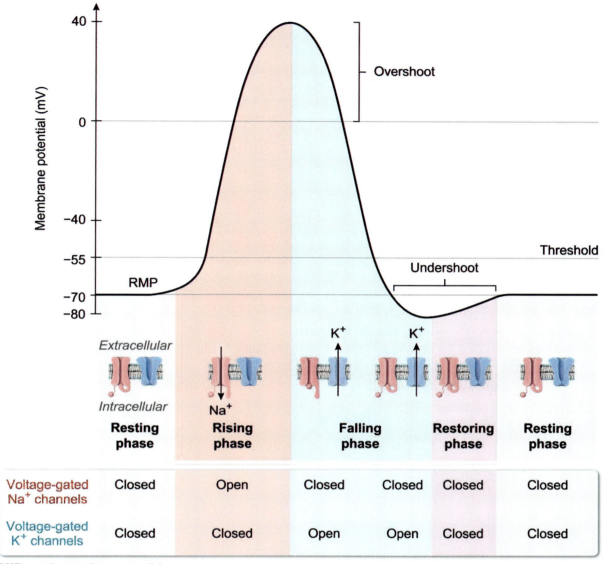

Figure 12.15 The action potential.

The excitability of a neuron changes during the action potential. As the rising membrane potential exceeds the threshold potential, all Na⁺ channels become committed to opening. As the membrane potential begins falling, Na⁺ channels are closed via the inactivation gates, a state incompatible with channel opening. With no Na⁺ channels available for activation, the neuron enters an **absolute refractory period** during which no amount of stimulation can elicit an action potential. This delayed return of excitability after the depolarization phase of an action potential ensures unidirectional flow along the axon.

After the absolute refractory period, the Na⁺ channels begin resetting to the resting state (ie, activation gates close and inactivation gates open). As this occurs, the neuron enters a **relative refractory period** that continues until the Na⁺ and K⁺ channels return to their resting states and the RMP is restored. During this period, newly reset Na⁺ channels can be recruited to open, provided that the depolarizing stimulus is stronger than normal. A stronger stimulus is necessary to overcome the influence of open K⁺ channels (early in the period), as well as the hyperpolarized membrane potential.

> **Concept Check 12.2**
>
> Identify the status (ie, open or closed) of voltage-gated Na⁺ and K⁺ channels for each of the following phases in a typical action potential: the rising portion of the overshoot, the falling portion of the overshoot, the falling portion of the undershoot, the rising portion of the undershoot.
>
> **Solution**
>
> *Note: The appendix contains the answer.*

12.2.03 Synaptic Transmission

Neurons communicate via junctions called **synapses** that are sites of actual or near contact between a neuron and another cell (either another neuron or a cell in a target tissue such as skeletal muscle). Communication between the two cells is directional, with the nerve impulse in the **presynaptic** (transmitting) neuron being communicated to the **postsynaptic** (receiving) cell. Synapses consist of the axon terminal of the presynaptic neuron and the plasma membrane of the postsynaptic cell, as well as the space between the two cells when the cells are not in direct contact.

In a synapse between two neurons, the nerve impulse is typically transmitted from the presynaptic axon terminal to a postsynaptic dendrite or soma, as shown in Figure 12.16.

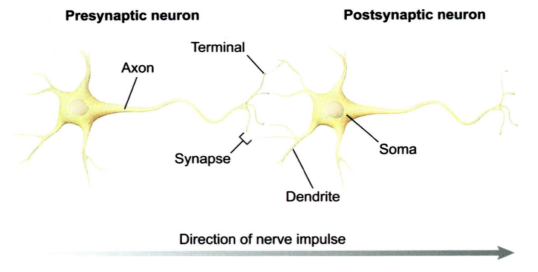

Figure 12.16 A synapse between two neurons.

Synaptic transmission of a nerve impulse between neurons can occur via an electrical or a chemical mechanism (Figure 12.17). At **electrical synapses**, electrical impulses are transmitted directly from one cell to the other via gap junctions. At **chemical synapses**, ligands called **neurotransmitters** are released from the presynaptic neuron into the **synaptic cleft** (ie, the space between the presynaptic axon terminal and the postsynaptic cell). The neurotransmitters bind postsynaptic membrane receptors, which often causes postsynaptic ligand-gated ion channels to open, facilitating ion movement into or out of the postsynaptic cell.

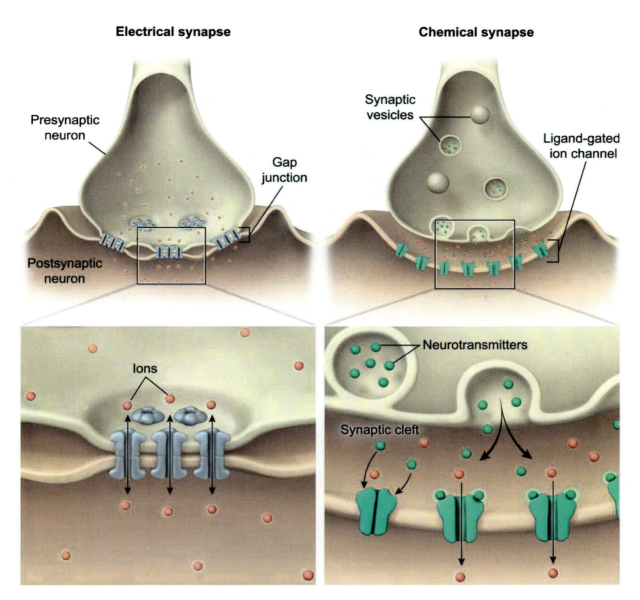

Figure 12.17 Electrical versus chemical synapses.

The signal from the presynaptic neuron is diminished or terminated in a chemical synapse when the concentration of neurotransmitter in the synaptic cleft is reduced. In some cases, this occurs through simple diffusion of the neurotransmitter away from the synapse. In other cases, neurotransmitter levels are reduced via reuptake into the presynaptic neuron or through enzymatic destruction.

12.2.04 Neurotransmitters

Neurotransmitters, the signaling molecules released from presynaptic neurons at chemical synapses, can elicit a variety of responses in postsynaptic cells. In nearly all cases, these responses are initiated by neurotransmitters binding to receptors on the postsynaptic cell membrane, which influences the membrane potential of the postsynaptic neuron. Neurotransmitters and the synapses at which they act are classified as excitatory or inhibitory based on their effect on the postsynaptic membrane potential.

Excitatory neurotransmitters have a *depolarizing* effect on the postsynaptic membrane (ie, cause the membrane potential to become more positive, often in response to positive ions such as Ca^{2+} entering the neuron), as shown in Figure 12.18. If the membrane potential of the postsynaptic neuron at an excitatory synapse exceeds a certain threshold (approximately −55 mV), an action potential is initiated in the postsynaptic neuron (see Concept 12.2.02).

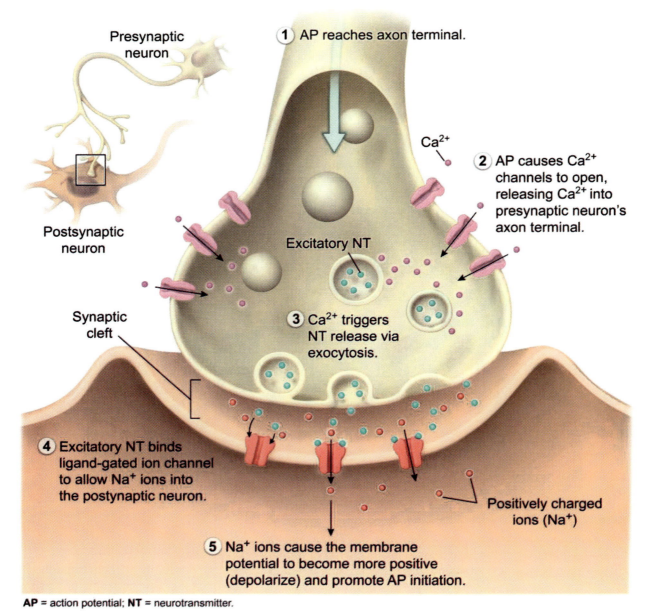

Figure 12.18 Synaptic transmission involving excitatory neurotransmitters.

The binding of **inhibitory neurotransmitters** to the postsynaptic neuron causes either an influx of negative ions (eg, Cl^-) or an efflux of positive ions (eg, K^+). In both cases, binding of the inhibitory neurotransmitter causes the cell's membrane potential to become more negative (ie, **hyperpolarize**), inhibiting action potential initiation, as shown in Figure 12.19.

Chapter 12: Nervous System

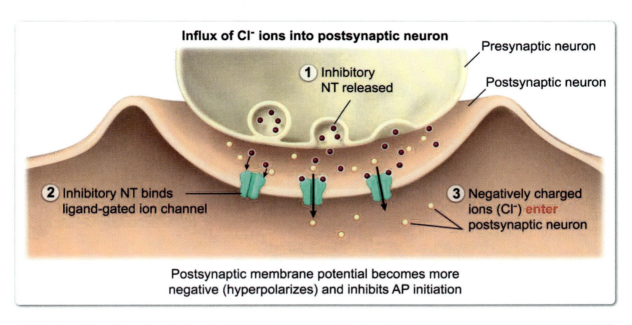

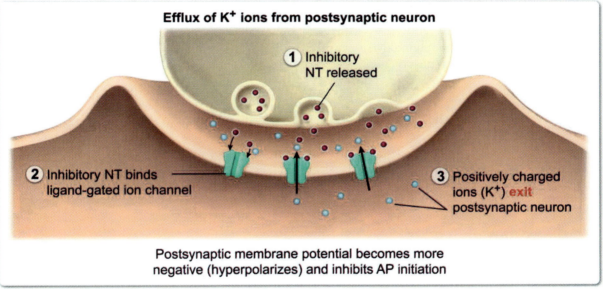

AP = action potential; NT = neurotransmitter.

Figure 12.19 Synaptic transmission involving inhibitory neurotransmitters.

As noted previously, most neurotransmitters act through binding to postsynaptic receptors. Table 12.1 presents several neurotransmitters and their functions.

Table 12.1 Common neurotransmitters and their functions.

	Neurotransmitter	Functions
Amino acids	Glutamate (Glu)	• Primary excitatory neurotransmitter of the central nervous system • Involved in learning and memory
	Gamma-aminobutyric acid (GABA)	• Primary inhibitory neurotransmitter of the brain
	Glycine (Gly)	• Primary inhibitory neurotransmitter of the spinal cord
Amines	Dopamine (DA)	• Involved in cognition, attention, movement, reward
	Serotonin (5-HT)	• Involved in sleep, appetite, mood
	Epinephrine	• Involved in sympathetic signaling in the autonomic nervous system
	Norepinephrine (NE)	• Involved in sympathetic signaling in the autonomic nervous system
	Acetylcholine (ACh)	• Involved in parasympathetic signaling in the autonomic nervous system • Released by motor neurons at NMJs of the somatic nervous system to excite skeletal muscle
Peptides	Endorphins	• Opiates produced by the body that modulate pain, as well as contribute to elevated mood following exercise

NMJ = neuromuscular junction.

12.2.05 Summation of Postsynaptic Potentials

As discussed in Concept 12.2.02, action potentials are "all-or-none" events, and, when triggered, an action potential is regenerated down an axon by the opening of new channels, such that the action potential strength is maintained. However, for an action potential to be triggered, the soma of the neuron must be depolarized to a great enough extent that the membrane potential in the trigger zone is more positive than the threshold potential.

In contrast to axons, dendrites and cell bodies do not generate action potentials. Instead, the postsynaptic cell membrane potential in these areas responds incrementally, and the response is called a **graded potential** because it is proportional to the strength of the stimulus from the presynaptic cell. Graded potentials dissipate as they travel from a synapse through the cell body. Graded potentials are called **excitatory** when they depolarize the postsynaptic neuron (bringing it closer to the threshold potential) and **inhibitory** when they hyperpolarize the postsynaptic neuron (taking it farther from the threshold potential).

Most postsynaptic neurons have multiple synapses, and a single presynaptic action potential typically produces only a postsynaptic graded potential (without an accompanying action potential). As a result, postsynaptic neuron behavior typically reflects the net influence of multiple presynaptic action potentials. The integration of multiple inputs from one or more presynaptic neurons is called **summation**. As shown in Figure 12.20, **spatial summation** is the integrated effect of multiple input signals from multiple presynaptic neurons, whereas **temporal summation** is the integrated effect from multiple input signals from a single neuron.

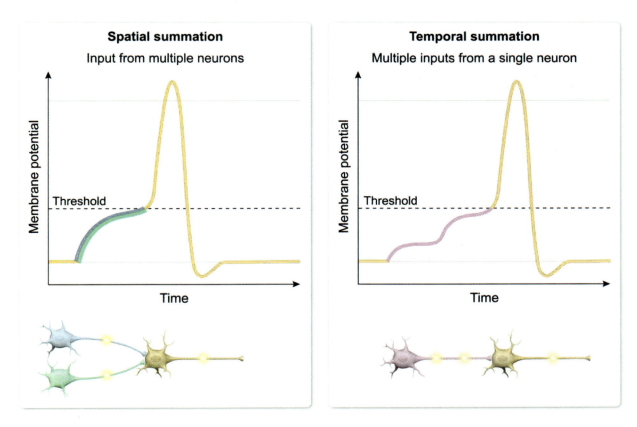

Figure 12.20 Spatial and temporal summation.

Some presynaptic neurons can release more than one type of neurotransmitter under the right circumstances; such neurons are called multi-transmitter neurons. In some cases, the different neurotransmitters are released from distinct vesicles (a phenomenon called co-transmission), either from the same axon terminal or from different axon branches. In other cases, two or more neurotransmitters are released from the same vesicle (a phenomenon called co-release).

Summation incorporates the overall influence of the number and types of neurotransmitters acting on a postsynaptic neuron. At an **excitatory synapse**, the overall effect of the various neurotransmitters to which the postsynaptic neuron is exposed is depolarizing. At an **inhibitory synapse**, the overall effect is hyperpolarizing.

Whether or not summation produces an action potential in a postsynaptic neuron is determined by the effect on the trigger zone. If a membrane potential more positive than the threshold potential (ie, a suprathreshold potential) is elicited at the trigger zone as a result of summation, an action potential will be produced. If summation produces only a graded potential or a suprathreshold potential that dissipates to a level below the threshold potential by the time it reaches the trigger zone, an action potential will not be generated.

Lesson 12.3

Nervous System Structure and Function

Introduction

The nervous system consists of multiple branches spread across several layers of organization. This complex organization reflects the efficient division of labor, as well as robust fine-tuning via antagonistic control. This lesson first presents the general organization of the nervous system before considering the general pathways by which the central and peripheral nervous systems communicate, the antagonistic divisions of the autonomic nervous system, and the participation of the nervous system in feedback regulation.

12.3.01 Organization of the Nervous System

The **nervous system**, which is responsible for the control and integration of all body systems, consists of the brain, spinal cord, and associated neurons and glial cells. The nervous system receives incoming information from external and internal environments, processes this information, and uses the processed information to coordinate purposeful responses. Through these actions, the nervous system coordinates movement, thought, and processes in the body that help maintain homeostasis.

The components of the nervous system can be divided and subdivided along several levels of organization, as shown in Figure 12.21. Anatomically, the **central nervous system (CNS)** consists of the cells of the brain and spinal cord, whereas the **peripheral nervous system (PNS)** consists of the nervous system components outside of the brain and spinal cord. Functionally, the CNS is the part of the nervous system that receives and processes sensory information and coordinates responses to this information throughout the body. The PNS is the part of the nervous system that relays information to and from the CNS and carries out responses.

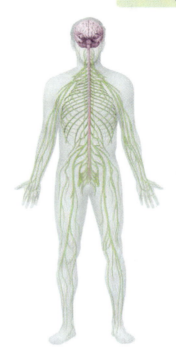

Figure 12.21 The nervous system is organized into central and peripheral components.

Analogous structures in the CNS and PNS have different names. Bundles of cell bodies are called **nuclei** within the CNS and **ganglia** within the PNS. Likewise, bundles of axons within the CNS are called **tracts**, whereas, in the PNS, such bundles are called **nerves**. Notably, neurons extending from the CNS to a ganglion are called **preganglionic** fibers, and neurons extending from a ganglion to a target tissue are called **postganglionic** fibers.

Within the CNS, neuron components are grouped to form **gray matter** (consisting of dendrites, cell bodies, and unmyelinated axons) and **white matter** (primarily myelinated axons). In the brain, gray matter is found on the surface of the cerebral cortex, with the many interconnecting axons of the white matter found on the interior. This arrangement is reversed in the spinal cord, with white matter on the surface and gray matter in the core.

The PNS can be further divided into **motor** and **sensory divisions** (sometimes called efferent and afferent divisions, discussed further in Concept 12.3.03). The motor division includes both the **somatic nervous system**, which innervates skeletal muscles, and the **autonomic nervous system (ANS)**, which innervates cardiac and smooth muscles, as well as glands. The enteric nervous system (see Concept 15.3.02) is also partly controlled by the ANS. The ANS is further divided into **sympathetic** and **parasympathetic divisions**, which are discussed in Concept 12.3.02. Figure 12.22 summarizes the PNS divisions.

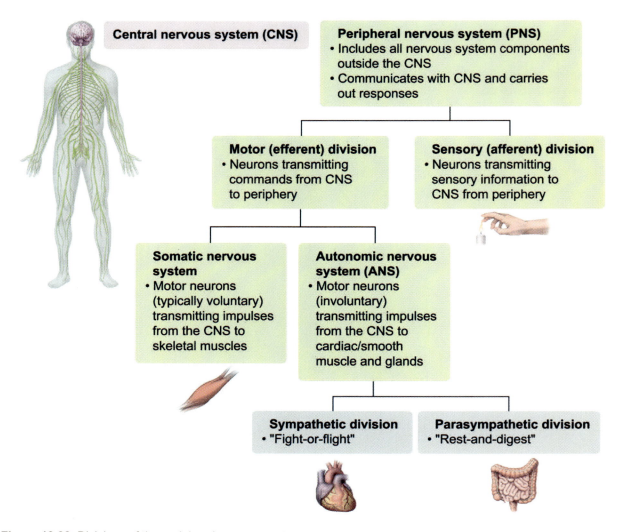

Figure 12.22 Divisions of the peripheral nervous system.

12.3.02 The Sympathetic and Parasympathetic Nervous Systems

The autonomic nervous system (ANS), itself part of the motor (efferent) division of the peripheral nervous system (PNS), can be divided into two largely antagonistic divisions, both regulating involuntary bodily functions:

- The **sympathetic division** (sympathetic nervous system) promotes "fight-or-flight" responses that prepare the body for action under stressful conditions. For example, the sympathetic nervous system activates processes that lead to increased heart rate, increased liver glucose release, and dilation of airways. At the same time, sympathetic activation inhibits nonessential activities such as digestion so that energy can be redirected toward addressing immediate stressors.
- The **parasympathetic division** (parasympathetic nervous system) promotes "rest-and-digest" responses under low-stress conditions. These responses include *increased* digestive functions and energy storage, normal urination patterns, and *decreased* blood pressure (ie, via lower heart rate and vasodilation).

Both sympathetic and parasympathetic responses are transmitted via preganglionic neurons arising from the central nervous system (CNS) and postganglionic fibers that synapse with the target tissue(s) (Figure 12.23). Preganglionic fibers are typically shorter than postganglionic fibers in sympathetic pathways, with the opposite being true in parasympathetic pathways. In both pathways, preganglionic neurons release acetylcholine, but the neurotransmitter released by the postsynaptic neuron varies, with norepinephrine typically released in sympathetic pathways and acetylcholine typically released in parasympathetic pathways.

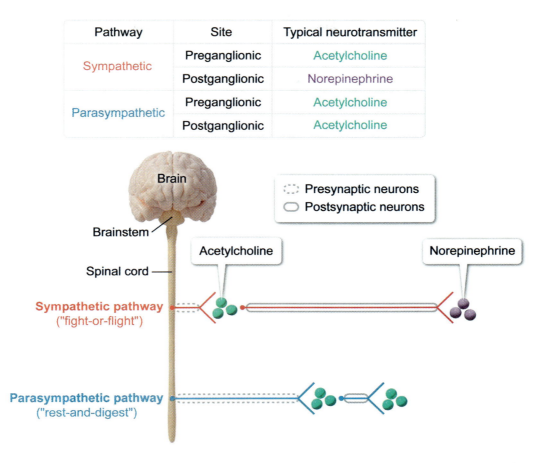

Figure 12.23 Pre- and postganglionic neuron lengths and neurotransmitters released in the sympathetic and parasympathetic nervous systems.

Figure 12.24 summarizes a number of the processes mediated by the parasympathetic and sympathetic nervous systems and highlights the differences in the locations from which parasympathetic and sympathetic preganglionic fibers arise from the CNS.

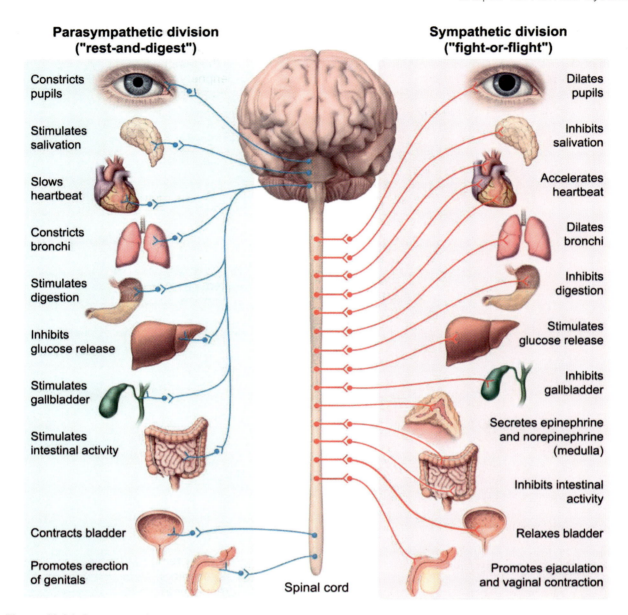

Figure 12.24 Parasympathetic and sympathetic divisions of the autonomic nervous system.

12.3.03 Afferent and Efferent Pathways

The terms afferent and efferent denote the direction a neural signal travels (toward and away from a point of reference, respectively). The terms afferent and efferent are used here to refer to neurons carrying information to and from the central nervous system (CNS), respectively.

The peripheral nervous system (PNS) communicates with the CNS via two groups of neurons (Figure 12.25):

- **Afferent (sensory) neurons** carry sensory information from the periphery (eg, pain, touch, pressure) to the CNS. The cell bodies of afferent neurons are located in ganglia on the dorsal (ie, posterior) part of the spinal cord, and afferent neurons synapse with neurons in the dorsal part of the spinal cord or in the medulla (ie, portion of brainstem).
- **Efferent (motor) neurons** carry motor commands from the brain to effector organs (eg, skeletal muscle, glands). Efferent neurons exit the ventral (ie, anterior) part of the spinal cord.

The nerves carrying information to and from the spinal cord are called spinal nerves. These nerves split into two branches, sometimes called roots, shortly before entering the spinal cord. Therefore, on each side of the spinal cord a dorsal root carries sensory information from the periphery to the spinal cord, and a ventral root carries motor commands from the spinal cord to the periphery. Within the spinal cord, the components of the communicating neurons are grouped into white matter (primarily axons) and dark matter (primarily cell bodies and dendrites).

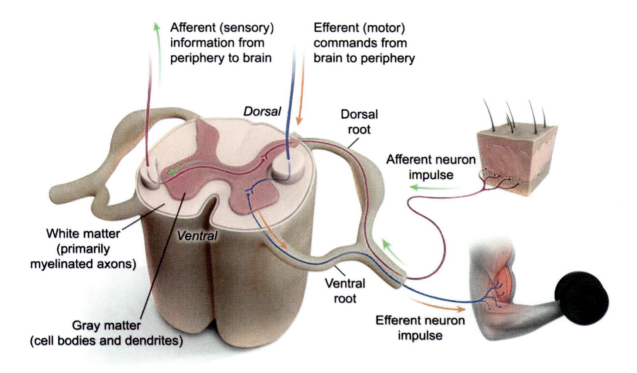

Figure 12.25 Afferent and efferent neurons.

Afferent neurons typically either interact with a specialized receptor (eg, for temperature, pain) or have ends (dendrites or "dendritic ends" of unipolar neurons) that serve as receptors. In many cases, an afferent neuron can activate a reflex response, mediated by an efferent neuron, as discussed in Concept 12.3.04.

12.3.04 Reflexes

Reflexes are involuntary responses to stimuli and may or may not require input from the brain. The specific neuronal pathway by which a stimulus directly causes the muscular or glandular effect associated with a particular reflex is called a **reflex arc**. Reflex arcs include a sensory (afferent) neuron, an effector (efferent) neuron, and, sometimes, an interneuron. A reflex arc begins with stimulation of a sensory receptor, which leads to an afferent electrical impulse that travels toward the spine or brain along a sensory nerve. This afferent impulse is then transmitted to an effector neuron in one of two ways:

- **Directly**, via a synapse between the afferent sensory neuron and the efferent effector neuron. This pathway is known as a **monosynaptic reflex arc**.
- **Indirectly,** through an interneuron between the sensory and the effector neuron. This pathway is known as a **polysynaptic reflex arc**.

Subsequently, electrical impulses travel along efferent neuron axons to stimulate a muscle fiber or gland, either directly or after synapsing with a postganglionic effector neuron. Some somatic motor reflexes are monosynaptic, whereas others are polysynaptic, as shown in Figure 12.26. Autonomic reflex arcs are always polysynaptic.

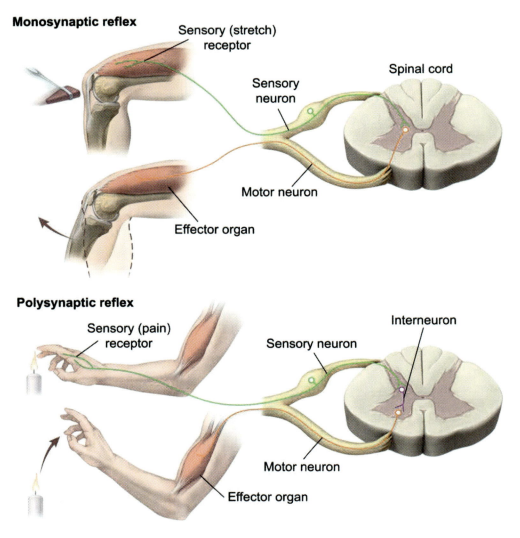

Figure 12.26 Monosynaptic and polysynaptic reflexes.

Reflexes can be modulated (ie, dampened or enhanced) by input from the brain. In the case of polysynaptic reflexes, this modulation takes the form of descending signals from higher areas in the central nervous system that act on preganglionic neurons in the reflex arc. Reflexes involving input from the brain are called supraspinal reflexes, whereas those mediated entirely within the spinal cord are called **spinal reflexes**.

12.3.05 Feedback Control

Nervous system responses play a vital role in the maintenance of homeostasis and can be viewed as operating as components in a **feedback loop**. In such a loop, afferent (sensory) nervous system pathways are activated in response to a deviation from a set point in a sensory pathway (eg, body temperature), providing a signal for the central nervous system, acting as a control center, to act on. The resulting efferent (motor) pathway then activates responses in effector organs that bring the regulated system back toward the baseline set point range. In this way, the feedback loop functions to maintain homeostasis, as depicted in Figure 12.27.

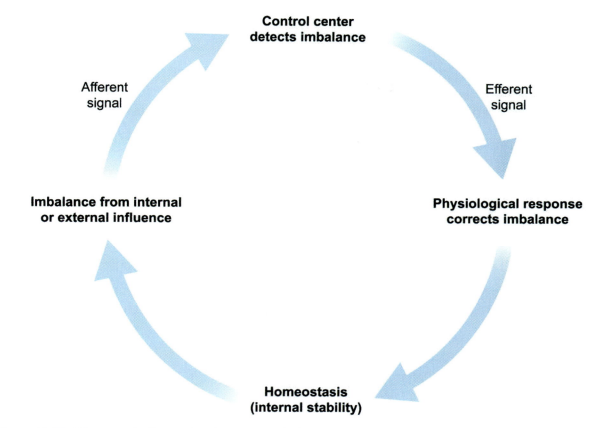

Figure 12.27 Afferent and efferent signaling as part of a feedback loop.

Feedback loops that function to maintain homeostasis are examples of negative feedback loops. In a negative feedback loop, the neural response to the activation of a neural receptor *inhibits* the initial stimulus in the same pathway. Negative feedback loops act to reduce the stimuli that move a system away from its normal range and thus help maintain homeostasis. As a result of this homeostatic tendency, nervous system feedback typically operates as part of a negative feedback loop.

Although negative feedback responses by the nervous system are much more common, in special circumstances, neural pathways can operate as part of a positive feedback loop in which activation of a sensory receptor triggers a response that moves a system *away* from homeostasis. For example, during childbirth (ie, parturition), uterine contractions increase pressure on the cervix, activating maternal pain receptors. Activation of these receptors leads to oxytocin release that, in turn, further stimulates painful uterine contractions, as discussed in Concept 10.4.03.

END-OF-UNIT MCAT PRACTICE

Congratulations on completing **Unit 6: Endocrine and Nervous Systems**.

Now you are ready to dive into MCAT-level practice tests. At UWorld, we believe students will be fully prepared to ace the MCAT when they practice with high-quality questions in a realistic testing environment.

The UWorld Qbank will test you on questions that are fully representative of the AAMC MCAT syllabus. In addition, our MCAT-like questions are accompanied by in-depth explanations with exceptional visual aids that will help you better retain difficult MCAT concepts.

TO START YOUR MCAT PRACTICE, PROCEED AS FOLLOWS:

1) Sign up to purchase the UWorld MCAT Qbank
   IMPORTANT: You already have access if you purchased a bundled subscription.
2) Log in to your UWorld MCAT account
3) Access the MCAT Qbank section
4) Select this unit in the Qbank
5) Create a custom practice test

Unit 7 Circulation and Respiration

Chapter 13 Circulation

13.1 Circulatory System

13.1.01	Cardiovascular Function
13.1.02	Blood Components
13.1.03	Oxygen Transport
13.1.04	Carbon Dioxide Transport
13.1.05	Blood Clotting
13.1.06	The Heart
13.1.07	Pattern of Blood Circulation
13.1.08	Blood Vessels
13.1.09	Blood Flow Regulation
13.1.10	Capillary Beds
13.1.11	Blood Pressure
13.1.12	Circulation and Thermoregulation
13.1.13	Endocrine and Nervous Control of Circulation

13.2 Lymphatic System

13.2.01	Lymphatic Vessels
13.2.02	Lymphatic System Function

Chapter 14 Respiration

14.1 Respiratory System Structure

14.1.01	Lung Structure
14.1.02	Mucociliary Escalator

14.2 Respiratory System Function

14.2.01	Mechanism of Respiration
14.2.02	Gas Exchange
14.2.03	Respiration and Thermoregulation
14.2.04	Control of Respiration
14.2.05	Role in Regulating pH

Lesson 13.1

Circulatory System

Introduction

The **circulatory system**, also known as the **cardiovascular system**, is composed of the **heart**, **blood vessels** (ie, vasculature), and **blood**. The circulatory system functions to transport materials throughout the body. This transport includes delivery of necessary substances obtained from the environment to the body's cells and movement of waste products away from cells to be eliminated from the body. The circulatory system also transports heat, molecules and cells that function in body defense, and signaling molecules involved in cell communication.

This lesson covers general cardiovascular system functions, composition and functional characteristics of blood, and structure and function of the heart. In addition, this lesson describes blood vessel structure and function, pattern and regulation of blood flow, and control mechanisms affecting circulation.

13.1.01 Cardiovascular Function

The cardiovascular system distributes blood between the heart and the rest of the body. This distribution is achieved through the pumping action of the heart and regulation of blood vessel diameter, which together create pressure differences that determine the extent to which blood flows to specific parts of the body.

Typically, blood flow distribution is well matched to the needs of tissues for delivery of oxygen, nutrients, water, electrolytes, and signaling molecules (ie, hormones) as well as for removal of metabolic wastes (eg, carbon dioxide, nitrogenous waste). The flow of blood also delivers defensive immune system cells and molecules to infected or injured tissues. In addition, the relative distribution of blood between the body's core and periphery (ie, skin) contributes to body temperature regulation.

The main functions of the cardiovascular system are summarized in Figure 13.1.

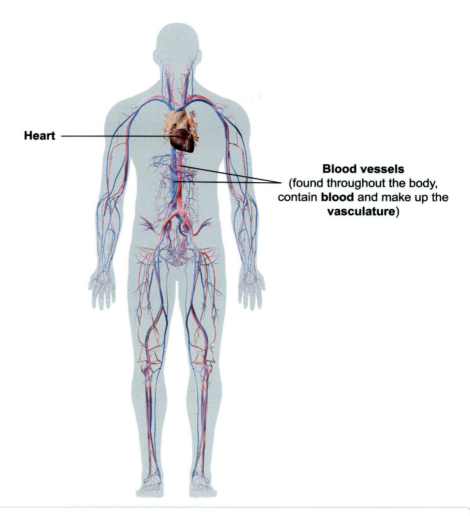

Cardiovascular system functions
- Transports oxygen, water, and nutrients from points of entry into the body to all cells
- Transports carbon dioxide and other metabolic wastes from all cells to points of exit from the body
- Transports hormones from endocrine cells to target cells
- Transports immune cells and molecules to sites of infection or injury
- Distributes heat to regulate body temperature

Figure 13.1 Primary functions of the cardiovascular system.

13.1.02 Blood Components

Blood is a fluid connective tissue composed of living cells (ie, formed elements) and a nonliving extracellular matrix. Formed elements originate in the bone marrow and include **erythrocytes** (ie, red blood cells [RBCs]) and **leukocytes** (ie, white blood cells [WBCs]). Blood also contains **thrombocytes** (ie, platelets), which are cell fragments derived from large bone marrow cells called megakaryocytes. RBCs function to transport oxygen and carbon dioxide, WBCs provide immune defense for the body (see Concept 20.1.01), and platelets play a crucial role in blood clotting.

RBC membranes possess molecules (ie, antigens) that are the basis for human blood groups (ie, blood types). Furthermore, most individuals' blood naturally contains antibodies that cause destruction of RBCs transfused (ie, received) from a blood donor with different RBC antigens than the recipient. There are many different groups of RBC antigens; however, antigens responsible for the Rh and ABO blood groups are of special importance because these antigens can cause potentially fatal transfusion reactions in recipients of mismatched blood. Antigens and antibodies are discussed in detail in Concepts 20.2.03 and 20.2.06.

Rh blood groups (ie, Rh$^+$, Rh$^-$) are based on the presence or absence of an antigen called the Rh factor, which follows an autosomal dominant inheritance pattern (see Lesson 7.2). ABO blood groups (summarized in Table 13.1) are determined based on the presence or absence of type A and type B RBC antigens, which are encoded by codominant alleles designated I^A and I^B (or simply A and B), respectively. In addition, a recessive allele of the ABO gene exists, designated i (or O). Individuals homozygous for this allele (ie, genotype ii) produce RBCs that lack type A and type B antigens, resulting in type O blood.

Table 13.1 Characteristics of ABO blood groups.

Blood group	Possible genotypes	Antigens present on red blood cells	Antibodies present in blood	Can donate blood to group	Can receive blood from group
A	$I^A I^A$, $I^A i$ (AA, AO)	Type A	Anti-B (binds to type B antigens)	A, AB	A, O
B	$I^B I^B$, $I^B i$ (BB, BO)	Type B	Anti-A (binds to type A antigens)	B, AB	B, O
AB (universal recipient)	$I^A I^B$ (AB)	Type A and type B	Neither anti-A nor anti-B	AB	A, B, AB, O
O (universal donor)	ii (OO)	Neither type A nor type B	Anti-A and anti-B	A, B, AB, O	O

Unlike the ABO system, in which anti-A and anti-B antibodies can be present in an individual's blood without prior exposure to type A and type B antigens, production of anti-Rh antibodies (ie, anti-D) requires that an Rh$^-$ individual be exposed to Rh$^+$ RBCs. Such exposure can occur when an Rh$^-$ mother gives birth to an Rh$^+$ baby. If this mother subsequently becomes pregnant with another Rh$^+$ offspring, the previously formed anti-Rh antibodies can cross the placenta and cause destruction of fetal RBCs (see Figure 13.2), which can lead to a potentially fatal condition called hemolytic disease of the fetus and newborn (ie, erythroblastosis fetalis).

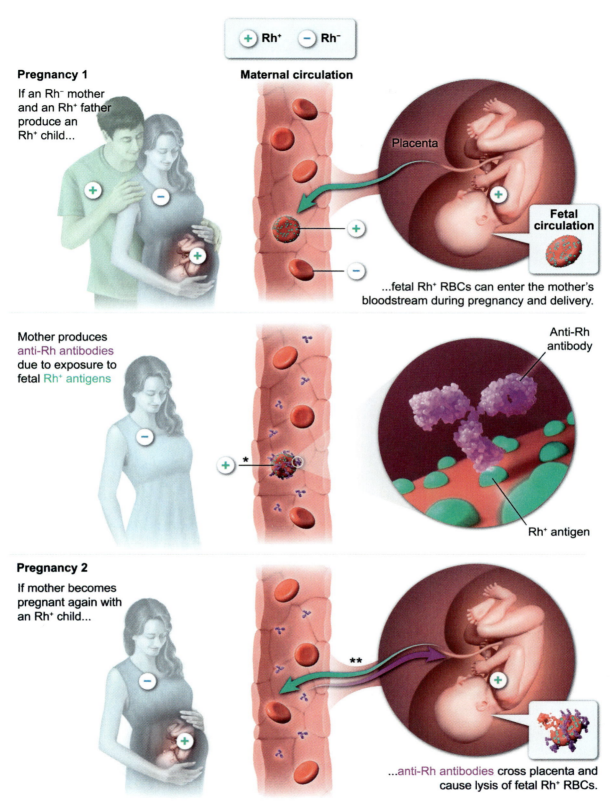

Figure 13.2 Situation leading to hemolytic disease of the fetus and newborn.

The relative abundance of each type of formed element in the blood differs, with the number of RBCs in a sample of blood greatly exceeding the number of WBCs and platelets. RBCs typically account for approximately 45% of a blood sample's total volume, whereas the combined volume of WBCs and platelets accounts for less than 1% of the total volume. The fraction (ie, percent) of a blood sample's total volume that consists of RBCs is called the **hematocrit**, with normal values ranging from approximately 37% to 52% in adults.

Plasma, the liquid matrix in which blood cells are suspended, accounts for roughly 55% of blood's volume. Plasma is approximately 90% water, and the remaining 10% is composed of a variety of dissolved substances. Figure 13.3 shows a breakdown of the general blood components.

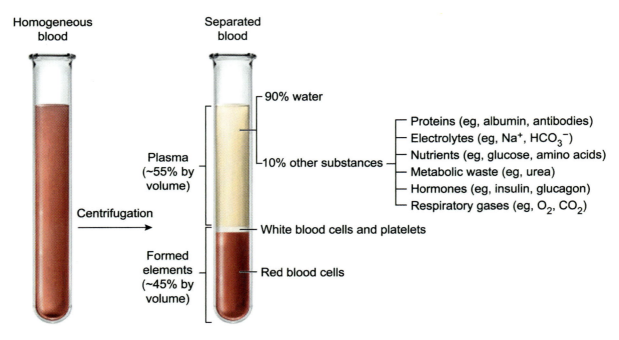

Figure 13.3 Components of blood.

Blood cells are produced by cell division of hematopoietic stem cells in the bone marrow. Most mature blood cells in circulation do not divide and must be continually removed and replaced as they reach the end of their life span. The spleen plays an important role in the removal of aged or damaged RBCs and platelets from circulation via phagocytosis by the abundant macrophages present in the spleen.

13.1.03 Oxygen Transport

Transportation of molecular oxygen (O_2) from the lungs to all cells of the body is a primary function of the cardiovascular system. Of the total O_2 carried by the blood, only about 2% is present as gas dissolved in the blood plasma. The remaining O_2 in the blood is reversibly bound to the protein **hemoglobin** and transported within red blood cells (RBCs).

As shown in Figure 13.4, a molecule of hemoglobin consists of four polypeptide chains (ie, subunits), each of which is attached to an iron-containing **heme group** that can reversibly bind one O_2 molecule. This configuration enables each hemoglobin molecule to transport up to four O_2 molecules.

RBCs are specialized for O_2 transport. Due to a small diameter and flattened, biconcave shape (see Figure 13.4), RBCs have a large surface area to volume ratio, which facilitates diffusion of O_2 into and out of RBCs. In addition, mature mammalian RBCs lack membrane-bound organelles (eg, nuclei, mitochondria), which is likely an adaptation (see Concept 8.2.03) allowing mature RBCs to hold more hemoglobin molecules, therefore increasing the cells' O_2 carrying capacity. Furthermore, RBCs' lack of mitochondria prevents aerobic cellular respiration, keeping RBCs from consuming the O_2 they carry.

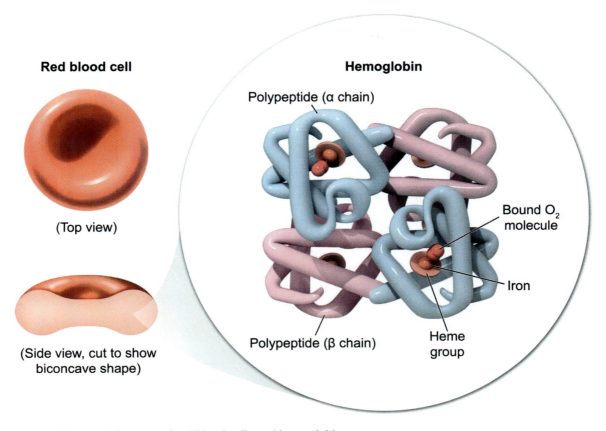

Figure 13.4 Structural features of red blood cells and hemoglobin.

In addition to RBC structural specializations, hemoglobin molecules within RBCs have specialized features that enhance O_2 transport. Hemoglobin is able to efficiently pick up O_2 at the lungs (as discussed in Concept 14.2.02) because hemoglobin's affinity for O_2 progressively *increases* as each O_2 is loaded onto hemoglobin. Likewise, hemoglobin is able to efficiently drop off O_2 at the tissues because hemoglobin's affinity for O_2 progressively *decreases* as each O_2 is unloaded.

The changes in hemoglobin's affinity for O_2 that occur during O_2 loading and unloading are due to **positive cooperativity** among hemoglobin's four subunits. Binding of an O_2 molecule to one subunit induces a conformational change in hemoglobin that makes binding of O_2 by the remaining subunits progressively easier. Similarly, the release of an O_2 molecule from one subunit of an oxygen-saturated hemoglobin molecule (ie, hemoglobin carrying four O_2 molecules) facilitates the progressive unloading of O_2 from the remaining subunits.

The graph shown in Figure 13.5 is called an **oxygen-hemoglobin dissociation curve (OHDC)**. This type of graph shows the fraction of hemoglobin present in a blood sample saturated with O_2 when the blood is exposed to different partial pressures of O_2 (ie, PO_2).

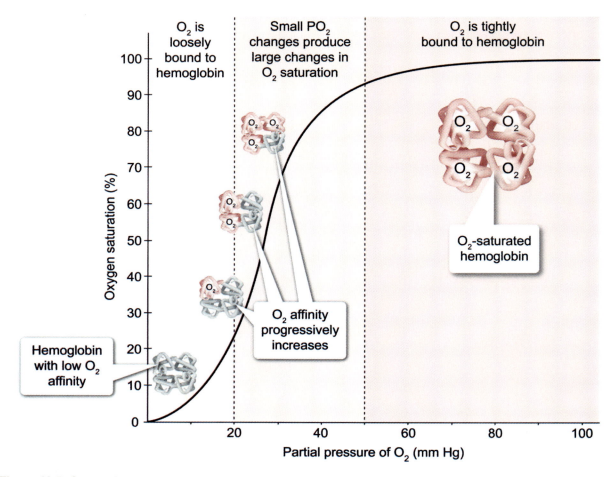

Figure 13.5 Oxygen-hemoglobin dissociation curve.

Positive cooperativity in O_2 binding among hemoglobin's subunits (ie, hemoglobin's variable O_2 affinity) causes the OHDC to be **sigmoidal** (ie, S-shaped). Hemoglobin without any bound O_2 has low affinity for O_2; therefore, as PO_2 increases from zero, the fraction of O_2-saturated hemoglobin increases slowly at first. As hemoglobin's O_2 affinity increases (due to positive cooperativity during O_2 loading), small PO_2 increases result in significant increases in the percentage of O_2-saturated hemoglobin. As O_2 saturation approaches 100% with increasing PO_2, the curve flattens (see Figure 13.5).

Various environmental factors also affect hemoglobin's O_2 affinity and can result in changes to the OHDC known as **left or right shifts** (Figure 13.6). These factors include temperature, carbon dioxide partial pressure (PCO_2), pH, and concentration of 2,3-bisphosphoglycerate (2,3-BPG), a molecule derived from an intermediate produced during glycolysis.

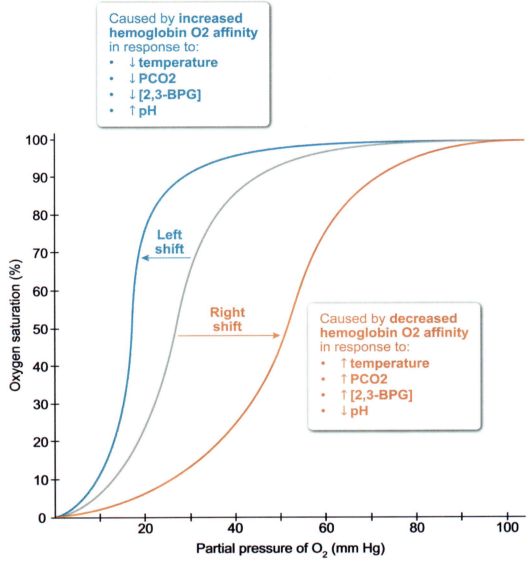

BPG = bisphosphoglycerate.

Figure 13.6 Factors causing shifts to the oxygen-hemoglobin dissociation curve.

As shown in Figure 13.6, hemoglobin's O_2 affinity increases (resulting in a left-shifted OHDC) in response to decreased temperature, PCO_2, or 2,3-BPG concentration, as well as increased pH. These conditions frequently characterize tissues at rest without an elevated demand for O_2.

Conversely, hemoglobin's O_2 affinity decreases (thereby facilitating O_2 unloading) in response to increased temperature, PCO_2, or 2,3-BPG concentration, as well as decreased pH. Such changes, which cause the OHDC to shift to the right, typically occur in tissues experiencing elevated metabolic activity (eg, skeletal muscle during exercise). This means that hemoglobin tends to unload O_2 with greatest efficiency to the tissues consuming O_2 most rapidly via aerobic metabolism. A right shift in the OHDC caused by increased PCO_2 and/or decreased pH is referred to as the **Bohr effect**.

RBCs produced during fetal development contain a specialized type of hemoglobin called **fetal hemoglobin (hemoglobin F)**. This type of hemoglobin contains polypeptides with an amino acid sequence that differs from that of polypeptides present in adult hemoglobin (hemoglobin A). The resulting structural differences between hemoglobin F and hemoglobin A cause hemoglobin F to have greater O_2

affinity than hemoglobin A. This difference in O_2 affinity allows O_2 to readily pass from maternal blood to fetal blood at the placenta, which is adaptive because the fetus must extract O_2 from maternal blood rather than from air (see Concept 10.4.02).

13.1.04 Carbon Dioxide Transport

Cells generating ATP via aerobic cellular respiration produce carbon dioxide (CO_2) as a metabolic waste product. This CO_2 is carried away from CO_2-producing cells and to the lungs for disposal (see Concept 14.2.02) via the circulatory system. There are three primary ways that CO_2 is transported in the blood, as illustrated in Figure 13.7:

- Approximately 7% of the CO_2 in the blood is carried as gas dissolved in the blood plasma. The poor solubility of CO_2 in water explains the minority of CO_2 transported in this manner.
- The remaining CO_2 in the blood is taken up by red blood cells (RBCs), in which approximately 23% of the blood's total CO_2 reversibly binds to hemoglobin. Unlike O_2, CO_2 does not bind to hemoglobin's heme groups; rather, CO_2 binds to the N-terminal amino acids of hemoglobin's four polypeptide chains, forming **carbaminohemoglobin**.
- The remaining CO_2 within RBCs (approximately 70% of the blood's total CO_2) is acted upon by **carbonic anhydrase**, an enzyme that catalyzes a reversible reaction between CO_2 and water. This reaction produces carbonic acid (H_2CO_3), which dissociates into hydrogen ions (H^+) and **bicarbonate ions (HCO_3^-)**. Upon formation, HCO_3^- ions exit RBCs via transport proteins and travel in the plasma. These transport proteins also bring chloride ions (Cl^-) into RBCs, preventing an imbalance of electric charge that would otherwise occur due to loss of HCO_3^-. This transfer is known as the **chloride shift**.

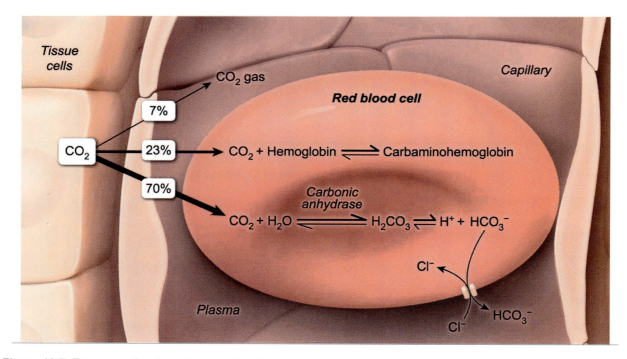

Figure 13.7 Transport of carbon dioxide in the blood.

When CO_2-rich blood reaches the lungs, CO_2 is transferred from the pulmonary capillaries to the lung air spaces (ie, alveoli) for disposal from the body via expiration. The steps of this transfer are depicted in Figure 13.8:

1. Because the carbon dioxide partial pressure (PCO₂) of pulmonary capillary plasma is higher than the PCO₂ of alveolar air, CO₂ gas diffuses from the blood into the alveoli, that is, from high to low partial pressure.
2. Loss of CO₂ from the plasma decreases plasma PCO₂, allowing CO₂ diffusion out of RBCs.
3. Diffusion of CO₂ out of RBCs causes carbaminohemoglobin within RBCs to release bound CO₂, which then diffuses from the RBCs into the plasma.
4. As CO₂ continues to exit RBCs, carbonic anhydrase in the RBCs converts H_2CO_3 into CO_2 and H_2O, reversing the reaction that occurred at the tissues. The resulting decrease in RBC H_2CO_3 concentration triggers the production of additional H_2CO_3 via the reaction between H^+ and HCO_3^-, which causes RBC HCO_3^- concentration to decrease.
5. Decreased RBC HCO_3^- concentration causes HCO_3^- to be pulled into the RBCs from the plasma, which further promotes the carbonic anhydrase–catalyzed reaction converting HCO_3^- to CO_2. This additional CO_2 diffuses out of the RBCs and ultimately into the alveoli.

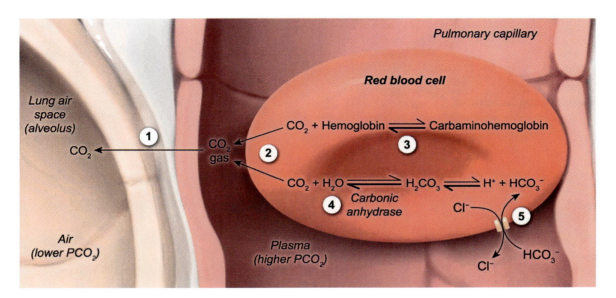

Figure 13.8 Steps involved in the transfer of carbon dioxide from the blood to the lungs.

13.1.05 Blood Clotting

Injuries that cause breaks in blood vessel walls result in a loss of blood from the circulatory system. To prevent excessive blood loss from such injuries, the circulatory system responds to damaged blood vessels by forming blood clots that function as plugs to stop additional blood from escaping through broken vessel walls.

A disease known as **hemophilia** occurs when this clotting process is impaired, which can lead to potentially life-threatening blood loss following minor injuries. Hemophilia is typically an inherited disorder that exhibits an X-linked recessive pattern of transmission from parents to offspring (see Concept 7.3.02).

The process by which blood clots are formed is called **hemostasis** and requires the activity of platelets (thrombocytes) in the blood. The steps that occur during hemostasis are summarized in Figure 13.9.

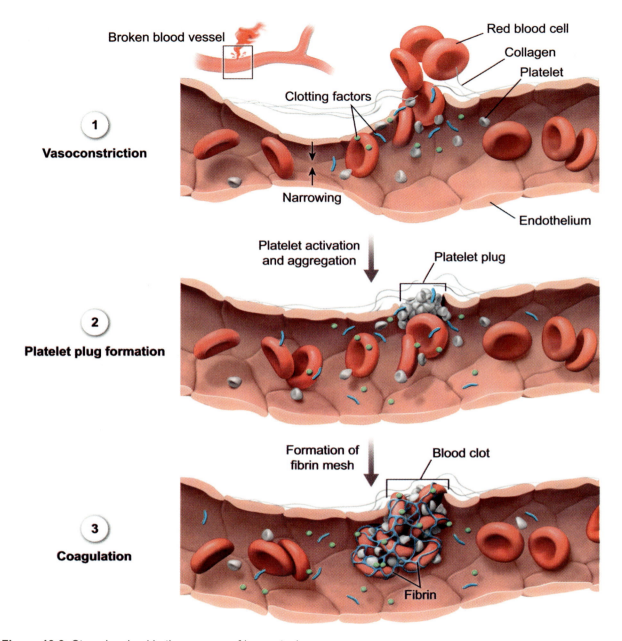

Figure 13.9 Steps involved in the process of hemostasis.

The first step in hemostasis is **vasoconstriction** (or vascular spasm), which causes narrowing of the broken blood vessel. Vasoconstriction occurs via contraction of smooth muscle cells in the wall of the blood vessel and reduces blood flow through the damaged vessel, slowing blood loss. The inner lining of a blood vessel is called the **endothelium**, and damage to endothelial cells causes the release of chemical signals (eg, endothelin) that help trigger vasoconstriction.

The second step in hemostasis involves the formation of a **platelet plug**. Platelets typically do not adhere to undamaged endothelial cells; however, platelets stick tightly to the collagen fibers hidden beneath the endothelium in undamaged vessels but exposed in damaged vessels. As platelets begin sticking to exposed collagen, they become activated, causing further stickiness. The platelets release chemical signals and aggregate, forming a plug by adhering to one another. This signaling process, which enhances vasoconstriction and activates even more platelets, is an example of positive feedback.

The final step in hemostasis is **coagulation**, which produces the actual blood clot. Coagulation involves a complex series of steps in which multiple clotting factors in the blood plasma are sequentially activated (see Figure 13.10 for a simplified model; detailed knowledge of clotting factors is beyond the scope of the exam).

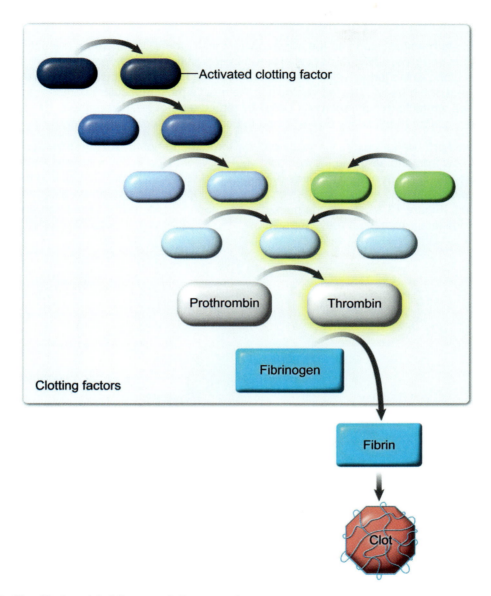

Figure 13.10 Simplified model of the coagulation cascade.

Near the end of the coagulation cascade, a clotting factor called **prothrombin** is converted into an active enzyme called **thrombin**, which cleaves soluble **fibrinogen** molecules into insoluble **fibrin** molecules. These fibrin molecules polymerize into fibrin strands that become cross-linked to form a mesh that traps platelets and red blood cells, producing a clot that stops additional blood loss from the broken blood vessel.

After the damaged blood vessel has healed, the blood clot undergoes **fibrinolysis**, in which the fibrin mesh is digested by an enzyme called **plasmin**, and the clot dissolves. Plasmin is produced via activation of an inactive precursor, a plasma protein known as **plasminogen**.

3.1.06 The Heart

The heart is a muscular structure that contains four chambers: two thin-walled **atria** and two thick-walled **ventricles**. The walls of the chambers are composed primarily of cardiac muscle. The left side and right side of the heart each consist of an atrium and a ventricle, and the two sides function as separate pumps that contract simultaneously to move blood throughout the body.

The atria function to transfer blood into the adjoining ventricles; therefore, the atria are not as thick-walled (muscular) as the ventricles, which must propel blood over significantly greater distances. The right ventricle pumps blood to the lungs, which are located near the heart, while the left ventricle pumps blood throughout the rest of the body. The left ventricle is the heart's most muscular chamber, an adaptation that allows the left ventricle to contract more forcefully than the other heart chambers, generating sufficiently high pressure to efficiently deliver blood to body regions located relatively far from the heart.

Specialized valves provide for unidirectional blood flow through the heart. **Atrioventricular (AV) valves** permit blood to flow from the atria into the ventricles but prevent blood in the ventricles from moving directly back into the atria. **Semilunar valves** allow blood to flow from the ventricles into the blood vessels (ie, arteries) that lead away from the ventricles but stop blood in these arteries from flowing back into the ventricles. Figure 13.11 shows the pathway that blood follows through the heart and the locations of the AV and semilunar valves that prevent backflow.

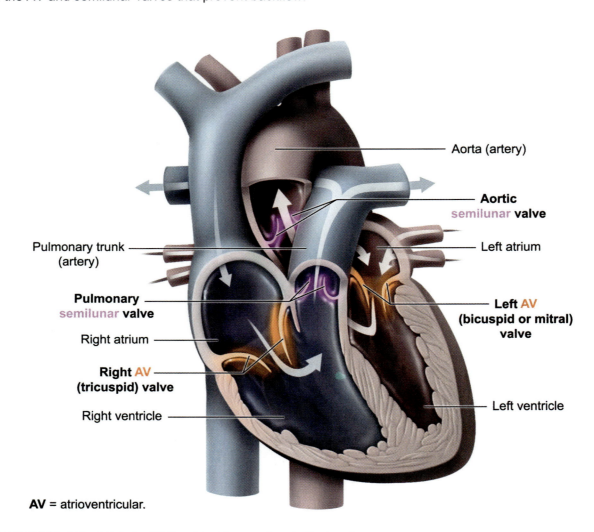

AV = atrioventricular.

Figure 13.11 Atrioventricular (AV) and semilunar valves in the heart.

Cardiac muscle cells making up the walls of the heart are connected end to end by structures called **intercalated discs**, which contain gap junctions and desmosomes. Gap junctions permit ions to pass directly from cell to cell, allowing action potentials (ie, series of membrane voltage changes responsible for triggering muscle cell contraction) to spread rapidly across the heart. Desmosomes prevent cardiac muscle cells from pulling away from each other during contraction.

The rapid transmission of action potentials occurring throughout the heart muscle via gap junctions and the strong connections that exist between cardiac muscle cells due to the presence of desmosomes allow the heart to act as a coordinated unit. The coordination of activity among cells is the basis for the heart's ability to function as an efficient pump. More information about cardiac muscle cell structure and function is found in Concept 17.2.02.

Cardiac muscle is **myogenic**, that is, the electrical signal that stimulates contraction of the heart comes from self-depolarizing cells within the heart muscle itself rather than from an outside source, such as the nervous system. The heart also contains specialized cells that form a **cardiac conduction system** (Figure 13.12). This system allows the contraction signal, which consists of a wave of depolarization, to be rapidly transmitted to particular regions of the heart in a specific sequence. In this way, the timing of contraction of different parts of the heart is controlled to ensure that the heart pumps blood effectively.

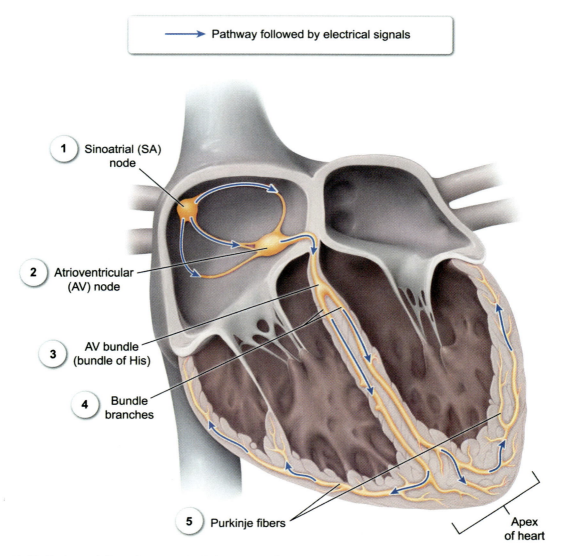

Figure 13.12 Pathway followed by electrical signals passing through the cardiac conduction system.

Events leading to coordinated heart contraction include the following:

1. The electrical signal stimulating heart contraction typically originates in a region called the **sinoatrial (SA) node** in the right atrium. This signal spreads from the SA node across both atria via gap junctions, triggering simultaneous contraction of the atria, and travels through cells of the conduction system to a region called the **AV node**.

2. The AV node is located in the inferior portion of the wall that separates the two atria (ie, interatrial septum). Cells making up the AV node possess structural adaptations that delay passage of the signal to the ventricles by approximately 0.1 seconds, which, as discussed later in this concept, contributes to the heart's coordinated pumping action.

3. After passing through the AV node, the contraction signal is carried through a portion of the conduction system called the **AV bundle** (bundle of His), which is the only structure that electrically links the atria to the ventricles.

4. The AV bundle divides into two **bundle branches**, which are conduction system pathways that carry the signal within the wall separating the ventricles (ie, interventricular septum) toward the heart's pointed inferior end (ie, apex).

5. The bundle branches divide to form **Purkinje fibers**, which carry the signal superiorly within the ventricular walls. The signal triggers ventricular contraction, which begins at the heart's apex and progresses superiorly toward the atria.

Operation of the heart's conduction system can be monitored using electrodes attached to the body's surface at specific locations. Data collected by these electrodes can be used to generate a graph with characteristic wave patterns. This graph, called an **electrocardiogram (ECG)**, provides a record of electrical activity within the heart as the heart beats. Figure 13.13 shows how specific components of an ECG correspond to depolarization or repolarization of specific regions of the heart.

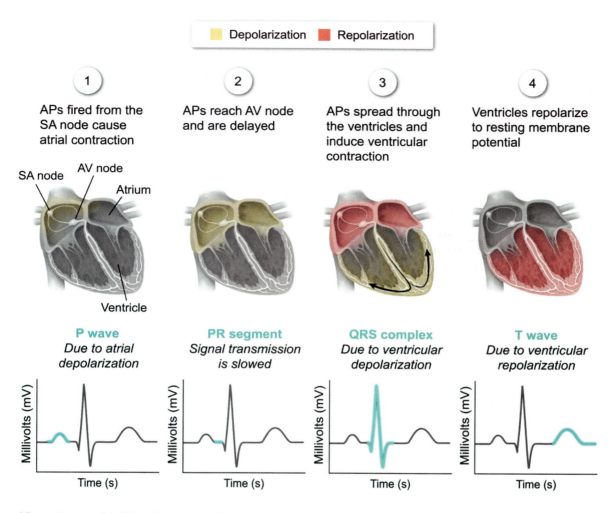

Figure 13.13 Electrocardiogram record of the heart's electrical activity.

The series of events that the heart undergoes to pump blood is called the **cardiac cycle** (see Figure 13.14). These events cause pressure changes within the heart that allow the heart to fill with blood and then eject the blood. The cardiac cycle involves two primary phases: diastole and systole. During **diastole**, cardiac muscle relaxes, which lowers pressure within the heart and allows blood to enter the heart. Conversely, during **systole**, cardiac muscle contracts (first in the atria and then in the ventricles), which increases pressure within the heart and causes blood to be pumped out of the heart into the attached arteries.

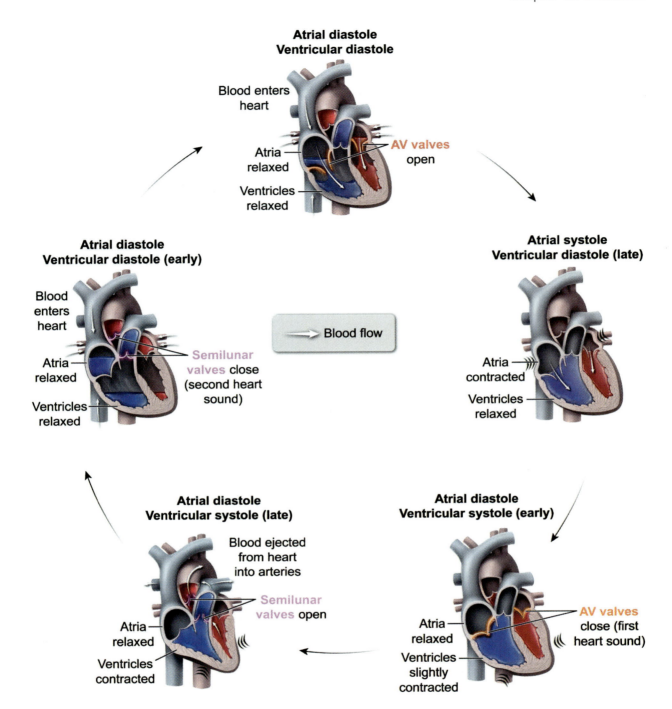

Figure 13.14 Events occurring during the cardiac cycle.

The pattern of activity in the cardiac conduction system underlies the pattern of muscle contraction that characterizes the cardiac cycle and results in efficient pumping of blood. For example, the delay in signal transmission mediated by the AV node allows the atria to finish contracting before the ventricles begin contracting, facilitating blood transfer from atria to ventricles. Likewise, transfer of blood out of the ventricles into arteries is efficient because ventricular contraction begins at the heart's apex and proceeds superiorly, thereby squeezing blood toward the heart's base (ie, broad superior end), where the arteries are attached.

Pressure changes within the heart cause opening and closing of heart valves during the cardiac cycle (Figure 13.14). Closing of these valves produces heart sounds ("lub-dup") characteristic of a beating heart. The first sound ("lub") occurs at the beginning of ventricular systole, when pressure in the contracting ventricles becomes greater than pressure in the relaxed atria, causing the AV valves to forcefully close. The second sound ("dup") occurs when the semilunar valves close at the beginning of ventricular diastole, when pressure in the ventricles drops below pressure in the arteries leading from the heart.

The purpose of the heart is to propel blood to all parts of the body. The heart's effectiveness at this task can be assessed by measuring **cardiac output (CO)**, which is defined as the amount of blood pumped from one ventricle (either left or right) in one minute. CO is affected by both **heart rate (HR)** and **stroke volume (SV)**. HR (ie, pulse) is the number of times the heart completes the cardiac cycle per minute (ie, number of heart beats per minute), and SV is the volume of blood pumped from the ventricle during a single contraction (ie, during ventricular systole). CO can be calculated as follows:

$$CO = HR \times SV$$

 Concept Check 13.1

A person's cardiac output (CO) is determined to be 5,100 mL/min. Every 0.8 seconds, the person's sinoatrial node generates a signal that passes through the cardiac conduction system. Assuming the person's heart is functioning as expected, what is this person's stroke volume?

Solution

Note: The appendix contains the answer.

13.1.07 Pattern of Blood Circulation

The left and right sides of the heart (each of which is composed of an atrium and a ventricle) function as independent pumps that propel blood through two separate circuits of blood vessels. The **pulmonary circuit** carries blood under lower pressure from the right side of the heart to the sites of gas exchange in the lungs and back to the left side of the heart. The **systemic circuit** transports blood pumped under higher pressure from the heart's left side to all parts of the body not served by the pulmonary circuit. Upon return to the heart, blood from the systemic circuit enters the heart's right side (ie, starting point of the pulmonary circuit).

In a healthy heart, blood passes through an atrium before entering a ventricle; consequently, blood is received by the heart into the atria and blood exits the heart through the ventricles. Specifically, oxygen-rich blood returning from the lungs via **pulmonary veins** enters the left atrium, which transfers this blood to the left ventricle to be pumped into the systemic circuit via the **aorta**. Likewise, oxygen-poor blood returning from the systemic circuit via the **venae cavae** enters the right atrium, where the blood is transferred to the right ventricle to be pumped back to the lungs via **pulmonary arteries** in the pulmonary circuit (see Figure 13.15).

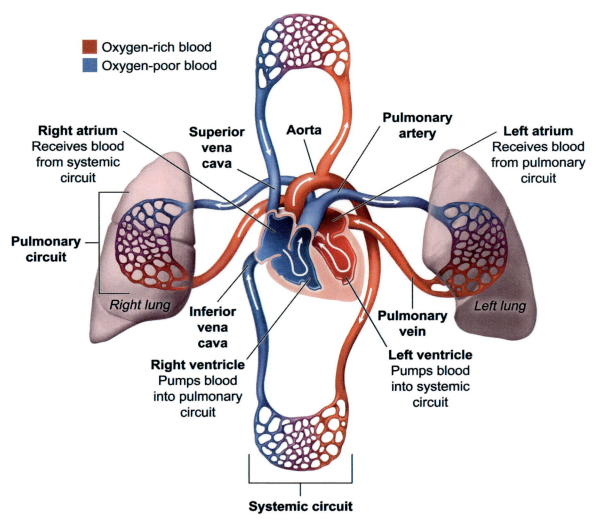

Figure 13.15 Pathway of blood through the pulmonary and systemic circuits.

 Concept Check 13.2

Identify and arrange in order the structures in the following list that blood in a pulmonary artery would pass through to reach the superior vena cava.

aorta, aortic semilunar valve, bicuspid valve, left atrium, left ventricle, pulmonary semilunar valve, pulmonary vein, right atrium, right ventricle, superior vena cava, tricuspid valve

Solution

Note: The appendix contains the answer.

13.1.08 Blood Vessels

Blood is transported through the body within an extensive system of blood vessels, collectively referred to as the vasculature. Different types of blood vessels, primarily **arteries**, **veins**, and **capillaries**, comprise the vasculature. Small branches that diverge from arteries are called **arterioles**, and small vessels that converge to form veins are called **venules**. Figure 13.16 shows the general arrangement of the different blood vessel types within the vasculature and the order in which blood flows through these vessels.

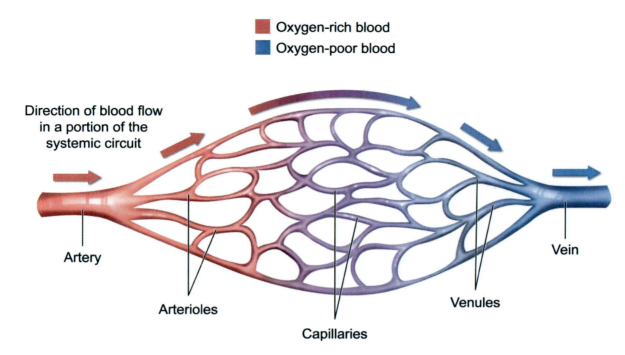

Figure 13.16 Sequential arrangement of different types of blood vessels in the vasculature.

All blood vessels have an internal lining called the **endothelium**, which is composed of a single layer of specialized epithelial cells (ie, endothelial cells). The endothelium provides a slick surface that reduces friction between the vessel wall and the blood, facilitating smooth blood flow through the vessels. Endothelial cells also secrete a variety of chemical signals (eg, endothelin, nitric oxide) involved in regulating various aspects of cardiovascular system function, including control of blood vessel diameter, blood clotting, blood vessel permeability, formation of new blood vessels, and inflammation.

Blood vessels other than capillaries typically have walls that, in addition to possessing an inner endothelial layer, have a middle layer containing smooth muscle cells and elastic fibers and an outer layer consisting of connective tissue, primarily collagen fibers. The relative thicknesses of these layers, which vary in different types of blood vessels (Figure 13.17), reflect the vessels' functional differences.

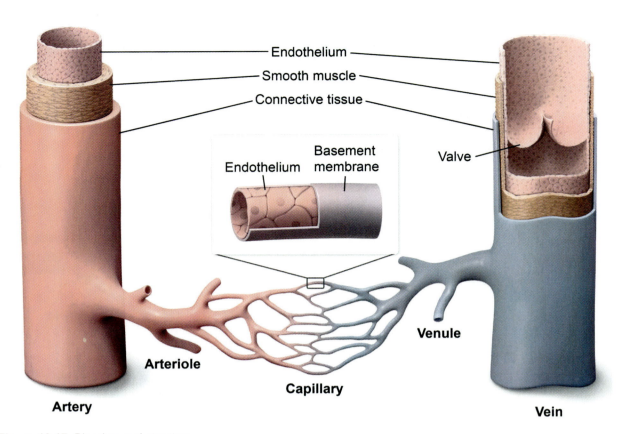

Figure 13.17 Blood vessel structure.

Arteries, which function to carry blood away from the heart, have relatively thick walls that contain abundant smooth muscle cells, elastic fibers, and collagen fibers. This wall structure enables arteries to withstand the pressure exerted on blood during contraction (ie, systole) of the heart's ventricles. Similarly, the wall structure allows arteries to stretch, which stores energy to maintain blood's forward motion during periods in which the heart is relaxed (ie, diastole). Arteries typically carry oxygen-rich blood; however, arteries of the pulmonary circuit (ie, pulmonary arteries) carry oxygen-poor blood to the lungs so the blood can pick up oxygen.

Veins carry blood toward the heart and have walls with relatively thin layers of smooth muscle and elastic tissue. The blood in veins is typically under low pressure; therefore, to prevent backflow of blood, many veins contain **valves** (see Figure 13.17). Furthermore, most veins transport oxygen-poor blood, an exception being veins that carry blood in the pulmonary circuit (ie, pulmonary veins). Because blood gains oxygen in the lungs, pulmonary veins carry oxygen-rich blood directly back to the heart from the lungs.

Capillaries are the smallest type of blood vessel. The walls of capillaries are extremely thin and consist of a single layer of endothelial cells attached to a basement membrane. The structure of capillary walls facilitates exchange of materials between blood and interstitial fluid in body tissues. This movement of materials enables blood to supply oxygen and nutrients to body tissues and to remove cellular waste products from the tissues.

13.1.09 Blood Flow Regulation

The circulatory system can efficiently distribute materials as needed to specific parts of the body (eg, organs) by regulating blood flow. By adjusting the amount of blood that flows to a particular body region, the body can vary the delivery of needed substances (eg, oxygen, nutrients, immune cells) to meet the changing demands for these substances in different parts of the body at different times.

Blood flows in the circulatory system due to differences in pressure, that is, blood (like all fluids) moves down a pressure gradient (ΔP) from areas of higher pressure to areas of lower pressure. During systole of the cardiac cycle, pressure in the circulatory system is greatest within the heart, which causes blood to exit the heart and enter the vasculature, where pressure is lower.

As blood moves through the vasculature, friction among blood cells and between the blood and blood vessel walls causes a continuous loss of pressure. Therefore, in the systemic circuit, blood pressure is highest in the aorta (ie, artery leading from the left ventricle) and lowest in the venae cavae (ie, veins leading into the right atrium), as depicted in Figure 13.18.

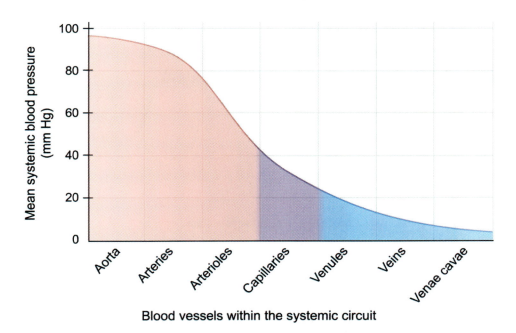

Figure 13.18 Blood pressure within different blood vessels of the systemic circuit.

Blood flow (Q) is defined as the volume of blood that moves through a structure (eg, organ) in a given amount of time, and friction within the circulatory system opposes blood flow. The total amount of friction encountered by blood as it passes through the circulatory system is referred to as the **total peripheral resistance (TPR)**. TPR is influenced by the viscosity (ie, thickness) of blood as well as by the length and internal (ie, lumen) diameter of blood vessels. TPR increases when blood viscosity or blood vessel length increases, or when blood vessel diameter decreases.

Because movement of blood through the circulatory system is driven by a pressure gradient (ΔP) and is opposed by resistance resulting from friction (ie, TPR), Q can be analyzed using the following equation (which is derived from Ohm's law):

$$Q = \frac{\Delta P}{\text{TPR}}$$

This equation shows that Q is directly proportional to ΔP (ie, Q increases as ΔP increases) and inversely proportional to TPR (ie, Q decreases as TPR increases).

Although blood viscosity and blood vessel length may change over long time frames (eg, lengthening of blood vessels as an infant grows to adulthood), these variables typically remain constant over short periods of time (ie, hours). Therefore, TPR (and therefore, Q) is regulated primarily via changes to blood vessel diameter (ie, vasoconstriction and vasodilation). Figure 13.19 shows a hypothetical example of how blood flow can be adjusted to meet the needs of particular body regions by changes in blood vessel diameter.

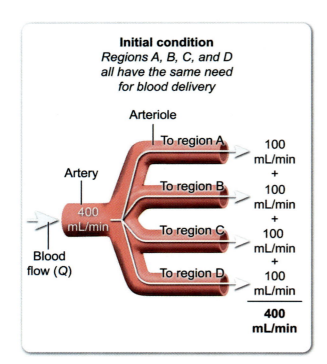 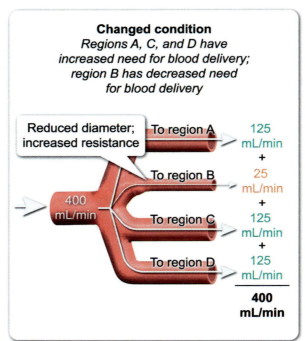

Figure 13.19 Regulation of blood flow via changes to blood vessel diameter.

As blood passes through the circulatory system, the greatest drop in blood pressure occurs as blood passes through arterioles (see Figure 13.18), which indicates that arterioles account for the majority of TPR. Consequently, arterioles are the primary sites at which TPR is regulated via vasoconstriction and vasodilation.

13.1.10 Capillary Beds

Capillaries are organized into networks called **capillary beds**. Each capillary bed typically receives blood from an arteriole and passes the blood to a venule, with blood flow (ie, microcirculation) through the capillary bed determined by the diameter of the supplying arteriole. In addition, entrances to some capillaries are encircled by smooth muscle cells that form structures called **precapillary sphincters**. Precapillary sphincters can constrict, causing blood to bypass the capillary bed (see Figure 13.20). In this way, blood flow through capillary beds can be closely matched to the metabolic activity level of the surrounding tissues.

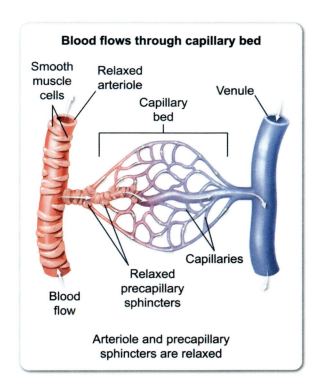 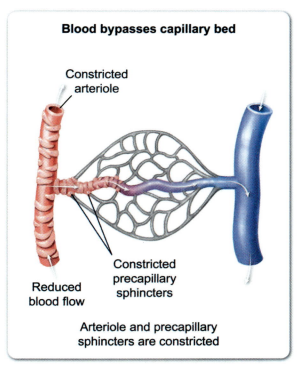

Figure 13.20 Regulation of blood flow through capillary beds.

After being pumped from the heart, blood generally passes through only one capillary bed before returning to the heart. However, in a few locations within the body (eg, intestines, pituitary gland, kidney), blood received from one set of capillaries is carried by one or more larger vessels to a second set of capillaries, through which this blood passes before returning to the heart via the systemic circuit. Such pathways that involve two capillary beds arranged in series are called **portal systems** and allow substances to be transported from one body region to another without being diluted or widely distributed.

Blood in capillaries moves at a slower velocity than blood in other types of vessels. This slow movement of blood increases the efficiency with which materials are exchanged across capillary walls between blood and interstitial fluid because the amount of time during which diffusion can occur is increased. Blood velocity is slowest in capillaries because the total (ie, combined) cross-sectional area of all capillaries is greater than the total cross-sectional areas of other blood vessel types (see Figure 13.21).

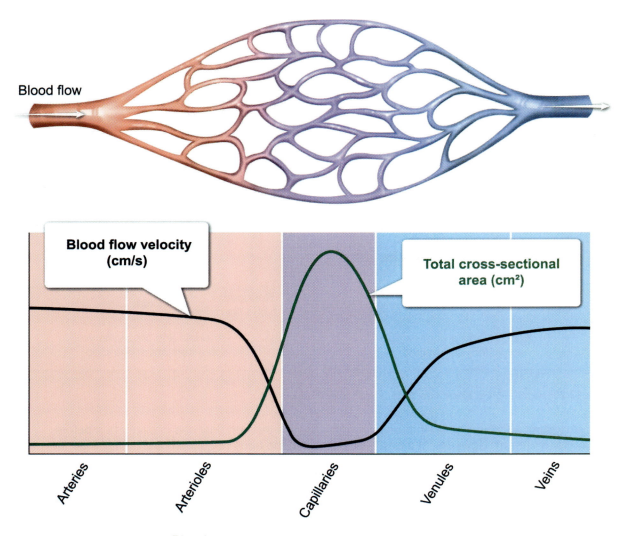

Figure 13.21 Blood flow velocity and total cross-sectional area of different types of blood vessels.

Exchange of materials between capillary blood and interstitial fluid occurs by diffusion, endocytosis, exocytosis, and bulk flow (ie, mass fluid transfer). Bulk flow is caused by **hydrostatic pressure (HP)** and **oncotic pressure (OP)** inside and outside capillaries. HP is the force a fluid exerts on a vessel wall (eg, blood pressure). OP (also known as colloid osmotic pressure) results from the presence of large molecules (eg, proteins) that do not freely cross the capillary wall, causing water to move from areas of lower OP to areas of higher OP by osmosis.

As shown in Figure 13.22, the OP and HP of the interstitial fluid, as well as the OP of the blood, remain essentially unchanged as blood passes through a capillary. However, due to friction, the HP of blood *decreases* from the arterial end to the venous end of a capillary. Consequently, differences in net pressure at opposite ends of a capillary bed cause fluid to move from the blood to the interstitial fluid (ie, filtration) at the arterial end of the capillary bed and cause fluid to move into the blood from the interstitial fluid (ie, absorption) at the venous end of the capillary bed (Figure 13.22).

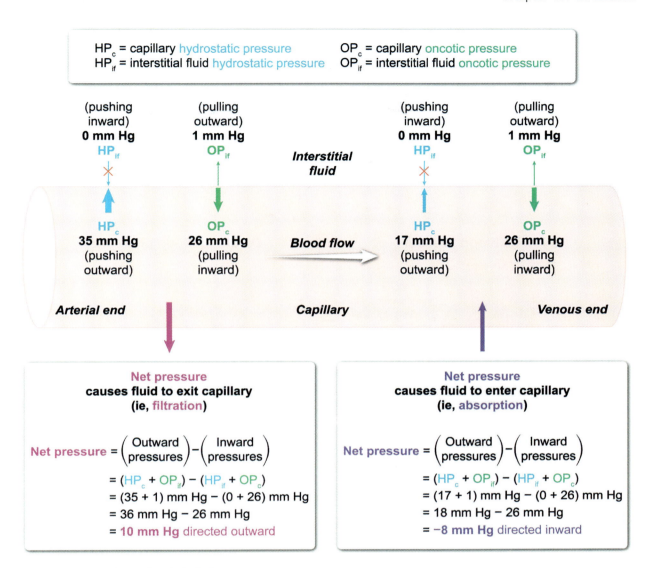

Figure 13.22 Fluid transfer via bulk flow.

Bulk flow of fluid out of and into capillaries plays an important role in determining the volume of fluid present in blood (ie, plasma volume) and in the interstitial space (ie, interstitial fluid volume). More fluid generally exits the blood via filtration into the interstitial space than is taken up by the capillaries via absorption. However, the **lymphatic system** functions to return excess interstitial fluid to the blood (as discussed in Concept 13.2.02); therefore, a higher filtration rate relative to absorption rate in capillary beds typically does not result in inappropriate accumulation of interstitial fluid, a condition known as **edema**.

13.1.11 Blood Pressure

As discussed in Concept 13.1.09, blood flow through the circulatory system occurs due to a pressure gradient resulting from the heart's activity. Specifically, blood flow is driven by contraction (ie, systole) of the heart's ventricles, and blood pressure develops as a result of the resistance to blood flow that exists within the system due to friction among blood cells and between blood vessel walls and the blood.

Arterial blood pressure resulting from ventricular systole, referred to as **systolic pressure**, causes elastic arteries near the heart to stretch, allowing these arteries to store energy. During ventricular relaxation (ie, diastole), the arteries that were stretched during systole recoil, releasing their stored energy and generating pressure by pushing on the blood. This **diastolic pressure**, although lower than systolic pressure, maintains forward blood movement in the vasculature because backward blood movement is prevented by the semilunar valves in the heart (ie, valves between the ventricles and the pulmonary trunk and aorta).

Systemic blood pressure in humans is typically measured using a **sphygmomanometer**, which consists of an inflatable cuff and a pressure gauge. Blood pressure readings are reported as a ratio of systolic to diastolic pressures (see Figure 13.23), typically in units of millimeters of mercury (mm Hg). Normal blood pressure for healthy adults at rest is generally defined as less than 120/80 mm Hg. Blood pressure that is higher than normal is known as **hypertension**, while blood pressure that is lower than normal is known as **hypotension**.

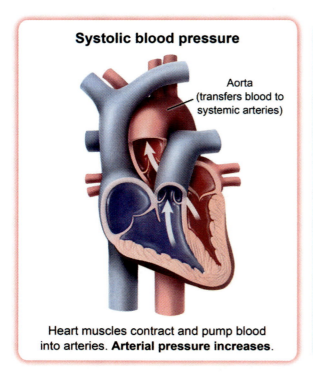
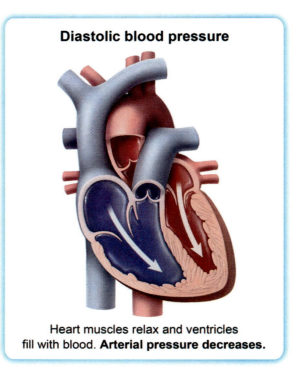

Activity	Blood pressure (mm Hg)
Resting	120 / 80
Running	155 / 73

Figure 13.23 Systolic versus diastolic blood pressure.

As shown in Figure 13.24, blood pressure within arteries rises and falls in a regular pattern, which is due to alternating systole and diastole of the heart. These pressure fluctuations result in a detectable **pulse** (ie, rhythmic stretching and recoiling of arteries). **Pulse pressure** represents the difference between diastolic and systolic pressures in an artery, and **mean arterial pressure (MAP)** represents the average pressure in an artery during one cardiac cycle. MAP is not exactly midway between diastolic and systolic pressures because diastole typically lasts longer than systole; therefore, the following equation is often used to calculate MAP:

$$\text{MAP} = \text{Diastolic pressure} + \frac{\text{Pulse pressure}}{3}$$

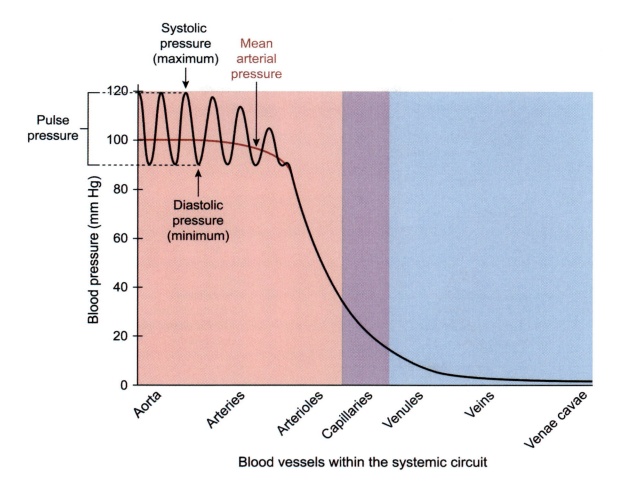

Figure 13.24 Parameters of blood pressure.

Arterial blood pressure varies based on blood flow into and out of arteries. If more blood enters than leaves the arteries in a given amount of time, blood volume in the arteries increases, causing blood pressure within the arteries to increase. Conversely, if more blood leaves than enters the arteries in a given amount of time, blood volume and blood pressure in the arteries decrease.

Consequently, increased cardiac output (which increases blood flow into the arteries) or increased peripheral resistance due to arteriole vasoconstriction (which decreases blood flow out of the arteries) tends to raise arterial blood pressure, as does an increase in total blood volume in the circulatory system. Figure 13.25 summarizes factors that affect arterial blood pressure.

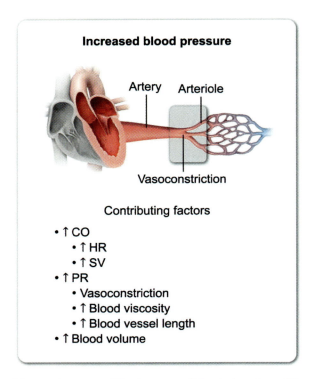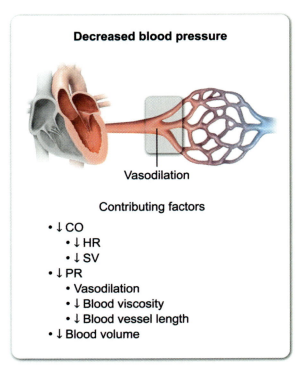

CO = cardiac output; HR = heart rate; SV = stroke volume; PR = peripheral resistance.

Figure 13.25 Factors that produce changes in blood pressure.

13.1.12 Circulation and Thermoregulation

Blood is composed mostly of water, which has a high specific heat capacity. Consequently, blood can carry a significant amount of heat. Distribution of heat within the body can be controlled via regulation of blood flow to specific body regions, allowing for **thermoregulation** (ie, maintenance of a physiologically appropriate internal body temperature).

When internal body temperature rises, the cardiovascular system facilitates thermoregulation by increasing blood flow through capillary beds near the surface of the skin via vasodilation, allowing heat to be dissipated to the environment. Conversely, when internal body temperature falls, decreased blood flow to capillary beds at the body's surface via vasoconstriction allows heat to be retained in the body's core. Figure 13.26 shows how blood flow to surface capillary beds is regulated via vasodilation and vasoconstriction of arterioles supplying blood to these capillaries.

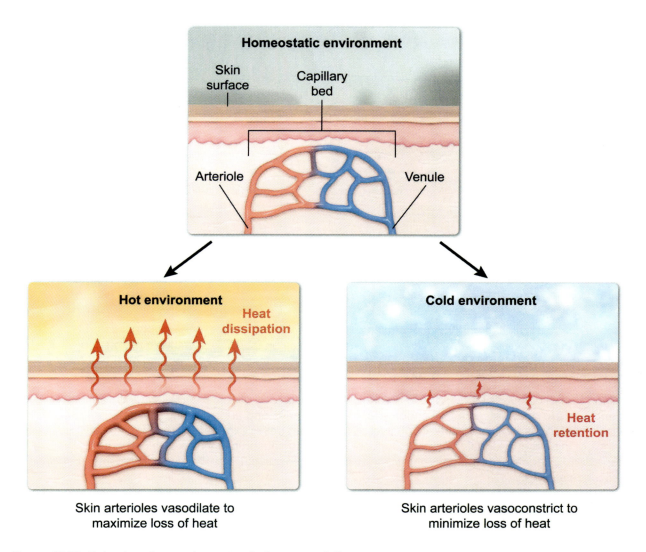

Figure 13.26 Role of cardiovascular system in thermoregulation.

13.1.13 Endocrine and Nervous Control of Circulation

The cardiovascular system is regulated by **extrinsic controls**, which are signals that originate in the nervous and endocrine systems and affect heart and blood vessel activity. In addition, local conditions (eg, extent of smooth muscle stretching, CO_2 concentration) within an organ or tissue can influence local blood flow by affecting the diameter of arterioles supplying blood to capillary beds in the region, a mechanism referred to as **autoregulation**.

As discussed in Concept 13.1.06, heart contraction is *not* initiated by signals outside the heart; however, the contraction *rate* and contraction *force* (ie, heart contractility) are regulated by extrinsic signals. Activity of the autonomic nervous system (ANS) plays an important role in regulating heart function. Specifically, activity of the **sympathetic division** of the ANS increases heart rate and contractility via release of **norepinephrine** from neurons. Conversely, activity of the **parasympathetic division** of the ANS decreases heart rate (but has little effect on contractility) via release of **acetylcholine**.

Hormones released by glands in the endocrine system also affect the function of the heart (see Lesson 11.2). **Epinephrine**, which is released by the adrenal medulla in response to sympathetic nervous system (SNS) activity, stimulates the heart to beat faster and more forcefully. Likewise, the hormone **triiodothyronine**, which is released from the thyroid gland, increases heart rate by acting on the heart muscle directly and by making the heart more responsive to epinephrine and norepinephrine. Table 13.2 summarizes the effects of extrinsic controls on the functioning of the heart.

Table 13.2 Effects of extrinsic controls on heart function.

Regulatory system	System component	Signal molecule	Effect on heart
Autonomic nervous system	Sympathetic division	Norepinephrine	Increased heart rate and contactility
Autonomic nervous system	Parasympathetic division	Acetylcholine	Decreased heart rate
Endocrine system	Adrenal medulla	Epinephrine	Increased heart rate and contactility
Endocrine system	Thyroid gland	Triiodothyronine	Increased heart rate and contactility

Blood vessel vasodilation and vasoconstriction are also subject to neural and hormonal control. This regulation of blood vessel diameter is complex and involves an interplay of multiple factors, some of which produce opposite effects (ie, vasodilation versus vasoconstriction) depending on the context in which signals are received.

Furthermore, blood vessels respond to a variety of chemical and physical stimuli via autoregulation. For example, **nitric oxide** released by endothelial cells in arteries and arterioles triggers vasodilation of these vessels in response to increased frictional force (ie, shear stress) experienced by endothelial cells during periods of increased blood flow. In addition, cell stretching caused by increased blood pressure in arteries and arterioles can trigger contraction of smooth muscle in the walls of these vessels, resulting in vasoconstriction.

Lesson 13.2
Lymphatic System

Introduction

The **lymphatic system** is composed of lymphoid organs, lymphatic vessels (ie, lymphatics), and fluid within the lymphatic vessels (ie, lymph). Lymphoid organs, such as the red bone marrow, thymus, spleen, and lymph nodes, play crucial roles in immune defense of the body, including in the production of cells (ie, lymphocytes) that constitute the immune system.

In addition to immune defense of the body (discussed in Lesson 20.1), the lymphatic system functions to move fluid and proteins that leak into the interstitial space from blood capillaries back to the blood. Furthermore, following the absorption of lipids from food in the digestive system, the lymphatic system transports lipids to the blood to be distributed throughout the body.

This lesson covers the structure and function of lymphatic vessels, movement of lymph in the body, and major functions of the lymphatic system.

13.2.01 Lymphatic Vessels

Lymphatic vessels carry lymph in one direction only, from the tissues toward the heart. Interstitial fluid initially enters the lymphatic system via blind-ended (ie, closed at one end) **lymphatic capillaries**, which are located among blood capillaries and tissue cells.

Like blood capillaries (see Concept 13.1.10), the walls of lymphatic capillaries are composed of a single layer of endothelial cells. However, unlike blood capillaries, lymph capillaries lack a continuous basement membrane, and lymphatic endothelial cells loosely overlap one another to form flaplike valves (Figure 13.27).

The flaplike valves in the walls of lymph capillaries permit easy entry of interstitial fluid (containing proteins, lymphocytes, and, potentially, pathogens) into lymph capillaries but prevent outward leakage. In addition, lymphatic endothelial cells are anchored to the surrounding extracellular matrix by protein filaments that help pull the overlapping cells apart to facilitate fluid entry.

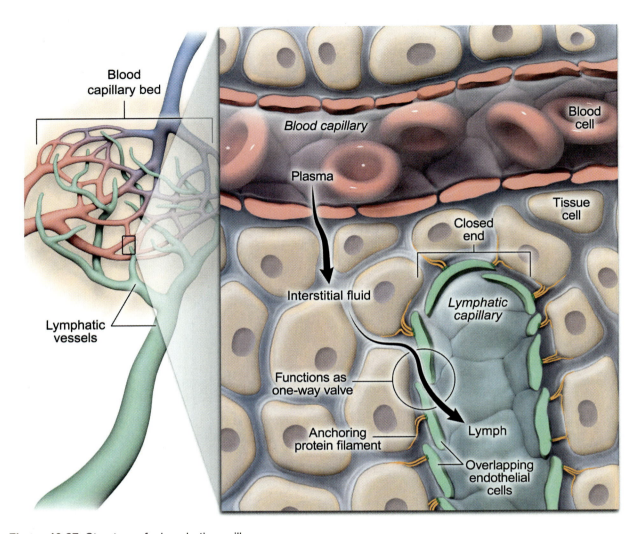

Figure 13.27 Structure of a lymphatic capillary.

After entering lymphatic capillaries, lymph moves through a series of lymph vessels that have progressively larger diameters and thicker walls containing smooth muscle. Lymphatic capillaries merge to form **collecting vessels**, which, as shown in Figure 13.28, contain **valves** (similar to those present in cardiovascular system veins) to prevent backflow of lymph.

Collecting vessels carry lymph to lymph nodes and eventually merge to form **lymphatic trunks**. Each lymphatic trunk carries lymph to one of two **lymphatic ducts** (ie, right lymphatic duct, thoracic duct), which empty the lymph into veins of the cardiovascular system near the heart.

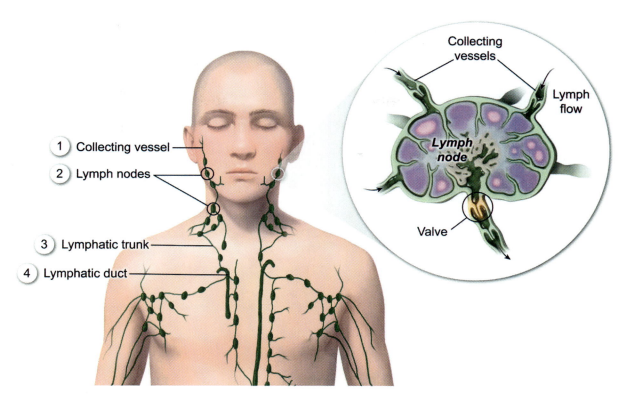

Figure 13.28 Sequence of structures through which lymph passes after entering lymphatic capillaries.

As shown in Figure 13.29, the **right lymphatic duct** receives lymph that drains from the right arm, the right side of the head, and the right side of the thorax. The **thoracic duct** drains lymph from all other body regions; consequently, the thoracic duct transfers the majority of the body's lymph to the blood.

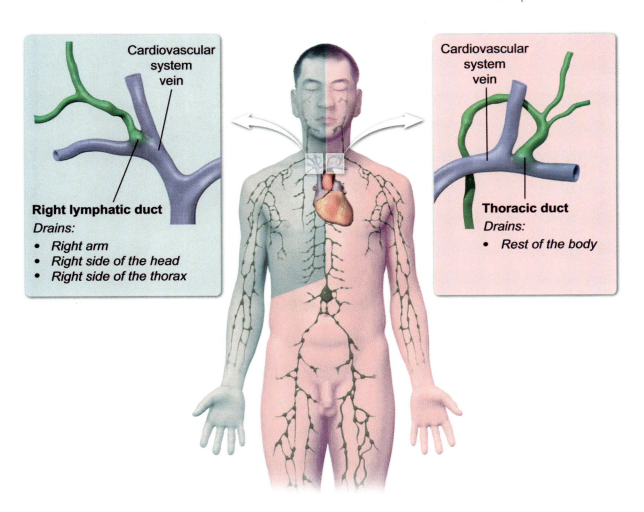

Figure 13.29 Lymphatic drainage.

Concept Check 13.3

The following diagram depicts steps in a pathway along which a plasma protein would likely travel after the protein has leaked into the interstitial space from a blood vessel. Complete the diagram by determining which of the following terms belongs in each blank box: "Blood capillary," "Cardiovascular system vein," "Collecting vessel," "Lymphatic capillary," "Lymphatic duct," "Lymphatic trunk," "Lymph node."

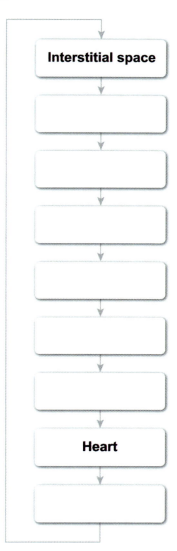

Solution

Note: The appendix contains the answer.

13.2.02 Lymphatic System Function

As discussed in Concept 13.1.10, bulk flow causes fluid to move out of blood capillaries into the interstitial space at the arterial end of a capillary bed and reenter the capillaries at the venous end of the capillary bed. However, less fluid typically reenters the capillaries than leaves the capillaries, which could result in harmful fluid accumulation in the interstitial space (ie, edema) if interstitial fluid was not continuously returned to the blood. Consequently, to maintain appropriate fluid balance, a primary lymphatic system function is to collect and transport excess interstitial fluid (along with constituent proteins) to the blood via the lymphatic vessels.

In addition to its role in balancing fluid, the lymphatic system functions to transport chylomicrons (ie, triglyceride-rich lipoproteins) to the blood from cells in the small intestine that absorb lipids during food digestion (see Concept 15.1.03). Chylomicrons are too large to enter blood capillaries, so they are taken into specialized lymphatic capillaries called **lacteals**, which are present within villi of the small intestine. The fluid within lacteals, which becomes incorporated into the lymph transported by the lymphatic system, is called **chyle**.

Figure 13.30 depicts the role of the lymphatic system in maintaining fluid balance in the body and in transporting lipids from the small intestine to the blood.

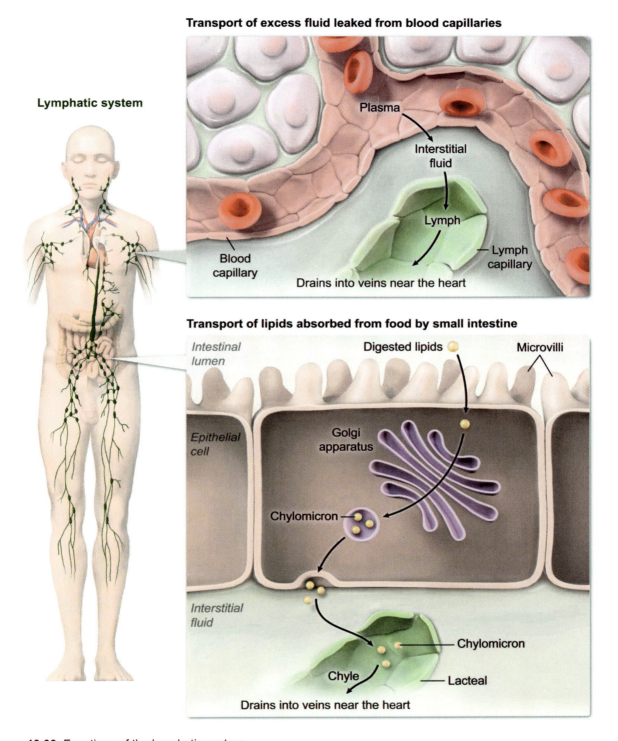

Figure 13.30 Functions of the lymphatic system.

Unlike the cardiovascular system, in which the heart is the dedicated pump, the lymphatic system lacks a dedicated pump to move lymph through lymphatic vessels. Instead, skeletal muscle contraction results in compression of lymphatic vessels, and the presence of valves within lymphatic vessels prevents backflow of lymph. Consequently, lymph is propelled in only one direction (ie, toward the heart) as skeletal muscle contraction occurs. In addition, the walls of lymphatic vessels larger than lymphatic capillaries contain smooth muscle, which contracts rhythmically to facilitate lymph movement.

Lesson 14.1
Respiratory System Structure

Introduction

The respiratory system includes the lungs and the structures that carry air to and from the lungs. This system is functionally divided into two zones. In the **respiratory zone**, exchange of respiratory gases (ie, oxygen and carbon dioxide) occurs. In the **conducting zone**, air is transported to and from the respiratory zone, and the air is conditioned to enhance respiratory zone function. This lesson describes the general structure of the components of the respiratory system, including those components that participate in protecting the system from potentially harmful materials in the environment.

14.1.01 Lung Structure

Figure 14.1 shows the major components of the respiratory system. The respiratory zone (ie, region responsible for gas exchange) is present within the lungs, which also contain some structures of the conducting zone (ie, passageways through which air is transported).

In addition to air-conducting structures within the lungs, the conducting zone includes the nose (ie, **nostrils**, **nasal cavity**), **pharynx**, **larynx**, **trachea**, and **primary bronchi** (singular: bronchus). Many components of the conducting zone, including the nose, larynx, trachea, and bronchi, are reinforced with cartilage. Cartilage helps maintain open airways by preventing collapse of these structures during breathing.

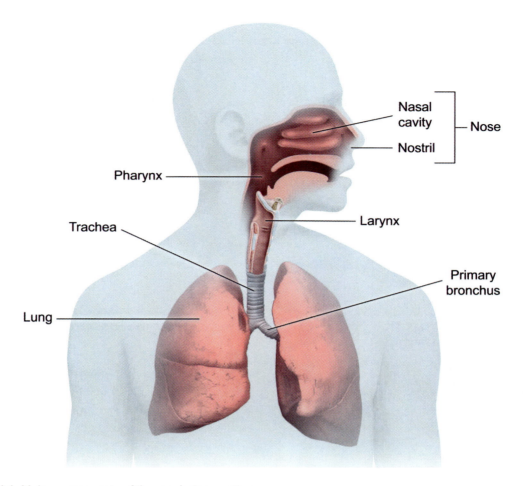

Figure 14.1 Major components of the respiratory system.

The primary bronchi connect the trachea to the lungs. Within the lungs, the primary bronchi repeatedly branch to form additional bronchi with smaller and smaller diameters and progressively less cartilage reinforcement. Within the lungs, air passageways with diameters less than approximately 1 mm are called **bronchioles**, the walls of which contain smooth muscle and lack cartilage entirely, allowing bronchiolar diameter to be adjusted to regulate airflow. Bronchioles continue branching to form smaller air passageways, eventually giving rise to **terminal bronchioles**, which constitute the final section of the conducting zone.

Terminal bronchioles connect the conducting zone to the respiratory zone, which consists of **respiratory bronchioles**, **alveolar ducts**, and **alveolar sacs** within the lung (see Figure 14.2). Respiratory bronchioles have scattered, thin-walled microscopic air pockets called **alveoli** (singular: alveolus) that bulge from the bronchioles' surface.

Respiratory bronchioles receive air from terminal bronchioles and transfer air into alveolar ducts, which are short tubes that have many individually attached alveoli and less smooth muscle than terminal bronchioles. Alveolar ducts carry air to alveolar sacs, which are dead-end structures composed of small clusters of interconnected alveoli. Alveolar sacs and their constituent alveoli contain no smooth muscle.

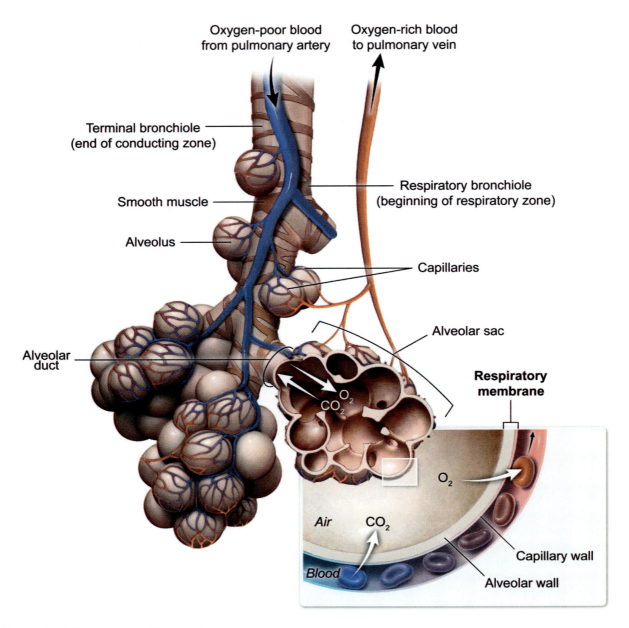

Figure 14.2 Components of the respiratory zone.

An average human lung in a healthy adult contains approximately 250 million alveoli, with an internal surface area (ie, surface area exposed to air in the lung) of approximately 50 m² (incredibly large when compared to an average skin surface area for the entire body of approximately 1.7 m²). The wall (ie, septum) of an alveolus consists of a single layer of epithelial cells. These cells include type I cells, which account for about 95% of the wall's structure and allow for gas exchange, and type II cells, which secrete **pulmonary surfactant**, a substance that reduces surface tension within the lung and facilitates lung inflation.

Alveoli are surrounded by elastic fibers and capillaries, with adjacent alveoli and capillary walls forming the **respiratory membrane** (see Figure 14.2). This membrane permits rapid exchange of gases between the blood and air in the lungs via simple diffusion.

The lungs are located within the thoracic cavity, the inferior boundary of which is formed by the **diaphragm**. Within the thoracic cavity, each lung is surrounded by a double-layered saclike membrane, or **pleura** (Figure 14.3). The inner layer of this sac, the **visceral pleura**, is attached to the exterior surface of the lung, and the outer layer, the **parietal pleura**, is attached to the thoracic cavity wall. The pleural layers slide easily past one another but typically do not separate because the slit-like **pleural cavity** between them contains a small volume of watery fluid (ie, **pleural fluid**) that keeps the layers in close contact via hydrogen bonding.

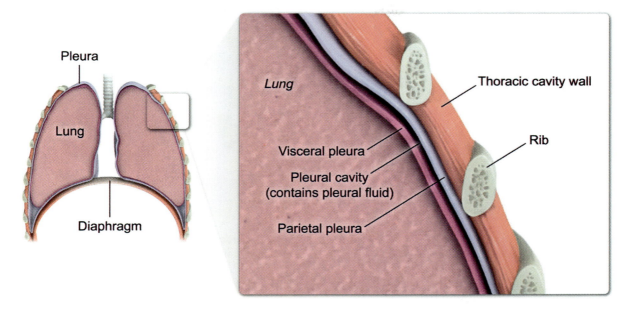

Figure 14.3 Double-layered structure of a pleura.

14.1.02 Mucociliary Escalator

Both the wall of an alveolus and the wall of a capillary are primarily composed of a single layer of flat epithelial cells. As a result, the respiratory membrane (ie, structure separating air from blood in the lung) is very thin, approximately 0.5 μm, and does not typically function as an effective barrier against entry of pathogens or other harmful substances (eg, toxic chemicals) into the body. Consequently, the respiratory system employs a variety of mechanisms by which pathogens and other potentially harmful materials are prevented from reaching the lung's alveoli or are removed from the alveoli upon arrival.

A primary means by which the lungs are protected from pathogens and other potentially harmful inhaled substances is a mechanism called **mucociliary clearance (MCC)**, which occurs via the **mucociliary escalator**. The mucociliary escalator consists of airway **mucus**, which traps foreign materials, and **cilia**, hairlike structures that continuously wave back and forth on the surface of specialized cells to sweep mucus toward the pharynx. Upon reaching the pharynx, the mucus is expectorated (eg, spit from the mouth) or swallowed (ie, mucus-entrapped pathogens are destroyed via acidic stomach secretions).

The mucociliary escalator is present in most regions of the respiratory system, including the nasal cavity, larynx, trachea, bronchi, and bronchioles. These regions are lined by an epithelium that consists of ciliated cells interspersed with secretory cells (eg, mucus-secreting **goblet cells**). In addition, the entrance to the nasal cavity contains hairs that help filter large particles from inhaled air. Pathogens and particles not eliminated by MCC can be removed from alveoli by **alveolar macrophages**, which engulf and destroy foreign materials via phagocytosis. Figure 14.4 depicts structures that help protect the respiratory system.

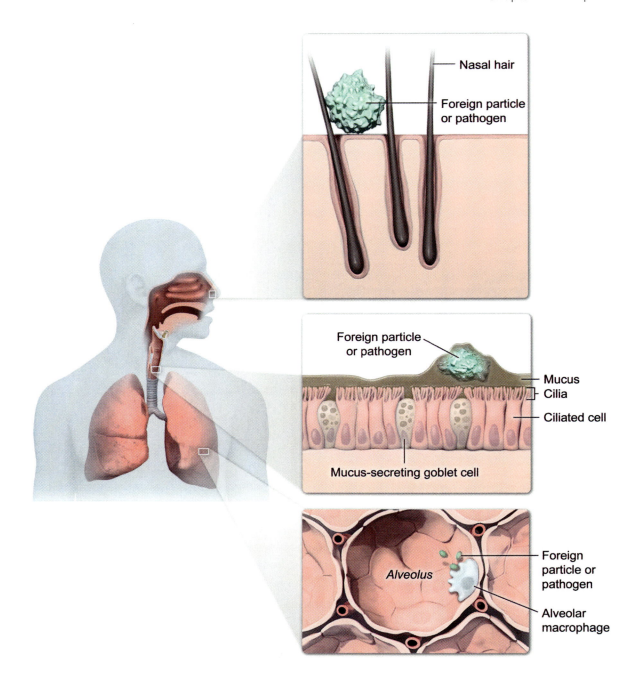

Figure 14.4 Protective structures in the respiratory system.

Chapter 14: Respiration

 Concept Check 14.1

Complete the following table by writing "yes" or "no" in each of the empty cells.

Structure	Reinforced with cartilage	Diameter regulated by smooth muscle	Contains ciliated lining and mucus
Nasal Cavity	Yes	No	
Larynx		No	
Trachea		Yes	
Bronchi		Yes	
Bronchioles			
Alveoli			

Solution

Note: The appendix contains the answer.

Lesson 14.2
Respiratory System Function

Introduction

The respiratory system functions to bring oxygen (O_2) into the body for distribution to cells and to remove carbon dioxide (CO_2) waste generated by cellular metabolism from the body. In addition, the respiratory system participates in regulation of body pH and body temperature, produces sounds used for vocal communication, and protects the body from inhaled pathogens and other harmful inhaled substances. This lesson describes mechanisms by which the respiratory system carries out its primary functions.

14.2.01 Mechanism of Respiration

Functions of the respiratory system entail movement of air into and out of the lungs via a process called **ventilation**, which consists of alternating phases of **inspiration** and **expiration**. Air movement to and from the lungs occurs due to differences in pressure between air in the lungs and air outside the body.

Within the lungs, changes in pressure that result in inspiration or expiration occur via activity of skeletal muscles coupled with the lungs' natural resiliency (ie, elasticity). Skeletal muscles that function primarily in ventilation include the dome-shaped **diaphragm** and the **intercostal muscles**, which are attached to the ribs (Figure 14.5).

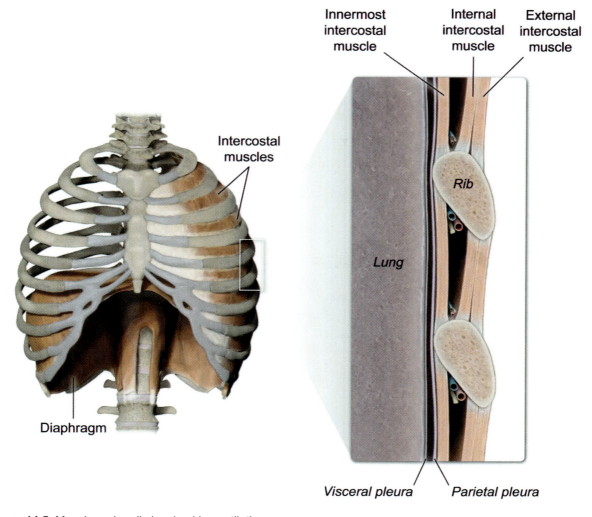

Figure 14.5 Muscles primarily involved in ventilation.

Inspiration occurs when pressure within the lungs' alveoli drops below atmospheric pressure. This decrease in alveolar pressure occurs when the volume of the thoracic cavity increases due to contraction (ie, flattening) of the diaphragm and, to a lesser extent, rib cage expansion via contraction of the external intercostal muscles.

Because pleural membranes connect the lungs to the thoracic cavity wall, thoracic cavity expansion causes lung expansion, resulting in decreased lung pressure in accordance with Boyle's law. Consequently, air moves from an area of higher pressure in the atmosphere to an area of lower pressure in the lungs.

Expiration occurs when the volume of the thoracic cavity decreases, which causes air pressure within the lungs to exceed atmospheric pressure, driving air from the respiratory system out of the body. Passive (ie, unforced) expiration takes place as the diaphragm relaxes and resumes a dome shape, whereas forced expiration is an active process that involves contraction of the internal intercostal muscles and abdominal muscles. The recoil of elastic fibers that stretch during inspiration facilitates expiration by generating a force that constricts alveoli. Figure 14.6 summarizes factors responsible for inspiration and expiration.

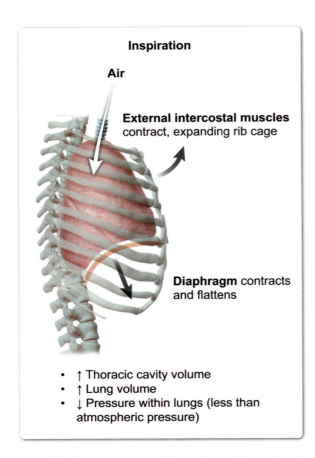 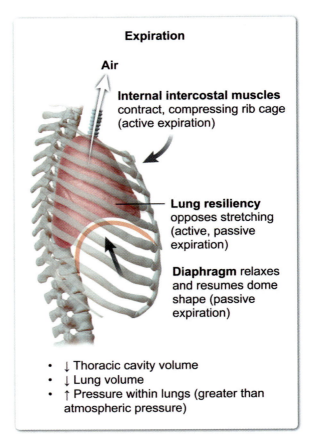

Figure 14.6 Factors responsible for inspiration and expiration.

Proper lung function is dependent on the maintenance of a thin layer of watery fluid called **alveolar lining fluid (ALF)**, which is found on the interior surface of alveoli. This fluid protects alveoli from desiccation (ie, drying out), provides immune defense of the lungs via alveolar macrophages and antimicrobial proteins, and mediates respiratory gas exchange. The interface between ALF and alveolar air results in the development of **surface tension**, which, along with the behavior of elastic fibers around alveoli, creates lung resiliency.

Pulmonary surfactant, a lipid-protein mixture secreted into ALF by type II alveolar cells, reduces surface tension, thereby lowering the amount of work needed to expand alveoli during inspiration (see Figure 14.7). A potentially fatal condition called neonatal respiratory distress syndrome can occur in premature infants who do not produce sufficient surfactant. This condition greatly increases the amount of energy these infants expend during inspiration.

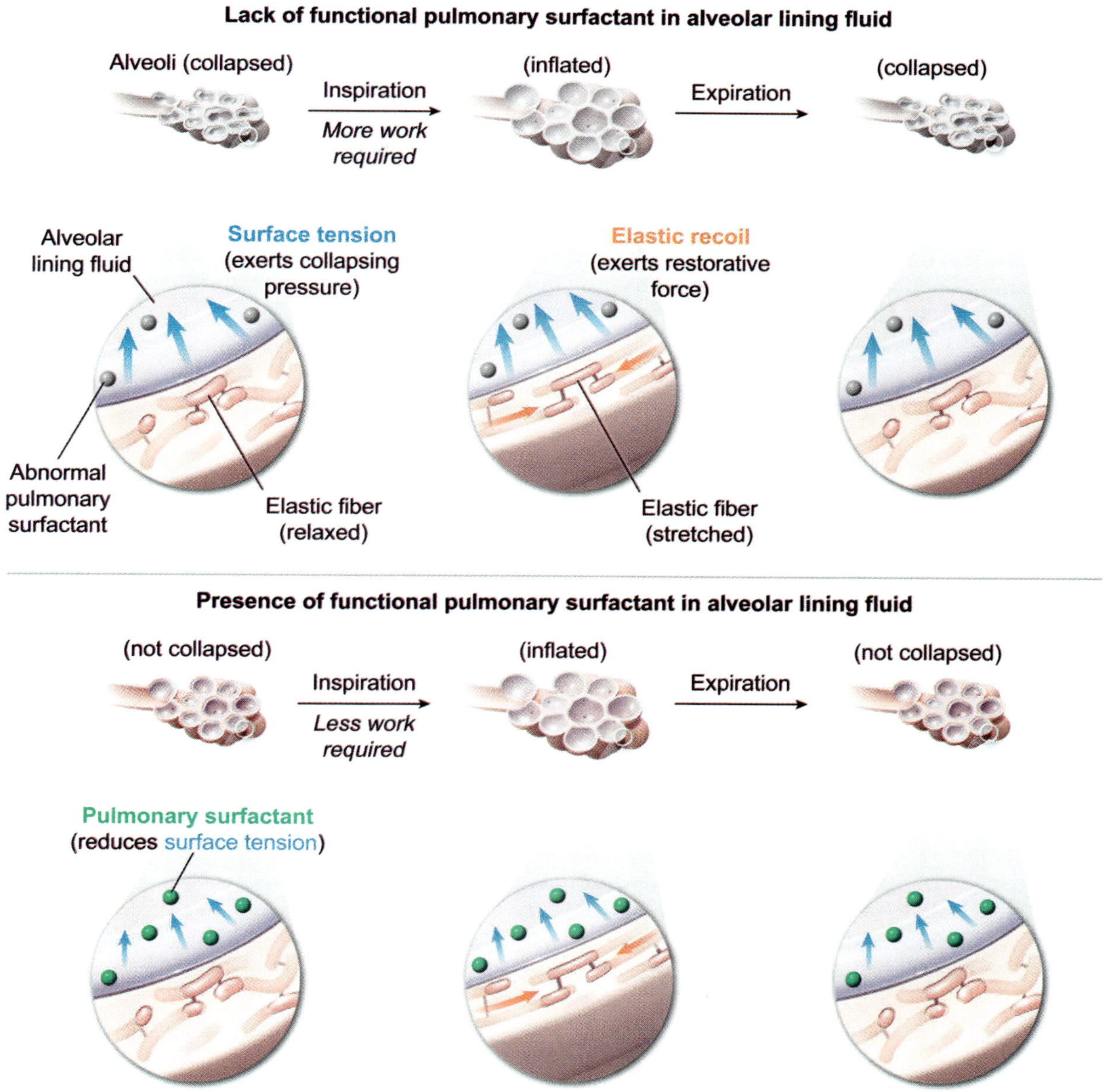

Chapter 14: Respiration

Lack of functional pulmonary surfactant in alveolar lining fluid

Alveoli (collapsed) → Inspiration (More work required) → (inflated) → Expiration → (collapsed)

Alveolar lining fluid

Surface tension (exerts collapsing pressure)

Elastic recoil (exerts restorative force)

Abnormal pulmonary surfactant

Elastic fiber (relaxed)

Elastic fiber (stretched)

Presence of functional pulmonary surfactant in alveolar lining fluid

(not collapsed) → Inspiration (Less work required) → (inflated) → Expiration → (not collapsed)

Pulmonary surfactant (reduces surface tension)

Figure 14.7 Effect of pulmonary surfactant on lung resiliency.

Some aspects of lung function can be evaluated via spirometry, a diagnostic test used to determine various lung volumes and capacities (see Figure 14.8). The following parameters can be directly measured via spirometry or indirectly calculated from other test results:

- **Tidal volume (TV)** is the amount of air that moves into or out of the lungs during one respiratory cycle while at rest (ie, during a single unforced inspiration and expiration cycle).
- **Inspiratory reserve volume (IRV)** is the amount of additional air that can be forcefully breathed in following a normal, quiet inspiration.
- **Expiratory reserve volume (ERV)** is the amount of additional air that can be forcefully breathed out following a normal, passive expiration.
- **Inspiratory capacity (IC)** is the total amount of air that can be forcefully breathed in following a normal, passive expiration (ie, IC = TV + IRV).

- **Vital capacity (VC)** is the maximum amount of air that can be forcefully breathed out following a forceful, maximal inspiration (ie, VC = TV + IRV + ERV).
- **Residual volume (RV)** is the amount of air present in the lungs following a forceful, maximal expiration (ie, RV = FRC − ERV).
- **Functional residual capacity (FRC)** is the amount of air present in the lungs following a normal, passive expiration (ie, FRC = ERV + RV).
- **Total lung capacity (TLC)** is the maximum amount of air that can be contained within the lungs (ie, TLC = TV + IRV + ERV + RV).
- **Anatomical dead space** is the total volume of the structures that constitute the respiratory system's conducting zone (which extends from the nose to the terminal bronchioles and contains inspired air that does not participate in alveolar gas exchange).

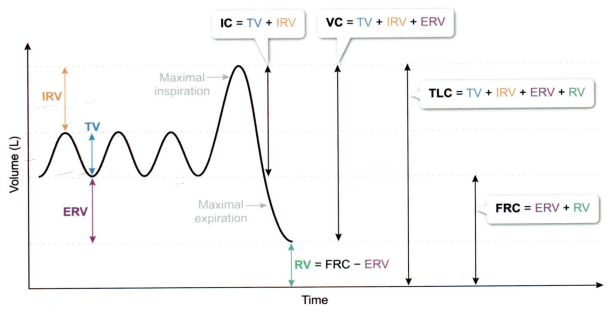

TV = tidal volume; IRV = inspiratory reserve volume; ERV = expiratory reserve volume; IC = inspiratory capacity; VC = vital capacity; RV = residual volume; FRC = functional residual capacity; TLC = total lung capacity.

Figure 14.8 Lung volumes and capacities.

Respiratory system function can be adversely affected by various diseases that impact ventilation and/or lung volumes and capacities. For example, **asthma**, which involves airway inflammation and bronchoconstriction (ie, narrowing of bronchi and bronchioles via contraction of smooth muscle), results in increased resistance to airflow. Asthma can also involve an overproduction of mucus, which can further impede airflow. Acute asthma attacks can be life-threatening due to greatly reduced gas exchange resulting from a lack of lung ventilation.

Chronic obstructive pulmonary disease (COPD) also causes reduced airflow and makes ventilation difficult. The main types of COPD are emphysema and chronic bronchitis. Emphysema, which is typically associated with cigarette smoking, results in damage to the lungs' alveoli. Emphysema leads to the enlargement of alveoli due to destruction of alveolar walls and also results in a loss of alveolar elasticity. These changes cause reduced alveolar surface area available for gas exchange and increased residual volume, which reduces ventilation efficiency. Figure 14.9 illustrates the effects of asthma and emphysema.

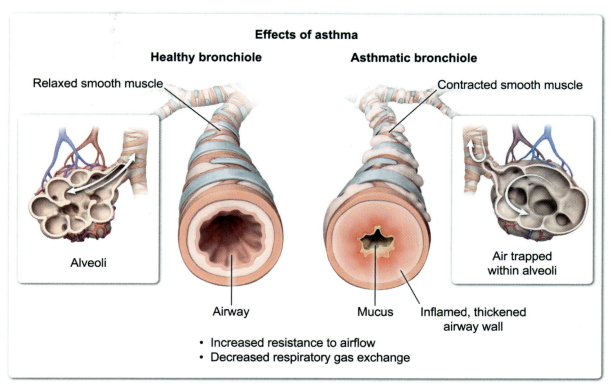

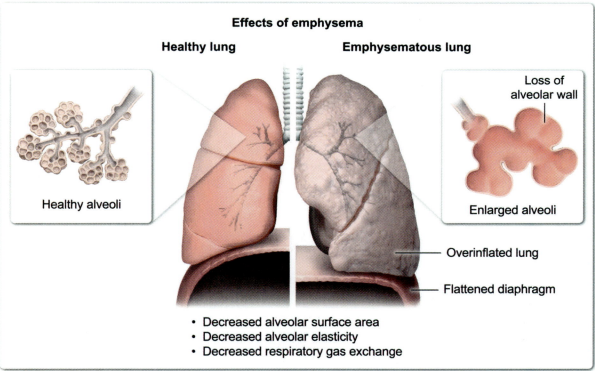

Figure 14.9 Effects of asthma and emphysema.

The structures through which air passes to reach the lungs (Figure 14.10) condition the air. The changes made to inhaled air help facilitate gas exchange and protect the lungs' delicate alveoli from damage due to desiccation or cold temperatures. As air passes through the nasal cavity, pharynx, and trachea, mucous membranes lining these structures transfer heat and moisture to the air. In addition, as discussed in Concept 14.1.02, the mucociliary escalator operates within the nasal cavity, larynx, trachea, bronchi, and bronchioles to remove particulates (eg, dust, pathogens) from the air before it reaches the alveoli.

The larynx, the opening to which is covered during swallowing by the **epiglottis** to prevent food or liquid from entering the airway, contains **vocal folds** (ie, true vocal cords). These folds of tissue can produce sounds via vibration as air passes between them during expiration. The length and tension of the vocal folds can be changed by contraction of muscles within the larynx to regulate the sound's pitch. The loudness of the sound is a function of airflow force between the vocal folds, with stronger airflow producing louder sounds. Airflow force is dependent on the activity level of the muscles responsible for forced expiration.

Forced expiration also enables the respiratory system to expel potentially harmful materials via coughing or sneezing. Coughing can be performed voluntarily and is also triggered reflexively by substances that irritate the airways (eg, larynx, bronchioles) in some manner. Sneezing functions to reflexively expel irritating substances from the nasal cavity.

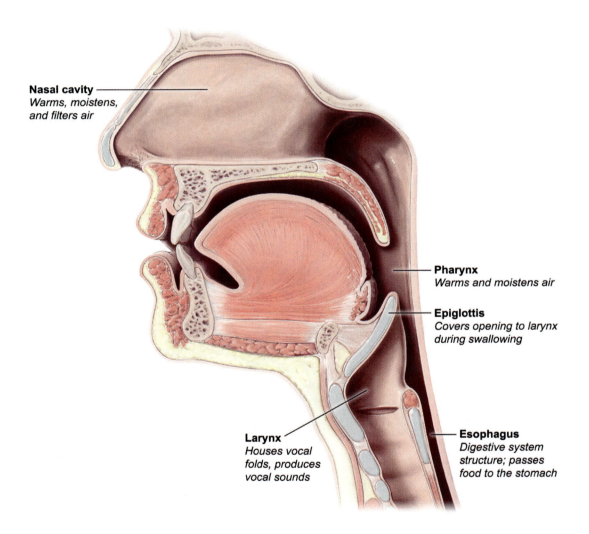

Figure 14.10 Components and functions of the upper respiratory tract.

Chapter 14: Respiration

14.2.02 Gas Exchange

Gas exchange between air within the lungs' alveoli and blood within pulmonary capillaries around the alveoli is driven by differences in partial pressures of respiratory gases (ie, O_2 and CO_2). Like other gases, O_2 and CO_2 move from regions of higher partial pressure to regions of lower partial pressure. As described by Henry's law, the amount of a gas that becomes dissolved in a solution with which the gas is in contact is directly proportional to the gas's partial pressure. Therefore, the dissolved gas content of blood moving through pulmonary capillaries varies based on partial pressures of the gases in alveolar air.

Blood is carried from the heart to the gas exchange sites (ie, alveoli) of the lungs via pulmonary arteries. Blood returns to the heart from the lungs via pulmonary veins and is pumped to the rest of the body (see Concept 13.1.07). As shown in Figure 14.11, the partial pressure of O_2 (ie, PO_2) in alveolar air is initially greater than the PO_2 of blood plasma carried past the alveoli within pulmonary capillaries. Consequently, O_2 diffuses from the alveolar air into the blood within the pulmonary capillaries until the blood's PO_2 matches the PO_2 of the alveolar air (ie, reaches dynamic equilibrium).

The partial pressure of CO_2 (PCO_2) in blood plasma within pulmonary capillaries is initially greater than the PCO_2 of alveolar air (see Figure 14.11), causing CO_2 to diffuse out of the blood into the air within the alveoli. The air within the alveoli is then removed from the body via expiration.

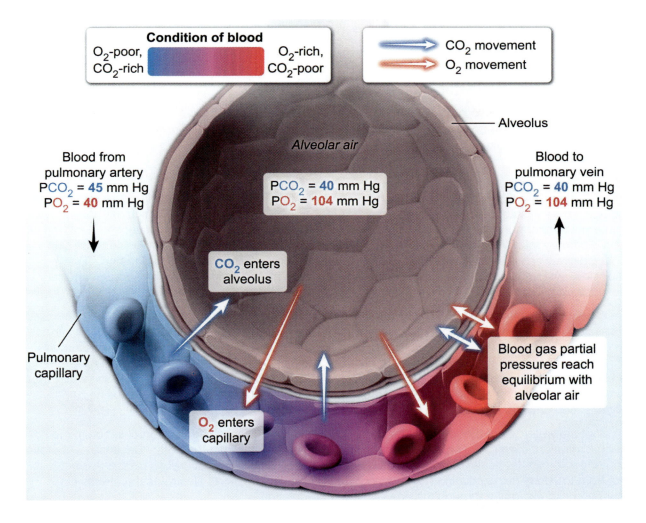

Figure 14.11 Gas exchange in the lung.

Gas exchange also occurs between blood in systemic capillaries and tissues in the body. Because metabolically active body tissues consume O_2 and produce CO_2 via cellular respiration, the PO_2 in tissues is typically lower than the PO_2 of blood in systemic capillaries. Furthermore, the PCO_2 in tissues is typically higher than the PCO_2 in systemic capillaries. As a result, O_2 diffuses from the blood into metabolically active tissues and CO_2 diffuses from metabolically active tissues into the blood, as shown in Figure 14.12.

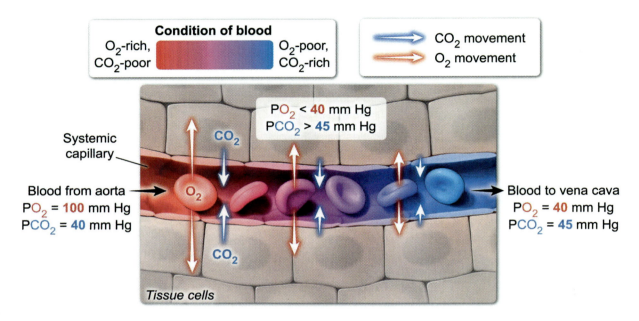

Figure 14.12 Gas exchange in a body tissue.

14.2.03 Respiration and Thermoregulation

The respiratory system contributes to **thermoregulation**, the overall process by which the body maintains an internal temperature within the normal physiological range. As discussed in Concept 19.2.03, the skin plays a primary role in thermoregulation via regulation of blood flow through capillaries near the skin's surface as well as via sweat production, which allows for body cooling through evaporation. The respiratory system similarly participates in thermoregulation by regulating blood flow through capillary beds located near the air-exposed internal surfaces of the nasal cavity and trachea.

The nasal cavity contains structures (ie, nasal conchae) that increase the surface area of the nasal mucosa (ie, respiratory epithelium), to which air entering and leaving the respiratory system via the nose is exposed. This extensive mucosa transfers heat and moisture (ie, water vapor) to inspired air, and vasodilation of arterioles supplying blood to capillary beds in the nasal cavity facilitates this air-warming and humidification process. When expiration occurs through the nose, some of the heat and water that were added to the inspired air are reclaimed as the warmer air exits the cooler nose, causing condensation of water back onto the nasal mucosa.

Not all of the water vapor that evaporates into inspired air can be reclaimed via condensation in the nose during expiration; therefore, ventilation results in a net loss of water from the body. Because evaporation is an endothermic (ie, heat-absorbing) process, a net loss of water via evaporation results in a net loss of heat from the body (ie, ventilation typically *cools* the body). Figure 14.13 illustrates the effects of inspiration and expiration through the nose on thermoregulation.

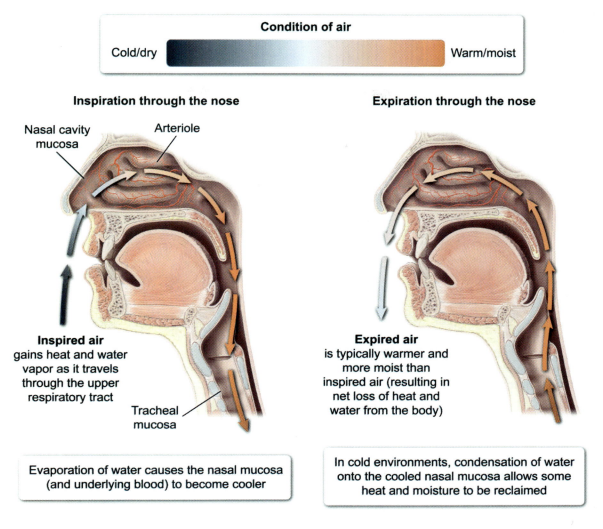

Figure 14.13 Effects of inspiration and expiration through the nose on thermoregulation.

Increased ventilation accompanied by decreased tidal volume (ie, **panting**) results in increased water evaporation from the respiratory system, thereby increasing heat dissipation (Table 14.1). Inspiration during panting occurs through the nose but may also occur through the mouth to facilitate cooling by allowing additional water to evaporate (ie, from the oral mucosa and tongue).

Body cooling via panting is maximized when inspiration occurs through the nose and expiration occurs through the mouth. This pattern of air flow prevents condensation (which causes heat to *return* to the body) from occurring in the nasal cavity.

Table 14.1 Mechanisms by which the respiratory system participates in thermoregulation.

Mechanism	Effect
Vasodilation of arterioles supplying blood to capillary beds (ie, in nasal cavity and trachea)	Warms inspired air
Evaporation of water from respiratory mucosa	Cools respiratory mucosa and underlying blood
Condensation of water vapor onto nasal mucosa	Returns heat to the body
Panting (ie, rapid, shallow ventilation)	Enhances evaporative cooling (ie, facilitates evaporation of water from respiratory mucosa)

14.2.04 Control of Respiration

As discussed in Concept 14.2.01, ventilation occurs due to the activity of skeletal muscles (eg, diaphragm, intercostal muscles), which contract upon receipt of signals from the central nervous system (ie, brain, spinal cord) via somatic motor neurons. As shown in Figure 14.14, the brain contains **respiratory centers**, which govern involuntary ventilation. These respiratory centers are located within the **pons** and **medulla** of the brainstem. The brain's **cerebral cortex** controls voluntary ventilation.

The respiratory center in the medulla is composed of a network of neurons that establishes the basic rhythm of involuntary respiration, which is characterized by a resting ventilation rate of approximately 12–16 breaths per minute. Neurons in the pons participate in respiratory control by communicating with and affecting the activity of the medullary respiratory center.

The respiratory centers receive input from a variety of sensory receptors, including **central chemoreceptors** located in the brainstem and **peripheral chemoreceptors** located in blood vessels that carry blood to the brain (ie, aortic arch and carotid arteries, see Figure 14.14). Integration of this sensory input allows the respiratory centers to adjust the rate and depth of ventilation to meet changing demands, which are typically brought about by changes in the body's activity level. In addition, input from higher brain centers (eg, cerebral cortex, hypothalamus) regulates respiratory center output and influences involuntary ventilation.

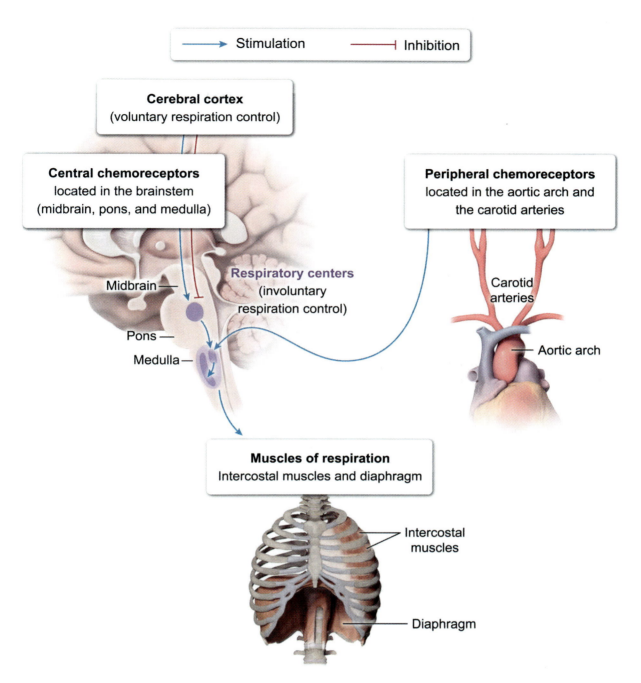

Figure 14.14 Structures involved in respiration control.

The body is able to meet changing demands for ventilation by monitoring the concentrations of CO_2, hydrogen ions (H^+), and O_2 in body fluids. Peripheral chemoreceptors monitor all three of these factors in arterial blood, and central chemoreceptors detect CO_2 levels in the brain's cerebrospinal fluid (CSF). More specifically, the central chemoreceptors monitor pH (ie, H^+ concentration) of the CSF, which becomes more acidic (ie, pH decreases) as CO_2 moves from the blood into the CSF by readily diffusing across the blood-brain barrier. The effect of CO_2 concentration on body fluid pH is discussed in greater detail in Concept 14.2.05.

Under normal conditions, arterial blood CO_2 concentration is the most important determinant of ventilatory rate and depth, primarily via the effect of blood CO_2 concentration on CSF pH. The partial pressure of CO_2 (PCO_2) in arterial blood is tightly regulated, typically being held at approximately 40 mm Hg due to negative feedback. When arterial blood PCO$_2$ rises, excess CO$_2$ is eliminated via increased ventilation resulting from increased stimulatory input from chemoreceptors. Conversely, if arterial blood PCO$_2$ is too low, CO$_2$ is allowed to accumulate in the blood through decreased ventilation resulting from decreased chemoreceptor stimulation.

The level of O$_2$ in arterial blood typically does not play a major role in regulating ventilation. Blood O$_2$ content becomes a primary factor in determining ventilatory rate only if the partial pressure of O$_2$ (PO$_2$) in arterial blood drops below approximately 60 mm Hg (ie, 20–40% lower than typical arterial blood PO$_2$). If arterial blood PO$_2$ drops below this point, decreased PO$_2$ stimulates increased ventilation by triggering increased stimulatory input to the respiratory centers by the peripheral chemoreceptors. Table 14.2 summarizes the effects of activated peripheral and central chemoreceptors on ventilation control.

Table 14.2 Effects of sensory inputs sent from activated chemoreceptors to respiratory centers.

Stimulus triggering chemoreceptor activity	Type of chemoreceptor	Chemoreceptor location	Ventilatory response
↓ pH of CSF (↑ PCO$_2$ of CSF)	Central	Throughout brainstem	↑ Rate and depth
↑ PCO$_2$ of blood	Peripheral	Aortic arch, carotid arteries	↑ Rate and depth
↓ pH of blood	Peripheral	Aortic arch, carotid arteries	↑ Rate and depth
↓ PO$_2$ of blood	Peripheral	Aortic arch, carotid arteries	↑ Rate and depth

CSF = cerebrospinal fluid.

Activity of the respiratory centers is also affected by other inputs. For example, pain and strong emotions, such as fear and excitement, affect ventilation (typically by increasing ventilation rate) via inputs from the amygdala and hypothalamus. In addition, the lungs contain **mechanoreceptors** (ie, stretch receptors) that are stimulated by lung inflation and send inhibitory signals to the respiratory centers. This inhibitory input from mechanoreceptors likely prevents lung overinflation.

The cerebral cortex controls voluntary respiration via motor neurons that bypass the brainstem's respiratory centers and directly stimulate inspiratory muscles. However, voluntary control of ventilation (eg, breath-holding) is limited because involuntary signals from the respiratory centers eventually override voluntary signals from the cerebral cortex as the PCO$_2$ in the blood and CSF rises.

Concept Check 14.2

The table shows blood test results obtained from a human subject with no underlying health problems who provided blood samples for analysis on three separate occasions. Complete the table by ranking the subject's likely ventilation rate at the time of sample collection as highest, intermediate, or lowest.

	Sample 1	Sample 2	Sample 3
Arterial blood PO_2	80 mm Hg	70 mm Hg	95 mm Hg
Arterial blood PCO_2	40 mm Hg	35 mm Hg	45 mm Hg
Arterial blood pH	7.40	7.46	7.35
Ventilation rate			

Solution

Note: The appendix contains the answer.

14.2.05 Role in Regulating pH

The respiratory system (along with the excretory system, as described in Concept 16.2.01) plays a major role in regulating the pH of body fluids. Specifically, ventilation provides a mechanism by which minute-to-minute changes in body fluid pH can be responded to and corrected, thereby maintaining acid-base balance (ie, pH homeostasis). The respiratory system can regulate pH in the body because carbon dioxide (CO_2), which is eliminated from the body via ventilation, has a significant impact on body fluid pH by reacting with water (H_2O) in body fluids to produce carbonic acid (H_2CO_3) as follows:

$$CO_2 + H_2O \rightleftharpoons H_2CO_3 \rightleftharpoons HCO_3^- + H^+$$

This equation shows that CO_2 and H_2O are in equilibrium with H_2CO_3, which in turn is in equilibrium with bicarbonate ions (HCO_3^-) and hydrogen ions (H^+). Consequently, in accordance with Le Châtelier's principle, if the CO_2 concentration in a body fluid increases, the concentrations of H_2CO_3 and H^+ in this fluid will also increase, resulting in decreased pH (because pH decreases as H^+ concentration increases). Likewise, if the CO_2 concentration in a body fluid decreases, less H_2CO_3 will be produced, resulting in lower H^+ concentration and higher pH.

The body regulates blood pH via the respiratory system by adjusting ventilation rate. Changes in ventilation rate cause changes in blood pH by affecting the extent to which CO_2 is removed from the blood via expiration. As shown in Figure 14.15, an increased ventilation rate raises blood pH (ie, makes the blood more alkaline) by transferring more CO_2 from the blood into the alveoli and eliminating that CO_2 from the body via expiration. Likewise, a decreased ventilation rate lowers blood pH (ie, makes the blood more acidic) by allowing more CO_2 to remain in the blood rather than being transferred to the alveoli and expired.

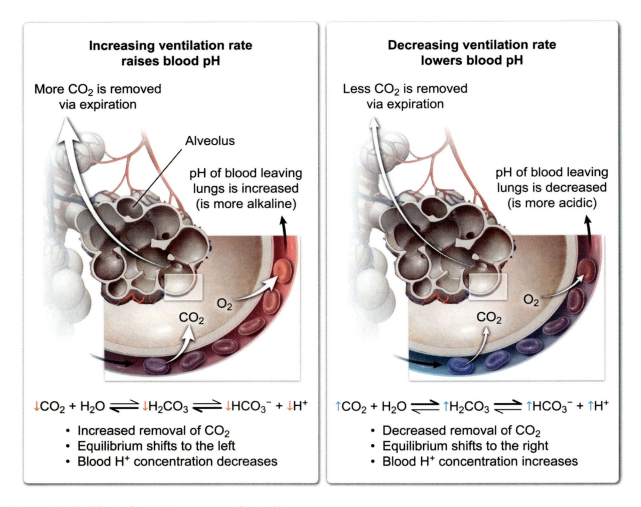

Figure 14.15 Effect of ventilation rate on blood pH.

Various factors (eg, diseases, medications, emotional state) that affect ventilation rate can lead to altered blood pH. Factors that result in **hyperventilation** (ie, rapid, deep breathing) typically produce respiratory alkalosis (abnormalllly high blood pH). Conversely, factors that result in **hypoventilation** (ie, slow, shallow breathing) typically produce respiratory acidosis (abnormally low blood pH). Figure 14.16 illustrates the effects of hyperventilation and hypoventilation on blood pH.

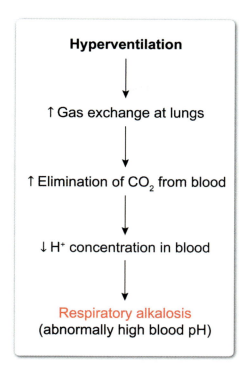 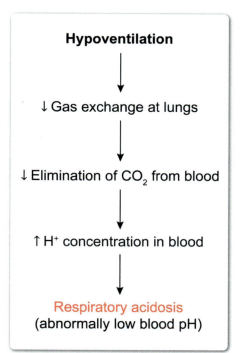

Figure 14.16 Causes of respiratory alkalosis and acidosis.

Chemical buffer systems within the body provide an important means by which the body resists changes in pH. Buffer systems help maintain pH homeostasis by releasing H⁺ in response to increased pH and binding H⁺ in response to decreased pH. The **bicarbonate buffer system**, which consists of the reversible reactions involving carbonic acid and bicarbonate ions previously described in this concept, is the primary buffer system that stabilizes pH in the body's extracellular fluids (including blood plasma).

As described in Concept 13.1.04, the enzyme **carbonic anhydrase** is present within red blood cells to catalyze the formation of carbonic acid, which dissociates to produce bicarbonate ions and hydrogen ions. The protein hemoglobin, which is present in abundance within red blood cells, functions as a buffer by binding these hydrogen ions, thereby preventing a significant decrease in pH. Hemoglobin molecules, along with other intracellular proteins and blood plasma proteins (eg, albumin), make up the **protein buffer system**, which helps stabilize the pH of both intracellular and extracellular fluids.

The **phosphate buffer system** is similar to the bicarbonate buffer system but uses hydrogen phosphate ions (HPO_4^{2-}) rather than bicarbonate ions. The concentration of phosphate in blood plasma is relatively low; therefore, the phosphate buffer system functions primarily within cells and in urine, locations in which phosphate concentration is higher.

END-OF-UNIT MCAT PRACTICE

Congratulations on completing **Unit 7: Circulation and Respiration**.

Now you are ready to dive into MCAT-level practice tests. At UWorld, we believe students will be fully prepared to ace the MCAT when they practice with high-quality questions in a realistic testing environment.

The UWorld Qbank will test you on questions that are fully representative of the AAMC MCAT syllabus. In addition, our MCAT-like questions are accompanied by in-depth explanations with exceptional visual aids that will help you better retain difficult MCAT concepts.

TO START YOUR MCAT PRACTICE, PROCEED AS FOLLOWS:

1) Sign up to purchase the UWorld MCAT Qbank
   IMPORTANT: You already have access if you purchased a bundled subscription.
2) Log in to your UWorld MCAT account
3) Access the MCAT Qbank section
4) Select this unit in the Qbank
5) Create a custom practice test

Unit 8 Digestion and Excretion

Chapter 15 Digestion

15.1 Alimentary Canal

15.1.01	Oral Cavity and Esophagus
15.1.02	Stomach
15.1.03	Small Intestine
15.1.04	Large Intestine
15.1.05	Peritoneum
15.1.06	Peristalsis

15.2 Accessory Digestive Organs

15.2.01	Liver
15.2.02	Gallbladder
15.2.03	Pancreas

15.3 Control of Digestion

15.3.01	Endocrine Control of Digestion
15.3.02	Enteric Nervous System

Chapter 16 Excretion

16.1 Excretory System Structure

16.1.01	General Excretory System Structure
16.1.02	Kidney Anatomy
16.1.03	Nephron Structure

16.2 Excretory System Function

16.2.01	Kidney Function
16.2.02	Urine Formation
16.2.03	Urine Elimination

Chapter 15: Digestion

Lesson 15.1

Alimentary Canal

Introduction

The components of the **alimentary canal**, also known as the **gastrointestinal tract**, include the oral cavity, esophagus, stomach, small intestine, and large intestine (Figure 15.1). Ingested food moves through these components, each of which has a distinct function, to undergo mechanical and chemical digestion. Nutrients are absorbed by epithelial cells lining certain portions of the canal, and indigestible materials and waste are excreted from the body at the distal end of the canal. This lesson details the alimentary canal components and functions.

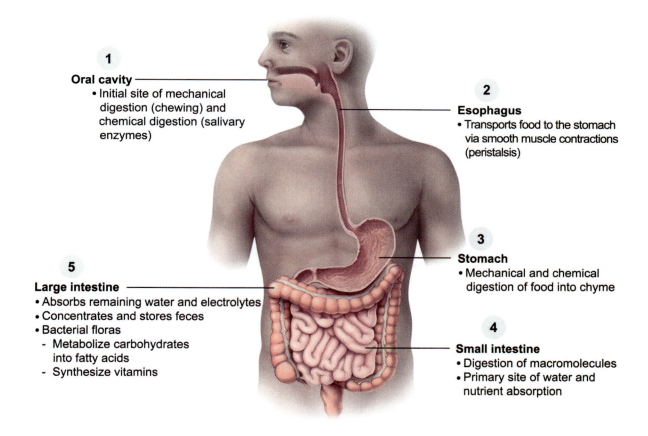

Figure 15.1 Components of the alimentary canal.

15.1.01 Oral Cavity and Esophagus

Initial mechanical digestion of food occurs via mastication (ie, chewing) by the teeth in the **oral cavity** (Figure 15.2). While mastication occurs, saliva and mucus secreted by salivary glands (exocrine glands located in the oral cavity) lubricate ingested food. Saliva, a fluid substance, contains enzymes for initial chemical digestion of some macromolecules. These enzymes include **lingual lipase**, which hydrolyzes triglycerides into free fatty acids, glycerol, monoglycerides, and diglycerides, and **salivary amylase**, which hydrolyzes the polysaccharide starch into the disaccharide maltose.

In the oral cavity, the muscles in the tongue and cheeks compress mechanically digested food into a bolus (ball) that can be swallowed. This bolus passes from the oral cavity to the pharynx (throat) and past the upper esophageal sphincter (sphincters are rings of muscle that control transit through tubular structures, see Concept 15.1.06). Next, the bolus enters the **esophagus**, a passageway for food to be carried to the stomach.

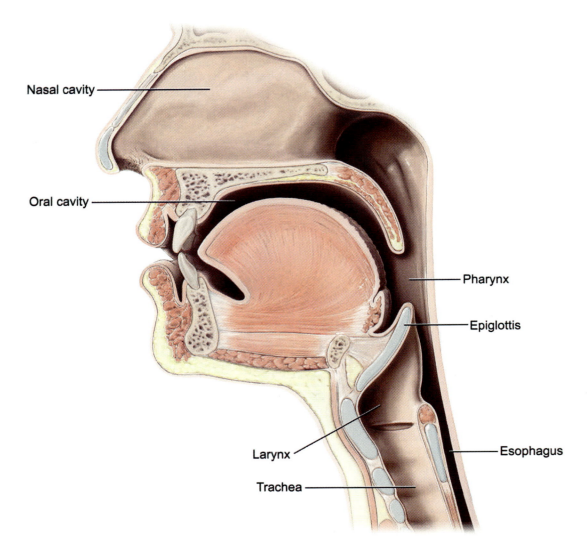

Figure 15.2 Anatomy of the oral cavity and esophagus.

15.1.02 Stomach

After passing through the esophagus, masticated food exits through the **lower esophageal sphincter** (sometimes called the cardiac sphincter because it is near the heart), a muscular ring located at the junction of the esophagus and the **stomach** that controls the passage of food into the stomach (Figure 15.3).

In addition to being the first opportunity for temporary food storage in the alimentary canal, the sac-like stomach functions in chemical and mechanical digestion of the food bolus. In the stomach, the food bolus is converted to **chyme**, a semifluid, partially digested mixture of water, hydrochloric acid (HCl), digestive enzymes, food nutrients (eg, proteins, carbohydrates, fats) and indigestible food components.

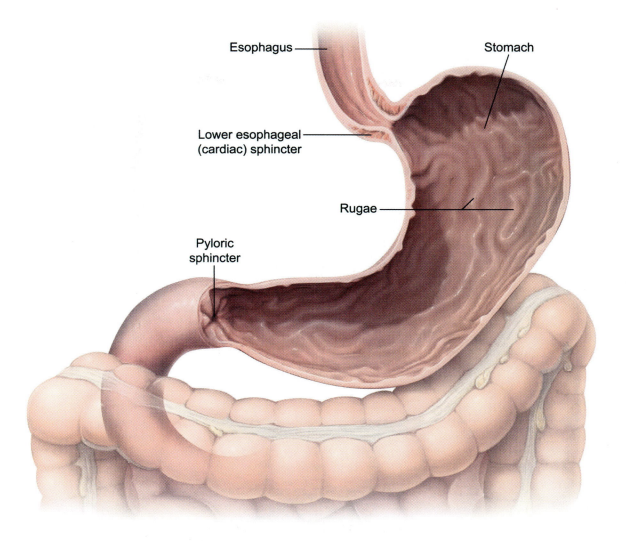

Figure 15.3 Anatomy of the stomach.

Different cells of the stomach secrete different products (Table 15.1). Some cells secrete products that form a solution known as **gastric juice**, which aids in stomach function. Other cells secrete products that allow the stomach to efficiently form chyme without harming the stomach itself. Cells of the stomach include G cells, parietal cells, chief cells, and mucous cells.

- **G cells** produce **gastrin**, a hormone that helps regulate secretion of hydrochloric acid by parietal cells. Hydrochloric acid creates an acidic environment (pH 1–3) in the stomach, which serves to activate digestive proteolytic enzymes, defend against pathogens, and denature proteins.
- **Parietal cells** produce **hydrochloric acid** and **intrinsic factor**, a glycoprotein that aids in the absorption of vitamin B_{12} in the ileum of the small intestine.
- **Chief cells** produce **pepsinogen**, an inactive form (ie, zymogen) of pepsin. When activated by the low pH of the stomach, the proteolytic enzyme pepsin breaks down polypeptides into smaller peptides. Chief cells also produce **gastric lipase**, which serves to hydrolyze lipids in the stomach.
- **Mucous cells** produce **mucus**, which forms a protective physical barrier for the stomach wall to guard against autodigestion (self-destruction) in the stomach's acidic, proteolytic environment. **Bicarbonate ions** released from stomach epithelial cells form a chemical barrier (ie, acid-neutralizing buffer) under the mucus barrier.

Table 15.1 Summary of secretions produced by various types of stomach cells.

Stomach cell type	Secretory product	Function of secretory product
G cells	Gastrin	Signals parietal cells to secrete hydrochloric acid
Parietal cells	Hydrochloric acid	Primary component of gastric juice; activates proteolytic enzymes, kills microbes, and denatures (unfolds) proteins
Parietal cells	Intrinsic factor	Aids in the absorption of vitamin B_{12} in the ileum
Chief cells	Pepsinogen	Zymogen (inactive form) of pepsin, which cleaves polypeptides into smaller peptides when activated by the low pH of gastric juice
Chief cells	Gastric lipase	Carries out hydrolysis of lipids in the stomach
Mucous cells	Mucus and bicarbonate	Protect the stomach walls from autodigestion by gastric juice, which contains acid and proteases

Once the food bolus has been converted to chyme, it passes through the **pyloric sphincter** into the **small intestine**.

15.1.03 Small Intestine

The **small intestine**, so named because it is smaller in diameter than the large intestine, is the site in which macromolecular digestion is completed, and it is the primary site from which substances useful to the body are absorbed. As shown in Figure 15.4, there are three subdivisions of the small intestine: the **duodenum**, the **jejunum**, and the **ileum**.

The **pyloric sphincter** controls the flow of chyme from the stomach into the duodenum, and digestion is largely, if not entirely, completed in the duodenum and jejunum. Nutrient absorption occurs throughout the small intestine, with the specific nutrients and amounts absorbed varying in the three small intestine subdivisions.

Chapter 15: Digestion

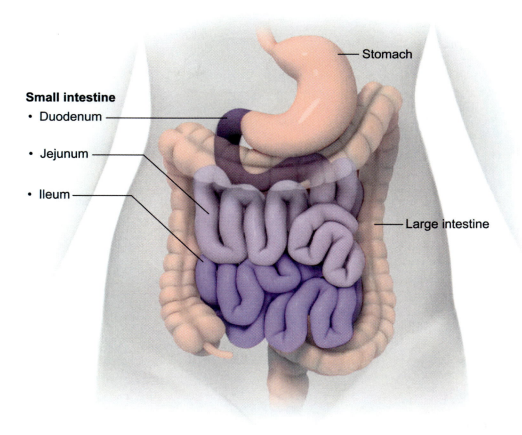

Figure 15.4 Anatomy of the small intestine.

In the duodenum, the presence of meal-derived fats within chyme and the acidity of chyme itself stimulate the release of bile and pancreatic secretions that digest fat and counteract the acid load (ie, raise the pH). **Bile** is a nonenzymatic solution produced by liver cells and stored in the gallbladder. When secreted into the duodenum, bile promotes the neutralization of acidic chyme arriving from the stomach and the mechanical digestion of fats (ie, emulsification). Likewise, the pancreas assists both in neutralizing chyme (by secreting bicarbonate) and in digesting fats (by secreting lipases).

Bile and pancreatic secretions enter the duodenum of the small intestine through the common bile duct (Figure 15.5).

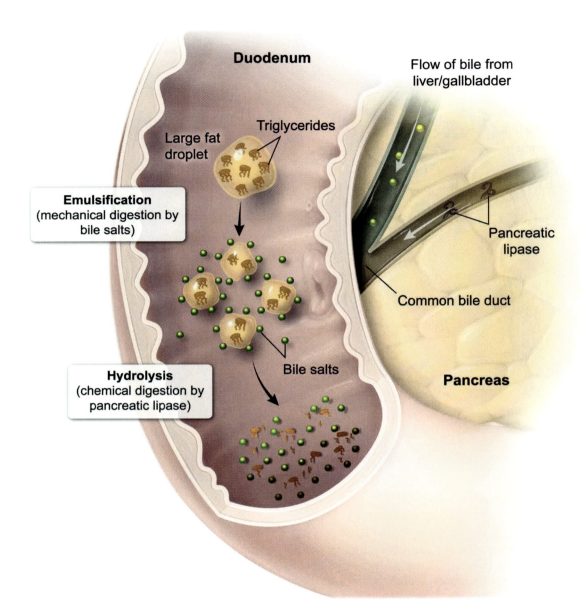

Figure 15.5 Digestion of lipids in the small intestine.

In addition to the bile duct's digestive secretions and the mechanical mixing that occurs in the small intestine to make nutrients available, the surface of the small intestine itself strongly influences nutrient absorption.

Large **circular folds** in the intestine's epithelial lining increase surface area and slow the movement of chyme through the intestinal tract. In addition to these circular folds, **villi** (fingerlike projections extending from the lining) and **microvilli** (smaller fingerlike extensions of individual absorptive cells) maximize the time and surface area available for nutrient absorption (Figure 15.6). The microvilli-covered epithelial surface where digestion and absorption of nutrients occur is called the brush border of the small intestine.

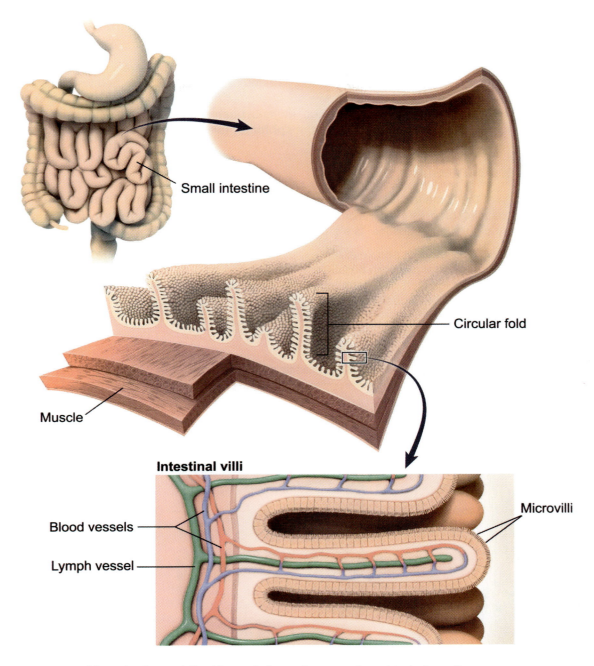

Figure 15.6 The small intestine is specialized to maximize surface area for nutrient absorption.

Nutrient absorption is also influenced by various intestinal enzymes. Complementing pancreatic lipases are lipases secreted from small intestinal cells. Proteolytic enzymes, which hydrolyze polypeptides, are synthesized and secreted from the microvilli-covered brush border and participate in protein digestion within the small intestine. **Dipeptidase** is a proteolytic enzyme that functions to directly digest proteins, while **enteropeptidase** serves to convert pancreatic trypsinogen in the small intestine into its active proteolytic form, **trypsin**. The small intestine also secretes enzymes that hydrolyze disaccharides (**disaccharidases**).

In addition to enzymes, the small intestine releases several hormones important in digestion. **Secretin** is a hormone released by the small intestine that functions to promote pancreatic enzyme and bicarbonate release into the duodenum and inhibit gastric acid secretion by parietal cells of the stomach, thereby regulating digestive tract pH. Secretin also slows digestive tract motility, allowing sufficient time for digestive enzymes to interact with chyme. **Cholecystokinin** (CCK) is another intestinal hormone that functions to promote pancreatic enzyme and bile release.

The small intestine absorbs nutrients and most of the water from the chyme it receives from the stomach. Fat digestion products (eg, free fatty acids, monoglycerides) are ultimately absorbed via microvilli of small intestine epithelial cells, as are mono-, di-, and trisaccharides from digested carbohydrates and small peptides and amino acids from digested proteins. Material unable to be digested in the small intestine passes to the large intestine through the **ileocecal sphincter** (also known as the ileocecal valve).

15.1.04 Large Intestine

The final portion of the alimentary canal is the **large intestine**, wider yet shorter than the small intestine but similarly consisting of three subdivisions: the **cecum**, **colon**, and **rectum** (Figure 15.7). The cecum, the first segment of the large intestine, is a small pouch connected to the appendix.

The colon is subdivided into four sections: the ascending colon, transverse colon, descending colon, and sigmoid colon, from proximal (near) to distal (far). The colon is responsible for reabsorption of electrolytes (eg, sodium ions, chloride ions) and water from indigestible material. As this reabsorption occurs, undigestible material is compacted into solid waste (feces), which is stored in the rectum prior to elimination from the body.

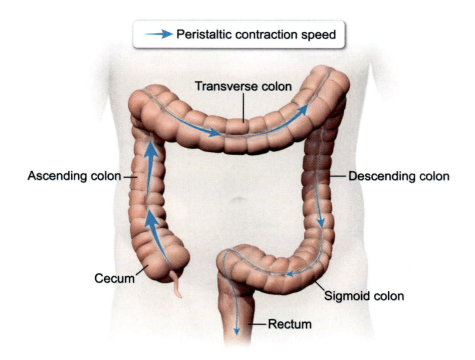

Figure 15.7 Large intestine anatomy.

The large intestine contains a diverse array of bacterial species (ie, **gut flora**) that aid in digestive processes. For example, some of these bacteria can process food items that would otherwise be indigestible, such as certain carbohydrates. Gut bacteria metabolize these undigested carbohydrates into short-chain fatty acids that can be absorbed and used by the body for energy. In addition, some bacterial species in the large intestine synthesize certain vitamins.

15.1.05 Peritoneum

The **peritoneum** lines the abdominal cavity and is made of two tissue layers. As shown in Figure 15.8, the **peritoneal cavity** is the space enclosed by the peritoneum. It is described as a potential space, or an area between two adjacent structures that may be pressed against one another.

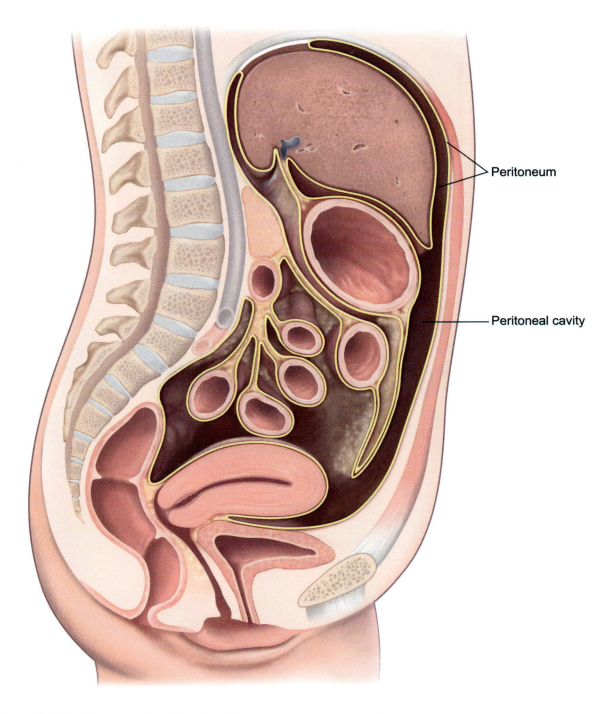

Figure 15.8 Peritoneum and peritoneal cavity.

15.1.06 Peristalsis

The movement of material through the alimentary canal relies on rhythmic muscular contractions rather than gravity. These wavelike muscular movements are termed **peristalsis** and occur due to the action of smooth muscle (Figure 15.9).

In addition to peristaltic waves, the movement of food through the digestive system is reliant on the relaxation of **sphincter muscles**. These rings of muscle (eg, cardiac sphincter, pyloric sphincter, ileocecal sphincter) divide the alimentary canal into segments with distinct functions. A sphincter is typically constricted so its center is closed but can relax to allow the natural passage of substances based on physiological requirements.

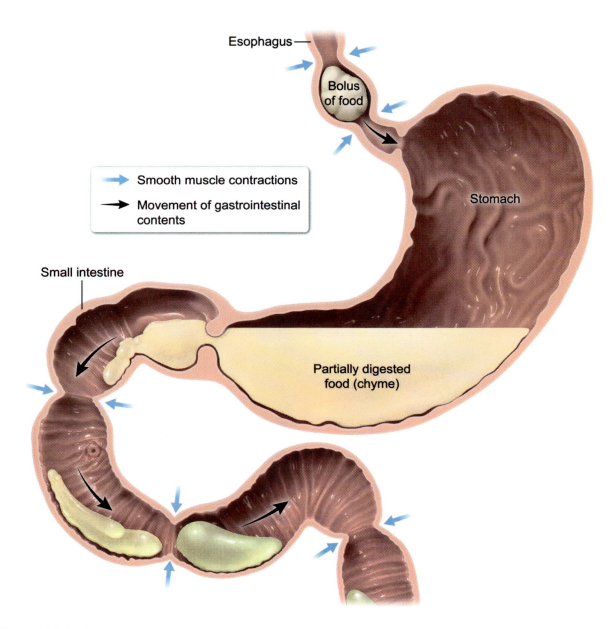

Figure 15.9 Peristalsis.

Concept Check 15.1

Small intestinal bacterial overgrowth (SIBO) is a disorder characterized by gastrointestinal discomfort and an abnormally high number of bacterial species present in the small intestine. It is thought that one type of SIBO is caused by a dysfunctional gastrointestinal sphincter muscle. Which sphincter is most likely dysfunctional in patients affected with this type of SIBO?

Solution

Note: The appendix contains the answer.

Lesson 15.2

Accessory Digestive Organs

Introduction

In addition to the organs of the alimentary canal, which are discussed in Lesson 15.1, there are several accessory organs important to the digestion and subsequent absorption of nutrients liberated by digestion. These organs include the **liver**, **gallbladder**, and **pancreas** (Figure 15.10) and are detailed in this lesson.

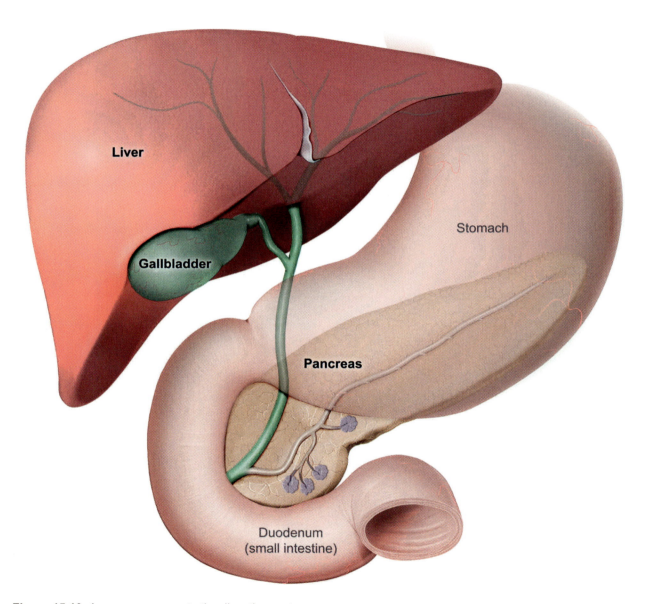

Figure 15.10 Accessory organs to the digestive system.

15.2.01 Liver

The **liver** is located in the upper right quadrant of the abdominal cavity, just below the diaphragm, and performs various functions (Figure 15.11).

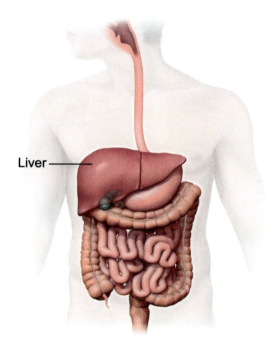

Liver functions
- Glucose homeostasis
- Bile production
- Breakdown and detoxificaion of drugs and waste products
- Macromolecule synthesis and storage

Figure 15.11 The liver is an accessory digestive organ.

Prior to food consumption and the initiation of digestion, the liver plays an essential role in whole-body glucose homeostasis, supplying glucose to the blood for uptake by the tissues of the body. Upon food ingestion, concentrations of small molecules derived from carbohydrate, protein, and/or fat digestion increase in the blood. These molecules are delivered to the liver for processing. In addition, insulin is released by the pancreas (discussed in more detail in Concept 15.2.03), which signals the liver to take up glucose and reduce the production of glucose from gluconeogenesis and glycogenolysis.

As the site of bile production, the liver is crucial for fat digestion (Figure 15.12). **Bile** (also discussed in Concept 15.2.02) is a solution released into the duodenum of the small intestine to help mechanically (ie, by physical, nonenzymatic means) digest lipids by breaking down large lipid globules into micelles (smaller droplets) during emulsification.

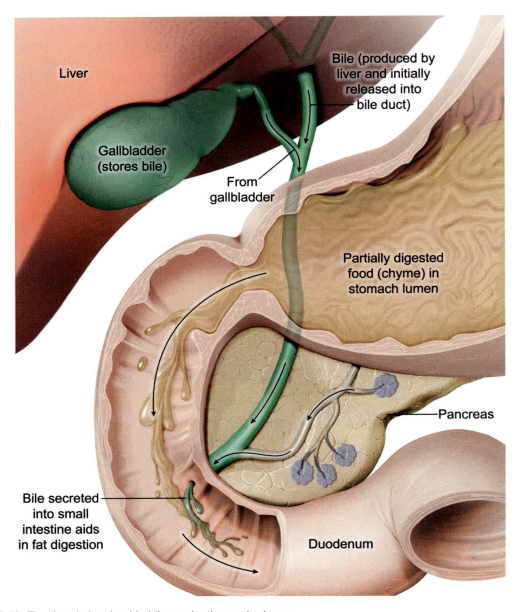

Figure 15.12 The liver is involved in bile production and release.

The liver is also an important organ in the breakdown and detoxification of many drugs and waste products. Drugs are exogenous compounds that may have toxic effects when present in the body at elevated concentrations. The body produces endogenous waste products (eg, bilirubin, ammonia) that must be modified in the liver to avoid adverse effects.

Macromolecules such as plasma proteins (eg, clotting factors and albumin), fats, ketone bodies, and cholesterol are produced by liver cells. Additionally, the liver stores several molecules (eg, glycogen), minerals (eg, iron), and vitamins.

15.2.02 Gallbladder

The gallbladder is an accessory digestive organ. It does not synthesize molecules (eg, digestive compounds, enzymes); rather, the gallbladder is the storage reservoir for bile produced by the liver.

Bile is an alkaline fluid that facilitates fat digestion. Cholesterol, bile acids, and bile pigments (eg, bilirubin) are contained within bile. Before bile acids are released from the liver, they are conjugated with additional compounds to form bile salts, which are amphipathic (ie, containing hydrophobic and hydrophilic regions), a property essential for fat digestion.

On a molecular level, hydrophobic regions of bile salts associate with fat globules, whereas hydrophilic regions associate with the aqueous environment. This allows bile salts to act as detergents and break down large lipid globules into smaller spherical structures called **micelles** (Figure 15.13). The formation of small micelles from larger lipid globules serves to increase lipid surface area for hydrolysis by lipases.

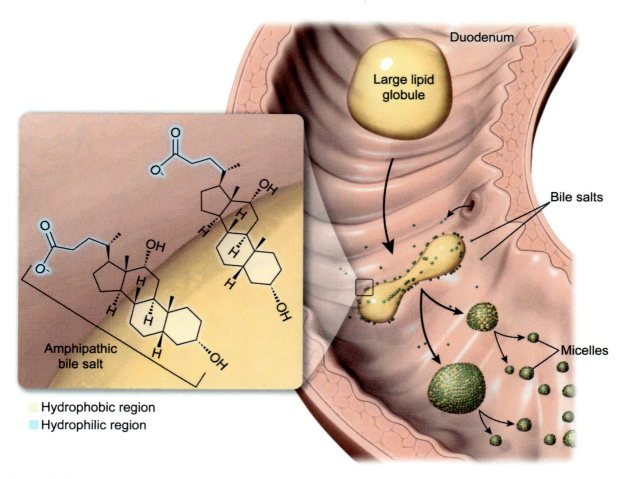

Figure 15.13 Emulsification of lipids by bile salts.

In the duodenum of the small intestine, the presence of meal-derived fats within chyme and the acidity of chyme itself stimulate bile release from both the liver and the gallbladder. Bile secretion into the duodenum promotes the neutralization of chyme and the physical digestion of fats (ie, emulsification). Emulsification is an example of mechanical digestion, a nonenzymatic process that physically breaks down food particles into smaller pieces.

15.2.03 Pancreas

The pancreas, composed of various cell types (eg, alpha cells, beta cells, delta cells), has **paracrine**, **exocrine**, and **endocrine** functions. Cells with paracrine function secrete substances that exert effects on neighboring cells, and cells with exocrine function secrete substances (eg, saliva, sweat, enzymes) through a duct and onto an epithelial surface. In comparison, cells with endocrine function secrete hormones into the bloodstream to cause an effect in a different part of the body.

Endocrine cells of the pancreas secrete insulin and glucagon, which are hormones involved in the regulation of blood glucose, into the bloodstream (Figure 15.14). Exocrine cells of the pancreas secrete digestive enzymes and bicarbonate into the small intestine to assist in digestive processes and to neutralize the acidity of chyme.

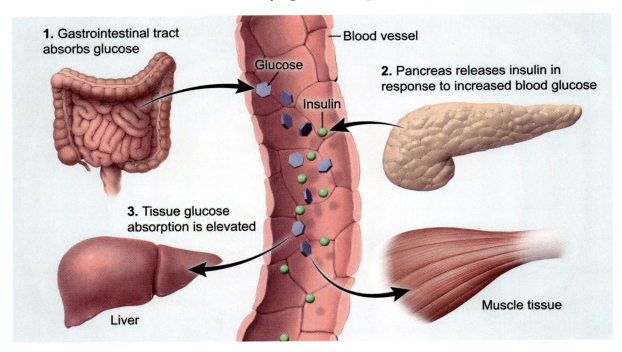

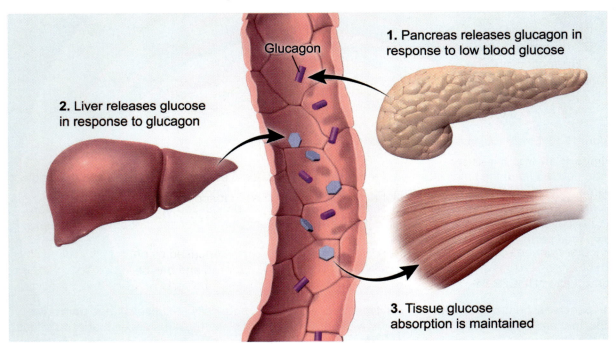

Figure 15.14 Involvement of the pancreas in the regulation of blood glucose.

The control of blood glucose is important in the maintenance of homeostasis. Food ingestion increases the blood glucose level and stimulates pancreatic beta cells to release **insulin**. Insulin is a glucose-regulating hormone that decreases the concentration of glucose in the blood. Insulin regulates glucose concentration in the blood by *promoting* glucose uptake by tissues (eg, adipose, muscle) and decreasing liver glucose production (ie, via gluconeogenesis, glycogenolysis), while at the same time promoting glucose storage in glycogen molecules (ie, glycogenesis).

Insulin sensitivity refers to the biological response elicited by a fixed quantity of insulin. An organism is considered insulin-sensitive if only a minimal amount of insulin is needed to induce a reduction in blood glucose levels. In contrast, insulin-resistant organisms need substantially more insulin to stimulate cells to take up the same amount of glucose from the blood (Figure 15.15).

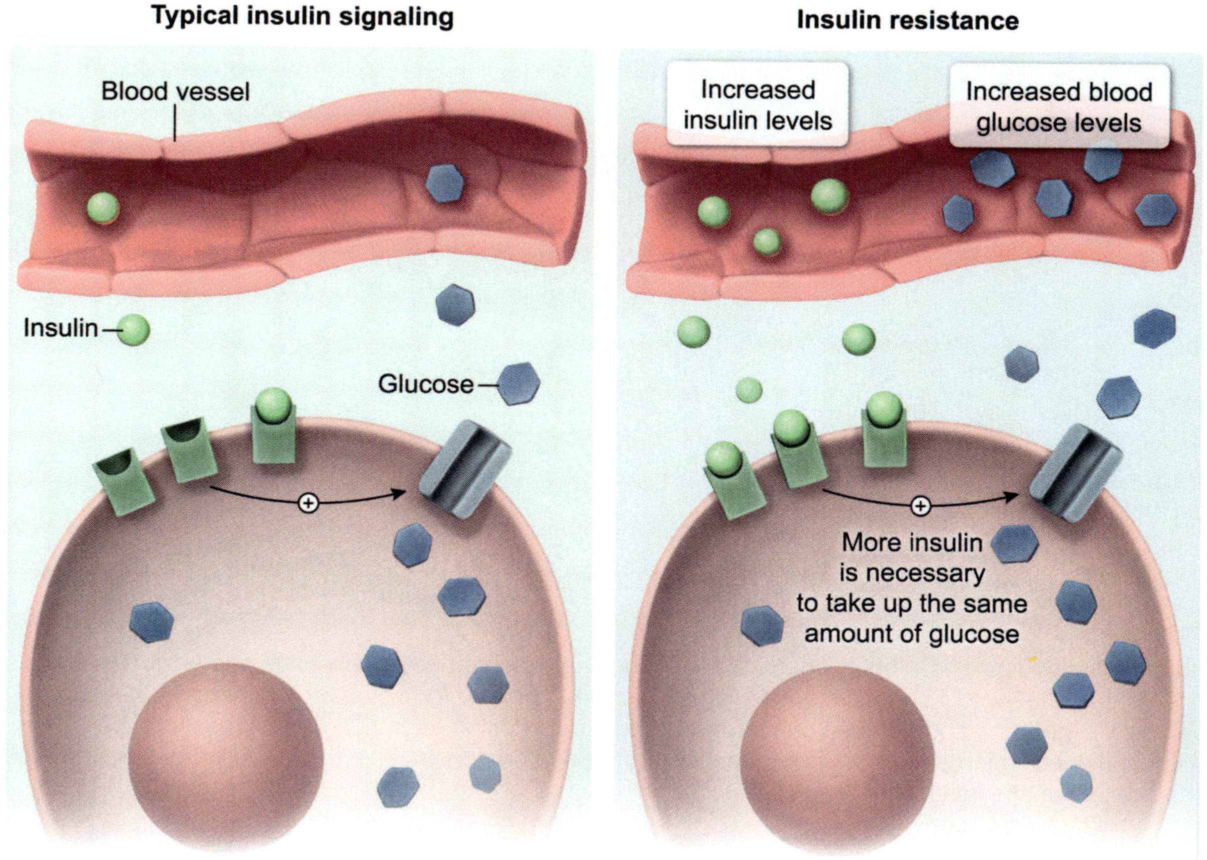

Figure 15.15 Insulin resistance.

Glucagon, a hormone released by alpha cells of the pancreas in response to low blood glucose levels, opposes the effects of insulin. Release of glucagon results in increased gluconeogenesis and glycogenolysis and decreased glycogenesis.

Another hormone important to digestion is **somatostatin**, which is released by pancreatic delta cells. This hormone has a generalized inhibitory effect on digestive function and has been shown to suppress insulin and glucagon release. Figure 15.16 summarizes the cells of the pancreas and their products.

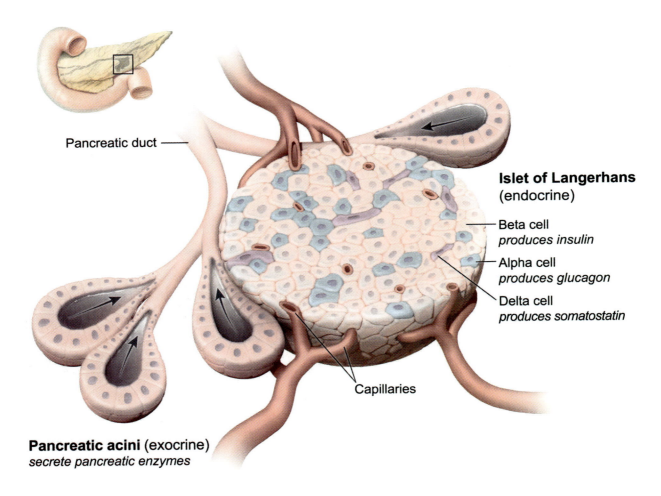

Figure 15.16 Cells of the pancreas and their products.

The exocrine pancreas assists in chemical digestion, which is carried out by acids and enzymes and involves the cleavage of chemical bonds within macronutrients to form simpler compounds that can be absorbed. The formation of these compounds occurs through the secretion of enzymes into the pancreatic duct, which empties into the duodenum. These enzymes aid in the digestion of chyme in the lumen of the small intestine.

Enzymes secreted by the pancreas include **pancreatic lipase**, which acts to chemically digest lipids, and **pancreatic amylase**, which facilitates polysaccharide hydrolysis to form disaccharides. The pancreas also secretes proteolytic digestive enzymes (eg, trypsinogen, chymotrypsinogen). A duodenal enzyme, enteropeptidase, converts the zymogen **trypsinogen** to its active form, **trypsin**. Trypsin activates other pancreatic zymogens (eg, converts **chymotrypsinogen** to **chymotrypsin**) and functions in continued peptide digestion.

Concept Check 15.2

Complete the table to match each accessory digestive organ (ie, "gallbladder," "pancreas," and "liver") with its function. Each organ may be used more than once, and some functions may have more than one related organ.

Accessory digestive organ(s)	Function
	Lipid digestion
	Secretion of trypsinogen and chymotrypsinogen
	Production of bile
	Detoxification of drugs and waste products
	Secretion of somatostatin
	Glucose homeostasis
	Storage of bile

Solution

Note: The appendix contains the answer.

Lesson 15.3
Control of Digestion

Introduction

Both digestion and the subsequent absorption of ingested nutrients are complex processes involving many different organs. Therefore, nutrient digestion and absorption are tightly regulated by both the endocrine system (via hormones) and the nervous system (via innervation). This lesson covers endocrine control of digestion and the enteric nervous system.

15.3.01 Endocrine Control of Digestion

The balance between intake and utilization of dietary nutrients is regulated by numerous factors, including hormones released from cells throughout the body (eg, intestine, adipose tissue, pancreas). Some of these hormones promote desire for food intake (ie, appetite), whereas others promote a feeling of satiety (ie, fullness or dietary satisfaction, Figure 15.17).

Leptin is one hormone that responds to changing energy availability to influence nutrient intake and metabolism. When the body is in an energy-rich state (eg, after a meal), the hormone leptin is released by white adipose tissue. Elevated adipose tissue stores are an indicator of elevated long-term energy stores and, in general, the greater the adipose tissue stores, the higher the leptin levels in the serum. Leptin release triggers feelings of satiety by communicating to the hypothalamus that energy stores are elevated, thereby suppressing appetite.

In contrast, a fasting or energy-poor state triggers release of the hormone **ghrelin** by gastric cells in the stomach wall. Ghrelin release triggers feelings of hunger by communicating to the hypothalamus that energy stores are diminishing, thereby increasing appetite and triggering food-seeking behavior.

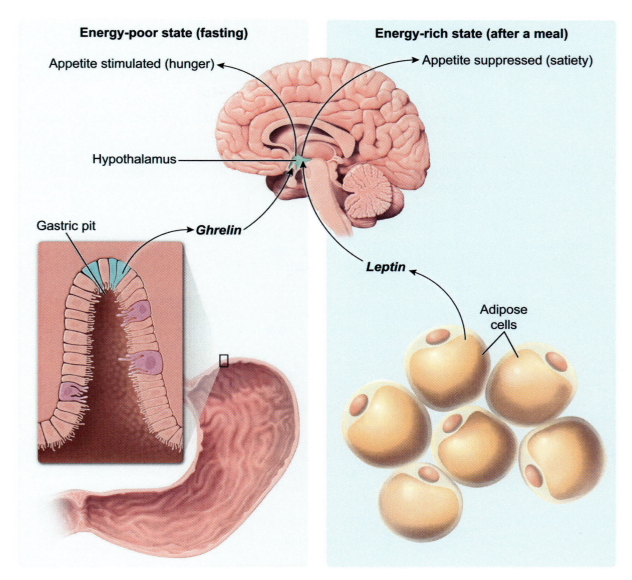

Figure 15.17 Ghrelin and leptin are hormones involved in appetite regulation.

Other hormones involved in the digestive system are discussed in other lessons and are summarized in Table 15.2.

Table 15.2 Hormones with functions in digestion.

Hormone	Site of secretion	Site of function	Function
Gastrin	G cells of the stomach wall	Parietal cells of the stomach wall	Signals parietal cells to release HCl; promotes stomach motility
Secretin	Duodenal epithelial cells	Pancreas, stomach	Promotes pancreatic enzyme and bicarbonate release into duodenum; inhibits HCl secretion by parietal cells
Cholecystokinin (CCK)	Duodenal epithelial cells	Pancreas, gallbladder, stomach	Promotes pancreatic enzyme and bile release; decreases stomach motility; promotes satiety
Insulin	Pancreatic beta cells	Adipose, muscle, liver	Promotes glucose uptake by tissues; decreases liver glucose production; promotes liver glycogen formation
Glucagon	Pancreatic alpha cells	Adipose, liver	Promotes increased gluconeogenesis and glycogenolysis; decreases glycogenesis; promotes fat release into bloodstream
Somatostatin*	Pancreatic delta cells	Pancreas	Generalized inhibitory effect on digestive function; suppresses insulin and glucagon release
Leptin	White adipose tissue	Hypothalamus	Triggers feelings of satiety; supresses appetite
Ghrelin	Endocrine cells of the stomach wall, pancreas	Hypothalamus	Triggers feelings of hunger; increases appetite

* Somatostatin is also produced by the hypothalamus, where its release inhibits growth hormone secretion by the anterior pituitary

15.3.02 Enteric Nervous System

The **enteric nervous system** (ENS) is a specialized division of the nervous system found within the digestive system. As shown in Figure 15.18, the ENS consists of the **submucosal nerve plexus** and the **myenteric nerve plexus**. These plexuses innervate the walls of the gastrointestinal tract and regulate digestive function by controlling digestive secretions and triggering peristalsis.

Chapter 15: Digestion

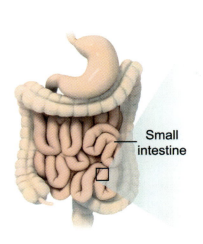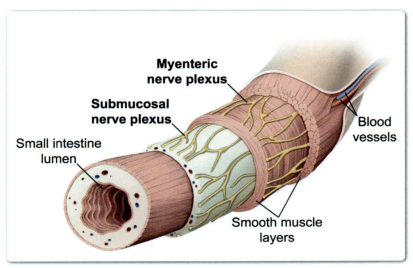

Figure 15.18 The enteric nervous system consists of the submucosal and myenteric nerve plexuses.

The ENS does not require input from the central nervous system; however, the autonomic nervous system can modulate ENS function. Specifically, the parasympathetic division of the autonomic nervous system stimulates digestive activity by modulating ENS function. In contrast, the sympathetic division of the autonomic nervous system inhibits digestive activity by decreasing both gut motility and blood flow to the gastrointestinal tract in the face of an immediate stressor.

Lesson 16.1
Excretory System Structure

Introduction

Excretion involves disposing of wastes from the body. While both the digestive system (see Chapter 15) and the skin system (see Chapter 19) are involved in excretion, this chapter focuses on the specific role the kidneys play in waste excretion and how this waste is removed from the body. This lesson covers the structures of the excretory system.

16.1.01 General Excretory System Structure

The **excretory system** is responsible for producing, storing, and excreting urine, the waste solution produced by the kidneys. Urine is produced as blood is filtered in the **kidneys** by individual subunits known as **nephrons**. Liquid urine is funneled from each kidney into paired tubes known as **ureters**, which carry the urine to the **bladder**. The bladder then stores urine for excretion, which occurs when urine travels through the **urethra** as it exits the body during urination. These excretory system structures are depicted in Figure 16.1.

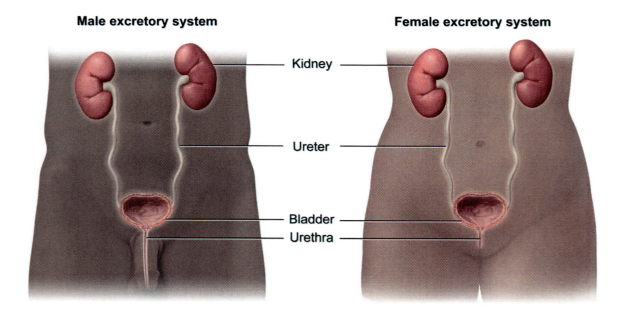

Figure 16.1 Structures of the excretory system.

16.1.02 Kidney Anatomy

The kidneys are anatomically divided into an outer **cortex** and an inner **medulla**, both of which contain portions of **nephrons**, the functional units of the kidney. Nephrons filter blood and selectively excrete or reabsorb the contents of the resulting filtrate (see Concept 16.2.02 for more information on nephron function).

Filtrate produced in nephrons exits the renal medulla at **renal papillae**, which are the tips of pyramid-shaped structures known as medullary pyramids. Filtrate flows from the papillae into spaces called

calyces, which come together to form the **renal pelvis**. From the renal pelvis, paired **ureters** transport filtrate to the bladder for storage prior to excretion from the body as urine (Figure 16.2).

Entering and exiting each kidney near the renal pelvis are the **renal artery** and **renal vein**. Blood flow to the kidneys is discussed further in Concept 16.1.03.

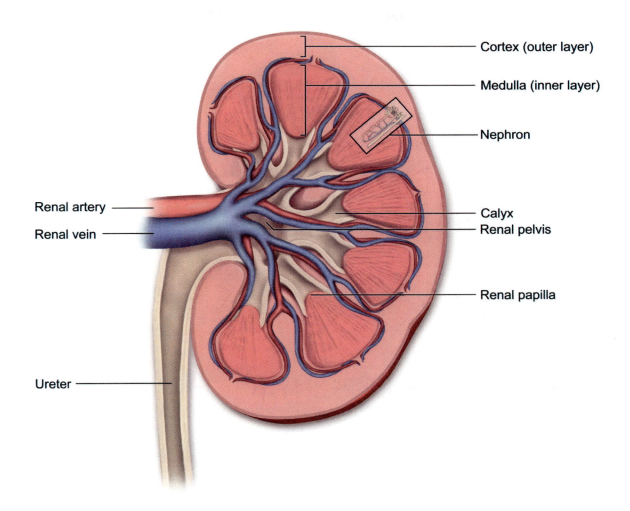

Figure 16.2 Kidney anatomy.

16.1.03 Nephron Structure

Within each kidney are numerous functional units known as nephrons (Figure 16.3), and each of these nephrons is composed of a glomerulus and associated tubular segments. The **glomerulus**, found in the renal cortex, is a ball-like network of capillaries that receives and filters blood from the renal arteries. This filtered blood, known as filtrate, enters the **tubular segments** of the nephron, where the concentration of nutrients, electrolytes, water, and other substances is adjusted to ultimately produce urine. The tubular segments can span both the cortex and the medulla.

Surrounding each nephron tubule are capillaries (the **peritubular capillaries**). During urine formation, the peritubular capillaries allow the exchange of water and other molecules between the filtrate and the blood. The capillaries around medullary tubular segments are collectively known as the **vasa recta**.

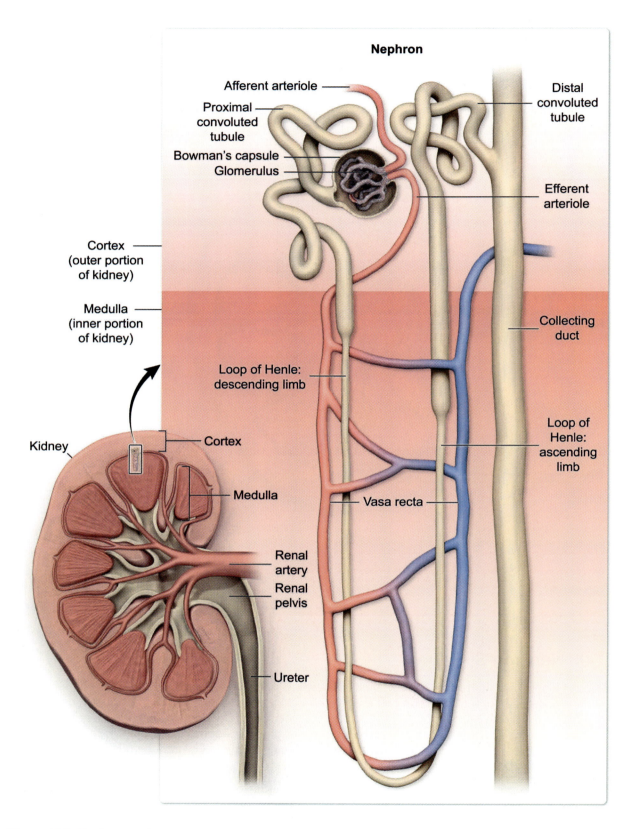

Figure 16.3 Nephron anatomy.

Blood enters each kidney via a renal artery, which branches into smaller arterioles. The arterioles entering each nephron are known as **afferent arterioles**, which branch further to form glomerular capillaries. Glomerular capillaries contain large pores, which are the first part of the three-part system (sometimes collectively referred to as the **filtration membrane**) by which blood passing through a glomerulus is filtered to produce filtrate.

Glomerular capillary pores are large and serve as a coarse filter that excludes blood cells and large proteins from the rest of the blood components, which pass freely through. Glomerular capillaries are lined by a basement membrane with closely apposed Bowman's capsule epithelial cells. The basement membrane and epithelial cells serve as the second and third layers of the filtration membrane, respectively, blocking all but the smallest proteins while allowing passage of water and many small solutes (eg, glucose, ions, urea).

Blood components too large to be filtered by the glomerular capillaries remain in circulation, exiting the glomerulus via **efferent arterioles**, as shown in Figure 16.4.

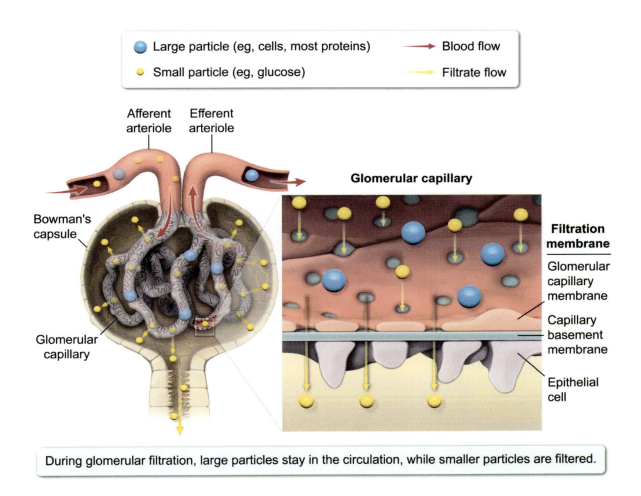

During glomerular filtration, large particles stay in the circulation, while smaller particles are filtered.

Figure 16.4 Blood filtration at the glomerulus is dependent on particle size.

Filtrate passing through the glomerular filtration system is collected by **Bowman's capsule**, a cup-shaped extension of the renal tubule that surrounds the glomerulus, and is delivered into a long tubule composed successively of the following segments:

1. **Proximal convoluted tubule**: The first of the tubular segments, the proximal convoluted tubule delivers filtrate from Bowman's capsule to the loop of Henle.

2. **Loop of Henle**: This segment is a hairpin structure composed of two limbs. The descending limb moves filtrate down into the medulla, while the ascending limb transports filtrate out of the medulla and back to the cortex. Some nephrons have long loops of Henle that extend into the inner region of the medulla. These nephrons are known as **juxtamedullary nephrons**. Other nephrons, called **cortical nephrons**, have shorter loops of Henle that extend only into the superficial portions of the medulla (Figure 16.5).
3. **Distal convoluted tubule**: This is a tubular segment that conveys filtrate from the loop of Henle to the collecting duct.
4. **Collecting duct**: Sometimes also referred to as the collecting tubule, this is the tube into which fluid from the distal convoluted tubule flows. One collecting duct may serve several nephrons.

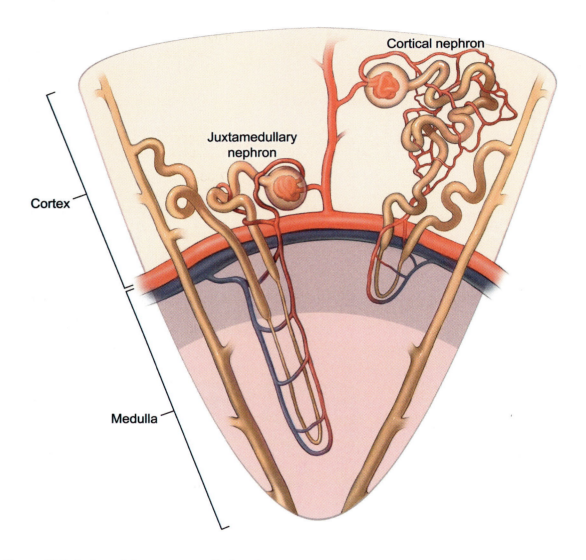

Figure 16.5 Juxtamedullary versus cortical nephrons.

Further details about the function of each of the nephron portions, the influence of hormones, and urine formation are discussed in Lesson 16.2.

Lesson 16.2
Excretory System Function

Introduction

Homeostasis, or a stable internal environment, is vital to maintain in the body. The excretory system (and the kidneys in particular) plays a large role in the maintenance of homeostasis through the regulation of fluid volume and composition, hormone production, and waste removal.

In addition, the functions of the excretory system are largely linked to the regulation of urine formation and excretion. This lesson details the various functions of the excretory system in addition to how urine is formed and eliminated from the body.

16.2.01 Kidney Function

By contributing to the regulation of osmotic balance, blood pressure, hormone production, waste removal, and blood pH, the kidneys play a large role in the maintenance of homeostasis in the body.

Blood pH (ie, H^+ concentration) is maintained within a narrow range (7.35–7.45), primarily through regulatory mechanisms in the kidneys and lungs, and is influenced by carbon dioxide (CO_2) levels. Most CO_2 transported in blood combines with water to form carbonic acid (H_2CO_3), which rapidly dissociates into bicarbonate (HCO_3^-) and H^+:

$$CO_2 + H_2O \rightleftharpoons H_2CO_3 \rightleftharpoons HCO_3^- + H^+$$

This reaction is known as the **bicarbonate buffer system** and shows that CO_2 and H^+ levels are related in the body. Because it is an equilibrium system, the bicarbonate buffer system adjusts when perturbed according to Le Chatelier's principle. Therefore, the system can aid in maintaining blood pH. Various homeostatic regulatory mechanisms (eg, changes in ventilation patterns in the lungs) are based on the bicarbonate buffer system.

Lower blood CO_2 concentrations decrease blood acidity (ie, increase pH) due to decreased H^+ production. Conversely, higher blood CO_2 concentrations increase blood acidity (ie, decrease pH) due to increased H^+ production. In physiological settings, the acid-base response to a loss of H^+ is functionally similar to a gain of HCO_3^- (ie, both result in increased HCO_3^-), and vice versa.

The kidneys typically reabsorb most of the bicarbonate entering the filtrate. However, when extracellular fluid (eg, blood plasma, interstitial fluid) becomes excessively alkaline (high pH), the kidneys excrete HCO_3^- and retain H^+, causing blood pH to decrease, as shown in Figure 16.6.

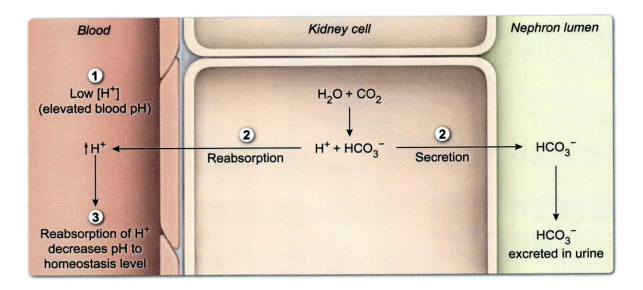

Figure 16.6 Role of the kidneys in acid-base balance when extracellular pH increases.

Conversely, when the extracellular fluid becomes excessively acidic (low pH), the kidneys excrete H^+ and retain HCO_3^-, causing blood pH to rise (Figure 16.7). Via these processes and others, the kidneys act with the lungs to maintain blood pH within the required narrow physiological range.

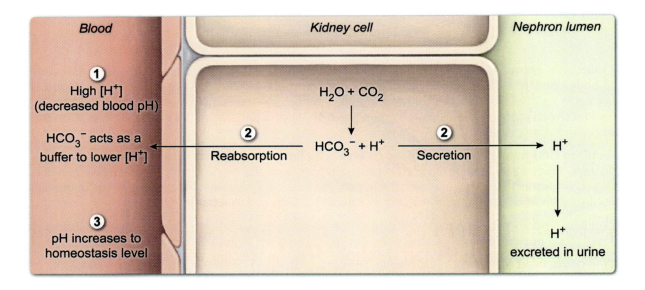

Figure 16.7 Role of the kidneys in acid-base balance when extracellular pH decreases.

The kidneys likewise play an essential role in regulating **blood pressure (BP)**, or the force blood exerts on blood vessel walls. BP decreases when blood vessels dilate (become wider) or when blood volume decreases. Hormones such as angiotensin II, aldosterone, and antidiuretic hormone (ADH; vasopressin) regulate BP by modulating reabsorption of water and salts by the kidneys (Figure 16.8).

Chapter 16: Excretion

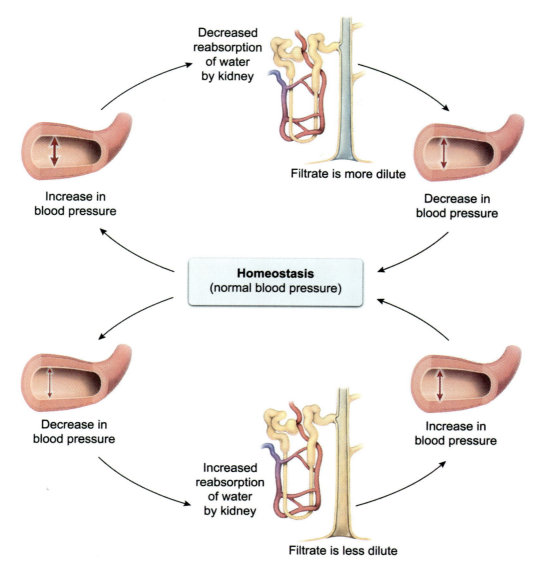

Figure 16.8 Regulation of blood pressure by the kidneys.

The **renin-angiotensin system (RAS)** is a multiorgan molecular cascade activated when BP (or blood volume) falls (Figure 16.9). A drop in BP causes juxtaglomerular cells in the kidneys to release renin, an enzyme that cleaves the plasma protein angiotensinogen to form angiotensin I. Angiotensin-converting enzyme then cleaves angiotensin I to form angiotensin II. Angiotensin II ultimately raises BP by inducing both the constriction of arterioles (increasing BP without changing blood volume) and the release of aldosterone from the adrenal cortex (increasing BP by increasing blood volume, as described next).

Aldosterone is released in response to RAS activation or to an increased serum level of K^+. Aldosterone acts on nephron distal tubules and collecting ducts to promote the reabsorption of Na^+ and the secretion of K^+. Increased reabsorption of Na^+ increases the osmolarity, or solute concentration, of the renal interstitial fluid. Elevated osmolarity promotes water reabsorption, which ultimately causes blood volume and BP to increase.

ADH is a hormone released by the posterior pituitary when BP falls or when blood osmolarity rises. ADH promotes water reabsorption by increasing the permeability of the distal tubule and collecting duct to water. Release of ADH also induces vasoconstriction, the narrowing of blood vessels. Both of these effects increase BP. The contributions of the RAS, aldosterone, and ADH to BP regulation are shown in Figure 16.9.

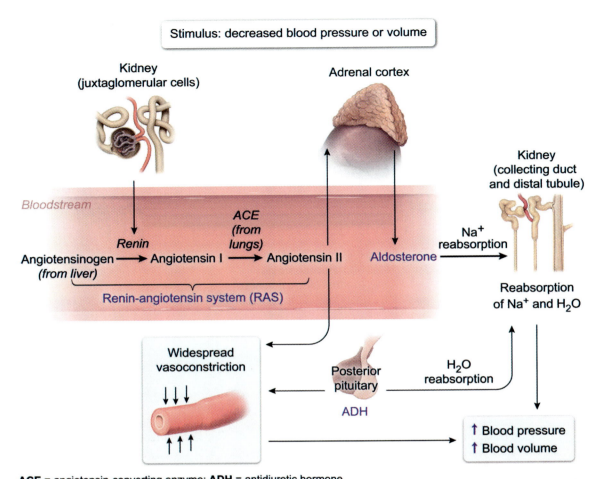

Figure 16.9 The renin-angiotensin system, aldosterone, and ADH act to increase blood volume and blood pressure.

The kidneys are vital to **osmoregulation**, or the process of regulating water and ion content, of the blood. During osmosis, water generally flows from *lower* to *higher* osmotic pressure environments across cell membranes (see Concept 5.2.03). **Osmotic pressure** is the minimum pressure that must be applied to prevent water movement by osmosis; in other words, osmotic pressure describes the extent to which water "wants" to cross a semipermeable membrane by osmosis. This pressure is the result of solute concentration differences on either side of the membrane: A *higher* solute concentration creates a *higher* osmotic pressure.

Osmosis occurs across capillary walls (ie, semipermeable membranes) into the interstitial fluid and vice versa due to differences in the concentrations of solutes (eg, plasma proteins such as albumin). This water movement across capillary walls also occurs because of osmotic pressure, which in the capillaries is referred to as oncotic pressure.

The kidneys help control blood osmolarity by balancing the intake of specific molecules with the excretion of specific molecules in urine. For example, increased water consumption decreases blood osmolarity. To maintain homeostasis, higher concentrations of solutes (eg, Na^+) are reabsorbed and more water is excreted by the kidneys, producing a larger volume of dilute urine. When blood osmolarity is high, the kidneys respond by increasing water reabsorption and decreasing solute reabsorption. In this way, the concentration of water and solutes in the blood stays relatively constant, as shown in Figure 16.10.

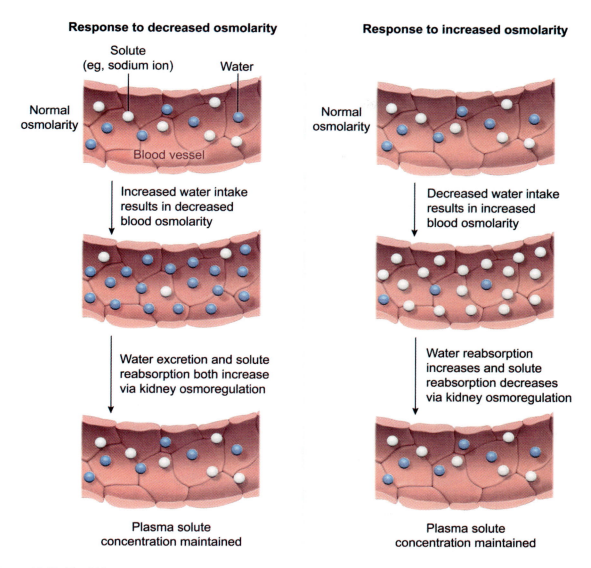

Figure 16.10 The kidneys participate in osmoregulation when blood osmolarity decreases (left panel) or increases (right panel).

In addition to being the target of several hormones (eg, aldosterone, ADH), the kidneys produce a hormone called **erythropoietin (EPO)**. EPO stimulates red bone marrow to produce erythrocytes (ie, red blood cells) when blood oxygen levels fall. This increased production of erythrocytes leads to a rise in the oxygen-carrying capacity of the blood. Conversely, high levels of oxygen or erythrocytes inhibit the production of EPO, promoting homeostasis.

The kidneys filter blood and produce liquid waste (**urine**) that is eventually excreted from the body. The body produces metabolic waste products such as creatinine from muscle metabolism and ammonia (NH_3) from metabolism of compounds containing nitrogen. Because ammonia can disrupt extracellular fluid pH, it is converted to urea in the liver before being delivered to the kidneys.

The kidneys remove metabolic wastes such as urea and foreign substances (eg, drugs, environmental toxins) by collecting these wastes in the filtrate and excreting them in the urine. If the kidneys are unable to excrete waste, the resulting waste accumulation becomes toxic and can cause serious medical complications (eg, increased blood nitrogen levels due to lack of urea excretion). The details of urine formation by nephrons are covered in Concept 16.2.02.

16.2.02 Urine Formation

Within each kidney are functional units known as **nephrons**, which perform blood **filtration** and concentrate urine. During urine formation, nephrons facilitate the removal of some solutes and waste products from the blood into filtrate (ie, fluid filtered into the nephron) as well as the **reabsorption** of useful solutes back into the blood from the filtrate (Figure 16.11). Waste products not filtered in the glomerulus are transferred from the blood to the filtrate in nephron tubules (**secretion**). **Renal clearance** is a measure of the ability of the kidneys to remove substances from the bloodstream to be excreted.

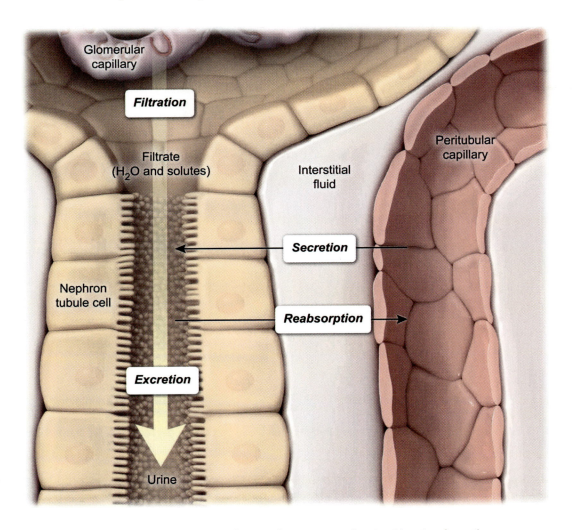

Figure 16.11 Filtration, secretion, and reabsorption are the processes involved in urine formation.

Blood enters the kidneys via the renal arteries, which branch into smaller arterioles and, ultimately, capillaries of the glomerulus. Glomerular capillaries are lined by a basement membrane that allows passage of water, many solutes (eg, glucose, ions, urea), and very small proteins into the filtrate. The filtrate is collected by Bowman's capsule, which surrounds the glomerulus. Materials not filtered by the glomerulus due to their larger size (eg, blood cells, large proteins) exit the glomerular capillaries. Additional details about blood flow and filtration in nephrons can be found in Lesson 16.1.

The degree of vascular constriction of afferent and efferent arterioles controls the rate of blood flow into and out of the glomerulus, thereby affecting the volume of fluid filtered through the kidneys per unit time, known as the **glomerular filtration rate (GFR)**. These effects are detailed in Figure 16.12.

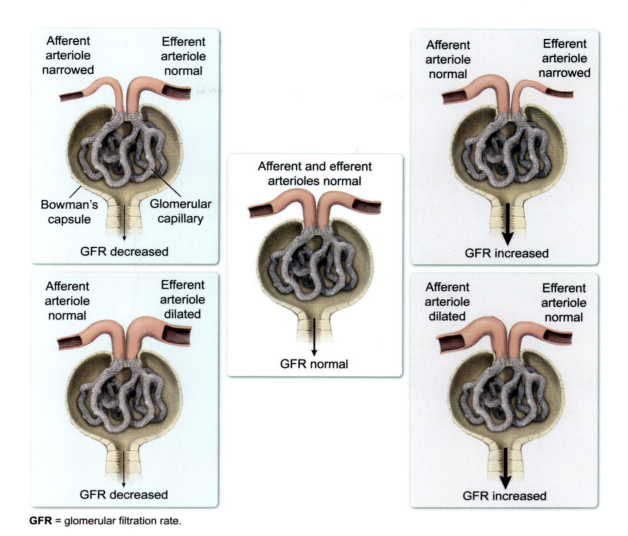

GFR = glomerular filtration rate.

Figure 16.12 Effect of dilation and constriction of afferent and efferent arterioles on glomerular filtration rate.

Within the glomerulus, several pressures affect the movement of fluid from the glomerular capillaries into Bowman's capsule. The hydrostatic pressure of glomerular blood (GBHP) tends to push fluid out of the capillaries, while the hydrostatic pressure of Bowman's capsule fluid (BCHP) tends to push fluid into the capillaries. Because GBHP is higher than BCHP, there is a driving force for fluid to flow from the capillaries into Bowman's capsule.

Oncotic pressure (see Concept 16.2.01) also affects fluid movement in the glomerulus. The oncotic pressure of the filtrate in Bowman's capsule (BCOP) tends to pull fluid from the capillaries, while the glomerular blood oncotic pressure (GBOP) tends to pull fluid into the capillaries. Because larger proteins (eg, albumin) are not filtered and remain in the glomerular capillaries, GBOP is higher than BCOP. Therefore, net oncotic pressure tends to decrease fluid flow out of the capillaries, opposing the net hydrostatic pressure. The pressures affecting fluid movement in the glomerulus are summarized in Figure 16.13.

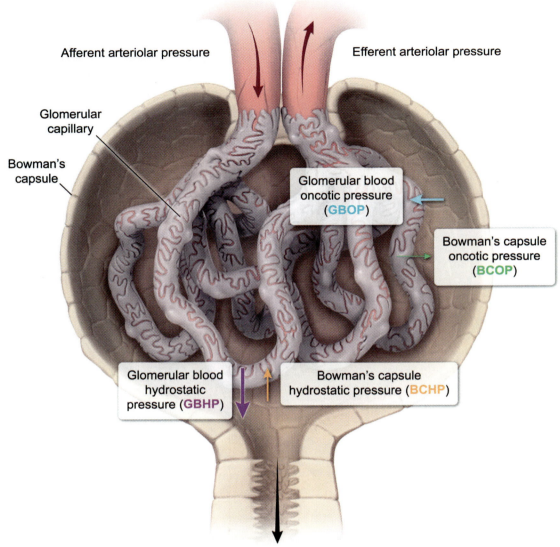

Figure 16.13 Pressures in the nephron glomerulus.

Filtrate travels from Bowman's capsule to the **proximal convoluted tubule (PCT)**, where important nutrients (eg, amino acids, vitamins, salts, glucose, water) are reabsorbed from the nephron and incorporated back into the blood through the peritubular capillaries surrounding the PCT. Waste products not filtered by the glomerular capillaries are actively secreted from the peritubular capillaries into the nephron.

Next, filtrate flows into the **loop of Henle**, a hairpin structure composed of two limbs. The descending limb extends down into the relatively salty medulla (inner portion of the kidney) and is highly *permeable* to water but *impermeable* to NaCl. The ascending limb, which travels from the loop's lowest point in the medulla back toward the cortex (outer portion of the kidney), is *impermeable* to water but *permeable* to NaCl. These differential permeability characteristics in the two limbs act together to form a concentration gradient within the loop of Henle that maximizes water reabsorption, as described next.

Water moves from areas of low solute concentration to areas of high solute concentration. Accordingly, because salt concentration in the medulla is *high*, water is passively reabsorbed (via osmosis) from the filtrate flowing through the descending limb into the salty medulla, where the water is taken up by blood vessels (vasa recta). As the water is passively reabsorbed via osmosis into the medulla, the filtrate becomes more concentrated.

The ascending limb transports filtrate from the medulla and into the cortex. NaCl is first passively reabsorbed into the medulla as the filtrate travels up the ascending limb. As the limb nears the cortex, NaCl is actively transported out of the filtrate and into the medulla, maintaining the medulla's high salt concentration and facilitating continued water reabsorption in the descending limb. Because water follows NaCl, the saltiness of the medulla promotes continued water reabsorption from the descending limb (and the collecting duct). This full **countercurrent multiplication system** is depicted in Figure 16.14.

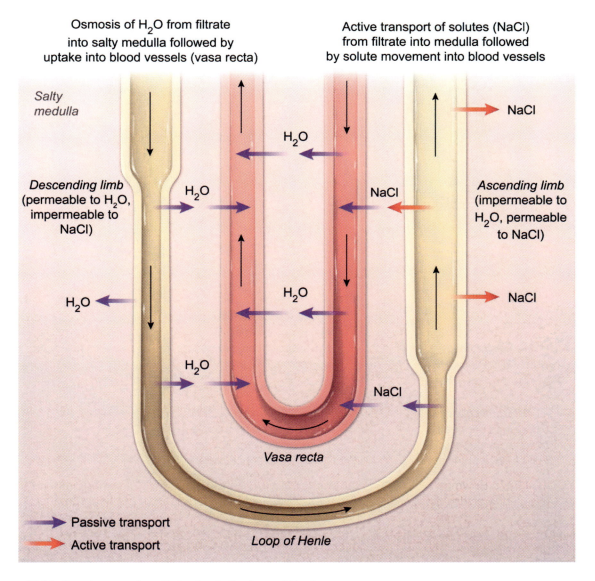

Figure 16.14 Countercurrent multiplication in the loop of Henle.

From the loop of Henle, filtrate enters the **distal convoluted tubule (DCT)**, where aldosterone and antidiuretic hormone (ADH; vasopressin) promote the reabsorption of water. Aldosterone is released from the adrenal cortex in response to increased serum potassium (K^+) and low blood pressure. In response to aldosterone, nephrons reabsorb sodium (Na^+) and secrete K^+; however, more Na^+ is reabsorbed than K^+ is secreted. This generates an ion gradient between the filtrate and the fluid surrounding the nephron, causing water to be reabsorbed into the interstitial fluid (and eventually the bloodstream) via osmosis.

When blood osmolarity increases (eg, through the actions of aldosterone), ADH is secreted from the posterior pituitary. ADH increases the permeability of the DCT and collecting duct to water, causing even more water to be reabsorbed. The actions of aldosterone and ADH promote an increased blood volume and blood pressure and cause urine to become more concentrated, decreasing the final excreted urine volume. Additional waste products are secreted into the DCT from the peritubular capillaries. Figure 16.15 summarizes the effects of aldosterone and ADH on urine formation.

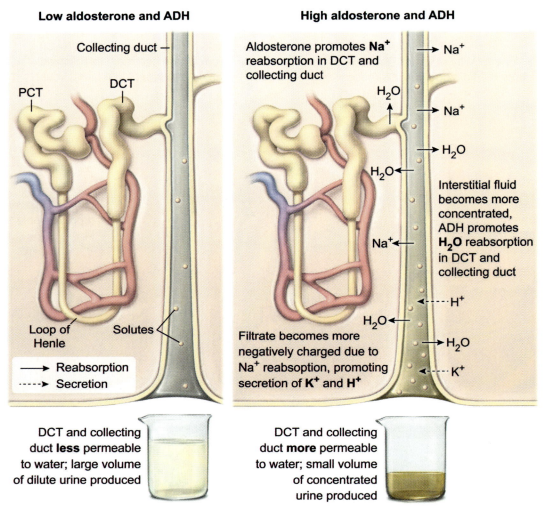

PCT = proximal convoluted tubule; **DCT** = distal convoluted tubule; **ADH** = antidiuretic hormone.

Figure 16.15 Effect of aldosterone and ADH on urine formation.

Finally, in the **collecting duct**, ADH and aldosterone act in the same way they do in the DCT, further concentrating urine and reducing its volume. Based on the state of the body, the amount of water reabsorbed varies; for example, when the body is hydrated, less water is reabsorbed and the volume of excreted urine is greater. Conversely, when the body is dehydrated, more water is reabsorbed, reducing urine volume and further concentrating the urine exiting the body. Reabsorbed water reenters the circulation via the peritubular capillaries.

Once concentration in the collecting duct is finalized, urine is emptied into the ureters for eventual excretion. Figure 16.16 summarizes the location and function of each section of the nephron.

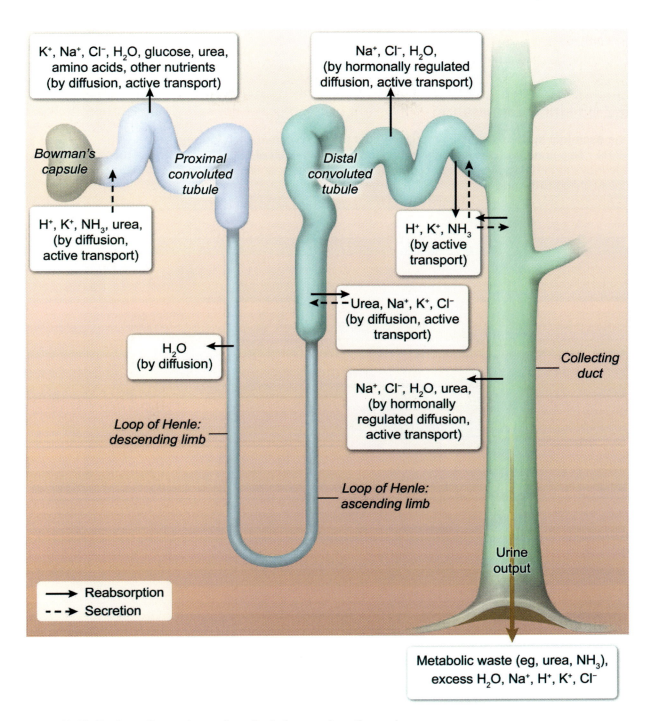

Figure 16.16 Reabsorption and secretion of substances along the nephron.

> ✓ **Concept Check 16.1**
>
> After purchasing a new reusable water bottle, a student finds that he is drinking much more water each day than he was previously. Which portion(s) of the nephron are most impacted by the student's lifestyle change? How would these portion(s) respond to the student's increased water intake?

> **Solution**
>
> *Note: The appendix contains the answer.*

16.2.03 Urine Elimination

In the excretory system, **urine** is produced in the nephrons within each **kidney**, as detailed in Concept 16.2.02. This urine from the kidneys is then funneled into each of the two **ureters**, the tubes connecting the kidneys to the **bladder**. Urine accumulates in the bladder until it exits the body via the urethra in a process called **urination** (Figure 16.17).

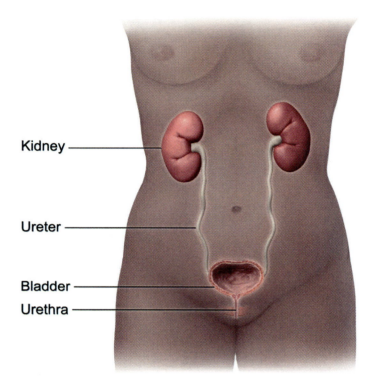

Figure 16.17 Components of the excretory system.

Urination is under the control of both smooth and skeletal muscle along the urinary tract (the system for removing urine from the body). Specifically, a layer of smooth muscle called the **detrusor muscle** lines the bladder and is relaxed while urine is stored in the bladder. The detrusor muscle is under involuntary control by the autonomic nervous system.

While urine is being stored, two sphincter muscles remain contracted so urine cannot exit the body. The proximal sphincter muscle is known as the **internal urethral sphincter (IUS)**. The IUS is composed of smooth muscle and is under involuntary control, like the detrusor muscle. The distal sphincter, or the **external urethral sphincter (EUS)**, is composed of skeletal muscle and is under voluntary control by the somatic nervous system (Figure 16.18).

During urination, activity within stretch receptors in the bladder leads to contraction of the detrusor muscle to push urine out of the bladder and into the urethra. At the same time, the IUS relaxes to open the urethra and allow urine through. For urine to be excreted from the body, IUS relaxation must be paired with voluntary relaxation of the EUS. Relaxation of the IUS alone will not allow urine excretion.

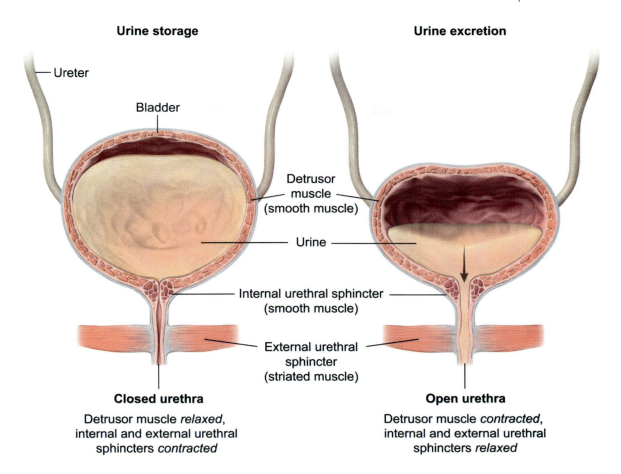

Figure 16.18 Urination requires contraction of the detrusor muscle and relaxation of the internal urethral sphincter and the external urethral sphincter.

END-OF-UNIT MCAT PRACTICE

Congratulations on completing **Unit 8: Digestion and Excretion**.

Now you are ready to dive into MCAT-level practice tests. At UWorld, we believe students will be fully prepared to ace the MCAT when they practice with high-quality questions in a realistic testing environment.

The UWorld Qbank will test you on questions that are fully representative of the AAMC MCAT syllabus. In addition, our MCAT-like questions are accompanied by in-depth explanations with exceptional visual aids that will help you better retain difficult MCAT concepts.

TO START YOUR MCAT PRACTICE, PROCEED AS FOLLOWS:

1) Sign up to purchase the UWorld MCAT Qbank
   IMPORTANT: You already have access if you purchased a bundled subscription.
2) Log in to your UWorld MCAT account
3) Access the MCAT Qbank section
4) Select this unit in the Qbank
5) Create a custom practice test

Unit 9 Musculoskeletal System

Chapter 17 Muscular System

17.1 General Muscle Characteristics

- 17.1.01 Myocytes
- 17.1.02 Myocyte Structure and Function Relationships
- 17.1.03 Muscle Excitation
- 17.1.04 Excitation-Contraction Coupling

17.2 Characteristics of Specific Muscle Types

- 17.2.01 Skeletal Muscle
- 17.2.02 Cardiac Muscle
- 17.2.03 Smooth Muscle

Chapter 18 Skeletal System

18.1 Skeletal System Structure

- 18.1.01 Bone Types
- 18.1.02 Bone Structure
- 18.1.03 Bone Matrix
- 18.1.04 Bone Cells
- 18.1.05 Joints
- 18.1.06 Cartilage

18.2 Skeletal System Function

- 18.2.01 Bone Function
- 18.2.02 Endocrine Control of Skeletal System

Lesson 17.1
General Muscle Characteristics

Introduction

Muscles are tissues composed primarily of cells specialized to produce force through the interaction of actin and myosin proteins. An instance of muscle force production is called a **contraction**. Contraction requires the pumping of ions and powering of the motor protein myosin, both of which increase the energetic requirements of muscle cells. Based on microscopic structural organization, muscle cells (ie, **myocytes**) can be described as striated (ie, cardiac and skeletal muscle) or smooth (Figure 17.1).

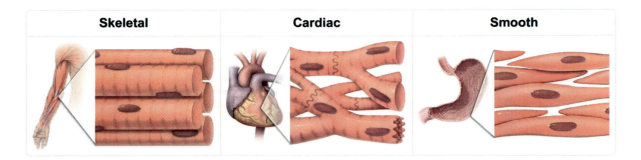

Figure 17.1 Types of muscle cells.

17.1.01 Myocytes

Myocytes are the predominant and defining cell type of muscles. Although myocyte structure varies according to the type of muscle in which myocytes are found, all myocytes share a few notable properties. First, all myocytes are specialized for force production through the interaction of the proteins actin and myosin. Second, force production by myocytes is dependent on elevation of intracellular calcium ion (Ca^{2+}) concentration. Finally, like most eukaryotic cells, myocytes are nucleated, although the precise number of nuclei per myocyte varies.

Certain structures common to eukaryotic cells are sometimes referred to by different names when discussing myocytes (which are often called fibers). For example, the plasma membrane of muscle cells is often called the **sarcolemma**, and muscle cells have a specialized type of endoplasmic reticulum called the **sarcoplasmic reticulum**. Myocyte cytoplasm is sometimes called **sarcoplasm**. Like the cytoplasm of other cells, sarcoplasm has a very low Ca^{2+} concentration at rest. The concentration of Ca^{2+} is much higher in the sarcoplasmic reticulum and in the extracellular fluid surrounding each myocyte (Figure 17.2).

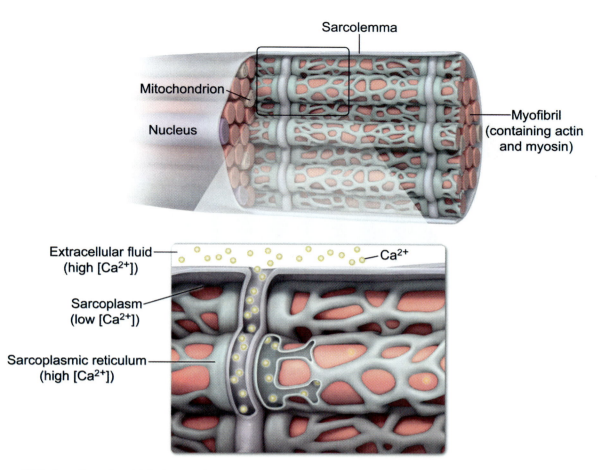

Figure 17.2 A portion of a striated myocyte.

17.1.02 Myocyte Structure and Function Relationships

All muscles produce force through the interaction of **actin** and **myosin** protein molecules. In muscle, actin molecules are linked together with other actin molecules to form filaments of actin; similarly, myosin molecules are linked together with other myosin molecules to form filaments of myosin.

Based on their relative thicknesses, actin filaments are also referred to as **thin filaments**, and myosin filaments are referred to as **thick filaments**. Myosin filaments contain a rod-like central region, composed of myosin tails, as well as heads that extend out from the rod-like region. Connecting the myosin heads to the myosin tails in the central region are thin arms, or hinges (Figure 17.3).

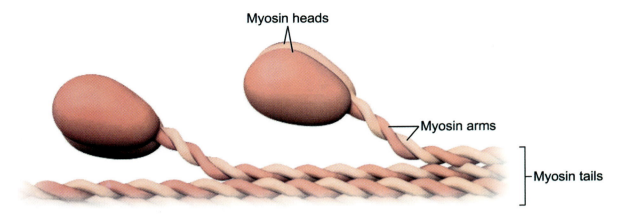

Figure 17.3 Myosin molecules in a thick filament.

Actin and myosin filaments are not organized in the same way in all muscles. In **striated** muscles, actin and myosin filaments are arranged parallel to one another in a way that produces regions of overlapping actin and myosin next to regions without overlap. These two types of regions regularly repeat, with actin filaments aligned with neighboring actin filaments and myosin filaments aligned with neighboring myosin filaments. When viewed under a microscope, this arrangement results in regularly spaced lighter and darker bands called **striations** (Figure 17.4).

In other muscles, actin and myosin filaments are organized differently and lack striations. Because these nonstriated muscles appear relatively homogeneous when viewed under a microscope, they are known as **smooth** muscles (Figure 17.4). Striated muscle cells are long cylindrical fibers, whereas smooth muscle cells, or fibers, are often described as spindle-shaped (ie, wide in the middle and tapered at the ends).

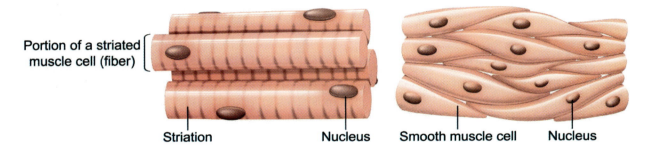

Figure 17.4 Striated versus smooth muscle.

During muscle contraction, myosin and actin filaments interact with one another (Figure 17.5). Specifically, myosin heads bind to and pull upon actin filaments. This pulling generates force along the myosin and actin filaments, and if this force exceeds any opposing forces, the actin filaments slide along the myosin filaments.

This sliding movement shortens the muscle cell along the length of the interacting thick and thin filaments. However, the length of each actin and myosin filament remains *unchanged*. Instead, it is the sliding of the two filament types along one another that causes the shortening, as shown in Figure 17.5. Striated muscle force production is greatest when actin-myosin overlap is maximal.

Relaxation

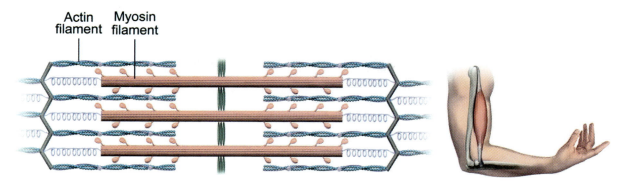

Contraction

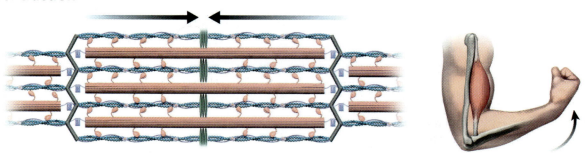

Figure 17.5 Arrangement of actin and myosin filaments during relaxation and contraction of striated muscle.

The banding pattern of striated myocytes is related to the presence of multiple functional units along the length of **myofibrils**, which are tiny cylindrical bundles of interacting thick and thin filaments surrounded by sarcoplasmic reticulum within a myocyte. These functional contractile units are called **sarcomeres** and are arranged in series along the length of a striated muscle cell.

Each sarcomere has thick filaments centered within it. Myosin heads extend out from most of the thick filaments but are absent in the middle of the sarcomere. Actin filaments extend from each end of the sarcomere but do not reach the middle. As a result of this configuration, sarcomeres contain several regions (Figure 17.6):

- The **I band** at each end of a sarcomere consists of only actin (thin) filaments.
- The **H band** in the center of the sarcomere consists of only myosin (thick) filaments.
- The **A band** fully encompasses the H band and extends further to include the regions of actin and myosin overlap; therefore, the A band marks the length of the myosin filaments.
- The **M line** marks the center of each sarcomere.
- The **Z line** is the boundary between neighboring sarcomeres. A large, elastic protein called **titin** is bound to the Z line and helps maintain the ordered arrangement of actin and myosin filaments within the sarcomere.

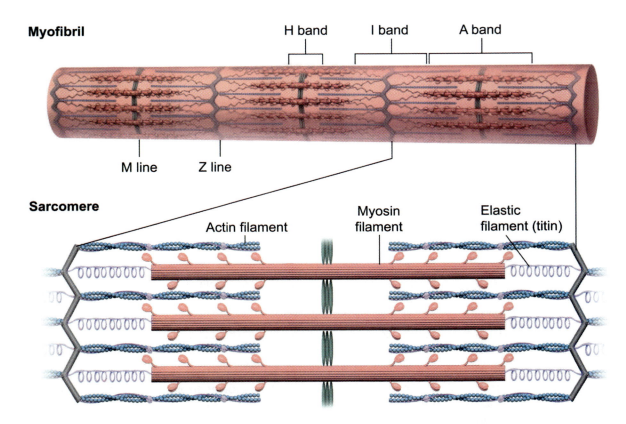

Figure 17.6 Components of a sarcomere.

In contrast to striated muscle cells, smooth muscle cells lack sarcomeres, and thick filaments in smooth muscle cells have myosin heads along their entire length. This allows optimal actin-myosin interaction (ie, optimal force production) in smooth muscle cells over a much greater length than in striated muscle cells. In addition, smooth muscle thick and thin filaments are not aligned in parallel with their respective neighboring filaments but are oriented in multiple directions.

Structural differences between striated and smooth muscle cells lead to important functional differences. Relative to smooth muscle cells, striated myocytes are specialized for relatively rapid, intermittent contractions (eg, beating of the heart or lifting of an object). Smooth muscle cells are specialized for slower and more sustained contractions (eg, constriction of a blood vessel or contraction of an intestinal segment, Figure 17.7). Smooth muscle contractions are much more energetically efficient than striated muscle contractions, as discussed later in this concept.

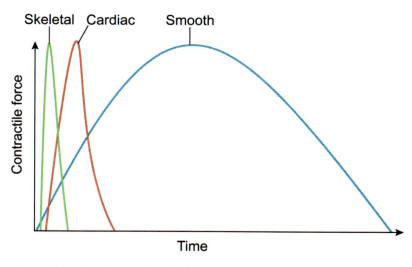

Figure 17.7 Relative rates of force development in skeletal, cardiac, and smooth muscle.

An energetic cost is associated with muscle contraction, and different strategies exist to match energetic supply and demand in striated and smooth muscles. The energetic costs of muscle contraction are primarily due to the movement of actin filaments by myosin and the maintenance of ionic gradients.

ATP hydrolysis by myosin ATPases alters the position of myosin heads, correctly positioning them for the force-generating interaction between actin and myosin during the power stroke of the cross-bridge cycle (described in more detail in Concept 17.1.04). Movement of ions across cellular membranes is necessary for muscle activation, force development, and relaxation. ATP-requiring pumps such as the plasma membrane Na^+/K^+-ATPase and the sarcoplasmic reticulum Ca^{2+}-ATPase (SERCA) use energy released during ATP hydrolysis to pump ions against their concentration gradients.

In striated muscles, more active, fatigue-resistant muscle cells exhibit more capillaries and greater concentrations of O_2-binding myoglobin and mitochondria. These characteristics reflect an increased reliance upon aerobic energy production to meet the energetic demands of repetitive muscle contraction. Like hemoglobin, myoglobin is a heme-containing protein that imparts a reddish color to cells in accordance with its concentration. The heme in myoglobin binds and releases oxygen according to cellular concentrations of oxygen and certain other molecules, thereby acting to stabilize oxygen supply for respiration.

Striated muscles that are only infrequently or briefly contracted can rely upon less energetically efficient, but much faster, means of replenishing cellular ATP stores. These faster alternatives include substrate-level phosphorylation via glycolysis or other high-energy phosphates (eg, creatine phosphate, also known as phosphocreatine, Figure 17.8).

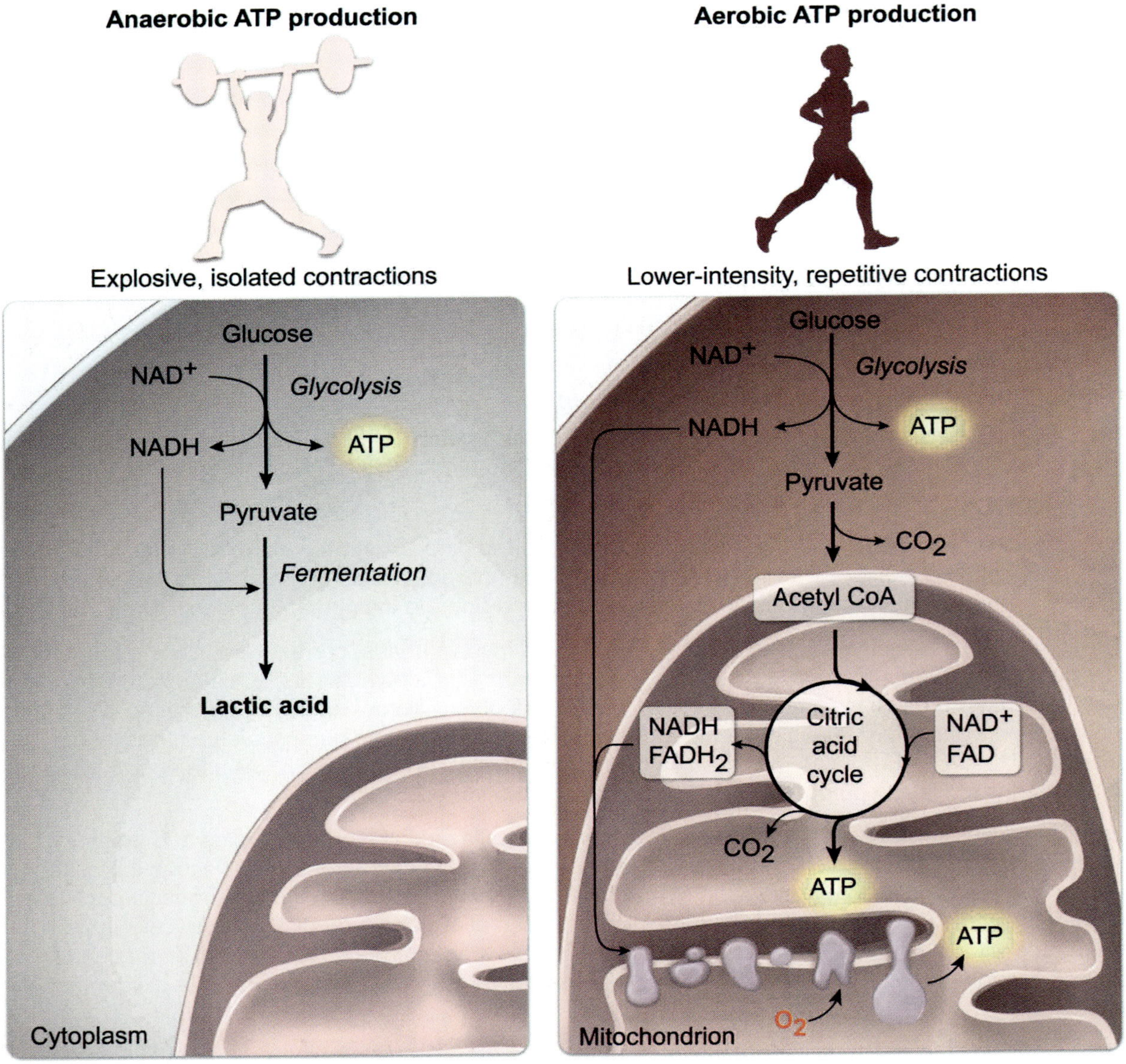

Figure 17.8 Anaerobic and aerobic ATP production in skeletal muscle cells.

In contrast to striated muscle, smooth muscle achieves energetic balance and avoids fatigue by minimizing energetic demands through efficient force production. As a result, smooth muscle typically has less myoglobin and fewer mitochondria than striated muscle, despite prolonged force production in smooth muscle.

17.1.03 Muscle Excitation

The force-producing interaction between actin and myosin filaments requires activation (ie, it is not always "on"), and this activation occurs differently in different muscle types. Usually, activation takes the form of an electrical signal, a wave of depolarization that spreads along and excites the myocyte membrane.

Differences in muscle activation are related to whether contraction of a muscle is under conscious control and whether myocytes within the muscle are electrically connected with each other. **Involuntary**

muscles (ie, smooth and cardiac) carry out automatic functions, while **voluntary muscles** (ie, skeletal) are involved in conscious movement or force production.

Some involuntary muscles have specialized intercellular connections called **gap junctions**. These connections allow the muscle to act as a **syncytium**, or a collection of connected cells effectively sharing cytoplasmic molecules via the pores formed by gap junctions (Figure 17.9). Gap junctions allow the contraction of all cells in a muscle at the same time by electrically connecting the cells. All cardiac muscle cells have gap junctions, allowing for synchronous contraction of the cells in each heart chamber, as do most smooth muscle cells (eg, intestinal smooth muscle cells).

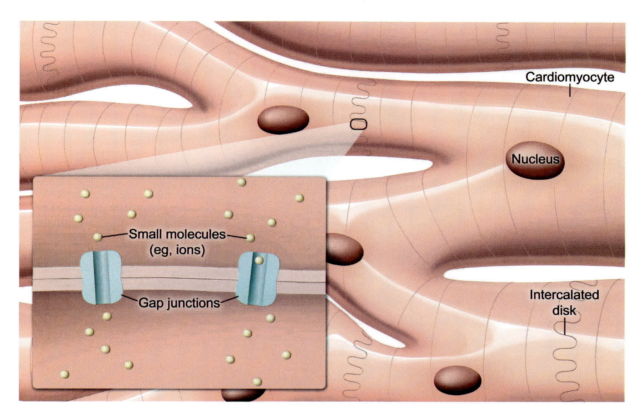

Figure 17.9 Gap junctions in cardiomyocytes (a functional syncytium).

As mediators of automatic, subconscious functions, involuntary muscles are innervated by nerve fibers of the autonomic nervous system (one branch of the peripheral nervous system). In cardiac muscle and certain smooth muscles, the action potentials that cause contraction occur without input from the nervous system (ie, some of the cells spontaneously depolarize). However, neural regulation influences the frequency and intensity of contractions. In other smooth muscles, a nervous impulse is necessary for contraction initiation.

In addition to neural regulation, involuntary muscle contractions can be regulated by certain hormones. For example, epinephrine can promote contraction or relaxation in smooth muscle depending on the type of receptor the epinephrine binds. In addition, some smooth muscle cells can be activated by chemical signaling, in which ligand binding to a receptor activates muscle contraction without depolarization.

Voluntary (ie, skeletal) muscles allow conscious control of body movement, position, and processes. Although defined by their ability to be voluntarily activated, voluntary muscles can also be activated involuntarily, such as during a reflex response. Likewise, during repetitive activities such as breathing or walking, rhythmic contractions can be involuntarily maintained as a special type of reflex movement but can also be influenced by conscious input. Similarly, conscious control of a few voluntary muscles mainly involves a brief interruption of prolonged contraction (eg, conscious relaxation of a sphincter muscle).

Skeletal muscles are innervated by the somatic nervous system. Although all synapses, including those in cardiac and smooth muscle, involve a connection between a neuron and a target cell, the interface between a somatic motor neuron and a skeletal muscle fiber is a specialized synapse known as a **neuromuscular junction (NMJ)**. An NMJ between a motor neuron and a skeletal muscle fiber is illustrated in Figure 17.10.

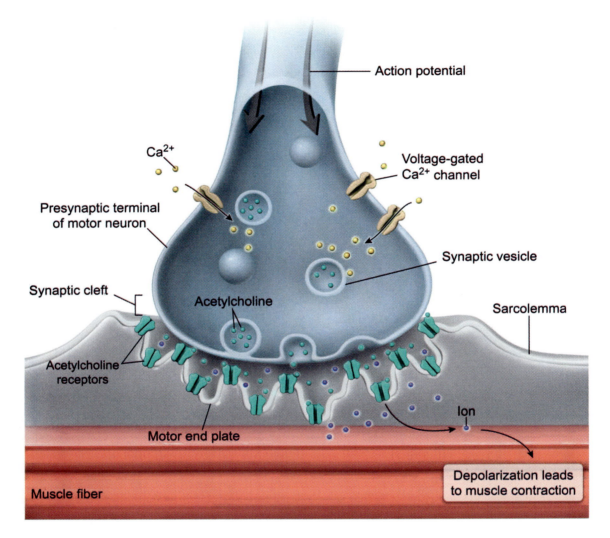

Figure 17.10 A neuromuscular junction.

An NMJ converts motor neuron action potentials into muscle fiber action potentials through a series of steps, as shown in Figure 17.10:

1. The arrival of an action potential at the axon terminal of a presynaptic motor neuron triggers release of acetylcholine (ACh) from synaptic vesicles into the synaptic cleft via exocytosis.

2. Some ACh diffuses across the synaptic cleft to bind ligand-gated ACh receptors (channel proteins) embedded within the sarcolemma of the postsynaptic muscle fiber (any unbound ACh either diffuses away or is degraded by the enzyme acetylcholinesterase).

3. ACh binding leads to channel opening.

4. Na⁺ flows down its electrochemical gradient and into the cell through the open channels, resulting in depolarization of the sarcolemma and generation of an action potential that propagates along the muscle fiber in all directions.

5. At certain locations along muscle fibers, the sarcolemma extends deep into the cells, forming invaginations known as transverse tubules (t-tubules). Spreading of the action potential into t-tubules brings the depolarizing current close to the sarcoplasmic reticulum, a specialized endoplasmic reticulum that plays an essential role in skeletal muscle contraction (described in more detail in Concept 17.1.04).

17.1.04 Excitation-Contraction Coupling

Excitation of a muscle cell triggers a series of steps that lead to calcium ion (Ca^{2+}) release and contraction in a process referred to as **excitation-contraction coupling** (Figure 17.11). In striated muscles (ie, skeletal and cardiac muscle), depolarization triggers the release of Ca^{2+} from the sarcoplasmic reticulum and from extracellular fluid into the cytosol. In skeletal muscle, Ca^{2+} released only from the sarcoplasmic reticulum is required for contraction, whereas in cardiac muscle, Ca^{2+} entering from outside the cell plays an important role as well.

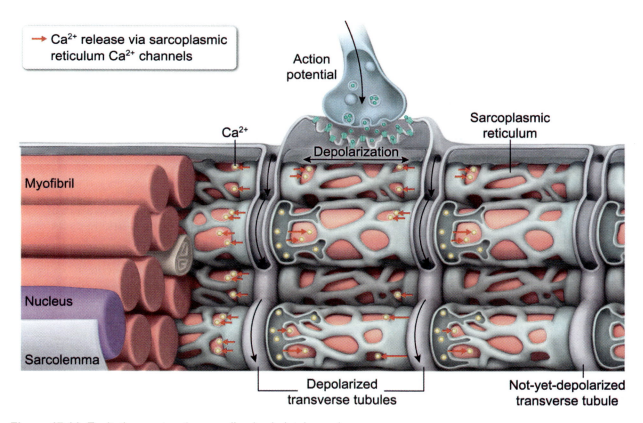

Figure 17.11 Excitation-contraction coupling in skeletal muscle.

The increase in intracellular Ca^{2+} triggers the **cross-bridge cycle**, in which force is produced by actin-myosin interactions (Figure 17.12). In striated muscles, the cross-bridge cycle occurs as follows:

1. At rest (ie, before intracellular Ca^{2+} rises), myosin heads are loosely associated (ie, weakly bound) with actin and bound to ADP and Pi. Actin filaments are bound to troponin and tropomyosin. **Troponin** is a small protein complex associated with **tropomyosin**, an elongated protein wrapped around actin filaments. At rest, tropomyosin prevents the strong actin-myosin binding required for force generation.

2. In response to increased cytosolic Ca2+ concentration during muscle excitation, Ca2+ binds troponin, causing a conformational change to tropomyosin described by the **sliding filament model**. This conformational change fully exposes myosin binding sites on actin filaments, allowing myosin heads to strongly bind actin filaments and form a **cross-bridge**.

3. Cross-bridge formation triggers the **power stroke**, in which myosin heads pull strongly on actin, and the eventual dissociation of Pi and ADP from myosin heads. If the force exerted by the myosin heads on actin is greater than any opposing forces, the actin filament will slide past the myosin molecule to which the actin is bound, and the muscle fiber will shorten.

4. Following the power stroke, new ATP molecules bind myosin, which leads to the temporary dissociation of actin and myosin cross-bridges.

5. ATP hydrolysis results in ADP and Pi once again being bound to myosin heads and weak binding between myosin heads and actin.

This process continues cycling until the intracellular Ca^{2+} concentration falls when muscle activation ceases, leading to muscle relaxation. The decline in Ca^{2+} is primarily due to pumping of Ca^{2+} back into the sarcoplasmic reticulum by sarcoplasmic reticulum Ca^{2+}-ATPase (SERCA).

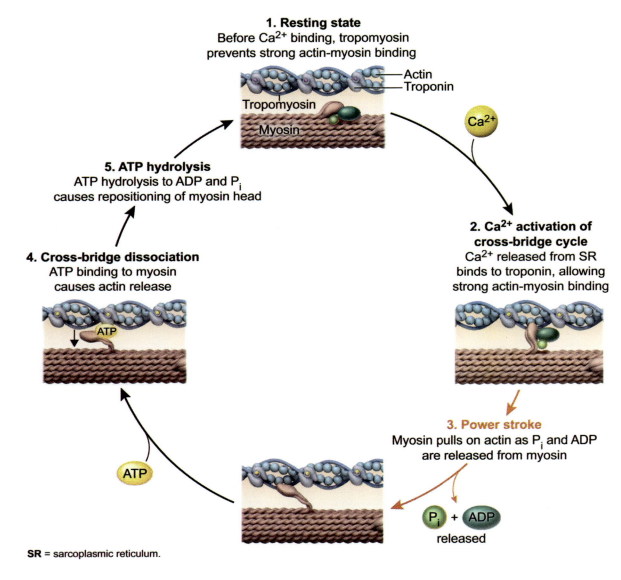

SR = sarcoplasmic reticulum.

Figure 17.12 The cross-bridge cycle of muscle contraction.

Smooth muscle excitation-contraction coupling resembles that in striated muscle, with a few important differences. In smooth muscle, Ca^{2+} binds to the protein **calmodulin (CaM)**, rather than troponin, to regulate the cross-bridge cycle, and myosin phosphorylation plays an essential role. The general process in smooth muscle occurs as follows (Figure 17.13):

1. Increased intracellular Ca^{2+} concentration leads to the formation of Ca^{2+}/CaM complexes.

2. Ca^{2+}/CaM complexes bind to the enzyme **myosin light chain kinase (MLCK)**, thereby increasing MLCK activity.

3. MLCK phosphorylates a protein called **myosin regulatory light chain (RLC)**, a part of the myosin molecules composing the thick filament.

4. RLC phosphorylation by MLCK facilitates the formation of actin-myosin cross-bridges and force development. As in striated muscle, cross-bridge cycling can continue as long as the intracellular Ca^{2+} concentration remains elevated.

5. Dephosphorylation of RLC by the enzyme **myosin light chain phosphatase** promotes eventual cross-bridge detachment. However, this detachment sometimes occurs very slowly after dephosphorylation, allowing smooth muscle to enter a state in which force is sustained due to continued actin-myosin interaction. This *latch* state is one means by which smooth muscle is able to maintain force in a very energetically efficient manner without fatiguing.

6. Smooth muscle relaxation occurs when the intracellular Ca^{2+} concentration is lowered by removal of Ca^{2+} from the cytosol into either the sarcoplasmic reticulum or the extracellular fluid. This reduces Ca^{2+} binding to CaM, a situation analogous to lowered intracellular Ca^{2+} concentration promoting striated muscle relaxation by reducing Ca^{2+} binding to the troponin complex.

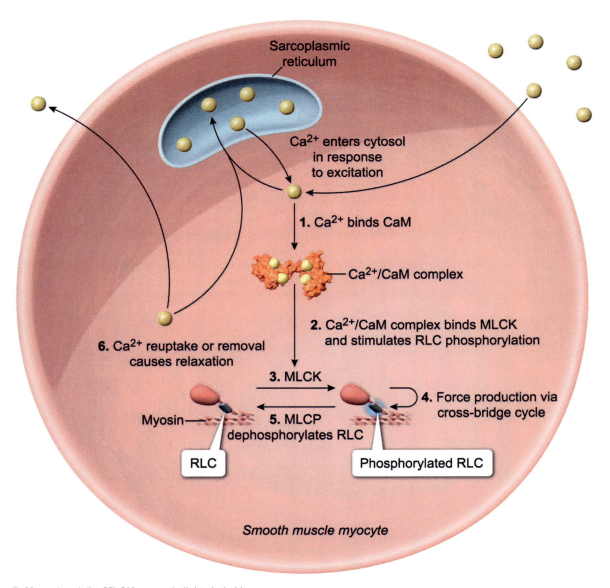

CaM = calmodulin; MLCK = myosin light chain kinase; MLCP = myosin light chain phosphatase; RLC = myosin regulatory light chain.

Figure 17.13 Smooth muscle excitation-contraction coupling.

Concept Check 17.1

In a relaxed myocyte, an approximately 20,000-fold difference in Ca^{2+} concentration typically exists across the sarcolemma, and a gradient approximately one-fifth as large exists between the sarcoplasmic reticulum and the sarcoplasm. Place labels for the sarcoplasmic reticulum, extracellular fluid, and cytoplasm along the *x*-axis to accurately depict the Ca^{2+} concentrations typically present in a myocyte and its environment prior to contraction.

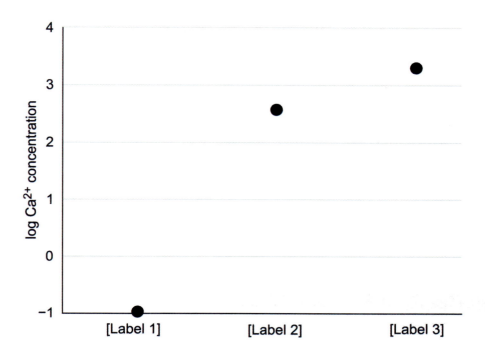

Solution

Note: The appendix contains the answer.

Lesson 17.2
Characteristics of Specific Muscle Types

Introduction

The force produced by muscle contraction can cause bodily movement or stabilization of position. In addition, this force can compress or expand tubular cavities to cause, modulate, or prevent physical transport of a substance (eg, air, blood). There are three broad classifications of muscle, **skeletal**, **cardiac**, and **smooth**, with each class specialized for specific functions (Figure 17.14).

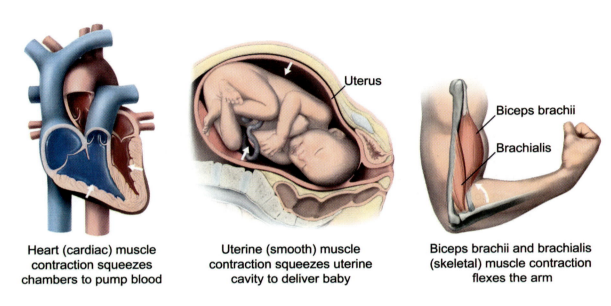

Heart (cardiac) muscle contraction squeezes chambers to pump blood

Uterine (smooth) muscle contraction squeezes uterine cavity to deliver baby

Biceps brachii and brachialis (skeletal) muscle contraction flexes the arm

Figure 17.14 Examples of the function of the three muscle types.

17.2.01 Skeletal Muscle

Skeletal muscle is found throughout the body, typically in muscles attached at each end to bone. When thin sections of skeletal muscle samples are viewed under a microscope, regular light and dark regions, known as **striations**, are apparent. Skeletal muscle cells are long and cylindrical fibers and are multinucleated (ie, contain multiple nuclei per cell). The microscopic organization of skeletal muscle fibers is discussed in further detail in Lesson 17.1.

Skeletal muscle fibers have specific mitochondrial enzymes and myosin ATPases that enable these fibers to contract with varying degrees of force, velocity, and endurance. Although skeletal muscles are **voluntary muscles**, the control of some of these muscles allows for conscious interruption or activation of contraction (eg, control of the diaphragm muscle during breathing or of certain sphincters composed of skeletal muscle).

Skeletal muscle is the only type of muscle innervated by the somatic nervous system. Generation of an action potential at the motor end plate of a neuromuscular junction (ie, the interface between a somatic motor neuron and a skeletal muscle fiber) causes depolarization to spread across the sarcolemma. Action potentials propagate along transverse tubules (t-tubules) just as they propagate along the sarcolemma, carrying the wave of depolarization deep into muscle fibers and resulting in the rapid and complete depolarization of the muscle fiber (Figure 17.15).

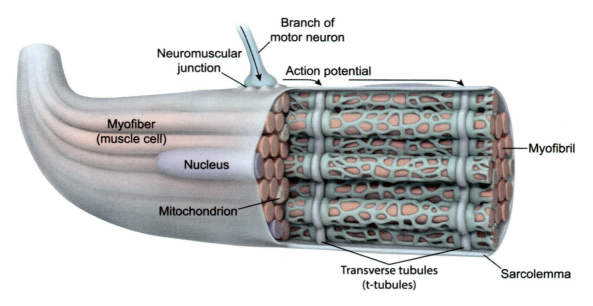

Figure 17.15 Depolarization in a skeletal muscle fiber.

Voltage-sensing calcium channels in t-tubules and neighboring calcium channels in the sarcoplasmic reticulum largely span the small space between the external (t-tubular) and internal (sarcoplasmic reticulum) membranes of striated muscle fibers (Figure 17.16). In skeletal muscle, voltage-sensing calcium channels detect depolarization and cause the calcium channels in the sarcoplasmic reticulum to open, thereby causing contraction to occur. The Ca^{2+} released during contraction is rapidly taken back up into the sarcoplasmic reticulum by the sarcoplasmic reticulum Ca^{2+}-ATPase (SERCA).

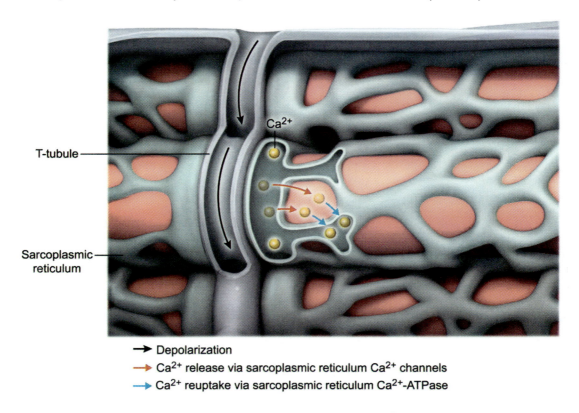

Figure 17.16 Depolarization-induced release and subsequent reuptake of Ca^{2+} in skeletal muscle.

The force produced when skeletal muscle contracts is dependent upon several factors. These factors include the number of muscle fibers activated, the frequency at which muscle fibers are activated, and the length of activated muscle fibers.

Groups of muscle fibers distributed throughout a muscle but controlled by the same motor neuron are called **motor units**. Certain motor units are more readily stimulated than others, resulting in a progressive accumulation of active motor units as the intensity of nervous stimulation increases. This process, called **motor unit recruitment**, increases the proportion of active muscle fibers in a muscle during a contraction, thereby increasing the force produced (Figure 17.17).

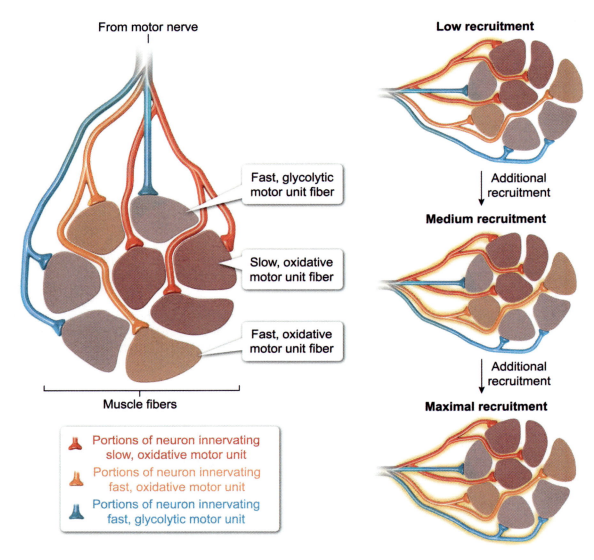

Figure 17.17 Skeletal muscle motor units and motor unit recruitment.

A single motor unit is composed of muscle fibers with identical activity levels; therefore, all the fibers in a motor unit are of a single fiber type. As muscle force production begins, motor units with low activation thresholds are activated first, when the level of neural stimulation is also low. These lowest threshold motor units activate *slow* (ie, slower-contracting), *oxidative* (ie, showing a preference for ATP generation via oxygen-requiring mitochondrial metabolism) muscle fibers.

As force production increases further, motor units with higher activation thresholds that innervate *fast* (ie, faster-contracting), oxidative fibers are recruited. As force production approaches the maximum, the final motor units recruited are those with the highest thresholds for activation, or the fast, *glycolytic* (ie, showing a preference for ATP generation via glycolysis) fibers.

As motor unit activation frequency increases, the interval during which Ca^{2+} can be pumped back into the sarcoplasmic reticulum and the muscle fiber can relax decreases. When a new burst of excitation activates Ca^{2+} release before relaxation from the previous stimulus has occurred, active force production occurs in a muscle fiber in which tension is already above the baseline. As a result, isolated **twitch** muscle contractions in response to individual action potentials fuse with one another, producing a "staircase" of twitches and greater total force than when each burst of stimulation occurs in isolation (Figure 17.18).

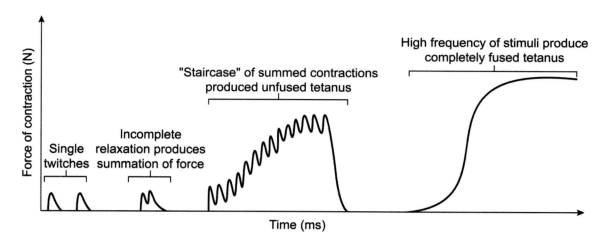

Figure 17.18 Individual muscle twitches can sum to produce a tetanus.

A contraction in which individual twitches fuse is called a **tetanic contraction** (or simply a **tetanus**). At high enough frequencies of stimulation, individual twitches are no longer discernible, and force production smoothly approaches a maximum plateau level with each burst of high-frequency stimulation. When individual stimulation peaks are discernible within a tetanus, the tetanus is described as **unfused**; a tetanus in which individual peaks are not discernible is called a **fused tetanus**, as shown in Figure 17.18.

The physical length of a muscle fiber as it contracts also affects the tension produced in the muscle. Force production is maximal when sarcomere lengths allow maximal actin-myosin overlap, and stretching a shortened muscle will increase force production up to the point at which overlap begins diminishing (ie, excessive stretching) or the fiber is damaged (Figure 17.19). Below the ideal sarcomere length, actin and myosin filaments from one side of the sarcomere crowd structures from the other side of the sarcomere (eg, Z line), resulting in hindered force production.

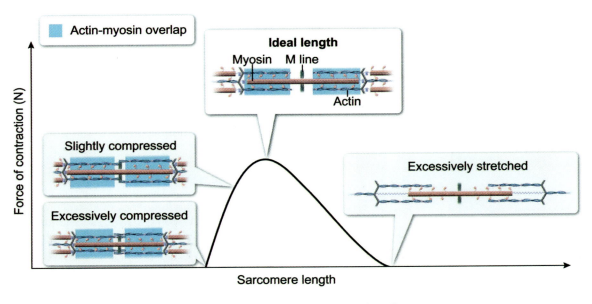

Figure 17.19 Relationship between force of contraction and sarcomere length.

Due to its large mass, contraction of skeletal muscle can exert a large influence on whole-body energy expenditure and thermoregulation. During exercise, increased skeletal muscle energy utilization can greatly increase the rate of whole-body energy expenditure, and the associated increase in heat production can challenge the body's thermoregulatory mechanisms. In addition, cold exposure can induce rapid, involuntary muscle contractions (ie, shivering) that can also markedly increase whole-body energy expenditure and heat production (**thermogenesis**, Figure 17.20).

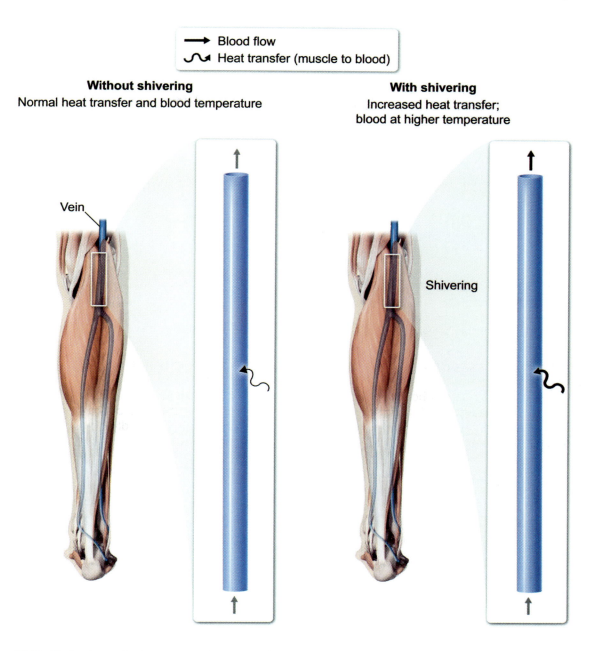

Figure 17.20 Skeletal muscle thermogenesis resulting from shivering.

Similar to the inevitable increases in energy expenditure and heat production, squeezing of blood vessels during skeletal muscle contractions is unavoidable. Veins are more affected by this squeezing than arteries because veins are more easily compressed.

This contraction-induced compression of veins, known as the **skeletal muscle pump**, can serve an important function: Squeezing blood from skeletal muscle toward the heart increases venous return (ie, the amount of blood returning to the heart). Because gravity causes pooling of blood in the lower extremities, the effect of the skeletal muscle pump is most pronounced during contractions of the leg muscles, as shown in Figure 17.21.

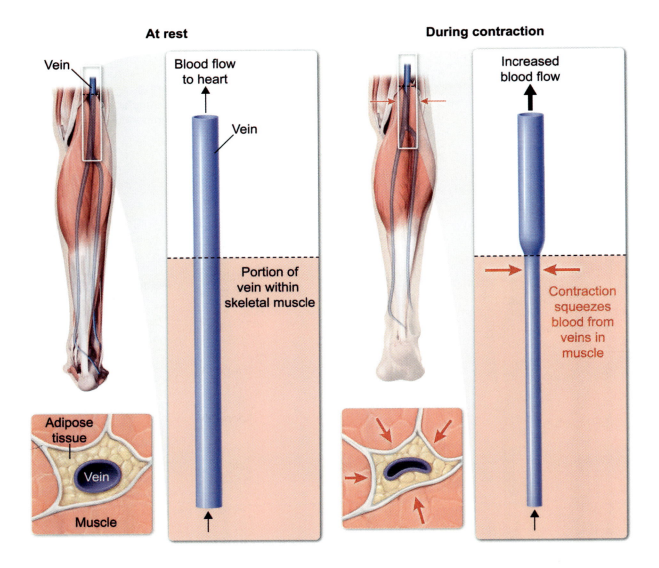

Figure 17.21 The skeletal muscle pump.

Skeletal muscle fibers can be grouped into three broad categories according to their characteristics (eg, speed of contraction, myoglobin content, enzyme activity). These categories and their characteristics are summarized in Table 17.1.

Table 17.1 General characteristics of typical skeletal muscle fiber types.

Fiber type	Slow oxidative (type 1)	Fast oxidative (type 2A)	Fast glycolytic (type 2X)
Speed of contraction	Slow	Medium	Fast
Best use	Endurance activity (eg, long-distance running, posture)	Moderate endurance activity (eg, medium-distance running)	Explosive movements (eg, weight lifting, jumping)
Resistance to fatigue	Fatigue resistant	Intermediate susceptibility to fatigue	Easily fatigable
Primary source of ATP	Aerobic respiration	Anaerobic glycolysis and aerobic respiration	Anaerobic glycolysis
Mitochondria	Plentiful	Plentiful to moderately plentiful	Few
Capillaries	Plentiful	Plentiful to moderately plentiful	Few
Myoglobin	Plentiful	Plentiful to moderately plentiful	Few
Appearance	Red	Intermediate	White

✓ Concept Check 17.2

Fill in the blanks in the sentences below with the words "fast," "glycolytic," "motor unit," "muscle fiber type," "oxidative," and "slow."

A _____ is a collection of skeletal myocytes voluntarily activated together and composed of a single _____. As the force produced by a voluntary muscle contraction gradually increases, any motor units with _____, _____ muscle fibers are activated first, followed by any motor units with fast, oxidative fibers. _____, _____ muscle fibers are recruited last, as the force produced approaches the maximum.

Solution

Note: The appendix contains the answer.

17.2.02 Cardiac Muscle

Cardiac muscle is found in the walls of the heart and is specialized to generate the powerful, coordinated contractions responsible for the continuous pumping of blood through the vessels of the circulation (Figure 17.22).

Microscopically, cardiac muscle appears striated like skeletal muscle, but cardiac muscle fibers are much shorter and can branch, and each cardiac muscle cell contains only one or two nuclei. In addition, cardiac muscle fibers are generally similar to each other, with abundant mitochondria and capillaries, reflecting the single function (ie, blood-pumping) and high metabolic demand of the heart.

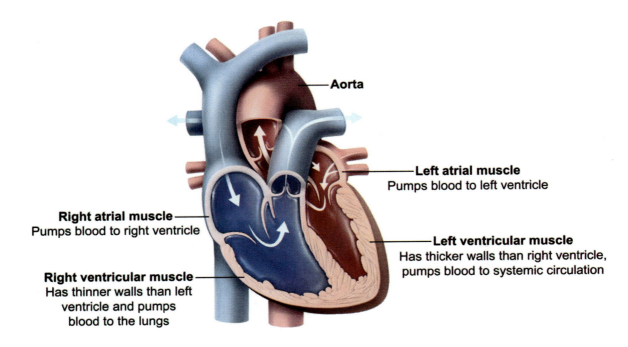

Figure 17.22 Cardiac muscle is found in the atria and ventricles of the heart.

Cardiac muscle cells are connected to adjacent cells via **intercalated discs**. These discs are regions of cell contact that contain both desmosomes (to prevent cells from separating during contraction) and gap junctions (to facilitate direct ion exchange for synchronized contraction), as shown in Figure 17.23.

Cardiac muscle is under involuntary control and is **myogenic**, meaning that it *does not require* nervous system input to contract. Instead, specialized cells spontaneously depolarize to produce electrical impulses that spread through the cardiac muscle via intercalated discs. However, cardiac muscle contraction can be *regulated* by neural and hormonal input. For example, parasympathetic signaling slows the rate of contraction, and sympathetic and hormonal signaling can increase heart rate (see Concept 13.1.13).

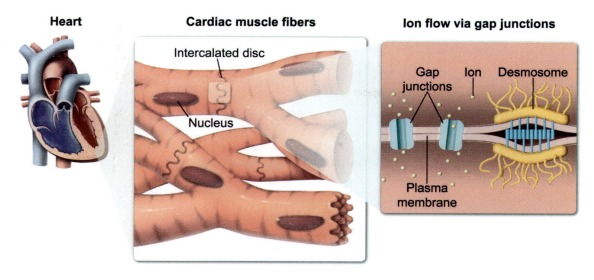

Figure 17.23 Intercalated discs in cardiac muscle.

As in skeletal muscle, t-tubules in cardiac muscle contain voltage-sensing Ca^{2+} channels near the Ca^{2+} channels in the sarcoplasmic reticulum. In cardiac muscle, however, depolarization-induced Ca^{2+} entry through the voltage-sensing Ca^{2+} channels is essential in triggering contraction. In addition, although some of the Ca^{2+} released during contraction is pumped back into the sarcoplasmic reticulum by sarcoplasmic reticulum Ca^{2+}-ATPases, release of Ca^{2+} back into the extracellular fluid is also an important contributor to the restoration of the baseline Ca^{2+} concentration after a cardiac contraction.

Because each heart chamber is a functional syncytium (ie, a collection of connected cells effectively sharing cytoplasm via the pores formed by gap junctions), all the muscle cells in each chamber are activated to contract with each heartbeat. As a result, recruitment of additional muscle fibers is not possible, unlike in skeletal muscle contraction. In addition, action potentials in cardiac muscle cells last much longer than in skeletal muscle cells. Consequently, the refractory period (see Concept 12.2.02) for each action potential extends until muscle relaxation has occurred, and frequency summation cannot occur.

However, cardiac muscle cells can greatly increase their force production when stretched, and contractions initiated at greater sarcomere lengths generate more force (up to a point). As in skeletal muscle cells, this general phenomenon is referred to as the length-tension relationship in the context of muscle fiber and sarcomere lengths. At the whole-heart level, greater atrial and ventricular blood volumes stretch cardiac muscle fibers, producing more forceful contractions and increased pumping of blood, a phenomenon called the Frank-Starling mechanism.

17.2.03 Smooth Muscle

Smooth muscle, so named because it lacks the microscopic striations found in skeletal and cardiac muscle, is found in internal organs and tissues (Figure 17.24). This muscle type exhibits a wide range of specialization to allow periodic or sustained contractions, typically to squeeze or compress a tubular structure (eg, gastrointestinal tract, blood vessel). Such contractions propel or prevent movement of materials (eg, chyme); smooth muscle contractions can also fine-tune flow (eg, blood in vessels).

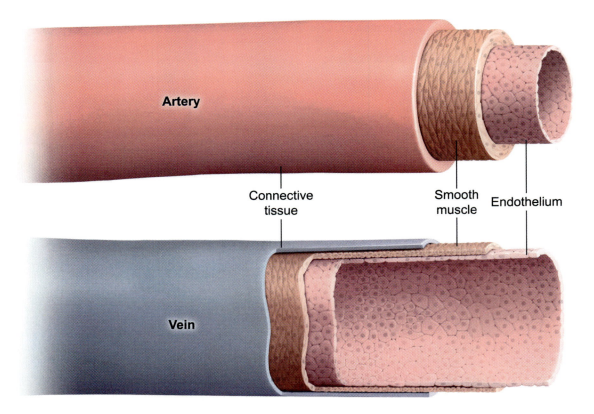

Figure 17.24 Smooth muscle in blood vessels.

Smooth muscle shares some characteristics with striated muscle, such as the essential role of cytoplasmic Ca^{2+} entry in causing contraction and force production via actin-myosin interaction, but important differences exist. For example, unlike striated muscles, the small myocytes in smooth muscle do not possess t-tubules to carry waves of depolarization deeper into the cells. Figure 17.25 depicts smooth muscle cells in relaxed and contracted states in blood vessels.

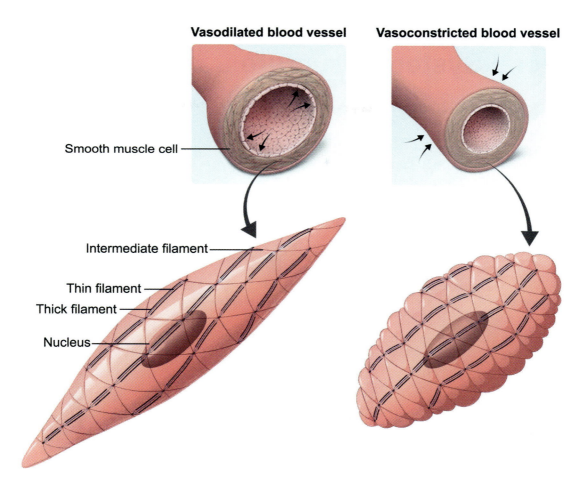

Figure 17.25 Smooth muscle cells.

Like the heart, smooth muscle is innervated by the autonomic nervous system and can be myogenic. Also similar to cardiac muscle, relaxation is facilitated by both Ca^{2+} reuptake into the sarcoplasmic reticulum as well as transport of Ca^{2+} back to the extracellular fluid. Some smooth muscles also resemble the heart in that gap junctions allow the cells in such muscles to act as a functional syncytium. In these cells, sharing of cytoplasm allows activation of all cells simultaneously.

Lesson 18.1
Skeletal System Structure

Introduction

Humans have an endoskeleton that is derived from the embryonic mesoderm. The skeletal system is composed of bones and the intimately associated tissues that cover bones or serve as bone linkages. The term *bone* can refer to a macroscopic unit (ie, one piece of the skeleton) or a tissue type with certain characteristic features. The human skeletal system is depicted in Figure 18.1.

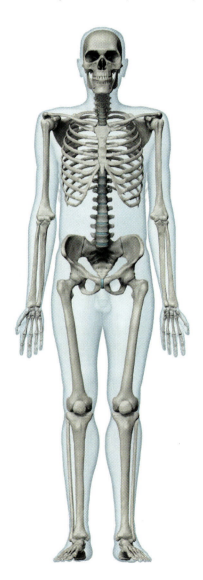

Figure 18.1 The human skeletal system.

This lesson introduces the structure of the skeletal system.

18.1.01 Bone Types

Bones and bone tissue can be classified based on characteristic traits, such as anatomical location, bone shape, and bone structure.

When classifying bones based on anatomical location, the skeleton is divided into axial and appendicular components. The bones of the **axial skeleton** (ie, skull, spine, ribs, and sternum) make up the long axis of the skeleton, and the bones of the **appendicular skeleton** (eg, femur, clavicles) are appended (ie, attached) to those of the axial skeleton (Figure 18.2).

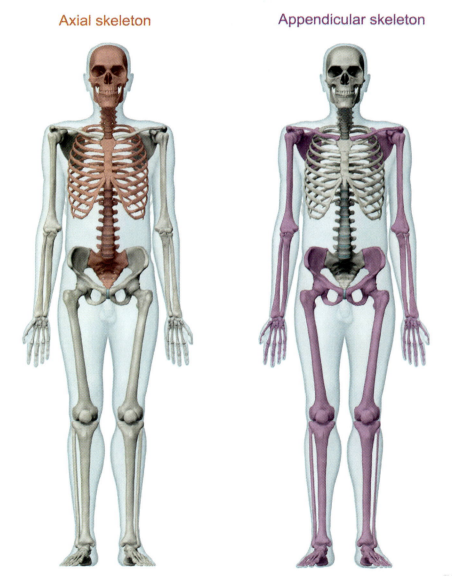

Figure 18.2 Classifying bones by anatomical location.

When classifying bones according to shape, they can be grouped into several types: **long** (eg, the ulna in the forearm), **short** (eg, the cylindrical bones of the fingers), **flat** (eg, skull bones), or **irregular** (eg, vertebrae in the spine). Examples of bone shapes are illustrated in Figure 18.3.

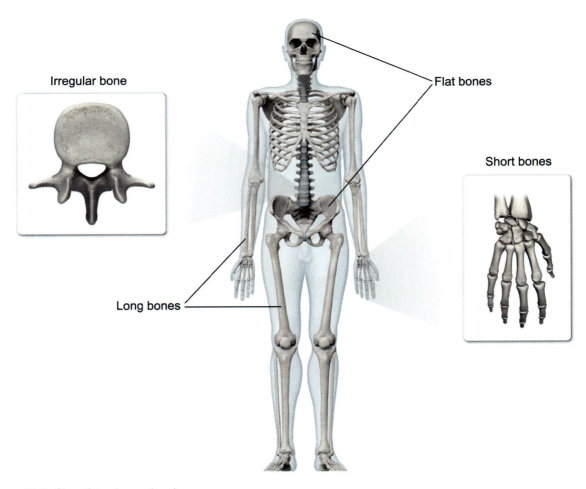

Figure 18.3 Classifying bones by shape.

Bone classification based on intrinsic structure (ie, hard surface bone versus less dense interior bone) is discussed in detail in Concept 18.1.02.

18.1.02 Bone Structure

Bone is a type of connective tissue and is composed of both living bone cells and nonliving materials secreted by bone cells into the extracellular space. The components of the bone extracellular matrix are discussed in detail in Concept 18.1.03.

Although a multitude of bone shapes exist, all bones share some common features. For example, all bones have the same fundamental intrinsic structure: an outer layer of hard, dense bone called **compact** or **cortical bone** and an inner, softer core known as **spongy bone**. In addition, long bones share a gross structure that includes the following features (shown in Figure 18.4):

- **Epiphyses** are rounded ends that make up joint surfaces and are covered by articular cartilage.
- The **diaphysis** is a hollow central shaft enclosing a **medullary cavity** filled with bone marrow.
- The **metaphyses** are regions where the diaphysis and epiphyses meet.
- The **epiphyseal (growth) plate**, a cartilaginous structure that lies between each epiphysis and metaphysis, is present only during childhood and serves as the site of longitudinal growth. When growth ceases, the growth plate is replaced with mature bone, forming the **epiphyseal line**.

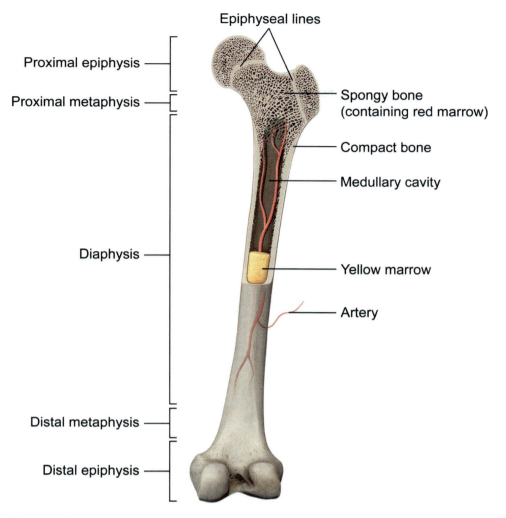

Figure 18.4 Structure of a long bone.

Compact bone is organized into structural units called **osteons**, or **Haversian systems**. Osteons are made up of **lamellae** (concentric rings of bone matrix) that surround a central **Haversian canal**, a cylindrical channel that runs parallel to the long axis of bone and through which blood vessels and nerves traverse.

Within the lamellar matrix are tiny spaces called lacunae, each containing a mature, mitotically inactive **osteocyte**, the most abundant non-marrow bone cell type. **Volkmann's (transverse) canals**, which run perpendicular to the long axis of bone, allow the passage of blood vessels and nerves between different Haversian canals. Microscopic channels called **canaliculi** allow osteocyte waste exchange and nutrient delivery (Figure 18.5).

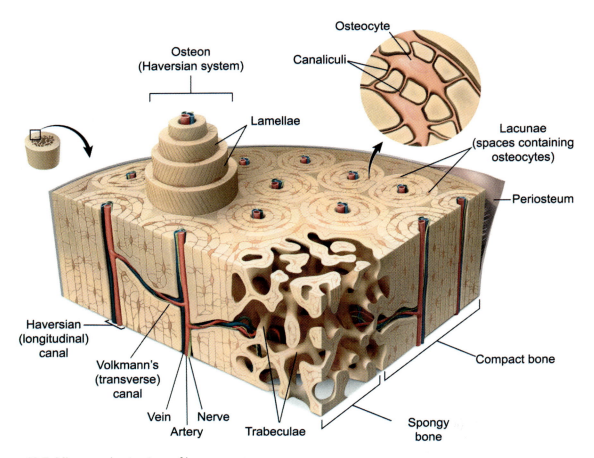

Figure 18.5 Microscopic structure of bone.

Spongy bone consists of a porous, interconnected network of irregular fine spikes of bone called **trabeculae** (spongy bone is sometimes called trabecular bone). Spongy bone is less dense than cortical bone, and in all bones of young children and certain bones of adults, this bone type contains marrow.

Bone marrow is composed of various types of cells, including fat cells (adipocytes) and precursors for red and white blood cells. Marrow that actively produces blood cells is called red marrow because the hemoglobin of red blood cell precursors imparts a reddish color. Inactive marrow is called yellow marrow and consists primarily of adipocytes. The two types of bone marrow are summarized in Table 18.1.

Table 18.1 Bone marrow characteristics.

Bone marrow type	Location	Function
Red	Spaces between the trabeculae of the spongy bone in flat and long bones (progressively replaced by yellow marrow with age)	Site of hematopoiesis (blood cell formation)
Yellow	Medullary cavity of the diaphysis of long bones	Site of fat (adipocyte) storage

Exposed bone surfaces (ie, those not covered by cartilage, tendon attachments, or ligaments) have a thin surface membrane layer called the **periosteum**. The periosteum provides some protection to bone and contains blood vessels, nerves, and a population of cells that can contribute to the repair of fractured bone. In this type of reparative bone formation (called **intramembranous ossification**), stem cells within the periosteum differentiate into osteoblasts (described in Concept 18.1.04) and secrete bone matrix.

The ends of some bones are covered with cartilage, and cartilage-producing cells (chondrocytes) within bones play an important role in bone growth during development. Cartilage and cartilage-mediated bone growth are discussed further in Concept 18.1.06.

Concept Check 18.1

Match the bone structure terms below with their description in the table.

Canaliculi

Haversian canals

Lacunae

Lamellae

Osteons (Haversian system)

Volkmann's canals

Structure	Description
	Structural units of compact bone
	Cylindrical channels that contain nerves and blood vessels and run *perpendicular* to the long axis of bone
	Tiny, osteocyte-containing spaces within the matrix surrounding Haversian canals
	Cylindrical channels that contain nerves and blood vessels and run *parallel* to the long axis of bone
	Tiny channels through which nutrient and waste exchange between osteocytes and the circulation occurs
	Concentric rings of bone matrix surrounding Haversian canals

Solution

Note: The appendix contains the answer.

18.1.03 Bone Matrix

The **bone matrix** is the extracellular material surrounding the cells in bone and is formed from bone cell secretions and other components from the blood. The matrix is composed of both inorganic materials, such as **calcium phosphate**, and organic materials, such as **collagen**. Collagen is secreted by bone cells, and calcium phosphate (a salt) forms by the precipitation of calcium (ie, Ca^{2+}) and phosphate (ie, PO_4^{3-}) from the bloodstream.

Precipitation of calcium phosphate salts onto collagen fibers in the matrix eventually results in the formation of **hydroxyapatite** crystals, which are responsible for the hardness of bone. Enzymes secreted by osteoblasts (Concept 18.1.04) catalyze the breakdown of calcium- and phosphate-containing compounds secreted from osteoblast vesicles, facilitating precipitation of Ca^{2+} and PO_4^{3-} released from these compounds into crystals of hydroxyapatite. Secreted collagen provides a site for hydroxyapatite crystals to bind. Figure 18.6 shows the bone matrix structure from macro- to nanoscale.

Chapter 18: Skeletal System

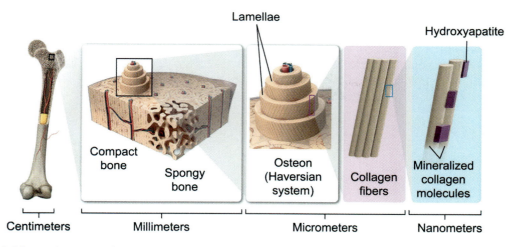

Figure 18.6 Macro- to nanoscale bone structure.

The bone matrix consists of both **mineralized** (ie, hard) and **unmineralized** (ie, soft) parts. The unmineralized bone matrix is called the **osteoid**. Specialized bone cells (discussed in Concept 18.1.04) mediate the transfer of calcium phosphate between the blood, osteoid, and mineralized bone matrix.

In certain circumstances, such as in individuals with the bone disease osteoporosis or in astronauts whose bones are no longer loaded by gravity, bone matrix mineralization can be reduced, as shown in Figure 18.7. If severe enough, such demineralization can reduce bone mass and strength to the point that bones become fragile and fracture more easily.

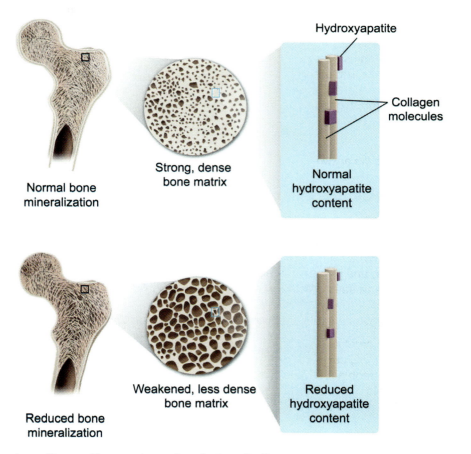

Figure 18.7 Structure of bone with normal or reduced mineralization.

18.1.04 Bone Cells

A whole bone (ie, one of the units of the skeleton) can be composed of several tissue types, including cartilage, marrow, and bone (ie, bone *tissue*), as depicted in Figure 18.8. The structural rigidity of the skeleton belies the dynamic nature of bones; for example, each day millions of new osteocytes and bone marrow cells are formed, and each year up to 10% of the skeleton is replaced.

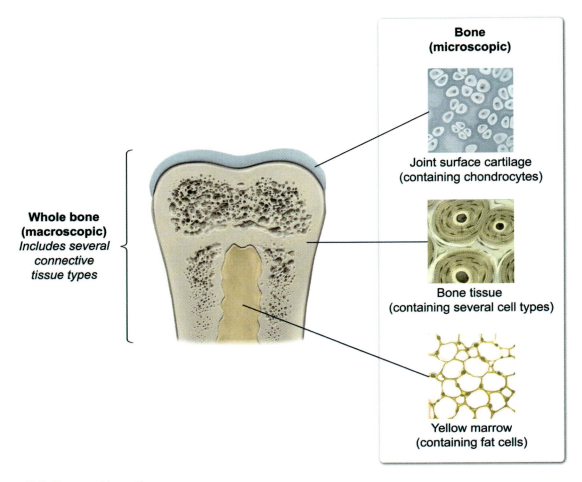

Figure 18.8 Bone and bone tissue.

Connective tissues, such as bone tissue, contain specialized cells that secrete and maintain an extensive extracellular matrix as well as other types of cells (see Concept 5.3.07). Bone tissue, which is sometimes connected by gap junctions, is made of several types of cells surrounded by bone matrix:

- **Osteoprogenitor cells** (sometimes called **osteogenic cells**) are mitotically active stem cells in bone that initially differentiate into osteoblasts.
- **Osteoblasts** are bone-forming cells that coordinate the incorporation of calcium and phosphate ions into bone. Osteoblasts secrete proteins that create the osteoid, or unmineralized bone matrix, into the extracellular space. Osteoblasts eventually become fully surrounded by bone matrix and differentiate into osteocytes.
- **Osteocytes** are mature (ie, fully differentiated) and mitotically inactive bone cells that maintain bone structure. Located within lacunae in the Haversian system (where they were originally osteoblasts), osteocytes can release signals to other bone cells to regulate compact bone remodeling.
- **Osteoclasts** are large cells that secrete proteolytic enzymes and acid that promote **resorption** (breakdown) of the organic and mineral components of bone.

Osteocytes, osteoclasts, and osteoblasts function to maintain the strength and integrity of the skeleton through **bone remodeling**. During this process, old bone is resorbed by osteoclasts and new bone is deposited by osteoblasts. Osteoblasts form successive concentric layers of bone (lamellae), and as the osteoid secreted by osteoblasts mineralizes, some osteoblasts become trapped within lacunae (spaces) in the lamellar matrix. Eventually, these trapped osteoblasts become either flattened bone-lining cells or osteocytes within the interior of bone. The process of bone remodeling is summarized in Figure 18.9.

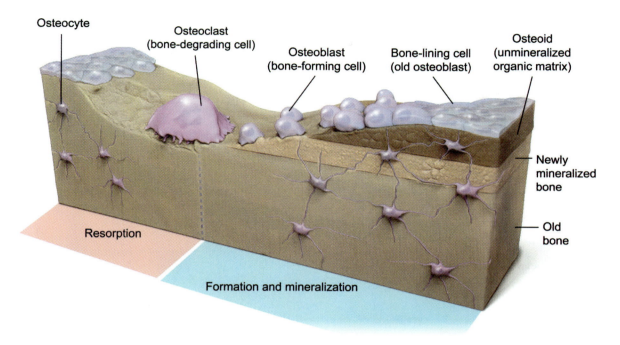

Figure 18.9 Bone remodeling.

In healthy individuals, the process of bone remodeling is tightly regulated to ensure that bone mass and density remain constant. To maintain this consistency, the rate of bone resorption by osteoclasts equals the rate of bone deposition by osteoblasts.

As discussed in Concept 18.1.02, certain bones contain bone marrow, which is made of various types of cells, including precursors for red and white blood cells (ie, erythrocytes and leukocytes, respectively). Such **bone marrow cells** are hematopoietic (ie, blood cell-producing) and remain active in some bones throughout life. Bone marrow and its role in the immune system are discussed in more detail in Concept 20.1.02.

18.1.05 Joints

Joints are specialized components of the vertebrate musculoskeletal system where two or more bones articulate (ie, interact). There are several types of joints, including immovable joints (eg, skull sutures) and freely movable joints (eg, hinge, ball-and-socket).

Freely movable joints, also known as **synovial joints**, consist of several structures that interact with bone and skeletal muscle, as shown in Figure 18.10:

- **Articular (hyaline) cartilage** surrounds the ends of bones within a joint, providing a smooth surface that absorbs compressive forces and reduces friction between interacting bones.
- A **joint cavity**, located between articulating bones, is lined by the synovial membrane (synovium), which produces specialized fluid known as **synovial fluid**. Synovial fluid lubricates joint surfaces and protects articular cartilage from excessive friction and damage.

- A **fibrous layer** of connective tissue extends across a joint from the periosteal membranes of the articulating bones. This fibrous layer lies superficial to the synovial membrane.
- **Fat pads** are adipose tissue structures that provide additional cushion between bones in some synovial joints (eg, knee and hip joints).
- Ligaments and tendons are dense connective tissue structures. Ligaments attach bones to other bones, and tendons generally attach bones to surrounding muscles.

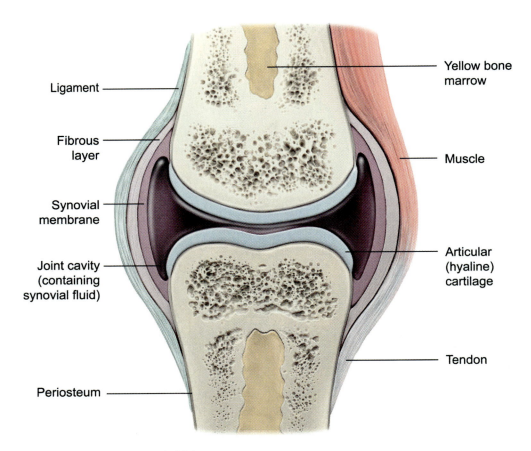

Figure 18.10 A freely movable (synovial) joint.

Ligaments and tendons are important connective tissue structures at joints (Figure 18.11), as they stabilize the positioning of bones and enable the application of muscle force across joints. In this latter role, muscle contraction transmits force to the muscle origin (the more proximal or less mobile end of a muscle) and insertion (the more distal or movable end of a muscle) points on the bones forming the joint.

Tendons are strong, fibrous bands of connective tissue that, in the context of joints, anchor muscle by attaching it to bony structures. Outside of joints, tendons can attach muscles to other structures (eg, the linea alba or other tendinous structures in the abdominal musculature). By transmitting the force generated by muscle contraction, tendons play an essential role in locomotion (movement).

Ligaments are strong bundles of connective fibers in joints that connect bones to other bones, thereby stabilizing a joint by holding the bony structures together. Ligaments also help stabilize internal organs and in rare cases can serve as muscle attachment sites (eg, part of the transverse abdominis muscle of the abdomen has its origin in a ligament near the hip/lower abdomen).

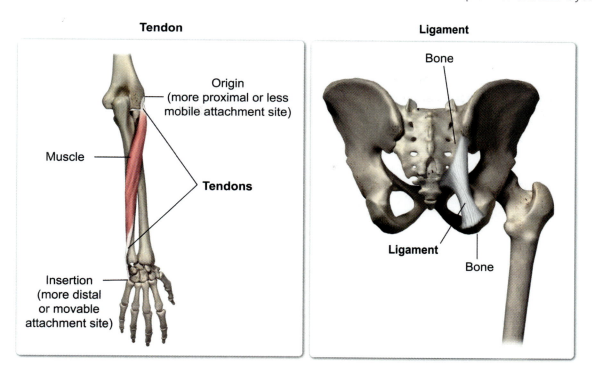

Figure 18.11 Tendons connect muscles to bones, and ligaments connect bones to other bones.

18.1.06 Cartilage

Cartilage is a connective tissue made up of chondrocyte cells. These cells secrete a specialized extracellular matrix called chondrin, which contains collagen fibers, proteoglycans, and water (Figure 18.12). The firm but flexible structure of cartilage is resistant to compression and stretching.

Most of the cartilage in the body is found in locations that require cushioning (eg, spine) and on the articulating surfaces of bone ends (ie, the bone surfaces that meet in a joint). Because cartilage lacks nerves and its own blood supply, this type of tissue must receive nutrients and oxygen via diffusion from surrounding fluids or vascularized areas.

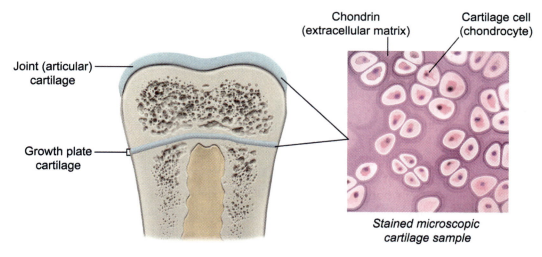

Figure 18.12 Cartilage location and structure.

Cartilage is classified as hyaline, elastic, or fibrous. **Hyaline cartilage** is the most abundant cartilage type in the body and is found in the ribs, nose, trachea, and larynx. The articular cartilage that surrounds the ends of bones within a joint is also hyaline cartilage. **Elastic cartilage** is enriched with elastic fibers and is found where flexibility is important (eg, the ear). **Fibrous cartilage** is structurally reinforced by a high content of collagen fibers, which contributes to its resistance to deformation (eg, when compressed between vertebrae).

Cartilage also plays a role in certain mechanisms of bone development. The process of **endochondral ossification** uses hyaline cartilage as a template for bone deposition by osteoblasts (as opposed to intramembranous ossification, in which osteoblasts in the periosteum secrete new bone in the absence of a cartilage template). During fetal development, the hyaline cartilage that composes the fetal skeleton is calcified (ie, replaced by bone). In children and adolescents, the epiphyseal (growth) plate of long bones is formed from hyaline cartilage and serves as the site of bone lengthening.

The cartilaginous epiphyseal plate is present only during childhood. When growth ceases, the growth plate is replaced with mature bone, forming the **epiphyseal line**, as shown in Figure 18.13.

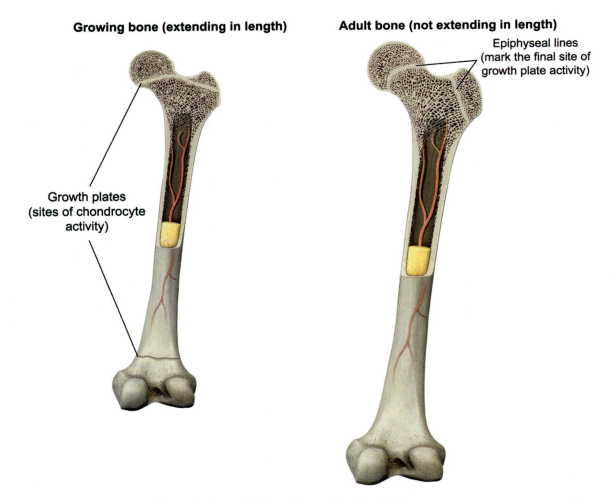

Figure 18.13 Replacement of growth plate with the epiphyseal line in adult bones.

Lesson 18.2
Skeletal System Function

Introduction

Bone plays multiple important roles in the body, and the storage and release of calcium and phosphate in bone help maintain whole-body calcium and phosphorus homeostasis. Specific hormones regulate bone growth and development as well as calcium and phosphorus release. This lesson details the functions of bone and how the skeletal system is regulated.

18.2.01 Bone Function

Vertebrates such as humans have an internal skeleton (endoskeleton) that consists of a bony vertebral column linked to other bones and tissues. The primary functions of the skeletal system are summarized in Table 18.2.

Table 18.2 Functions of the skeletal system.

Structural support	Bones provide a framework of structures to which other bones and tissues can attach.
Physical protection	Certain bones shield internal organs from physical trauma (eg, the skull protects the brain, the rib cage protects the lungs and heart, the vertebrae protect the spinal cord).
Mobility	By providing sites of attachment and joints across which muscles can exert force, the skeleton enables body movement.
Mineral and energy storage	Bones store and mobilize minerals (primarily phosphate and calcium) that add considerable strength to the skeleton. Additionally, yellow marrow present in some bones serves as a reservoir of stored energy in the form of fat.
Blood cell production (hematopoiesis)	Although cells in a variety of tissues are capable of producing blood cells during development or in response to certain stressors, bone marrow is typically the major site of blood cell production after birth.

18.2.02 Endocrine Control of Skeletal System

Bone structure and function are influenced throughout life by hormones. Hormones regulate bone mass primarily through their influences on bone growth, bone density, and handling of the body's calcium stores, 99% of which are in the form of calcium phosphate-containing crystals in bone.

Endocrine regulation of the skeletal system occurs in the context of a complex system of hormones that regulates numerous processes in the body (Table 18.3). In many cases, the hormones involved act upon other endocrine tissues as well as target cells in bone. For example, in addition to individual influences on bone, both **thyroid hormone** and **estradiol** can augment growth hormone secretion. Despite this complexity, a few general themes exist.

For example, during childhood and adolescence, **growth hormone** and thyroid hormone increase bone mass by promoting the synthesis of new bone necessary for linear bone growth. Abnormally high levels of either hormone can cause increased bone growth, whereas low growth hormone or thyroid hormone levels are associated with suboptimal bone growth. Some of the effects of thyroid hormone are likely due to its promotion of growth hormone synthesis.

Vitamin D, in addition to being a nutrient (ie, a fat-soluble vitamin) consumed in the diet, can be synthesized in the body and act as a steroid hormone that regulates bone growth and development. Vitamin D undergoes successive modifications in the liver and kidneys that increase its activity. The most active form of vitamin D is called vitamin D_3, or **calcitriol**. Calcitriol promotes absorption of calcium from ingested food and reabsorption of calcium in the kidneys. Inadequate vitamin D levels during childhood can cause rickets, a disease in which bone structure is abnormal.

Table 18.3 Hormonal modulators of bone mass via influences on linear growth and development and bone density.

Hormone	Effect on linear bone growth and development	Effect on adult bone density
Growth hormone	+	+
Thyroid hormone	+	-
Vitamin D	+	+
Estradiol	+	+
Testosterone	+	+
Parathyroid hormone	-	-
Calcitonin	+	+

The **sex hormones** estradiol (a type of estrogen) and testosterone are also associated with linear bone growth. The impact of both sex hormones on growth is likely due in part to their stimulation of growth hormone secretion. Testosterone can be converted to estradiol, so some (but not all) bone responses to testosterone also occur indirectly through estradiol. Estradiol contributes to the adolescent growth spurt but, in parallel, also promotes the slow conversion of chondrocytes to a senescent (aging) phenotype (see Concept 10.5.02). When senescence eventually occurs, the growth plates close and linear bone growth ends.

Although linear bone growth eventually ends, bone remodeling and the need to regulate blood calcium levels continues throughout life, and hormonal regulation of these processes remains vital. A major hormonal player in such regulation is parathyroid hormone, which indirectly stimulates bone resorption by stimulating osteoblasts to secrete factors promoting osteoclast activity. Parathyroid hormone also influences bone by stimulating loss of phosphate via urine excretion. Another hormone, calcitonin, is thought to prevent excessive blood calcium concentrations by limiting bone resorption.

Growth hormone and sex hormones continue to promote increased bone density throughout life, as does vitamin D. In contrast, there is a general inverse relationship between thyroid hormone levels and bone mineral density in adulthood, and excessive thyroid hormone concentrations can overstimulate bone resorption relative to bone deposition, leading to pathologically reduced bone density.

Chapter 18: Skeletal System

 Concept Check 18.2

Considering the mechanisms of action upon bone, classify each of the following hormones as promoting net deposition or net loss of bone when their secretion is increased in adults: parathyroid hormone, thyroid hormone, growth hormone.

Solution

Note: The appendix contains the answer.

END-OF-UNIT MCAT PRACTICE

Congratulations on completing **Unit 9: Musculoskeletal System**.

Now you are ready to dive into MCAT-level practice tests. At UWorld, we believe students will be fully prepared to ace the MCAT when they practice with high-quality questions in a realistic testing environment.

The UWorld Qbank will test you on questions that are fully representative of the AAMC MCAT syllabus. In addition, our MCAT-like questions are accompanied by in-depth explanations with exceptional visual aids that will help you better retain difficult MCAT concepts.

TO START YOUR MCAT PRACTICE, PROCEED AS FOLLOWS:

1) Sign up to purchase the UWorld MCAT Qbank
   IMPORTANT: You already have access if you purchased a bundled subscription.
2) Log in to your UWorld MCAT account
3) Access the MCAT Qbank section
4) Select this unit in the Qbank
5) Create a custom practice test

Unit 10 Skin and Immune Systems

Chapter 19 Skin System

19.1 Skin Structure

    19.1.01   Epidermis
    19.1.02   Dermis
    19.1.03   Pilosebaceous Unit
    19.1.04   Skin Capillaries

19.2 Skin Function

    19.2.01   Skin Barrier
    19.2.02   Skin's Role in Osmoregulation
    19.2.03   Skin's Role in Thermoregulation
    19.2.04   Regulation of Skin Function

Chapter 20 Immune System

20.1 Immune System Components

    20.1.01   Cells of the Immune System
    20.1.02   Bone Marrow
    20.1.03   Thymus
    20.1.04   Lymph Node
    20.1.05   Spleen

20.2 Immune System Function

    20.2.01   External Innate Defenses
    20.2.02   Internal Innate Defenses
    20.2.03   Adaptive Immunity
    20.2.04   Antigen Presentation
    20.2.05   Cell-Mediated Immunity
    20.2.06   Humoral Immunity
    20.2.07   Immune Response Integration
    20.2.08   Autoimmunity

Lesson 19.1

Skin Structure

Introduction

The skin, along with its accessory structures (eg, nails, hair, oil glands, sweat glands), make up the body's **integumentary system** (Figure 19.1). Skin consists of multiple layers, which are arranged to form two main structural elements (ie, main skin layers), the **epidermis** and **dermis**. The dermis rests on a deeper layer of connective tissue called the **hypodermis** (ie, subcutaneous layer), which is not technically part of the skin but contributes to the function of the integumentary system. This lesson outlines important integumentary system structures and their respective locations in the skin.

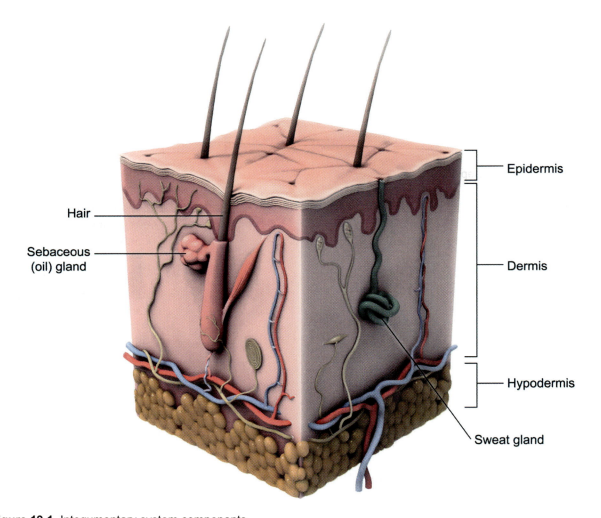

Figure 19.1 Integumentary system components.

19.1.01 Epidermis

The uppermost (ie, most superficial) layer of the skin is the **epidermis**. The epidermis is a **stratified squamous epithelium** that consists of multiple types of epithelial cells, including keratinocytes, melanocytes, Langerhans cells, and Merkel cells.

The epidermis is divided into five structurally distinct layers (strata). From deepest to most superficial, these layers include the **stratum basale**, **stratum spinosum**, **stratum granulosum**, **stratum lucidum** (present only in skin found on the palms of the hands and soles of the feet), and **stratum corneum** (Figure 19.2).

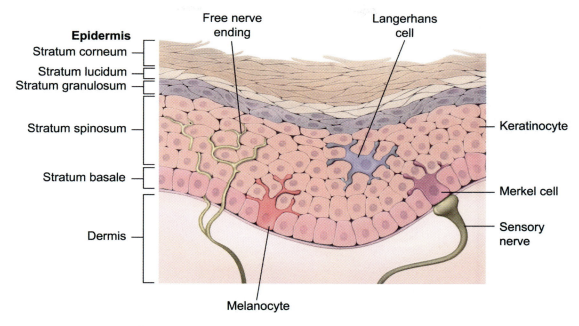

Figure 19.2 Layers of the epidermis.

The stratum basale consists of a single row of stem cells that continually divide to give rise to new stem cells. Of the two daughter cells produced from each mitotic division in the stratum basale, one cell remains in the basal layer (to continue proliferating) and the other begins differentiating into a mature keratinocyte.

As constant cell division in the stratum basale pushes cells outward, these cells mature and are sequentially organized to form the other layers of the epidermis, as shown in Figure 19.3. During their maturation, keratinocytes fill with **keratin proteins**, flatten, lose their organelles, and ultimately die. The stratum corneum (ie, outermost layer of the epidermis) is composed of 20–30 layers of dead, keratin-filled cells, which are constantly shed from the surface of the skin, being replaced by deeper, younger keratinocytes.

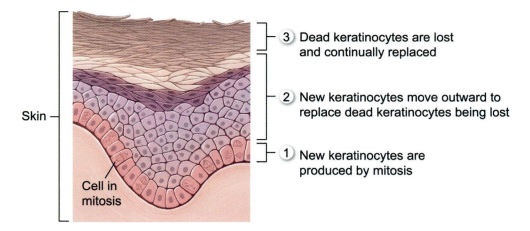

Figure 19.3 Replacement of dead cells lost from the surface of the skin.

19.1.02 Dermis

The layer that contributes most to the thickness of skin is the **dermis**, which is located between the epidermis and the hypodermis. The skin's blood vessels, lymph vessels, and sensory nerve fibers are located within the dermis, as are sweat glands, sebaceous (oil) glands, and hair follicles.

The dermis is composed of connective tissue, contains abundant fibroblast cells, and consists of two main layers: the **papillary layer** and the **reticular layer** (see Figure 19.4). The papillary layer lies just beneath the epidermis and consists of loose (areolar) connective tissue, which contains collagen fibers and elastic fibers produced by fibroblasts.

The reticular layer, which makes up about 80% of the thickness of the dermis, is the deeper of the two dermal layers. The reticular layer is composed of dense irregular connective tissue, which contains thick bundles of collagen fibers along with elastic fibers. The **hypodermis**, which is composed primarily of **adipose tissue**, is located just below the dermis.

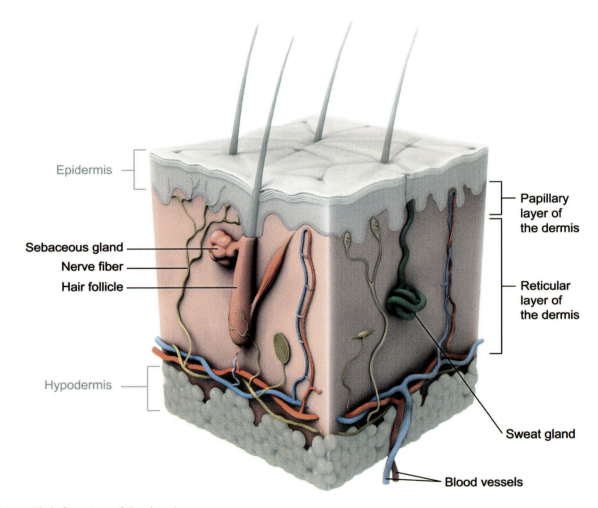

Figure 19.4 Structure of the dermis.

19.1.03 Pilosebaceous Unit

Hair, hair follicles, and sebaceous (oil) glands are accessory structures associated with the skin. A **hair follicle** is a living organ that produces a hair, whereas the hair itself is composed primarily of dead, keratinized cells. Each hair follicle is connected to a **sebaceous gland**, which secretes an oily lipid substance called **sebum** through a duct into the hair follicle and ultimately onto the surface of the skin, where it contributes to the barrier function of the skin, as described in Lesson 19.2.

In addition, an **arrector pili muscle**, made up of a bundle of smooth muscle cells, is attached to each hair follicle. Contraction of arrector pili muscles in an area of skin causes the hairs in that area to stand upright (ie, piloerection), which can play a role in regulation of body temperature in animals with abundant hair. A hair follicle, along with its associated sebaceous gland and arrector pili muscle, is referred to as a **pilosebaceous unit** (Figure 19.5).

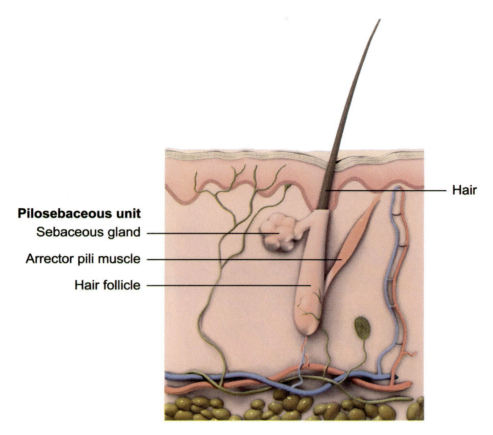

Figure 19.5 Components of a pilosebaceous unit.

19.1.04 Skin Capillaries

Unlike the epidermis, which contains no blood vessels, the dermis of the skin is **vascularized** (ie, contains blood vessels). **Arteries** that supply blood to the skin are present in the hypodermis, and branches from these arteries carry blood through the reticular layer of the dermis. These branches (ie, smaller arteries) eventually give rise to **arterioles**, which supply blood to **capillary loops** present in the papillary layer of the dermis. Blood drains from these capillary loops via **venules**, which join to form **veins** that carry blood away from the skin through the hypodermis (see Figure 19.6).

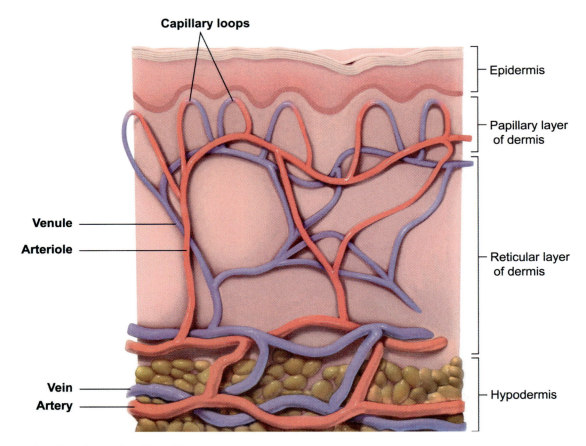

Figure 19.6 Blood vessels of the skin.

Lesson 19.2

Skin Function

Introduction

The complex structural organization of the skin allows it to carry out multiple important functions. These functions include providing protection against various potentially harmful environmental factors as well as participating in the maintenance of homeostasis within the body. This lesson explores the various functions that are performed by the skin.

19.2.01 Skin Barrier

A primary means by which the skin functions to protect the body is by serving as a barrier between the body's external and internal environments. This function is performed largely by the skin's epidermis, particularly by the outermost layer of the epidermis, the **stratum corneum**. The dead **keratinocytes** (ie, specialized cells) that make up the stratum corneum are filled with keratin proteins that serve to increase the skin's resistance to abrasion and chemicals.

The stratum corneum also functions to inhibit water loss from the body. Keratins and other proteins within the keratinocytes, lipids between these cells, and **sebum** secreted onto the surface of the epidermis make the stratum corneum highly water resistant. Damaged skin or skin that has lost sebum (eg, through frequent handwashing) will undergo excessive water loss through the stratum corneum (see Figure 19.7).

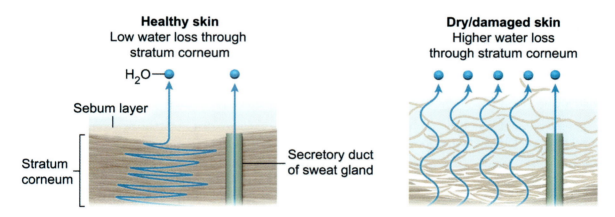

Figure 19.7 Role of stratum corneum in prevention of excessive water loss.

The epidermis also serves as a barrier to pathogens and ultraviolet radiation. The tightly packed keratinocytes of the epidermis help to physically block pathogen entry into the body. Furthermore, **Langerhans cells**, which are especially abundant in the stratum spinosum, perform surveillance to detect invading microorganisms. These phagocytic cells are active participants in the body's immune response, which is discussed in Lesson 20.2.

In addition, **melanocytes** in the stratum basale produce the pigment **melanin**, which absorbs ultraviolet radiation. After being synthesized, melanin granules (melanosomes) are transferred to nearby keratinocytes (see Figure 19.8), where the granules are arranged to shield the keratinocytes' nuclei from the DNA-damaging effects of exposure to ultraviolet radiation (discussed in Concept 1.3.02).

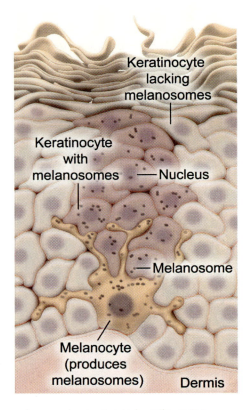

Figure 19.8 Transfer of melanosomes from melanocytes to keratinocytes.

In addition to the epidermis performing barrier functions, the dermis and hypodermis also play roles in protecting the body. **Collagen fibers** present in the dermis cause the skin to be strong and resist tearing, and **elastic fibers** contribute to the skin's resilience. The dermis contains phagocytic **dendritic cells**, which, like Langerhans cells in the epidermis, function in the body's immune response. In addition, adipose tissue in the hypodermis cushions the body and serves as an insulating layer to reduce heat loss.

Various environmental factors can result in harm to the body. Table 19.1 summarizes components of the skin that function to protect the body from these potentially harmful factors.

Table 19.1 Skin components that protect the body from potentially harmful environmental factors.

Potentially harmful factor	Layer of skin providing protection	Protective skin component
Abrasion, chemicals	Epidermis (stratum corneum)	Dead keratinocytes, keratin
Excessive water loss	Epidermis (stratum corneum)	Dead keratinocytes, keratin, lipids
Pathogens	Epidermis (stratum corneum)	Dead keratinocytes
	Epidermis (stratum spinosum)	Langerhans cells
	Dermis	Dendritic cells
Ultraviolet radiation	Epidermis	Melanocytes, melanin
Excessive heat loss	Hypodermis	Adipose tissue
Physical impact	Hypodermis	Adipose tissue
	Dermis	Collagen, elastic fibers

19.2.02 Skin's Role in Osmoregulation

Homeostasis refers to the ability of an organism to maintain a relatively stable physiological environment in response to external or internal changes. One important aspect of homeostasis involves maintaining appropriate water and solute concentrations in the body via the process of **osmoregulation**.

Although the kidneys are the body's main osmoregulatory organs (see Concept 16.2.01), skin also functions in an osmoregulatory capacity. **Sweat glands** (sudoriferous glands) in the dermis of the skin secrete **sweat**, which is a mixture of water, salts, trace nitrogenous waste, and antimicrobial proteins (Figure 19.9). The epidermis, which functions as a water barrier, prevents most direct water loss through the skin; however, sweat is deposited on the surface of the epidermis via ducts that travel through the dermis and epidermis. These ducts allow sweat to bypass the epidermal barrier and be released into the environment as necessary.

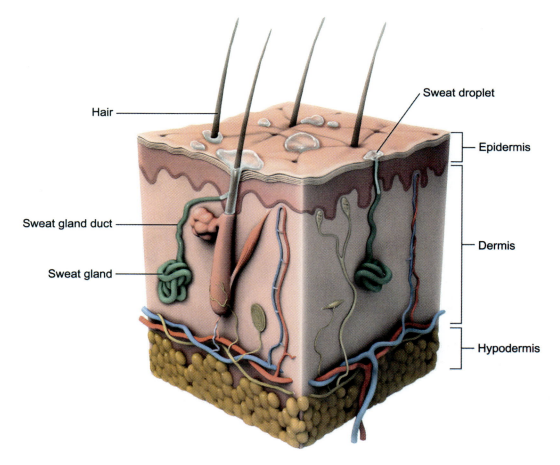

Figure 19.9 Secretion of sweat.

19.2.03 Skin's Role in Thermoregulation

Thermoregulation, the maintenance of body temperature within the normal physiological range, is a major homeostatic function of skin. Thermoregulatory centers in the hypothalamus monitor body temperature by processing information from internal and surface (skin) thermoreceptors. When a significant change in body temperature is detected, the hypothalamic center coordinates the physiological responses necessary to reset the temperature back within the normal range.

When body temperature is above normal, **vasodilation** (widening) of skin arterioles increases blood flow to skin capillaries and maximizes heat loss through the skin (Figure 19.10). Increased heat loss occurs because blood, which is warmed in the body core, transfers heat to the cooler skin surface as it passes through capillaries near the surface. Vasodilation occurs when the smooth muscle surrounding blood vessels relaxes.

The process of sweating (ie, **perspiration**) occurs when sweat is secreted onto the skin's surface by sweat (sudoriferous) glands. Heat loss and subsequent cooling occur due to **evaporation** of the water in the sweat, an endothermic process that absorbs heat from the body.

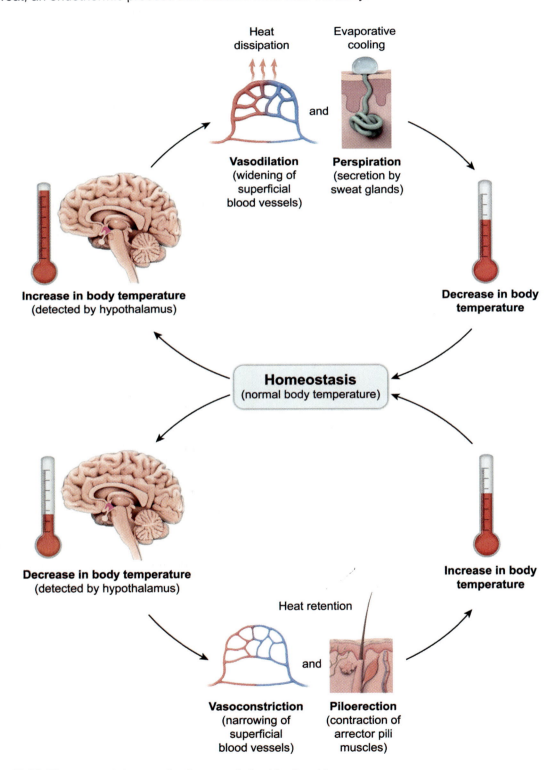

Figure 19.10 Thermoregulatory mechanisms carried out by the skin.

Chapter 19: Skin System

When body temperature is below normal, **vasoconstriction** (narrowing) of skin arterioles minimizes heat loss by diverting warm blood away from capillaries near the skin's surface and toward blood vessels in the interior of the body (Figure 19.10). Vasoconstriction occurs when the smooth muscle surrounding blood vessels contracts.

Adipose tissue in the hypodermis functions as an insulating layer between the body's interior and the outside environment and aids in the retention of core body heat. In cold environments, sympathetic signaling causes contraction of arrector pili muscles (attached to hair follicles in the skin), which causes **piloerection** (ie, hairs standing upright). Piloerection, which impedes heat loss by trapping warm air near the skin's surface, is an inefficient means of heat retention in humans due to insufficient body hair.

> ☑ **Concept Check 19.1**
>
> Fill in the blanks in the sentences below with the words "evaporation," "facilitate," "hinder," "insulation," "piloerection," "sweat," "vasoconstriction," and "vasodilation."
>
> When body temperature is too low, _____ occurs and arterioles within the skin undergo _____ to _____ the loss of heat from the body. In addition, adipose tissue beneath the skin provides _____ to help retain body heat. Conversely, when body temperature is too high, _____ of skin arterioles and _____ of _____ from the skin _____ transfer of heat from the body to the environment.
>
> **Solution**
>
> Note: The appendix contains the answer.

19.2.04 Regulation of Skin Function

The ability of the skin to participate in thermoregulation and osmoregulation depends on controlling vasodilation, vasoconstriction, and sweat gland function within the skin. These functions are regulated by the nervous system and hormonal inputs, as summarized in Table 19.2.

When body temperature increases above a set point established by the hypothalamus, vasodilation and activation of sweat glands occur as a result of stimulation of arterioles and sweat glands via the sympathetic nervous system. Although postganglionic sympathetic nerve terminals typically release the neurotransmitter norepinephrine, these skin responses to elevated temperature are triggered instead by the release of acetylcholine. In addition, emotionally stressful situations can cause sweat secretion as a result of the production of the fight-or-flight hormone epinephrine by the adrenal medulla.

Vasoconstriction of arterioles in the skin, which can occur in response to cold environmental temperatures and emotionally stressful situations, is regulated by the sympathetic nervous system via the release of norepinephrine.

Table 19.2 Factors affecting homeostatic functions of the skin.

Stimulus	Responding structures	Chemical signal	Effect
High body temperature	Superficial arterioles	Acetylcholine	Vasodilation
	Sweat glands	Acetylcholine	Secretion of sweat
Low body temperature	Superficial arterioles	Norepinephrine	Vasoconstriction
Emotional distress	Superficial arterioles	Norepinephrine	Vasoconstriction
	Sweat glands	Epinephrine	Secretion of sweat

Lesson 20.1

Immune System Components

Introduction

The immune system is responsible for protecting the body from pathogens such as viruses, bacteria, fungi, and parasites, as well as from defective body cells (eg, cancer cells) and toxins. The immune system is composed of cells that carry out the immune response and lymphoid organs and tissues, structures in which immune system cells form, mature, and interact with pathogens. This lesson details these immune system components and their locations in the body.

20.1.01 Cells of the Immune System

Leukocytes are white blood cells responsible for producing an immune response to defend the body. Various types of leukocytes exist, and they reside in different parts of the body, including in blood and body tissues. Table 20.1 summarizes characteristics of leukocytes in the immune system.

Table 20.1 Characteristics of immune system cells.

Cell type	Appearance	Distinctive feature	Location
Neutrophil		Visible cytoplasmic granules and segmented nucleus	Blood and other body tissues
Lymphocyte		Lacks cytoplasmic granules	Blood and lymphoid tissues
Monocyte		Lacks cytoplasmic granules	Blood
Macrophage		Lacks cytoplasmic granules (derived from monocyte)	Body tissues
Eosinophil		Visible cytoplasmic granules	Blood and other body tissues
Basophil		Visible cytoplasmic granules	Blood and other body tissues
Mast cell		Visible cytoplasmic granules	Body tissues
Dendritic cell		Cell membrane has numerous branchlike extensions	Body tissues

Immune cells can be categorized based on the structure of their cytoplasm. Some types of leukocytes (eg, neutrophils, eosinophils, basophils) have protein-containing cytoplasmic granules visible in stained cell preparations. Such leukocytes are referred to as **granulocytes**. Other types of leukocytes (eg, lymphocytes, monocytes) lack visible cytoplasmic granules and are called **agranulocytes**.

20.1.02 Bone Marrow

The **red bone marrow** is a soft, spongy **lymphoid organ** located within bones that gives rise to all types of blood cells, including leukocytes involved in the immune response (see Figure 20.1). Cells originating in the bone marrow begin as multipotent **hematopoietic stem cells**, which can differentiate into **lymphoid progenitor cells** or **myeloid progenitor cells**.

Lymphoid progenitor cells go on to become lymphocytes, including B lymphocytes (B cells), T lymphocytes (T cells), and natural killer (NK) cells. Most remaining cells of the immune system, including monocytes, macrophages, neutrophils, eosinophils, basophils, mast cells, and some types of dendritic cells, are derived from myeloid progenitor cells. Myeloid progenitor cells also give rise to cells not directly involved in immunity, including erythrocytes (red blood cells) and platelet-forming megakaryocytes.

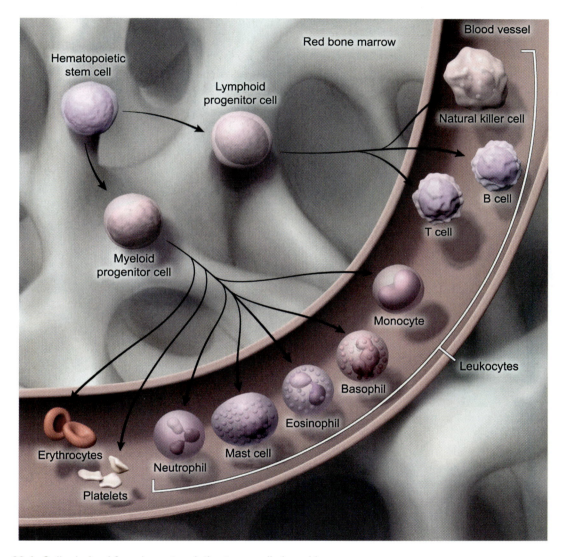

Figure 20.1 Cells derived from hematopoietic stems cells in red bone marrow.

In addition to producing all blood cells, the bone marrow functions in lymphocyte maturation. Specifically, B cells begin their maturation process in the bone marrow before migrating to other lymphoid organs, such as the spleen and lymph nodes, where B cell maturation is completed. Bone marrow is discussed further in Concept 18.1.02.

20.1.03 Thymus

The **thymus** is a lymphoid organ located just superior to the heart, between the lungs (Figure 20.2). The thymus, which is most active prior to puberty, achieves its largest size during adolescence and gradually shrinks thereafter, being replaced over time by adipose tissue and fibrous connective tissue.

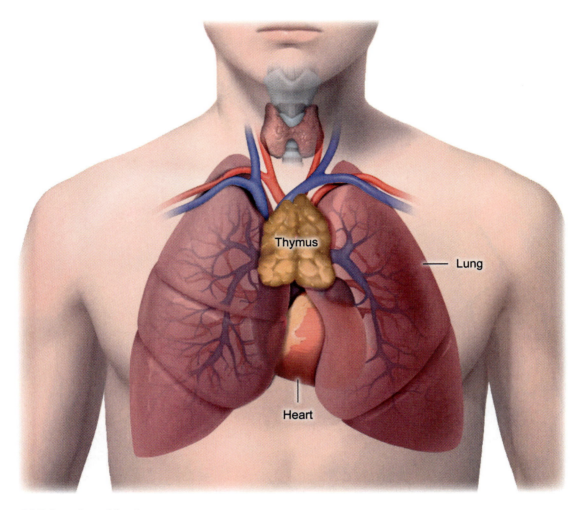

Figure 20.2 Location of the thymus.

During embryonic development, a population of lymphoid progenitor cells that ultimately differentiate into functional T cells migrates from the bone marrow to the thymus, the site of T cell maturation. During this maturation process, self-reactive T cells that could potentially mount an immune response against the body's own healthy cells are inactivated or induced to undergo apoptosis (ie, programmed cell death), as shown in Figure 20.3. Consequently, the remaining T cells typically exhibit **self-tolerance**, that is, they do not recognize the body's own healthy cells as foreign and therefore do not attack these healthy body cells.

Chapter 20: Immune System

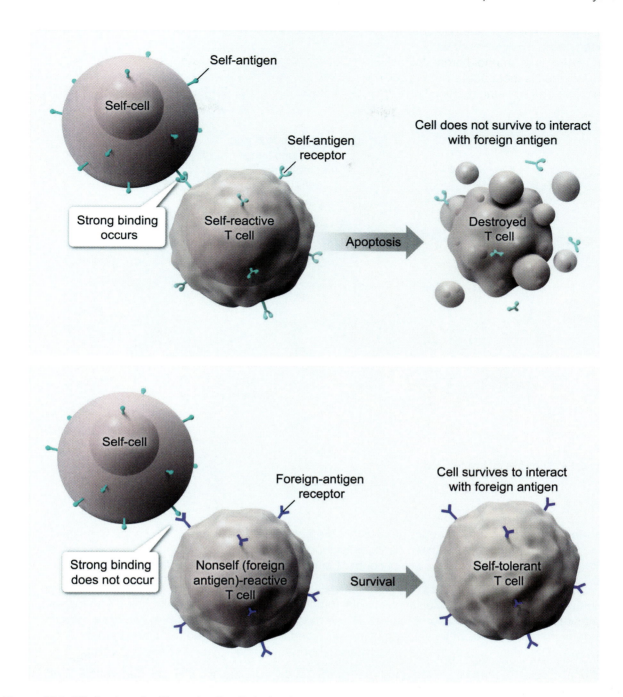

Figure 20.3 Elimination of self-reactive T cells in the thymus.

20.1.04 Lymph Node

Lymph nodes are organs of the lymphatic system, which is described in more detail in Lesson 13.2. As depicted in Figure 20.4, each lymph node contains lymphoid tissue and is enclosed in a dense fibrous capsule. Lymph nodes are clustered along lymph vessels and filter lymph as it flows through the lymphatic system.

Lymph nodes contain macrophages that identify pathogens and other foreign materials in the lymph and destroy these materials via phagocytosis. In addition, dendritic cells migrate to lymph nodes after engulfing pathogens in other tissues. These dendritic cells then participate in the activation of B and T lymphocytes in the lymph nodes.

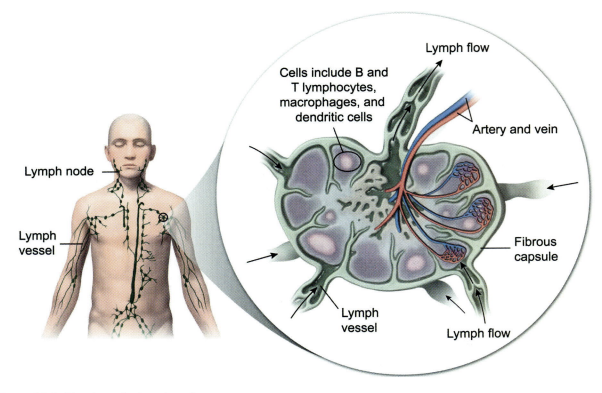

Figure 20.4 Structure of a lymph node.

20.1.05 Spleen

The **spleen**, located in the left upper quadrant of the abdomen (see Figure 20.5), is the largest lymphoid organ. Like lymph nodes, the spleen contains lymphoid tissue surrounded by a capsule and functions in the filtration of a body fluid. However, in contrast to lymph nodes, the spleen filters blood rather than lymph.

Within the spleen, there are two distinguishable components: red pulp and white pulp. The red pulp contains red blood cells and macrophages. These macrophages engulf and remove foreign material and pathogens from the blood, while also removing old, abnormal, or damaged red blood cells from circulation. The white pulp of the spleen contains abundant B and T lymphocytes that carry out surveillance of the blood to identify bloodborne pathogens and mount specific immune defenses against them.

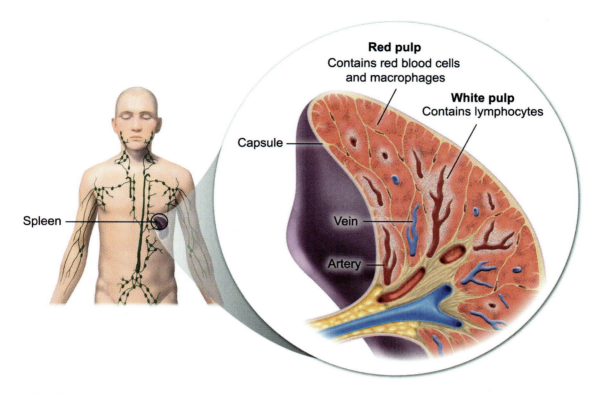

Figure 20.5 Structure of the spleen.

Concept Check 20.1

The diagram below depicts the differentiation of cells in the bone marrow into blood cells and other immune system cells. Complete the diagram and the sentences following the diagram by filling in the blanks with the appropriate cell names.

NK = natural killer.

_____ undergo maturation in the thymus, whereas _____ mature within the bone marrow before migrating to the lymph nodes and spleen. Cells that migrate to the lymph nodes after engulfing pathogens in body tissues are _____. In addition to lymphocytes, other cells present in large numbers within the spleen include _____ and _____.

Solution

Note: The appendix contains the answer.

Lesson 20.2

Immune System Function

Introduction

The immune system functions to protect the body from diseases by defending against potentially harmful agents such as pathogens, toxins, and cancer cells. The body employs various types of external (surface) barriers to keep foreign materials from entering the body. If these barriers fail to prevent the entry of harmful agents, the body responds with internal defenses, which include rapid, nonspecific (**innate**) defenses and slower, specific (**adaptive**) defenses.

The innate defenses are nonspecific in that they are not directed at particular pathogens, whereas the adaptive defenses are specifically targeted. Furthermore, activation of nonspecific defenses does not result in **immunological memory** (ie, a substantial increase in the ability of the defenses to respond to future attacks by the same pathogen), whereas activation of specific defenses does result in such memory. Consequently, specific defenses adapt and become more efficient at eliminating specific pathogens over time upon repeated exposure to the same pathogen.

This lesson covers immune system functions, including innate defenses and various types of acquired immunity.

20.2.01 External Innate Defenses

The body's **innate defenses**, which include the external (surface) structures that provide the body's first line of defense, are present at birth. The skin and mucous membranes that collectively form the body's external surface act as physical barriers to pathogen entry into the body and secrete various substances that enhance this barrier function. The structural basis for the skin's barrier function is discussed in Concept 19.2.01.

Some secretions produced by the skin and mucous membranes create environments unfavorable to pathogens at the body's surface (see Figure 20.6). Sweat and sebum (oil) produced by the skin cause the skin to have a slightly acidic pH, which inhibits bacterial growth. Gastric juice, secreted by the mucous membrane that lines the stomach, is very acidic and contains protein-digesting enzymes, resulting in an environment within the stomach that destroys most swallowed pathogens.

Various defensive proteins and peptides also contribute to the external barriers. **Mucin** is a protein that dissolves in water to form a sticky fluid called **mucus**, which helps protect the body by trapping pathogens, particularly in the respiratory, digestive, and urogenital tracts. The enzyme **lysozyme**, which kills bacteria by disrupting bacterial cell walls, is secreted into saliva, tears, and respiratory mucus. In addition, **defensins**, peptides produced by epithelial cells and various immune cells, can destroy a variety of pathogens, including bacteria, viruses, and fungi. Similarly, sweat contains **dermcidin**, a peptide that kills bacteria and fungi.

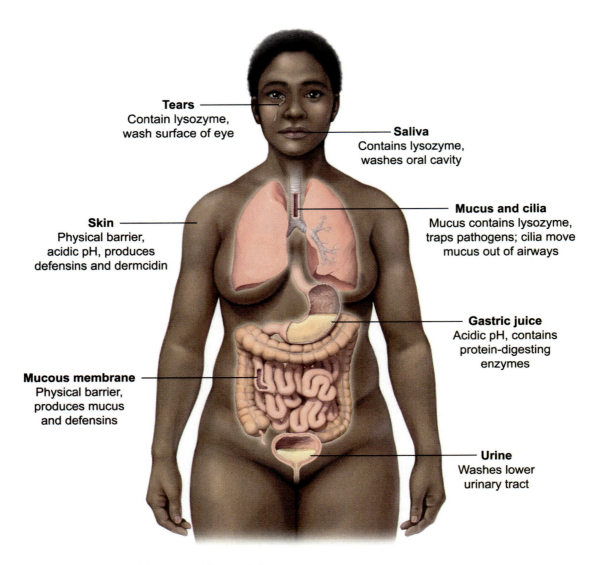

Figure 20.6 Components of the external innate defenses.

The physical and chemical defenses that operate at the body's surface are supplemented by mechanical mechanisms that help move pathogens away from the body. **Cilia**, located on cells of the upper respiratory tract, sweep mucus (and trapped pathogens) upward toward the mouth, where pathogens can be swallowed and destroyed by gastric juice. In addition, tears and saliva help wash pathogens from the surface of the eyes and oral cavity, respectively, and the outward flow of urine from the body helps remove pathogens from the lower urinary tract.

20.2.02 Internal Innate Defenses

Pathogens that evade the body's external innate defenses (described in the previous concept) are quickly met by internal innate defenses. Like external defenses, internal innate defenses are not directed at specific pathogens, but rather respond to a broad range of microorganisms, potentially harmful foreign substances, and abnormal (eg, cancerous) body cells.

Some body cells possess receptors called **pattern recognition receptors (PRRs)**, which bind to classes of molecules called **pathogen-associated molecular patterns (PAMPs)**. PAMPs are typically present on microorganisms and viruses but not on healthy human cells. Binding of PAMPs to PRRs results in the activation of internal innate defenses.

Two types of leukocytes that play significant roles in innate immunity are macrophages (derived from monocytes) and neutrophils. Macrophages and neutrophils both engage in **phagocytosis**, a process by which pathogens are engulfed by defensive cells and destroyed.

Pathogens or other materials engulfed by phagocytes (ie, phagocytic cells) are enclosed within vesicles (phagosomes) into which hydrogen ions are pumped. The resulting acidified phagosomes fuse with lysosomes to form phagolysosomes, which contain **hydrolytic enzymes**. The engulfed materials within the phagolysosomes are digested and destroyed by these enzymes, as shown in Figure 20.7.

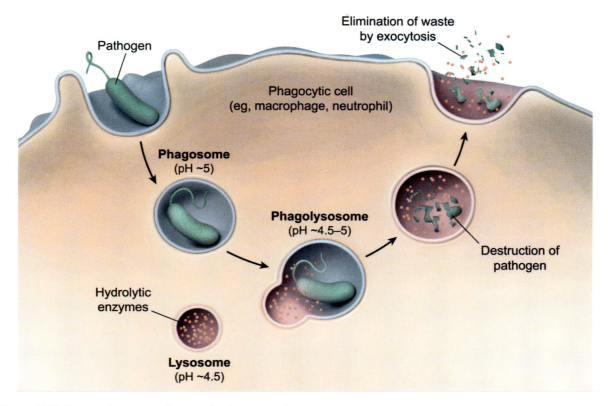

Figure 20.7 Destruction of a pathogen via phagocytosis.

In addition to destroying pathogens via phagocytosis, macrophages and neutrophils both perform other defensive functions. For example, macrophages participate in the activation of lymphocytes involved in adaptive immune responses, and neutrophils can undergo **degranulation**, a process by which granulocytes secrete the contents of various cytoplasmic granules via exocytosis. Degranulation results in the release of a variety of defensive substances, including antimicrobial molecules such as defensins, as well as cytokines (chemical signals) that help regulate the immune response.

Other leukocytes, including natural killer (NK) cells, mast cells, basophils, and eosinophils, also contribute to innate defenses. Although NK cells are not phagocytic, they can induce virus-infected cells and cancer cells to destroy themselves via apoptosis. Furthermore, mast cells, basophils, and eosinophils are granulocytes that participate in multiple immune system activities, including the **inflammatory response**, defense against parasites, and allergic reactions. Mast cells promote inflammation by releasing histamine, which triggers vasodilation and increased capillary permeability, as depicted in Figure 20.8.

Chapter 20: Immune System

1. Inflammatory chemicals (including histamine) are released in response to tissue injury.

2. Histamine promotes vasodilation and increases capillary permeability.

3. These changes facilitate delivery of plasma proteins and phagocytes (neutrophils and macrophages) to the infection site.

4. As inflammatory signaling subsides, macrophages clear the area of dead cells and debris.

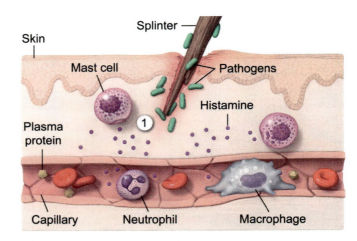
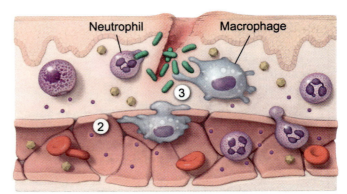
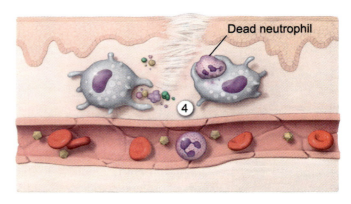

Figure 20.8 Events of the inflammatory response.

The actions of various antimicrobial proteins enhance the internal innate defenses. **Interferons** are proteins produced by virus-infected cells that interfere with viral replication in other cells and participate in regulating leukocyte activity. **Complement** (ie, a collective term referring to the complement system) is composed of more than 30 different proteins. Upon activation, complement increases phagocyte efficiency and helps regulate the inflammatory response. Activated complement also forms membrane attack complexes (MACs), which insert into pathogen cell membranes, forming pores that cause pathogen lysis.

20.2.03 Adaptive Immunity

In contrast to the innate defenses, which are nonspecific, the mechanisms responsible for **adaptive immunity** (ie, **cell-mediated immunity** and **humoral immunity**) are specific, in that they are directed at particular pathogens. Each different pathogen possesses distinct **antigens**, which are specific molecules (typically proteins or polysaccharides) that trigger an immune response by being recognized as foreign by the cells of the adaptive immune system.

Adaptive immunity depends on the activity of lymphocytes, namely T lymphocytes (T cells) and B lymphocytes (B cells). T and B cells have surface receptors that bind to specific antigens, and each of these cells possesses a great number of antigen receptors. All receptors on a single cell have the same specificity, so each T cell and B cell can recognize and respond to only one particular antigen. However, due to the large number of lymphocytes with unique receptors in each individual, the total number of different antigens to which a person can respond is very great.

Naïve lymphocytes (ie, mature lymphocytes that have yet to encounter antigens to which they can respond) must be activated before they can fight an invader. This activation happens only to lymphocytes that possess receptors that bind the invader's antigens. Consequently, only a small fraction of naïve lymphocytes in an individual are activated by a particular pathogen. Upon activation, each of these responsive lymphocytes proliferates to form a large number of cells identical to itself (ie, a clone). This process of activating and replicating only cells capable of responding to a particular antigen is called **clonal selection** (Figure 20.9).

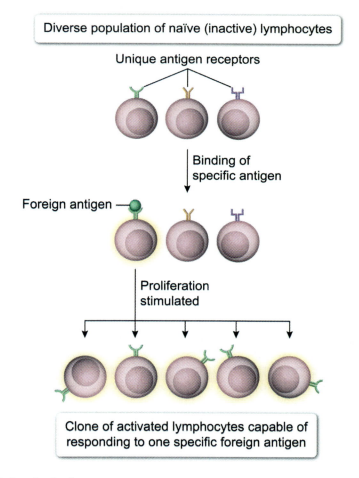

Figure 20.9 Process of clonal selection.

Following the proliferation of a particular lymphocyte, most cells in the resulting clone differentiate into **effector cells** that immediately function in combating the pathogen that stimulated clone production. The remaining cells of the clone differentiate into long-lived **memory cells**, which are held in reserve. The initial activity of the adaptive immune system to a new pathogen is called a **primary immune response**, and subsequent responses to the same pathogen are called **secondary immune responses**.

Memory cells are the basis for immunological memory, which allows specific immune responses to be adaptive. The presence of previously formed memory cells causes specific defenses to be faster and more effective in response to second or subsequent infections by a particular pathogen, as shown in Figure 20.10.

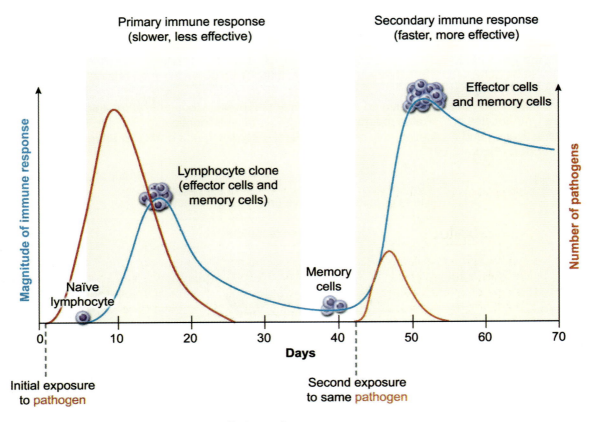

Figure 20.10 Effect of memory cells on specific immunity.

20.2.04 Antigen Presentation

Adaptive immune responses require lymphocyte activation, a process dependent on recognition and binding of foreign antigens by B-cell and T-cell receptors. Antigen receptors on B cells can recognize and bind unaltered, isolated antigens. However, T-cell receptors recognize antigens only when the antigens are processed and displayed on the surface of cells via membrane proteins called **major histocompatibility complex (MHC) proteins**. The process by which antigens are displayed in association with MHC proteins on cells is called **antigen presentation**.

There are two classes of MHC proteins. Class I MHC proteins (MHC I) are present on all nucleated cells, whereas class II MHC proteins (MHC II) are located only on **antigen-presenting cells (APCs)**, which include dendritic cells, macrophages, and B cells. MHC I molecules typically display fragments of antigens produced within the cell presenting them. These antigens can be normal (self) antigens in a healthy cell or foreign antigens, such as viral proteins produced in a cell during a viral infection.

Conversely, MHC II proteins typically display fragments of foreign antigens that were taken into APCs via phagocytosis (Figure 20.11).

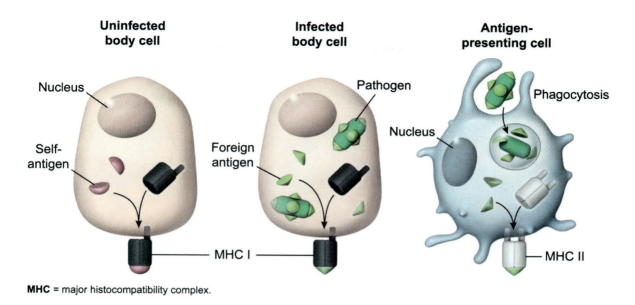

MHC = major histocompatibility complex.

Figure 20.11 Role of MHC proteins in antigen presentation.

20.2.05 Cell-Mediated Immunity

The adaptive immune system defends against intracellular pathogens (including viruses and certain bacteria), cancer cells, and foreign cells (eg, cells of transplanted organs) primarily through **cell-mediated immunity**. This type of immunity depends on the function of multiple types of T cells (so named because they mature in the thymus). In particular, T cells called **cytotoxic T (T_C) cells** (also known as CD8+ cells) function as effector cells directly responsible for destroying infected or abnormal body cells.

T cell activation requires interactions between naïve T cells and antigen-presenting cells (APCs). Such interactions typically occur in lymph nodes or the spleen. A type of T cell called a **helper T (T_H) cell** (or CD4+ cell) becomes activated when the T_H cell is presented with a foreign antigen, in association with MHC II, on the surface of an APC (eg, dendritic cell). If the antigen receptors of the T_H cell can recognize and bind specifically to the presented antigen, the T_H cell becomes activated and replicates to form a clone of the activated T_H effector cells, as depicted in Figure 20.12.

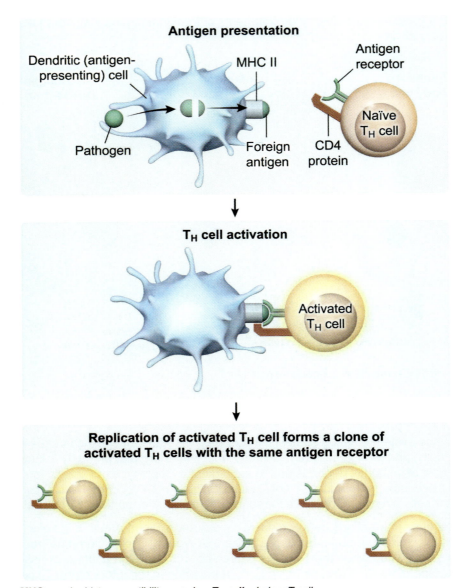

Figure 20.12 Activation and replication of T$_H$ cells.

Clones of T$_H$ effector cells participate in activating naïve T$_C$ cells that possess receptors capable of binding the specific foreign antigen that originally stimulated T$_H$ cell activation, which is presented (along with MHC I) to the T$_C$ cells by APCs. T$_H$ cells stimulate the APCs to present additionally required (ie, co-stimulatory) molecules to the T$_C$ cells, completing T$_C$ cell activation. The activated T$_C$ cells are then stimulated by cytokines (secreted by the T$_H$ cells) to divide and form a clone of T$_C$ effector and memory cells.

When T$_C$ effector cells subsequently encounter body cells displaying the targeted antigen (on MHC I), the T$_C$ cells release cytotoxins to induce these abnormal cells to undergo apoptosis (Figure 20.13).

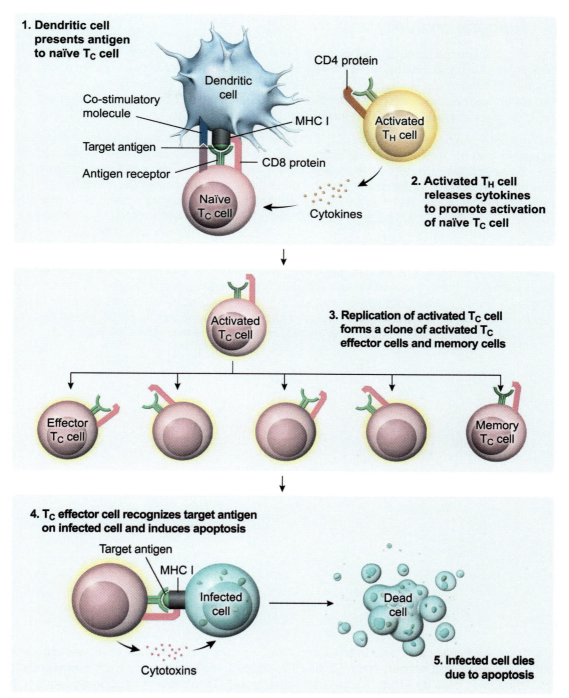

Figure 20.13 Steps in cell-mediated immunity.

In addition to T_H and T_C cells, some T cells differentiate into **regulatory T (T_{Reg}) cells**, which function to inhibit the immune response. T_{Reg} cells play an important role in preventing overactive immune responses, such as those directed against self-antigens (ie, autoimmune responses), which inflict damage on healthy tissues. This topic is discussed in greater detail in Concept 20.2.08.

20.2.06 Humoral Immunity

In addition to cell-mediated immunity, which results from the activity of cytotoxic T cells, adaptive immunity involves **humoral immunity**, which requires the participation of **B cells** (ie, lymphocytes that mature in the bone marrow). Humoral immunity ultimately depends on the function of specialized protein molecules called **antibodies**, which are secreted by activated B cells. For this reason, humoral immunity is also referred to as **antibody-mediated immunity**.

A humoral immune response begins with activation and proliferation (ie, clonal selection) of naïve B cells that possess antigen receptors capable of binding specific foreign antigens present on an invading pathogen. B-cell antigen receptors have essentially the same structure as secreted antibodies; however, these antibody-like receptor molecules are anchored to the B cell membrane. Some antigens (ie, T cell-independent antigens) induce B cell activation without helper T (T_H) cell involvement. However, most antigens (ie, T cell-dependent antigens) provoke B cell responses via a mechanism that requires T_H cell participation.

B cells function as antigen-presenting cells, and in doing so, B cells facilitate their own activation by T_H cells. When B-cell receptors bind their targeted antigens, the bound antigens can be taken up by B cells via endocytosis. After being internalized, the antigens are processed and displayed on the B cells' surface by class II major histocompatibility complex (MHC II) proteins, which allows the antigens to be presented to nearby T_H cells. In turn, these T_H cells secrete cytokines that stimulate replication and differentiation of the antigen-activated B cells into **plasma cells** and memory cells (Figure 20.14).

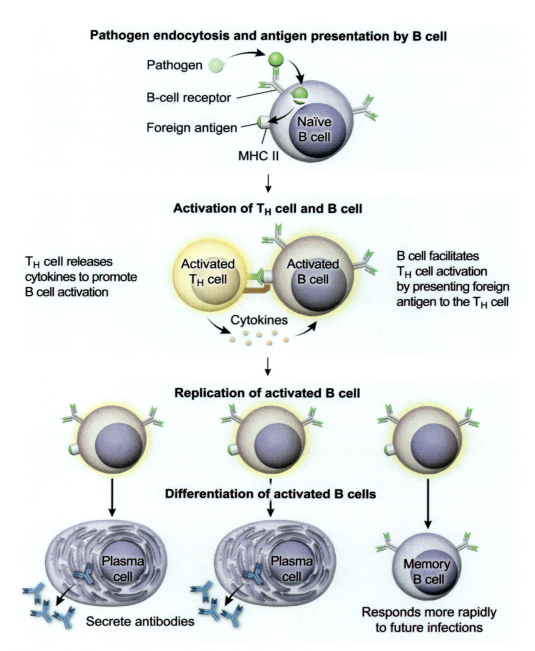

Figure 20.14 Steps in T-cell dependent activation of B cells.

Plasma cells, which contain abundant rough endoplasmic reticulum, are specialized to secrete large numbers of antibody molecules (ie, immunoglobulins [Ig]). Each antibody consists of four polypeptide chains, including two identical heavy chains and two identical light chains.

Each of these chains possesses a variable region and a constant region, and the chains are linked by disulfide bonds to produce a flexible Y-shaped protein. The tips of the two arms of the Y, which are formed by the heavy and light chain variable regions, are identical. These tips on the antibody serve as specific antigen-binding sites, as shown in Figure 20.15.

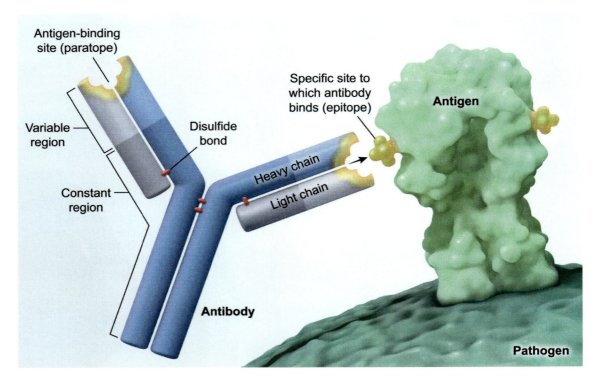

Figure 20.15 Antibody structure and interaction between antibody and antigen.

Antigens are large molecules that typically possess numerous specific sites to which antibodies can bind. These sites are referred to as **antigenic determinants**, or **epitopes**, whereas antigen-binding sites on antibodies are called **paratopes** (Figure 20.15). All antibodies secreted by a particular plasma cell have the same antigen specificity (ie, bind the same epitope), and this specificity is identical to that of the antigen receptors present on the original naïve B cell that was activated to form the plasma cell.

There are five general classes of antibodies: IgG, IgA, IgM, IgE, and IgD. These classes have different structural and functional characteristics that allow different classes to play specialized roles in varied aspects of the humoral immune response. Antibodies typically function in defense against extracellular threats, such as bacteria, fungi, toxins, parasites, and free viruses. However, antibodies do not destroy such invaders directly, but rather inactivate and mark foreign antigens for destruction by other components of the immune system. Various mechanisms of antibody defensive functions are summarized in Table 20.2.

Chapter 20: Immune System

Table 20.2 Mechanisms by which antibodies function in defense of the body.

Mechanism	Description	Effect
Neutralization	Antibodies bind to and block specific functional sites on viruses or toxins	Viruses and toxins are prevented from entering cells and causing damage
Pathogen clumping (or precipitation of soluble antigens)	Antibodies simultaneously bind to antigens on multiple pathogens	Clumped pathogens (or precipitated antigens) are efficiently engulfed and destroyed by phagocytes
Opsonization	Antibodies coat pathogens by binding to surface antigens	Antibody-coated pathogens are efficiently engulfed and destroyed by phagocytes
Complement activation	Antigen-antibody complexes on pathogen surface activate complement proteins	Activated complement proteins facilitate phagocytosis, regulate the inflammatory response, and cause pathogen lysis
Antibody-dependent cellular cytotoxicity	Antibodies bound to abnormal body cells or parasites trigger effector cells (eg, natural killer cells, neutrophils, eosinophils) to release cytotoxic molecules	Cytotoxic molecules released by effector cells induce apoptosis or cause lysis of target cells

20.2.07 Immune Response Integration

Different pathogens (eg, bacteria, viruses, parasites) present different challenges to the body's defenses. However, successful defense against any pathogen typically requires an integrated immune response that involves both innate and adaptive immune mechanisms operating in a coordinated manner (see Figure 20.16 and Table 20.3).

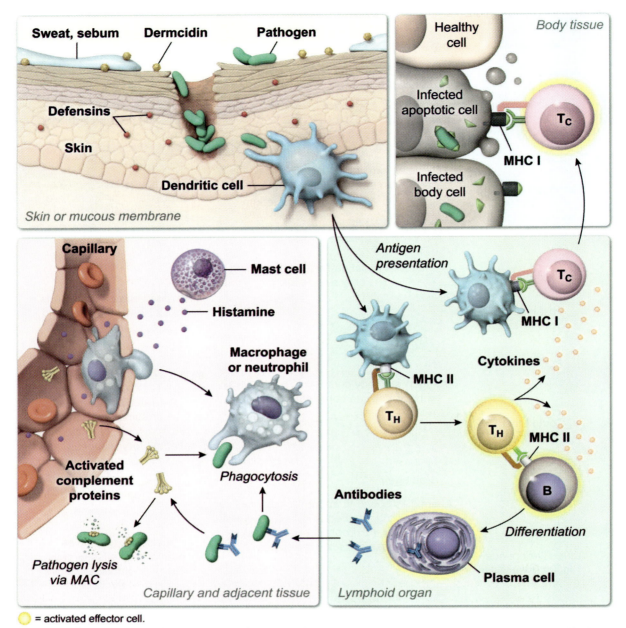

Figure 20.16 Integration of defensive mechanisms.

Table 20.3 Roles of components shown in Figure 20.16.

Location	Component	Role
Skin or mucous membrane	Skin	Functions as physical barrier to pathogens
	Pathogen	Causes disease
	Sweat, sebum	Acidic pH inhibits pathogen growth
	Dendritic cell	Kills pathogens via phagocytosis, participates in T_H and T_C cell activation in lymph node
	Defensins	Kill pathogens
	Dermcidin	
Lymphoid organ	Activated T_H cell	Secretes cytokines to activate T_C and B cells
	Plasma cell	Secretes antibodies
Body tissue	Activated T_C cell	Induces infected body cell to undergo apoptosis
Capillary and adjacent tissue	Mast cell	Secretes chemicals that trigger inflammatory response
	Histamine	Promotes vasodilation and increased capillary permeability
	Macrophage or neutrophil	Kills pathogens via phagocytosis
	Activated complement proteins	Facilitate phagocytosis, form membrane attack complex, promote inflammation
	Antibody	Neutralizes pathogens, facilitates phagocytosis, activates complement, triggers antibody-dependent cellular cytotoxicity

T_C cell = cytotoxic T cell; T_H cell = helper T cell.

Macrophages and dendritic cells play key roles in both innate immunity (ie, as active phagocytes) and adaptive immunity (ie, as antigen-presenting cells), and they therefore serve to functionally link the innate and adaptive defenses. Likewise, antibodies produced via an adaptive immune response can facilitate innate immunity by activating complement and by coating pathogens (ie, acting as **opsonins**), which enhances phagocyte function.

In addition to the integration that exists between innate and adaptive defensive mechanisms, the combined participation of both cell-mediated and antibody-mediated (ie, humoral) adaptive immunity is essential for defeating many pathogens. For example, the activity of cytotoxic T cells, as well as antibodies, is necessary to eliminate pathogens (eg, viruses) that reside both inside and outside of body cells.

> ☑ **Concept Check 20.2**
>
> The following flowchart depicts steps in an integrated adaptive immune response directed against an invading pathogenic virus. Complete the flowchart by determining which of the following phrases belongs in each blank box: "Antigen presentation," "B cell activation," "Cytotoxic T cell activation," "Destruction of virus-infected body cell," "Helper T cell activation," "Neutralization of extracellular viruses," "Phagocytosis of virus."
>
>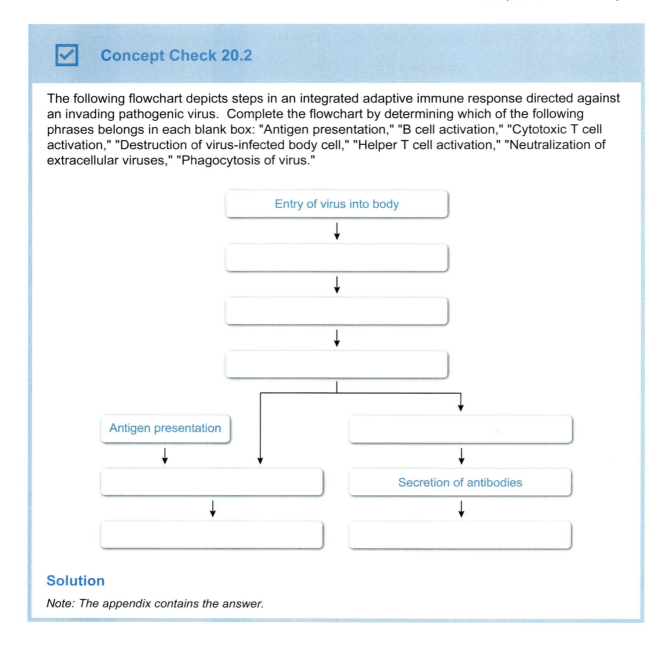
>
> **Solution**
> Note: The appendix contains the answer.

20.2.08 Autoimmunity

To function properly, an individual's immune system must be able to distinguish between **self-antigens** (ie, autoantigens), which are normal components of the individual's own healthy cells, and **foreign antigens**. As illustrated in Concept 20.1.03, autoreactive lymphocytes (ie, T and B cells that possess antigen receptors capable of strongly binding autoantigens) are typically eliminated or rendered inactive during immune system development. However, in some cases the immune system fails to tolerate autoantigens, and instead mounts an immune response against these antigens, a situation referred to as **autoimmunity**.

During an autoimmune response, autoreactive antibodies (ie, autoantibodies) and autoreactive cytotoxic T (T_C) cells cause destruction of healthy cells in the body, which can result in an autoimmune disease (eg, multiple sclerosis, type 1 diabetes mellitus). Regulatory T (T_{Reg}) cells that function to suppress the immune response are important for limiting autoimmune responses (see Figure 20.17). Therefore, T_{Reg} dysfunction is commonly observed in individuals with autoimmune diseases.

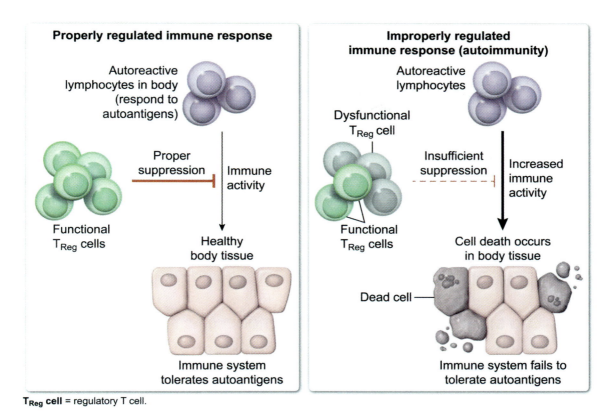

Figure 20.17 Role of regulatory T cells in the prevention of autoimmunity.

END-OF-UNIT MCAT PRACTICE

Congratulations on completing **Unit 10: Skin and Immune Systems**.

Now you are ready to dive into MCAT-level practice tests. At UWorld, we believe students will be fully prepared to ace the MCAT when they practice with high-quality questions in a realistic testing environment.

The UWorld Qbank will test you on questions that are fully representative of the AAMC MCAT syllabus. In addition, our MCAT-like questions are accompanied by in-depth explanations with exceptional visual aids that will help you better retain difficult MCAT concepts.

TO START YOUR MCAT PRACTICE, PROCEED AS FOLLOWS:

1) Sign up to purchase the UWorld MCAT Qbank
   IMPORTANT: You already have access if you purchased a bundled subscription.
2) Log in to your UWorld MCAT account
3) Access the MCAT Qbank section
4) Select this unit in the Qbank
5) Create a custom practice test

Appendix
Concept Check Solutions

You will find detailed, illustrated, step-by-step solutions for each concept check in the digital version of this book.

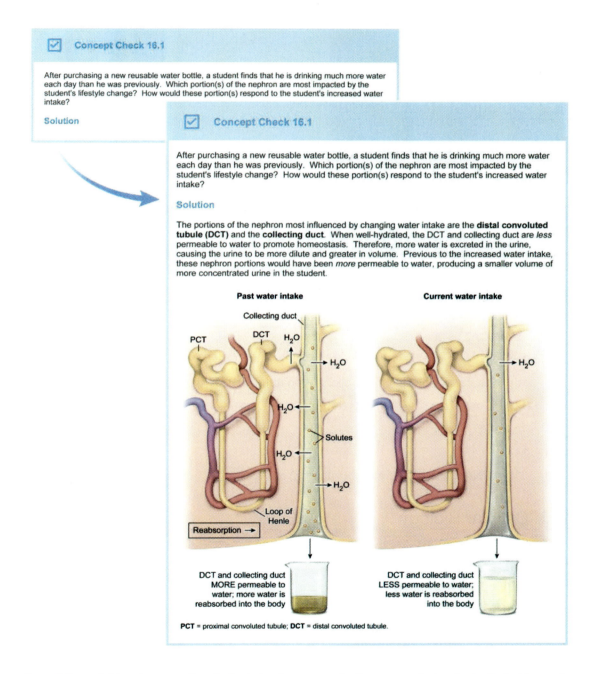

In this section of the print book, you will only find short answers to the concept checks included in each chapter. Please go online for an interactive and enhanced learning experience with visual aids.

Unit 1. Molecular Biology

Chapter 1. DNA Structure, Synthesis, and Repair

Lesson 1.1

    1.1   764 nucleotides.

Chapter 2. Gene Expression

Lesson 2.3

    2.1   mRNA 1

    2.2   No effect

Unit 2. Biological Research Techniques

Chapter 3. Designing and Interpreting Experiments

Lesson 3.1

    3.1   Blood type: nominal categorical; pH: continuous quantitative; test grade: ordinal categorical; reaction time: continuous quantitative; cell count: discrete quantitative

Lesson 3.2

    3.2   type II; decreases; increases; type I

Chapter 4. Biotechnology

Lesson 4.1

    4.1   5' - GCATT - 3'

Lesson 4.2

    4.2   nondenaturing; protein interactions; denaturing; molecular weights

Lesson 4.3

    4.3   A and D

Lesson 4.4

    4.4   1) miRNA, siRNA, and morpholinos. 2) more likely.

Unit 3. Cellular Biology

Chapter 5. Eukaryotic Cells

Lesson 5.2

    5.1   *Prior to transport*: red – extracellular surface of plasma membrane; green – cytoplasmic surface of plasma membrane; blue – extracellular. *After transport*: red – inside endosome; green – exterior of endosome; blue – inside endosome.

Lesson 5.3

5.2 By evaluating the protein's DNA coding information for a signal sequence or by evaluating the protein sequence for a signal peptide

5.3 Lysosomes

Lesson 5.4

5.4 Stages 2-5

5.5 Tumor suppressor gene

Chapter 6. Prokaryotes and Viruses

Lesson 6.1

6.1 Lipoteichoic acid, teichoic acid: gram-positive; Lipid A, LPS, outer membrane: gram-negative; NAG, NAM, peptide crossbridge, peptidoglycan: both.

Lesson 6.2

6.2

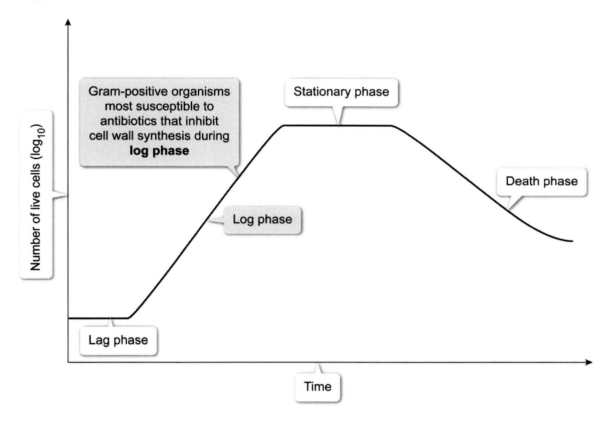

Lesson 6.3

6.3 The *Shigella* cells already contained an F plasmid (ie, were F$^+$).

6.4 *lac* gene expression would be repressed, and the organism would be unable to catabolize lactose.

Lesson 6.4

6.5 3′ UUUACACACGGCUUUACAACU 5′

Unit 4. Genetics and Evolution

Chapter 7. Genetics

Lesson 7.2

7.1 Male: hh; female: Hh

Chapter 8. Evolution

Lesson 8.1

8.1 34%

Unit 5. Reproduction

Chapter 9. Reproductive Systems

Lesson 9.2

9.1 A) hypothalamus, B) anterior pituitary gland, C) FSH, D) inhibin, E) LH, F) testosterone, G) GnRH.

Lesson 9.3

9.2 GnRH, FSH, and LH levels would remain high, low progesterone.

Chapter 10. Pregnancy, Development, and Aging

Lesson 10.1

10.1 1) Sperm would likely not reach the oocyte due to impaired motility. 2) Sperm would not penetrate the zona pellucida to reach the oocyte's plasma membrane. 3) Additional genetic information would be introduced, leading to a nonviable polyploid zygote.

Lesson 10.2

10.2 Ectoderm: adrenal medulla, anterior pituitary gland, oral epithelium, Schwann cells, sympathetic ganglia. Mesoderm: capillary endothelium, dermis of skin, red blood cells, rib cartilage. Endoderm: alveoli of lung, reproductive tract epithelium, urinary bladder epithelium.

Lesson 10.3

10.3 1) True; morphogen gradients are often layered. 2) False; lateral inhibition generally leads to divergent cellular identities. 3) True; once a developmental window has passed, cells lose the ability to respond to certain inductive signals.

Lesson 10.4

10.4 Umbilical vein, ductus arteriosus, umbilical arteries

Lesson 10.5

10.5 A senescent cell may re-enter an active state if the enzyme telomerase becomes active, restoring telomeres at chromosome ends.

Unit 6. Endocrine and Nervous Systems

Chapter 11. Endocrine System

Lesson 11.2

    11.1 TRH and TSH would be elevated. Abnormally low metabolism levels and low body temperature.

    11.2

Hormone	Location of hormone production	Stimulus for hormone release	Hormone target	Hormone function
Antidiuretic hormone (ADH)	Hypothalamus (released from posterior pituitary)	Reduced blood volume, increased blood osmolarity	Kidney nephron tubules	Increases H_2O reabsorption, increasing blood volume and decreasing blood osmolarity
Aldosterone	Adrenal cortex	Low blood pressure	Kidney nephron tubules	Promotes Na^+ reabsorption and K^+ secretion to increase H_2O reabsorption, leading to increased blood volume and blood pressure
Atrial natriuretic factor (ANF)	Atrial myocytes	Increased atrial stretch (sign of high blood pressure)	Kidney	Increases glomerular filtration rate, promotes Na+ and H_2O excretion, and inhibits aldosterone, causing decreased blood volume and blood pressure

Chapter 12. Nervous System

Lesson 12.1

    12.1 Regions 1 and 2: receive information at synapses; region 3: transmits information at synapses.

Lesson 12.2

    12.2 Rising portion of overshoot: voltage-gated Na^+ channels are open. Falling portion of overshoot: voltage-gated K^+ channels are open. Falling portion of undershoot: voltage-gated K^+ channels are open. Rising portion of undershoot: both channels are closed.

Unit 7. Circulation and Respiration

Chapter 13. Circulation

Lesson 13.1

    13.1 68 mL/beat

    13.2 pulmonary vein, left atrium, bicuspid valve, left ventricle, aortic semilunar valve, aorta

Lesson 13.2

    13.3   lymphatic capillary, collecting vessel, lymph node, lymphatic trunk, lymphatic duct, cardiovascular system vein, blood capillary

Chapter 14. Respiration

Lesson 14.1

    14.1

Structure	Reinforced with cartilage	Diameter regulated by smooth muscle	Contains ciliated lining and mucus
Nasal Cavity	Yes	No	Yes
Larynx	Yes	No	Yes
Trachea	Yes	Yes	Yes
Bronchi	Yes	Yes	Yes
Bronchioles	No	Yes	Yes
Alveoli	No	No	No

Lesson 14.2

    14.2   Sample 1: intermediate, sample 2: lowest, sample 3: highest.

Unit 8. Digestion and Excretion

Chapter 15. Digestion

Lesson 15.1

    15.1   Ileocecal sphincter

Lesson 15.2

    15.2

Accessory digestive organ(s)	Function
Liver, gallbladder, pancreas	Lipid digestion
Pancreas	Secretion of trypsinogen and chymotrypsinogen
Liver	Production of bile
Liver	Detoxification of drugs and waste products
Pancreas	Secretion of somatostatin
Liver, pancreas	Glucose homeostasis
Gallbladder	Storage of bile

Chapter 16. Excretion

Lesson 16.2

16.1 Distal convoluted tubule, collecting duct. Portions would be less permeable to water.

Unit 9. Musculoskeletal System

Chapter 17. Muscular System

Lesson 17.1

17.1 Label 1: cytoplasm, label 2: sarcoplasmic reticulum, label 3: extracellular fluid.

Lesson 17.2

17.2 Motor unit; muscle fiber type; slow; oxidative; fast; glycolytic.

Chapter 18. Skeletal System

Lesson 18.1

18.1

Structure	Description
Osteons (Haversian system)	Structural units of compact bone
Volkmann's canals	Cylindrical channels that contain nerves and blood vessels and run *perpendicular* to the long axis of bone
Lacunae	Tiny, osteocyte-containing spaces within the matrix surrounding Haversian canals
Haversian canals	Cylindrical channels that contain nerves and blood vessels and run *parallel* to the long axis of bone
Canaliculi	Tiny channels through which nutrient and waste exchange between osteocytes and the circulation occurs
Lamellae	Concentric rings of bone matrix surrounding Haversian canals

Lesson 18.2

18.2 Parathyroid hormone: net loss of bone; thyroid hormone: net loss of bone; growth hormone: net deposition of bone.

Unit 10. Skin and Immune Systems

Chapter 19. Skin System

Lesson 19.2

    19.1   piloerection; vasoconstriction; hinder; insulation; vasodilation; evaporation; sweat; facilitate.

Chapter 20. Immune System

Lesson 20.1

    20.1

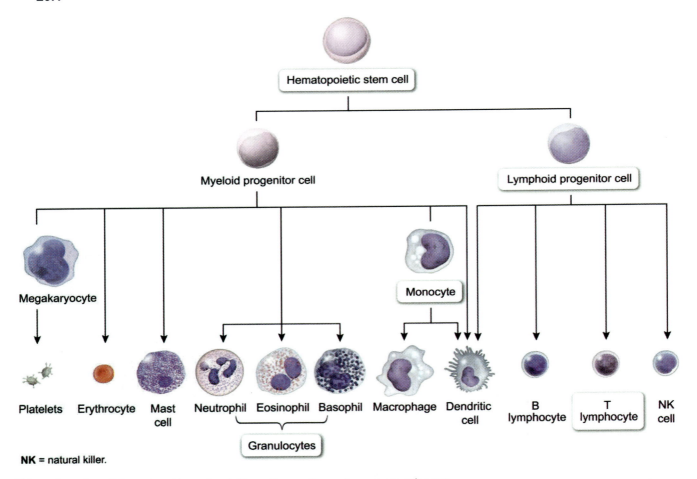

T lymphocytes; B lymphocytes; dendritic cells; erythrocytes; macrophages

Lesson 20.2

20.2

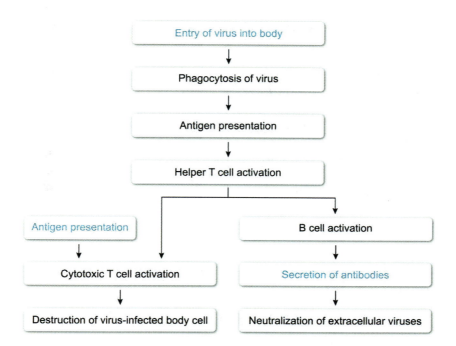

Index

A

A band, 559
ABO antigens, 452, *See also* blood types
absolute refractory period, 433
accessory digestive organs, 524
acetylcholine (ACh), 393, 438, 443, 479, 564, 611
acrosome, 323, 336
actin, 184–85, 187, 198, 556–59, 561–62, 565–67, 573
action potentials, 424, 426–27, 429, 438, 463, 563, 570, 579
activation gates, 429, 433
active transport, 160–61, 163, 165, 167, 171
adaptive immunity, 623, 628, 631, 633
adaptive radiation, 310–11
adenosine triphosphate (ATP), 158, 163–64, 167, 182, 229–31, 323, 577
adipocytes, 400, 586
adipose tissue, 410, 532, 592, 604, 608, 614
adrenal cortex, 344, 380, 389, 391–92, 396, 408, 414, 543, 549
adrenal glands, 372, 391
adrenal medulla, 344, 346, 357, 383, 391, 393–94, 396, 415, 479, 611
adrenocorticotropic hormone (ACTH), 388–90, 414
aerobic respiration, 230–3, *See also* cellular respiration
afferent neurons, *See* sensory neurons
agranulocytes, 612
albumin, 118, 509, 526, 544, 547
aldosterone, 392–93, 412, 415, 542–45, 549–50
alimentary canal, 513–14, 520, 522
allantois, 341, 360
alleles, 206, 278–81, 283, 286, 288–90, 299–301, 307, 351
all-or-none, 429
alternative hypothesis, 80–81
alternative mRNA splicing, 34–35
alveoli, 345, 458–59, 489–91, 495–96, 498, 500–501, 507
amino acids, 28, 37–42, 44, 46, 48–51, 53, 55, 215, 293, 298, 373, 392 438, 520
aminoacyl-tRNA synthetases, 40–41
anaerobic metabolism, 228
androgens, 324, 326, 393, 407–8
angiotensin, 392, 542–43
annealing, 7–8, 94
anterior pituitary gland, 319, 325–26, 333–34, 375, 382, 384–85, 388–91, 401, 408, 414
antibiotic resistance, 102, 233–35
antibiotics, 102, 138, 216–17, 223, 227–28, 233–35, 238–39
antibodies, 116, 118, 361, 365, 451–52, 628–30, 633
anticodon, 38–40, 44
antidiuretic hormone (ADH), 386–87, 414, 542–43, 545, 549–50
antigen presentation, 624–25
antigen-presenting cells (APCs), 624–26, 628, 633
antigens, 119, 451–52, 623–25, 628, 630, 634
antisense, 29, 116, 129
aorta, 362, 467, 471, 476
apoptosis, 201–3, 207, 330, 357–58, 369, 614, 621, 626
appendicular skeleton, 583
aquaporins, 161–62, 386
archaea, 148, 211–12, 218
archenteron, 343
arrector pili, 605, 611
arteries, 210, 462, 466–71, 475–77, 480, 575, 605
arterioles, 469, 472, 478–80, 502, 539, 543, 605
asexual reproduction, 3, 225
astrocyte, 421, 423
atrial natriuretic factor (ANF), 412–13
atrioventricular (AV) node, 464, 466
atrioventricular (AV) valves, 462
atrium, 361–62, 386, 412, 462, 464–67, 471, 578
autocrine signaling, 354, 379
autoimmunity, 634–35
autonomic nervous system (ANS), 393, 438, 440–42, 444, 479, 535, 552, 563, 581
autophagy, 180–81, 368
autoradiography, 112–114, 118
autoregulation, 479–80

autosomes, 280, 292
axial skeleton, 583
axons, 418–20, 426, 428, 433–34, 438–39, 441, 445, 564
axon hillock, 418, 426
axon terminal, 386, 418, 424

B

β-endorphins, 388, 390–91, 414
bacilli, 213
bacteria, 98–99, 102, 105, 148, 211–13, 218–24, 227–28, 232–33, 235, 239, 241, 243, 250, 252, 619
bacteriophages, 98, 241, 252, 257–60
Barr body, 352
basal body, 188, 221
basement membrane, 209, 470, 481, 539, 546
base pairing, 6–7, 12, 30, 37–38, 40, 43–44, 48, 55, 89, 109, 113
basophils, 612–13, 621
B cells, 613, 616, 623–24, 628–29, 634
bicarbonate, 413, 416, 458, 507, 509, 515–16, 519, 528, 534, 541
bicarbonate buffer system, 509, 541
bicuspid valve, 462
bile, 413, 416, 517, 520, 525–27, 534
bile duct, 517–18
bilirubin, 526–27
binary fission, 182, 193, 212, 224–27, 240, 258
biometry, 291–92
bioremediation, 135–36
biostatistics, 291
bipolar neurons, 419
birth, 332, 359, 363–64, 409, 452, 596, 619, *See also* parturition
bladder, 318, 536–37, 552
blastocoel, 339
blastocyst, 339–42, 348, 351, 357
blastomere, 339, 348
blood, 377–78, 84, 397, 450–52, 454–55, 458–59, 462, 481, 501–2, 505–8, 529, 536–37, 544–46, 575, 578–79
   cells, 454, 471, 475, 539, 546, 586, 596, 613
   coagulation, 461
   components, 451, 454, 539
   osmolarity, 386–87, 543–45, 550
   pressure, 383, 385, 392–93, 442, 471–72, 474–78, 541–44, 550
   types, 131, 155, 451–52
   volume, 386, 392–93, 414–17, 467, 471, 477, 542–44
blood-brain barrier, 421
blood vessels, 395–96, 460, 462, 467–71, 474, 480, 543, 575, 580, 585, 605–6, 611
bolus, 514, 516
bone, 210, 403–4, 570, 582, 596–97, 613
bone marrow, 451, 454, 584, 586, 591, 596, 613–14, 628
bone matrix, 404, 585, 587–90
bone remodeling, 591–91, 597
bottleneck effect, 303–4
Bowman's capsule, 539, 546–48
brain, 383, 390, 396, 410, 419, 438, 440–41, 444–46, 504,
bronchi, 488–89, 491, 498, 500
bronchioles, 396, 489, 491, 498, 500
bulbourethral (Cowper's) gland, 318, 320

C

calcitonin, 402–4, 415, 597
calcitriol, 398, 406, 416, 597
calcium, 398, 402, 404–6, 588, 590, 596–97
calmodulin, 567
calyx, 536–37
canaliculi, 585, 588
cancer, 15, 49, 199, 202–4, 206, 369, 619, 621, 625
capacitation, 336
capillaries, 468–70, 472–75, 478, 486, 490–91, 537, 544, 546–48, 550, 577–78, 610–11
capsid, 241, 251–52, 257, 260, 265
carbohydrates, 47, 170, 175, 177, 514, 520, 525
cardiac cycle, 465–67, 471, 476
cardiac muscle, 190–91, 462–63, 563, 565, 578–79
cardiac output (CO), 467
cardiac sphincter, 514, 522, *See also* lower esophageal sphincter
carrier, 288
carrier proteins, 160
cartilage, 210, 488–89, 587, 590–91, 593–94
catecholamines, 378, 393–96, 400, 415
cDNA (complementary DNA), 103–4, 106, 115–16
cecum, 520
cell-cell junctions, 190–91
cell-mediated immunity, 623, 625, 627–28

cell cycle, 19, 193–95, 199–202, 207, 369
cell death, 15, 199, 201, See also apoptosis
cell division, 23–24, 185, 193, 195, 201–2, 204, 212, 227, 270, 275, 277, See also mitosis, meiosis
cell membrane, 160, 162, 212, 424, 428, 544, See also plasma membrane
cell potency, 348–49, 367, 613
cell signaling, 154, 156, 158, 203, 206, 345, 348, 354–56, 358, 372–73, 379–80, 418, 450
cell theory, 145–47, 149, 193, 250
cell wall, 148–49, 163, 189, 198, 212, 214–15, 217–18, 227–28, 233–34,
cellular respiration, 181–83, 229–30, 455, 458, 502
central dogma of molecular biology, 27–28, 237
central nervous system (CNS), 345, 419–21, 423, 440–44, 446, 504, 535
centriole, 185, 188, 323, 336
centromere, 25, 195–97
centrosome, 185, 196–97
cerebral cortex, 383, 441, 504, 506
cerebrospinal fluid (CSF), 421, 505
cervix, 328, 363–64, 387, 447
channel protein, 159–60, 424, 426, 429–31, 433–34, 564, 579
chemical synapses, 434–35
chief cells, 515–16
cholecystokinin (CCK), 413, 416, 520, 534
cholesterol, 153, 155, 168, 176, 378, 526–27
chorion, 340, 342
chorionic villi, 340, 360
chromatids, See sister chromatids
chromatin, 23–25, 58–60, 172, 196, 351–52, 368
chromosomes, 3–4, 9, 13–15, 20, 49, 194–97, 200, 204, 211–12, 237–38, 270, 272–75, 278, 288–90, 294–95, 316–17, 352
chylomicrons, 486
chyme, 413, 514, 516–18, 520, 527–28, 530,
chymotrypsin, 530
cilia, 149, 187–89, 491, 620
circulation, 340, 359–61, 373, 378, 397, 451, 539, 550, 578, 588, 616
cleavage furrow, 198–99
clonal selection, 623, 628
cocci, 213

codominance, 281
codon, 37–38, 40–41, 43, 45, 49, 51–52, 293
collagen fibers, 189, 460, 469–70, 588, 593–94, 604, 608
collecting duct, 393, 540, 543, 549–50
colon, 520
complement system, 622
complementation, 124–25
complete dominance, 281
concentration gradient, 160–66, 355, 424–25, 428, 548, 561
confidence interval, 84–85
conjugation, 220, 233, 239–40
connective tissues, 208–10, 343, 357, 451, 469, 584, 590, 592–93, 602, 604
control group, 72, 76
control variable, 72–73
corona radiata, 332, 336
corpus luteum, 334–35, 340, 409
correlation coefficient, 74–75
cortical bone, 584, 586
cortical reaction, 336
corticosteroids, 392
corticotropin-releasing hormone (CRH), 389–91, 414
cortisol, 372, 392, 415
cortisone, 392, 415
countercurrent multiplication, 549
Cowper's gland, See bulbourethral gland
CRISPR-Cas9, 121–22
cristae, 182
cross-bridge cycle, 565–66
crossing over, 271, 273–75, 277, 289–90, See also recombination
cyclin-dependent kinase (CDK), 200–201, 207
cyclin proteins, 200–1
cytokines, 368–69, 621, 626, 628
cytokinesis, 185, 195, 198–99, 272
cytoplasm, 32, 149, 237, 260, 378, 581, 612
cytoskeleton, 156, 183–84, 189
cytosol, 149, 156, 158, 168, 174, 183, 229–30, 565, 567, See also cytoplasm
cytotoxic T cells (T_C), 625–27, 634

D

defensins, 619, 621
denaturation, 7–8, 93, 95
dendrites, 418–20, 438, 441, 445
dendritic cells, 608, 613, 615–16, 624–25, 633

dependent variable, 72, 76–77, 80, 82, 86
depolarization, 426, 429, 463–64, 562–63, 565, 570–71, 580
dermis, 602, 604–5, 608–9
desmosomes, 190, 463, 578
detrusor muscle, 552–53
diabetes mellitus, 401
diaphragm, 490, 494–95, 504
diaphysis, 584
diastole, 465, 470, 475–76
diastolic pressure, 476
differentiation, 321, 342, 350, 354, 356, 367, 404, 412, 415–16, 628
diffusion, 160–61, 166, 171, 354–55, 360, 378, 425, 435, 454, 473–74, 490
dipeptidase, 519
diploid, 270, 277, 317, 321, 330, 332, 337
direct hormones, 375, 391
distal convoluted tubule, 540, 543, 549
DNA (deoxyribonucleic acid), 3–8, 23–25, 27, 37, 48–49, 59, 113–14, 121–22, 149, 194–95, 239–41, 250, 293–94
DNA damage, 17, 19, 207, 296
DNA fingerprinting, 132–33
DNA libraries, 105–6
DNA ligase, 12–13, 18, 20, 99–100, 104
DNA methylation, 60, 338, 352, 354
DNA polymerase, 11–13, 17–20, 29, 48, 90–91, 100, 104, 204
DNA proofreading, 17
DNA repair, 15, 17, 19, 48, 206–7, 293
DNA replication, 9–10, 12–13, 17–18, 27, 29, 60, 172, 195, 200–201, 293, 296, *See also* replication
DNA sequencing, 89, 91, 95, 102, 133, *See also* Sanger sequencing
dominant, 280–82, 284
dopamine, 390–91, 414, 438
dorsal root, 445
double-blind study, 72
duodenum, 398, 413, 416, 516–17, 519, 525, 527, 530, 534
dynein, 185, 187

E
ectoderm, 342–44, 354, 356
efferent neurons, *See* motor neurons
elastic cartilage, 594
electrical gradient, 166, 424–25
electrochemical gradient, 166–67, 25, 565
electrolytes, 450, 520, 537
electron transport chain, 156, 229
electrophoresis, 90, 106, 108, 113–14, 117–18, *See also* polyacrylamide gel electrophoresis (PAGE)
ELISA, *See* enzyme-linked immunosorbent assay
embryogenesis, 48, 58, 324, 330, 332, 341, 351
embryonic cleavage, 338, 348
endochondral ossification, 594
endocrine glands, 373, 376, 383, 403, 413, 415
endocrine signaling, 373, 379–80
endocytosis, 167–69, 171, 179, 259–60, 474, 628
endoderm, 342–43, 346
endolysosome, 179
endomembrane system, 175–76, 178, 183, 260–61
endometrium, 328, 335, 340, 409, 415
endonucleases, 18, 98, 121, 127, 132
endoplasmic reticulum, 46, 172, 174–76, 178, 180, 259, 556, 565
endospores, 219, 222–23
endosymbiotic theory, 182
endothelial cells, 460, 469–70, 480–81
enhancers, 61–62, 353–54
enteric nervous system, 534
enteropeptidase, 519, 530
enzyme-linked immunosorbent assay (ELISA), 116–17
enzymes, 7, 9, 58–59, 63, 176, 182–83, 513, 519, 526–27, 530, 567
eosinophils, 612–13, 621
ependymal cells, 421
epiblast, 342
epidermis, 367, 602–3, 607–9
epididymis, 318, 323
epigenetics, 60, 206, 352, 354
epiglottis, 500
epinephrine, 381, 383, 393, 415, 438, 479–80, 563, 611
epiphyses, 584, 594
epiphyseal line, 584, 594
epiphyseal plate, 584
epithelial cells, 191, 208–10, 329, 357, 385, 413, 490–91, 520, 534, 539, 602
epitope, 630
equilibrium potential, 425
erythrocytes, 397, 451, 545, 591, 613, *See also* blood cells, red blood cells
erythropoietin (EPO), 397–98, 416, 545
esophagus, 513–14

estradiol, 329, 409, 596–97
estrogen, 328–29, 332, 334–35, 340, 359, 365, 389–90, 393, 407, 409, 415
euchromatin, 24–25, 58–59
evolution, 182, 208, 250, 299
excitation-contraction coupling, 565
excitatory neurotransmitters, 465–66
excretion, 392, 402, 405, 536
exocrine glands, 376, 513
exocytosis, 169–71, 179, 259, 261, 377, 386, 474, 564, 621
exon, 32–34, 103
exonuclease, 17–18
experimental design, 71–72, 76, 79, 291
expiration, 458, 494–96, 500–3, 507
expiratory reserve volume (ERV), 497
expressivity, 282
external urethral sphincter (EUS), 552–53
extracellular matrix (ECM), 189–90, 323, 451, 593

F

facilitated diffusion, 161, 163, 171
fallopian tubes, See uterine tubes
fast glycolytic fibers, 573, 577
fast oxidative fibers, 572, 577
fat pads, 592
fats, 152–53, 182, 210, 375, 390, 392, 414, 513–14, 517, 520, 526–27, 596, See also lipids
feedback loops, 363, 365, 388, 446–47
fermentation, 229–31
fertilization, 308, 317, 323, 328, 330, 332, 334–38, 340, 347, 359, 409
fetal circulation, 360–63, 458
fibrin, 461
fibrinogen, 461
fibrous cartilage, 494
filtrate, 392–93, 412, 536–37, 539–41, 545–49
filtration, 474–75, 546, 616
fimbriae, 219–20, 328, 336
first messenger, 380
FISH, See fluorescence in situ hybridization
fitness, 297–98, 306–7, 309–11
flagella, 149, 187–89, 219–21, 323
fluid mosaic model, 153–54
fluorescence in situ hybridization (FISH), 115
follicle, 329, 332, 334, 409
follicle-stimulating hormone (FSH), 325–26, 333–34, 388–90, 408–9, 414, 416
follicular phase, 333–34
foramen ovale, 362–63
forensics, 127, 131
frameshift mutations, 52, 293–94, 298
functional residual capacity (FRC), 498

G

gallbladder, 413, 517, 524, 526–27, 534
gametes, 48, 270–73, 283, 289, 296, 304, 317–18, 321, 328, 332, 336
gametogenesis, 325, 338
gamma-aminobutyric acid (GABA), 159, 438
ganglia, 347, 441, 444
gap junctions, 190–91, 420–21, 434, 463–64, 563, 578–79, 581, 590
gas exchange, 341, 467, 488, 490, 496, 498–99, 501–2
gastric juice, 515–16, 619–20
gastrin, 413, 416, 515–16, 534
gastrointestinal tract, See alimentary canal
gastrula, 343
gastrulation, 342–43
gene cloning, 99, 103
gene expression, 24, 27, 89, 103, 115, 244, 246–48, 256, 351, 353–55, 378
gene mapping, 290–91
gene regulation, 23, 58, 60–61, 64–65, 244, 248, 350
genes, 3–4, 57–58, 99, 101–3, 121, 123–24, 206–7, 238–39, 244–46, 278–79, 281, 288–90, 292, 351–53, 408
gene flow, 304, 307
gene pool, 281, 299, 302–4, 306
gene therapy, 130–31, 139
genetic crosses, 282
genetic drift, 302–3, 311
genetic leakage, 308–9
genetic linkage, 289–90
genome, 4, 57, 60, 105–6, 212, 222, 237, 243, 250, 253, 256, 263–64
genomic imprinting, 60, 351–52
genotype, 27, 279–82, 284, 286, 300–02
germ cells, 15, 296
germ layers, 342–45
germline mutations, 48, 296–97
gestation, 328, 359–61
ghrelin, 410, 532–34
glial cells, 346, 418, 420–21, 440
glomerular blood oncotic pressure (GBOP), 547
glomerular filtration rate (GFR), 412, 546–47
glomerulus, 537, 539, 546–47

glucagon, 394, 399–400, 528–29, 534
glucocorticoids, 380, 389–92, 396, 415
gluconeogenesis, 392, 399–400, 525, 529, 534
glucose, 165, 247, 392, 394, 399–400, 410, 525, 529, 531
glycogenesis, 394, 399–400, 529, 534
glycogenolysis, 394, 399–400, 525, 529
glycolysis, 229–30, 561, 573
glycoprotein, 155, 189, 252, 260–61, 336
glycosylation, 47, 176–77
goblet cells, 491
Golgi apparatus, 175–80, 259, 323, 377
gonadotropin, 325–26, 414
gonadotropin-releasing hormone (GnRH), 325–26, 333, 389–90, 408–9, 414
G protein, 154, 157–58, 380
G protein–coupled receptor (GPCR), 157–58, 380, 400
graded potentials, 438
gram-negative organisms, 214–17
gram-positive organisms, 214–16
Gram staining, 214
granulocytes, 612, 621
granulosa cells, 329, 332
gray matter, 441
growth hormone (GH), 375, 384, 388, 390–91, 401, 414–15, 534, 596–97
growth hormone–releasing hormone (GHRH), 384, 391

H
H band, 559
hair follicles, 604–5, 611
haploid, 238–39, 270–73, 275, 277, 317–18, 321, 328, 330, 332, 337
Hardy-Weinberg equilibrium, 300–2
Haversian canal, 585
Haversian system, 585, 588, 590
heart, 450, 462–65, 467, 470–71, 473, 475–76, 479–81, 487, 501, 575, 578–79, 581
helicase, 10–11, 13
helper T cells (T_H), 625–26
hematopoiesis, 454, 591, 586
hemoglobin, 50, 360, 454–58, 509, 561, 586
hemostasis, 459–61
heterochromatin, 24–25, 58
heterozygous, 279, 281, 284, 302, 352
histones, 23–25, 58–59, 212, 237
homeostasis, 381, 383, 385, 387, 440, 446–47, 507, 509, 541, 544, 607, 609

homologous chromosomes, 271–75, 277, 281, 290
homologous recombination, 20–21, 271, 273–75, 277, 282, 289–90. See also crossing over
homozygous, 124, 279, 281–82, 284, 302, 307
horizontal gene transfer, 225, 233–34, 237, 239–41
hormones, 340, 372–73, 375, 388, 402–3, 407, 409–11, 413–14, 416, 479, 519, 527–29, 532–34, 542–43, 545, 596–97
human chorionic gonadotropin (hCG), 340, 359
humoral immunity, 623, 628
hyaline cartilage, 591, 594
hybrid breakdown, 308
hybrid inviability, 308
hybridization, 7, 94, 109–11, 113–15, 121, 127, 129, 255
hybrid sterility, 308
hydrochloric acid (HCl), 413, 416, 514–16
hydrogen bonding, 6–7, 109, 491
hydrostatic pressure (HP), 474, 547
hydrostatic pressure of glomerular blood (GBHP), 547
hydroxyapatite, 404, 588
hyperpolarization, 430, 436, 438–39
hypoblast, 342
hypodermis, 602, 604–5, 608
hypothalamic-pituitary-adrenal (HPA) axis, 389, 392
hypothalamic-pituitary-gonadal (HPG) axis, 325–26, 332–34, 408–9
hypothalamic-pituitary-thyroid (HPT) axis, 390, 402
hypothalamus, 325–26, 363, 365, 382–86, 388–90, 401–2, 408, 410, 414–16, 504, 506, 532, 534, 609, 611

I
I band, 559
ileocecal sphincter, 520, 522
ileum, 515–16
immune response, 607–8, 612–14, 621, 623–24, 627–28, 630–31, 634
immunity, 613, 621, 624–25, 633
immunoglobulins (Ig), 361, 365, 629–30
immunohistochemistry, 119
implantation, 335–36, 340, 342, 359, 409
inactivation gate, 429–31, 433
inbreeding, 307

incomplete dominance, 281
independent assortment, 272–73, 289
independent variable, 72, 76, 80–81, 83, 86
inducible operons, 245–46
inflammation, 217, 469, 621–22
inhibin, 326–27, 334, 409, 416
inhibitory neurotransmitters, 436–37
innate immunity, 621, 633
inner cell mass (ICM), 339, 341–42, 349
inspiration, 494–96, 502–3
inspiratory capacity (IC), 497
inspiratory reserve volume (IRV), 497
insulin, 399–401, 525, 528–29, 534
integrase, 262–64
intercalated discs, 463, 578–79
intercostal muscles, 494–95, 504
intermediate filaments, 184–85, 190
internal urethral sphincter (IUS), 552–53
interneurons, 419, 445
interstitial fluid, 379, 393, 470, 473–75, 481, 485–86, 541, 543–44, 550
intestines, 518, 532
intramembranous ossification, 597, 594
intrinsic factor, 515–16
intron, 32–34, 61, 103
involuntary muscle, 562
ion channels, 159, 420, 428–29, 434

J
jejunum, 516
joint, 591–92, 596
juxtacrine, 354
juxtamedullary nephron, 540

K
keratin, 185, 187, 603, 607
keratinocyte, 602–3, 607–8
kidneys, 163, 397–98, 405–6, 412–13, 415–16, 536, 548, 552, 597, 609
kinesin, 185, 187
kinetochore, 196
Krebs cycle, 229, 399

L
labor, 363–65, *See also* parturition
lactation, 328, 359, 365
lacunae, 585, 590–91
lagging strand, 11–13
lamellae, 585, 591
Langerhans cells, 602, 607–8
large intestine, 513, 520
larynx, 488, 491, 594
lateral inhibition, 356–57
leading strand, 11
leptin, 410, 416, 532–34
leukocytes, 451, 591, 612–13, 621, *See also* white blood cells (WBCs)
Leydig cells, 319, 326, 408
ligament, 210, 587, 592–93
lipase, 513, 515–16, 519, 527
lipids, 154–56, 169–70, 175–76, 180, 183, 217, 515–16, 518, 525, 527, 530–31, 604, 607
lipopolysaccharide (LPS), 216–17
liver, 361, 372, 375, 392, 399–401, 524–27, 534, 545, 597
loop of Henle, 539–40, 548–49
lower esophageal sphincter (LES), 514
lungs, 361–63, 392, 396, 454–55, 458–59, 462, 467, 470, 488–91, 494, 501, 506, 541–42
luteal phase, 334
luteinizing hormone (LH), 325–27, 333–34, 388–90, 408–9, 414
lymph, 481–83, 486–87, 615–16
lymphatic vessels, 481–82, 485–87
lymph nodes, 481–82, 613, 615–16, 625
lymphocytes, 481, 612–13, 616, 621, 623–24, 628, 634
lymphoid organs, 412, 481, 612–14, 616
lymphoid progenitor cells, 613–14
lysogenic replication cycle (lysogeny), 258–59, 262
lysosomes, 175, 179–80, 621
lysozyme, 619
lytic replication cycle, 257–58

M
M line, 559
macrophages, 421, 613, 615–16, 621, 624, 633
major histocompatibility complex (MHC), 624–26, 628
mammary glands, 328, 365, 375, 387, 390, 414
mast cells, 613, 621
mean, 79
median, 79
megakaryocyte, 451, 613
meiosis, 212, 270, 289–90, 296, 321, 330, 408, *See also* cell division
melanocyte, 345–46, 357, 602, 607–8
melatonin, 411, 416

membrane, 118, 149, 151, 153–56, 159–69, 175, 221, 252, 424, 427–28, 430, *See also* cell membrane, plasma membrane
membrane attack complexes (MACs), 622
membrane potential, 167, 424, 426–30, 432
membrane transport, 160, 163–64, 167, 170–71, 424
memory B and T cells, 624, 626, 628
menstruation, 328, 335, 340
Merkel cells, 602
mesoderm, 342–43, 345–46, 582
messenger RNA (mRNA), 4, 27–29, 32, 37–39, 42–45, 60, 65, 103, 106, 127–28, 173–74, 245
metabolism, 183, 206, 229, 250, 298, 385, 390, 394, 402, 415, 457
metaphyses, 584
microarrays, 114–15
microfilaments, 184–85, 187, 189, 198
microglia, 421
microRNA (miRNA), 64–65, 127–29
microtubules, 184–88, 196, 221, 323
microvilli, 185, 518, 520
mineralocorticoids, 392, 415
missense mutation, 49–50, 293, 298
mitochondria, 156, 181–82, 229, 292, 323, 336, 419, 455, 561–62, 577–78
mitochondrial DNA, 182, 280, 293
mitosis, 24, 185, 193–98, 200–1, 275–77, *See also* cell division
mode, 79
molecular clock, 311–12
monocytes, 612–13
morphogens, 355, 358
morpholinos, 129–30
morula, 339
motor neurons, 438, 504, 506, 564, 570, 572
motor proteins, 184–86
motor units, 572–73
mucociliary escalator, 491, 500
mucous cells, 515–16
mucous membranes, 209, 500, 619
multipolar neurons, 419
muscle cells, 412, 445, 463, 556, 558–60, 562, 565–66, 570, 572–73, 577, 578–79, *See also* myocytes, muscle fibers muscle contraction, 466, 500, 558, 561, 563, 566, 570, 573
muscle fibers, 565–66, 572, 579
muscles, 495, 500, 514, 522, 556–58, 562–63, 570, 572–73, 592–93, 596
mutagens, 48, 293, 295–96
mutations, 48–51, 53–55, 60, 124, 206, 295–96, 301, 311
myelination, 420–21, 423
myocytes, 556–57, 559–60, 562, 580, *See also* muscle cells
myofibrils, 559
myogenic, 463, 578, 581
myoglobin, 561–62, 576–77
myometrium, 328
myosin, 184–85, 187, 556–62, 565–67, 570, 573
myosin light chain kinase (MLCK), 567
myosin light chain phosphatase, 567
myosin regulatory light chain, 567

N

Na^+/K^+-ATPase, *See* sodium-potassium pump
nasal cavity, 488, 491, 499–500, 502–3
natural killer (NK) cells, 613, 621
natural selection, 233–34, 297, 301, 305–6, 309
negative control, 76
negative correlation, 74
negative feedback, 246, 326, 333–34, 381–82, 387, 407, 409, 447, 506
nephrons, 386, 412, 536–37, 539–40, 543, 545–46, 548, 550, 552
nerves, 420, 441, 445, 587–88, 593
neural crest, 345–346, 357
neural tube, 345, 348, 356–57
neuroglia, 418, 421
neurohormones, 383, 385–386, 388
neuromuscular junction (NMJ), 438, 564, 570
neurons, 170, 346, 418–22, 424, 426, 433–36, 438–39, 441–46, 504, 564
neurotransmitters, 159, 383, 390, 434–35, 437–39, 443
neurulation, 342, 345–47
neutrophils, 612–13, 621
next-generation sequencing, 89, 92
nitrogenous bases, 4–7, 12
nodes of Ranvier, 420
noncoding RNA (ncRNA), 64–65, 127, 204
nonsense mutations, 51, 293, 298
nonspecific immunity, 619–20
norepinephrine, 378, 383, 393, 415, 438, 443, 479–80, 611

northern blotting, 114
nuclear envelope, 172, 175–76, 185, 196
nucleic acids, 3–5, 19, 37, 48, 89, 108–9, 113, 129, 252
nucleolus, 42, 173, 196–97
nucleosome, 23–24
nucleotides, 3–5, 9, 12–13, 17–18, 29–30, 32, 37–38, 40–41, 48–49, 51–53, 110, 293
nucleus, 28, 31–32, 40, 42, 58, 149, 172–74, 212, 259, 262–63, 292, 323, 378, 380
null hypothesis, 80–83, 291
Nuremberg code, 139–40

O
Okazaki fragments, 12
oligodendrocytes, 420–21
oncogenes, 206
oncotic pressure, 474, 544, 547
oocyte, 317, 323, 328–29, 336–37, 351
oogenesis, 328, 330–32, 337, 409, 415
open reading frame (ORF), 43, 49, 51–52
operon, 244–48
oral cavity, 513–14
origin of replication, 9–10, 224
osmolarity, 543
osmoregulation, 163, 544–45, 609, 611
osmosis, 162, 393, 474, 543–44, 548–50
osmotic pressure, 163, 544
osteoblasts, 404, 587–88, 590–91, 594
osteoclasts, 402, 404, 590–91
osteocytes, 585, 588, 590–91
osteoid, 589
osteons, 585
osteoprogenitor (osteogenic) cells, 590
outbreeding, 307
ovarian cycle, 329, 332–35, 409
ovaries, 328–30, 36, 389, 415–16
ovulation, 330, 332, 334, 336, 359, 389–90, 409, 414
oxygen, 229–32, 360–61, 395–96, 450, 454, 488, 494, 545, 561
oxygen-hemoglobin dissociation curve (ODHC), 455–57
oxygen saturation (SpO_2), 361
oxytocin, 363, 365, 386–88, 414, 447

P
pancreas, 398–99, 401, 410, 413, 415, 517, 524–25, 527–30, 534
pancreatic amylase, 398, 530
pancreatic lipase, 398, 519, 530
papillary layer, 604–605
paracrine, 354–55, 379, 527
parasympathetic nervous system (PNS), 345–46, 393, 419–20, 423, 438, 440–44, 479, 535
parathyroid glands, 403–4, 407, 415
parathyroid hormone (PTH), 403–7, 415, 597
parietal cells, 413, 416, 515–16, 519, 534
partial pressures, 455–57, 459, 501–2, 505–6
parturition, 342, 363–64, 387–88, 390, 447, See also birth
passive transport, 160–61, 165–66, 171
pathogen-associated molecular patterns (PAMPs), 620
pathogens, 233, 361, 481, 491, 607, 612, 615–16, 619–21, 623–24, 631, 633
pattern-recognition receptors (PRRs), 620
pedigree, 285–86
penetrance, 282
penis, 318, 328
pepsin, 515–16
pepsinogen, 515
peptide bonds, 42, 44
peptide hormones, 326, 373, 377–78, 380–81, 384, 386, 388, 397, 399, 402–3, 412–16
peptidoglycan, 212, 214–18, 222–24
periosteum, 587, 594
peripheral nervous system (PNS), 345, 357, 419–20, 423, 440, 442, 444, 563
peristalsis, 522
peritoneal cavity, 521
peritoneum, 521
peroxisomes, 183
phages, See bacteriophages
phagocytes, 202, 607, 621–22, 633
phagocytosis, 167–68, 454, 491, 615, 621, 625
phagolysosomes, 179, 260, 621
pharynx, 488, 491, 499, 514
phenotype, 27, 121, 124, 279–82, 284, 303, 309
phosphate buffer system, 509
phosphodiester bond, 12, 18–19, 89–90
phospholipid bilayer, 151, 153–54, 160, 252, 380, 424
phospholipids, 151–53, 176, 212
phosphorylation cascade, 158
piloerection, 605, 611

pilosebaceous unit, 604–5
pilus, 149, 220, 239
pinocytosis, 167–68
pituitary gland, 383, 385, 388, 402, 414–15
pituitary hormones, 388–90, 413–14
placenta, 336, 339–41, 359–65, 415, 452, 458
plants, 133, 136, 149, 163, 198, 265
plasma, 454, 458–59, 475
plasma cells, 628–30
plasma membrane, 151, 153, 155–57, 168–70, 175, 178, 188–90, 215–16, 230, 252, 260–61, 336, 380, 400, 419, 424–25, See also cell membrane
plasma proteins, 402, 461, 526, 544
plasmids, 99–103, 105–6, 212, 234, 238–40
plasmin, 461
plasminogen, 461
platelets, 451, 454, 459–61, See also thrombocytes
pleura, 490–91
ploidy, 270, 277
point mutation, 49, 52, 206, 293–94
polyacrylamide gel electrophoresis (PAGE), 117, 120
polycistronic, 245
polymerase chain reaction (PCR), 8, 93–98, 102–3, 106, 113, 115–16, 132
polyubiquitination, 63–64
population, 84–86, 124, 225, 228, 234, 297, 299–305, 307, 309
portal systems, 384, 388, 473
portal veins, 384
positive control, 76–77
positive cooperativity, 455–56
positive correlation, 74
positive feedback, 363, 387, 460
posterior pituitary gland, 363, 365, 385–88, 390, 417, 543, 550
postganglionic, 441, 443, 611
postsynaptic, 434, 438–39
postzygotic barrier, 308
post-translational modification, 47, 176
potassium, 164, 393, 428, 549
power stroke, 561, 566
preganglionic fibers, 441–43
pregnancy, 317, 328, 332, 334, 336, 390, 409
pressure, 363, 387, 444, 447, 465, 467, 470–71, 494–95, 547–48
pressure gradient, 471, 475
presynaptic neuron, 434–35, 439

prezygotic barrier, 308
primary active transport, 163
primary immunity, 624
primary sex determination, 316, 352
primase, 11, 13
primers, 29, 90, 93–94, 104
prions, 265–66
progenitor cells, 296, 350, 356
progesterone, 328, 332, 334, 340, 359, 365, 390–91, 409, 415
programmed cell death, See apoptosis
prokaryotic cell walls, 214
prokaryotic flagella, 187, 221
prokaryotic gene regulation, 237
prokaryotic genome, 237, 244
prokaryotic ribosomes, 174, 182
prolactin, 365, 375, 388, 390–91, 414
prolactin inhibitory factor (PIF), 390–91, 414
proliferative phase, 335
promoter, 29, 61, 123, 244
prophage, 258
prostate gland, 318, 320
proteasome, 64
proteins, 3–4, 13, 27–29, 37–39, 40–43, 45, 63, 113, 116–19, 153–56, 175–78, 222, 245, 251–52, 377, 514–15, 519–20
protein buffer system, 509
protein synthesis, 27, 46, 63, 121, 174, 176, 400, 414
protein trafficking, 47, 177–78, 252, 261
proteolytic enzymes, 201, 515–16, 519, 530, 590
prothrombin, 461
proto-oncogenes, 206
proximal convoluted tubule, 539, 548
puberty, 321, 323, 330, 332, 365, 408–9, 412, 614
pulldown assay, 119
pulmonary arteries, 361, 467, 470, 501
pulmonary circuit, 467, 470
pulmonary surfactant, 490, 496–97
pulmonary trunk, 362, 476
pulmonary veins, 361, 467, 470, 501
Punnett square, 283–84
p-values, 81
pyloric sphincter, 516, 522

Q
quantitative PCR (qPCR), 96–97

R
radiography, 90, 111–13

range, 79
reabsorption, 398, 405–6, 416–17, 520, 543–44, 546, 551, 597
reactive oxygen species (ROS), 48, 183, 229
reading frame, 38, 43, 52, 293
real-time PCR, 96
receptors, 156–59, 378–80, 383, 418–19, 435, 437, 445, 447, 563–64, 620, 623–26, 628, 630
receptor tyrosine kinase (RTK), 158
recessive, 124, 280–81, 284, 286, 288, 307
recombination, 99, 225, 240–41, 273–75, 290–91
rectum, 520
red blood cells (RBCs), 155, 163, 347, 397, 451–52, 454–55, 457–59, 461, 509, 613, 616, See also erythrocytes
reflex arc, 445–46
reflexes, 372, 446
refractory period, See absolute refractory period, relative refractory period
regeneration, 367
regulatory T cells (T_{Reg}), 627, 634
relative refractory period, 433
renal arteries, 537, 539, 546
renal clearance, 546
renal papillae, 536
renal pelvis, 537
renal veins, 537
renin, 392
renin-angiotensin system, 392–93, 543–44
replication, 5, 7, 9–10, 12–15, 17, 19, 24, 224–25, 238, 241, 252–54, 265–66
repolarization, 426–27, 464
reproduction, 147, 193, 224–25, 227, 305, 316, 318, 332, 407
reproductive anatomy, 209, 316–18, 320–21, 324, 332, 347
reproductive success, 297, 306, 309
residual volume (RV), 498
respiratory acidosis and alkalosis, 508–9
respiratory centers, 504, 506
resting membrane potential (RMP), 424–26, 428, 430–33
resting state, 431, 433
restriction enzymes, 98–100, 105, 132
reticular layer, 604–5
retrotransposons, 243, 263–64
retroviruses, 262–63
reverse transcriptase (RT), 104, 115–16, 204, 262, 264

ribonucleotides, 4–6
ribosomes, 4, 27, 32, 37, 42–47, 149, 173–74, 176–77, 182, 212, 260
ribs, 494, 583, 594, 596
risk-benefit consideration, 137
RNA (ribonucleic acid), 3–6, 27, 29, 37–38, 40, 42, 113–15, 118, 121, 250, 252
RNA end processing, 32
RNAi (RNA interference), 127, 129
RNA polymerase, 29–31, 40, 42, 60, 246–47, 253
RNA primer, 11–13
RNA splicing, 33–34, 128–29
RNA viruses, 252–55, 262
rRNA (ribosomal RNA), 27, 29, 42, 64, 148, 173, 211
RT-PCR, 115–16
RT-qPCR, 115

S

salivary amylase, 513
saltatory conduction, 420
Sanger dideoxy method, See Sanger sequencing
Sanger sequencing, 89–90, 93
sarcolemma, 556, 564–65, 70
sarcomere, 559–60, 573–74, 579
sarcoplasm, 556
sarcoplasmic reticulum, 556, 559, 561, 565–67, 569, 571, 573, 579, 581
satellite cells, 423
satiety, 385, 410, 413, 532, 534
scatterplots, 75
Schwann cells, 346–47, 420, 423
SDS-PAGE, See polyacrylamide gel electrophoresis (PAGE)
sebaceous gland, 604
sebum, 605, 607, 619
secondary active transport, 165
secondary immunity, 624
second messengers, 158, 380–81, 400
secretin, 413, 519–20, 534
secretory pathway, 178
secretory phase, 335, 340
selective permeability, 160, 424
self-antigens, 627, 634
self-renewal, 349–50, 367
self-tolerance, 614
semen, 318, 320–21
semiconservative, 9, 95
semilunar valves, 462, 467, 476
seminal glands, 318, 320

seminiferous tubules, 319–21, 323
senescence, 203–4, 209, 367–69, 597
sensory neurons, 419, 441, 444–45, 604
Sertoli (nurse) cells, 319, 324, 326, 408–9
sex cells, 272
sex chromosomes, 280–81, 288, 292, 316–17
sex determination, 288, 316, 324, 408
sex hormones, 317–18, 325, 328, 332, 407–10, 597, See also estrogen, testosterone
sex-linked inheritance, See X-linked inheritance, Y-linked inheritance
sexual reproduction, 3, 225, 270–73, 277, 296, 317
short interfering RNA, See small interfering RNA (siRNA)
signaling cascade, 158, 380
signal amplification, 380
signal peptide, 46, 174
signal sequence, 46, 174
significance level, 81–82, 84–86
silent mutation, 51–52, 293, 298
single-blind study, 72
single-stranded DNA (ssDNA) viruses, 253
single-stranded DNA-binding proteins, 10, 13
single-stranded RNA (ssRNA) viruses, 253
sinoatrial (SA) node, 464
sister chromatids, 20, 195–97, 270, 272, 274
skeletal muscle, 396, 438, 441, 444, 487, 494, 504, 556, 564–65, 570–72, 574–76, 578–79
skeletal muscle pump, 575–76
skeleton, 582–83, 590–91, 596
skin, 209, 342–43, 345, 478, 502, 602–11, 619
skull, 583, 596
sliding filament model, 566
slow oxidative fibers, 572, 577
small guide RNA (sgRNA), 121
small interfering RNA (siRNA), 64, 127–29
small intestine, 398, 413, 486, 513, 515–20, 525, 527–28, 530
small nuclear RNA (snRNA), 33
smooth muscle, 470, 472, 479–80, 482, 487, 489, 552, 558, 560–64, 567–68, 579–81, 605, 610–11
sodium, 164, 392, 412, 428, 520, 549
sodium-glucose linked transporter (SGLT), 165, 167

sodium-potassium pump (Na^+/K^+-ATPase), 164–65, 167
somatic cells, 49, 206, 270, 296
somatic mutations, 49, 296–97
somatic nervous system, 419, 438, 441, 552, 564, 570
somatic neurons, See sensory neurons
somatostatin, 390–91, 399, 401, 410, 415, 529, 531, 534
Southern blotting, 106, 108–10, 132
species, 304–5, 307–11
specific immunity, 623–24
sperm, 270–71, 275, 296, 308, 317–18, 320–21, 323, 328, 332, 336–37, 408
spermatocyte, 321
spermatogenesis, 318–19, 321–23, 326, 337, 389–90, 408–9, 414–15
sphincters, 514, 522, 552, 563, 570
spinal cord, 419, 438, 440–41, 444–46, 504, 596
spinal reflexes, 446
spindle fibers, 196–97
spine, 445, 583
spirilla, 213
spirochete, 213
spleen, 454, 481, 613, 616–17, 625
spliceosomes, 33–34
splicing, See alternative mRNA splicing, RNA splicing
SpO_2, See oxygen saturation
spongy bone, 584, 586
spores, See endospores
SRY, 287, 324, 332, 408
standard deviation (SD), 79
standard error (SE), 85
start codon, 38, 43–44, 245
statistical power, 83–84, 86
statistical significance, 81, 83, 85
statistics, 73–74, 77, 79–80, 84–87
stem cells, 321, 330, 349–50, 367, 369, 587, 590, 603
steroid hormones, 319, 326, 329, 332, 373, 378, 380–81, 389, 392, 398, 407, 415, 597
stomach, 410, 413, 416, 513–17, 519–20, 532, 534, 619
stop codon, 38, 43, 45, 51, 55, 293
stress response, 222, 368, 372, 383–85, 389, 391, 396–97, 415
striated muscle, See cardiac muscle, skeletal muscle
striations, 558, 570, 579

stroke volume (SV), 467
summation, 438–39
surface tension, 490, 496
sweat, 502, 609–11, 619
sweat glands, 602, 604, 609, 611
symbiosis, 232
sympathetic nervous system (SNS), 383, 393–94, 438, 441–44, 479, 535, 611
synapse, 418, 422, 434–38, 442, 445, 564
synapsis, 271, 274, 277
synaptic cleft, 434
synaptonemal complex, 274–75
syncytium, 563, 579
synovial fluid, 591
systemic circuit, 361–62, 467–68, 471, 473
systole, 465, 467, 470–71, 475–76
systolic pressure, 475–76

T

T cells, 613, 616, 623–29, 634–35
telomerase, 204–5
telomeres, 4, 14–15, 25, 203–5, 238, 368
tendons, 210, 587, 592–93
testcross, 284–85
testes, 318–19, 321, 324–25, 389, 408, 415–16
testosterone, 317, 319, 324, 326, 332, 389–90, 408–9, 414, 597
tetanus, 573
tetrad, 274
thermogenesis, 574–575
thermoregulation, 385, 390, 478–79, 502–4, 574, 609, 611
thick filament, 559–60, 567
thin filament, 559–60
threshold, 426, 436
threshold potential, 429, 438–39
thrombin, 461
thrombocytes, 451, 459, *See also* platelets
thymosin, 412, 416
thymus, 412, 416, 481, 614–15, 625
thyroid gland, 375, 390, 401–4, 414–15, 479
thyroid hormones, 375, 378, 390, 401–2, 414–15, 596–97
thyroid-stimulating hormone (TSH), 375, 388, 390, 414
thyrotropin, 390
thyrotropin-releasing hormone (TRH), 390, 414
thyroxine (T_4), 390, 402, 416
tidal volume (TV), 497, 503
tissue regeneration, 367

tissues, 208–10, 345–46, 367
titin, 559
tonicity, 163
topoisomerase, 11, 13
total lung capacity (TLC), 498
total peripheral resistance (TPR), 471–72
trabeculae, 586
trachea, 488–89, 491, 499–500, 502, 594
traits, 133, 138, 279, 283, 285–86, 288, 291–92, 297, 305, 309, 393
transcription, 4–7, 24, 27–32, 37, 40, 43, 59, 65, 244–48, 252–53, 256, 262
transcription factors, 29, 60–62, 206–7, 246, 351, 353, 380
transduction, 233, 239, 242, 258
transfer RNA (tRNA), 3, 27, 29, 38–45, 64
transformation, 101, 103, 233, 239–41
translation, 27–28, 31–32, 37, 52, 60, 65, 127–28, 174, 177, 252, 254, 256
translocation, 49, 294–95
transposons, 48, 243–44, 263, 296
transverse canals, *See* Volkmann's canals
transverse tubules, 565, 570–71, 579–80
treatment group, 72
tricuspid valve, 462
trigger zone, 426, 438–39
triglycerides, 176, 486, 513
triiodothyronine (T_3), 390, 402, 416, 479
trophoblast, 339–40
tropic hormones, 375, 384, 390
tropomyosin, 565–66
troponin, 565–67
trypsin, 519, 530
trypsinogen, 530–31
t-tubules, *See* transverse tubules
tumor suppressor genes, 206–7, 369
twitch, 573
type I and II errors, 82–83

U

ubiquitination, 47, 63
umbilical cord, 341, 360–62
unipolar neurons, 419, 445
urea, 539, 545–46
ureters, 536–37, 550, 552
urethra, 318, 536, 552
urine, 536–37, 540–41, 544–46, 550, 552, 597
urination, 536, 552–53
uterine contractions, 363, 387, 414, 447
uterine cycle, 332, 334–35, 340, 409
uterine tubes, 328, 334, 336, 339

uterus, 328, 332, 335–36, 339–40, 363, 409

V
vagina, 318, 328, 335, 363–64
validity, 86
variables, 71–72, 74–76, 80–81, 86
vasa recta, 537, 549
vasculature, 450, 468–69, 471, 476
vas deferens, 318
vasoconstriction, 383, 392, 395, 460, 471–72, 478, 480, 543, 611
vasodilation, 395, 412, 442, 471–72, 478, 480, 502, 609–11, 621
vasopressin, *See* antidiuretic hormone (ADH)
vector, 130, 139
veins, 468, 470–71, 482, 575, 605
vena cava, 361, 471
ventilation, 494–95, 498, 502, 504–7
ventilation rate, 504, 508
ventral root, 445
ventricles, 362, 462, 464–67, 470–471, 475–76, 578
venules, 469, 472, 605
vertebrae, 346, 583, 596
vesicles, 167–69, 175–80, 185, 187, 323, 336, 377–78, 386, 439, 564, 588, 621
villi, 486, 518
viral envelope, 252, 260–61
viral genomes, 241, 250–52, 256–57, 260, 262, 296
viral replication, 253, 255–56, 260
viral structures, 251
virion, 254, 257–58, 260–61
viruses, 121, 130, 147, 171, 206, 251, 296, 612, 620, 633
vitamins, 133, 135, 232, 398, 406, 515–16, 520, 526, 548, 597
Volkmann's canals, 585
voltage-gated channels, 428–29, 431, 571
voluntary muscles, 563, 570

W
waste products, 179, 228, 360, 450, 513, 520, 526, 531, 536, 545–46, 548
water balance, 163
western blotting, 118–19
white blood cells (WBCs), 451, 454, *See also* leukocytes
white matter, 441, 446
wild-type, 279
wobble pairing, 40

X
XIST, 352
X-linked inheritance, 288–89, 352, 459

Y
Y-linked inheritance, 289
yolk sac, 341–42

Z
Z line, 559
zona pellucida, 323, 332, 336
zygote, 296, 317, 332, 336–39, 348, 359
zymogen, 515–16, 530